Lecture Notes in Mathematics

Edited by A. Dold and B. Eckmann

1032

Ordinary Differential Equations and Operators

A Tribute to F. V. Atkinson

Proceedings of a Symposium held at Dundee, Scotland
March – July 1982

Edited by W. N. Everitt and R. T. Lewis

Springer-Verlag
Berlin Heidelberg New York Tokyo 1983

Editors

W. N. Everitt
Department of Mathematics, University of Birmingham
P.O. Box 363, Birmingham B15 2TT, England

R. T. Lewis
Department of Mathematics
University of Alabama in Birmingham
Birmingham, AL 35294, USA

AMS Subject Classifications (1980): 34-06, 35-06

ISBN 3-540-12702-X Springer-Verlag Berlin Heidelberg New York Tokyo
ISBN 0-387-12702-X Springer-Verlag New York Heidelberg Berlin Tokyo

Printing and binding: Beltz Offsetdruck, Hemsbach/Bergstr.
2146/3140-543210

DEDICATION

These Proceedings of the

Symposium on Ordinary Differential Equations and Operators

(Science and Engineering Research Council, United Kingdom)

held at the

University of Dundee, Scotland, UK

April, May and June 1983

are dedicated to

F. V. ATKINSON

of the University of Toronto, Canada

by his friends and colleagues

in recognition of his manifold and remarkable

contributions to the theory of ordinary and partial

differential equations, and of his warm and

endearing personality

PREFACE

During the months of March, April, May, June and July 1982 the Mathematics Committee of the Science and Engineering Research Council of the United Kingdom supported a Symposium on Ordinary Differential Equations and Operators; the Symposium was held in the Department of Mathematics at the University of Dundee, Scotland. These Proceedings form a permanent record of the contributions made to the Symposium, and other contributions from colleagues who were invited to submit manuscripts in view of the special nature of this volume.

These Proceedings are dedicated to F.V. (Derick) Atkinson by his many friends and colleagues in recognition of his mathematical contributions to the theory of differential equations and to mark his formal retirement from the University of Toronto, Canada. All the individual papers printed in this volume are dedicated by the authors to Derick Atkinson.

Following this preface there is a list of contents, and a list of names and addresses of all colleagues who took part in the Symposium. The main overseas visitors for the Symposium were F.V. Atkinson, František Neuman, R.T. Lewis, Tung Chin-Chu and Anton Zettl.

The main financial support came from the Mathematics Committee of SERC; there was a special grant from the Royal Society of London to enable Professor Tung Chin-Chu, Chinese Academy of Sciences, Beijing, PR China to attend the Symposium.

The Symposium could not have been mounted without the help and interest of the permanent staff of the Mathematics Committee of SERC. Professor F.F. Bonsall, University of Edinburgh, agreed to act for the Committee as an adviser for the Symposium; I am grateful to him for his help in this capacity.

I am greatly indebted to my colleague Commander E.R. Dawson R.N. who again stepped out from retirement to help with the organisation of the Symposium.

The University of Dundee generously supported the work of the Symposium.

I express keen gratitude to all colleagues who attended the Symposium, took part in the programme of lectures and discussions, and contributed manuscripts to this volume.

Likewise special thanks are due to the secretarial staff in the Department of Mathematical Sciences in the University of Dundee, and the Department of Mathematics in the University of Birmingham; in particular Caroline Peters, Pat Chapman and Carole Regan.

The move from the University of Dundee to the University of Birmingham provides me with an opportunity to thank the Editors of the series Lecture Notes in Mathematics, and the Staff of Springer-Verlag in Heidelberg for all their help and advice over the years 1972 to 1982 when the work of Conferences and Symposia held at the University of Dundee was reported on in Lecture Notes in Mathematics.

Finally I offer special thanks to my co-editor of these Proceedings, Roger Lewis, for all his help and advice over an extended period. The final form and content of this volume owes much to his wise and effective participation.

W.N. Everitt

Address list of authors and speakers

C D Ahlbrandt:	Department of Mathematics, University of Missouri, COLUMBIA, Missouri, U.S.A.
F V Atkinson:	Department of Mathematics, University of Toronto, TORONTO, Ontario, Canada, M5S 1A1
J V Baxley:	Department of Mathematics, Wake Forest University, WINSTON-SALEM, North Carolina 27109, U.S.A.
C Bennewitz:	Matematiska Institutionen, Uppsala Universitet, Thunbergsvägen 3, S-75228 UPPSALA, Sweden
P A Binding:	Department of Mathematics and Statistics, University of Calgary, Calgary, Alberta, Canada, T2N 1N4
K J Brown:	Department of Mathematics, Heriot-Watt University, Riccarton, Currie, EDINBURGH, EH14 4AS
P J Brown:	Department of Mathematics and Statistics, University of Calgary, CALGARY, Alberta, Canada, T2N 1N4
R C Brown:	Department of Mathematics, University of Alabama, UNIVERSITY, Alabama 35486, U.S.A.
L Collatz:	Eulenkrugstrasse 84, 2000 HAMBURG 67, Germany
E R Dawson:	Rhufwada, Westfield Terrace, Newport-on-Tay, DUNDEE, Scotland, U.K.
M S P Eastham:	Department of Mathematics, Chelsea College, Manresa Road, LONDON, SW3 6LX
W D Evans:	Department of Pure Mathematics, University College, P.O. Box 78, CARDIFF, CF1 1XL
W N Everitt:	Department of Mathematics, University of Birmingham, P. O. Box 363, BIRMINGHAM. B15 2TT
A M Fink:	Department of Mathematics, Iowa State University, AMES, Iowa 50011, U.S.A.

C T Fulton: Department of Mathematics, Pennsylvania State University, University Park, PENNSYLVANIA 16802, U.S.A.

T N T Goodman: Department of Mathematics, The University, DUNDEE, DD1 4HN, Scotland, U.K.

M W Green: Department of Mathematics, The University, DUNDEE, DD1 4HN, Scotland, U.K.

D B Hinton: Mathematics Department, University of Tennessee, KNOXVILLE, Tennessee 37916, U.S.A.

D S Jones: Department of Mathematics, The University, DUNDEE, DD1 4HN, Scotland, U.K.

R M Kauffman: Department of Mathematics, Western Washington University, BELLINGHAM, WA 98225, U.S.A.

I W Knowles: Department of Mathematics, University of Alabama, BIRMINGHAM, Alabama 35294, U.S.A.

A M Krall: McAllister Building, Pennsylvania State University, UNIVERSITY PARK, Pennsylvania 16802, U.S.A.

K. Kreith: Mathematics Department, University of California, DAVIS, California 95616, U.S.A.

V V Kučerenko Department of Mathematics, The University of Moscow, MOSCOW, U.S.S.R.

T Kusano: Department of Mathematics, Iowa State University, AMES, Iowa 50011, U.S.A.

M K Kwong: Department of Mathematics, Northern Illinois University, DEKALB, Illinois, U.S.A.

R T Lewis: Department of Mathematics, University of Alabama, BIRMINGHAM, Alabama 35294, U.S.A.

N G Lloyd: Department of Pure Mathematics, The University College of Wales, PENGLAIS, Aberystwyth SY23 3BZ, Wales

J W Macki: Department of Mathematics, University of Alberta, EDMONTON, Alberta, Canada, T6G 2E1

J B McLeod: Wadham College, Oxford University, 24-29 St. Giles, OXFORD, OX1 3LB, England

A B Mingarelli: Department of Mathematics, University of Ottawa, OTTAWA, Ontario, K1N 9B4, Canada

J W Neuberger: Department of Mathematics, North Texas State University, DENTON, Texas 76201, U.S.A.

F Neuman: Mathematical Institute, Czecholovak Academy of Sciences, Janáckovo nám 2a, 66295 BRNO, Czechoslovakia

T T Read: Department of Mathematics, Western Washington University, BELLINGHAM, WA 98225, U.S.A.

D A Sánchez: Department of Mathematics, University of New Mexico, ALBUQUERQUE, NM 87131, U.S.A.

R A Saxton: Department of Mathematics, Brunel University, UXBRIDGE, Middlesex, UB8 3PH, England

B D Sleeman: Department of Mathematical Sciences, University of Dundee, DUNDEE, DD1 4HN, Scotland

P D Smith: Department of Mathematical Sciences, University of Dundee, DUNDEE, DD1 4HN, Scotland

R A Smith: Department of Mathematics, University of Durham, South Road, DURHAM, DH1 3LE, England

K Soni: Department of Mathematics, University of Tennessee, KNOXVILLE, Tennessee, U.S.A.

C A Swanson: Department of Mathematics, University of British Columbia, VANCOUVER, British Columbia, Canada

Tung Chin-Chu: Graduate School, Chinese Academy of Sciences, P.O. Box 3908, BEIJING, P R China

J Walter: Institut für Mathematik, Templergraben 55, D-5100 AACHEN, Germany

A D Wood: School of Mathematical Sciences, N.I.H.E., Ballymun Road, DUBLIN 9, Ireland

S D Wray: Department of Mathematics, Royal Roads Military College, F.M.O., VICTORIA, British Columbia, Canada, VOS 1BO

A Zettl: Department of Mathematics, Northern Illinois University, DEKALB, Illinois, U.S.A.

Symposium on Ordinary Differential Equations and Operators

University of Dundee

March to July 1982

List of participants

1. Science and Engineering Council Visitors

(a) Main Overseas Visitors

F.V. Atkinson (Toronto, Canada)
R.T. Lewis (Birmingham, Alabama, U.S.A.)
F. Neuman (Brno, Czechoslovakia),
A. Zettl (DeKalb, Illinois, U.S.A.)

(b) U.K. Visitors

K.J. Brown (Heriot-Watt)
M.S.P. Eastham (Chelsea College, London)
W.D. Evans (Cardiff)
N.G. Lloyd (Aberystwyth)
J.B. McLeod (Oxford)
R.A. Saxton (Brunel)
R.A. Smith (Durham)
A.D. Wood (Cranfield)

(c) Other Overseas Visitors

C. Bennewitz (Uppsala, Sweden)
S.D. Wray (Victoria, Canada)

2. Royal Society Visitor

Tung Chin-Chu (Beijing, PR China)

3. Other Overseas Visitors

C.D. Ahlbrandt (Columbia Missouri, U.S.A.)
I.W. Knowles (Birmingham, Alabama, U.S.A.)
V.V. Kučerenko (Moscow, USSR)
A.B. Mingarelli (Ottawa, Canada)
D. Sánchez (New Mexico, U.S.A.)
K. Soni (Knoxville, U.S.A.)

4. University of Dundee

E.R. Dawson
W.N. Everitt
T.N.T. Goodman
D.S. Jones
B.D. Sleeman

C O N T E N T S

Inversion in the Unit Sphere for Powers of the Laplacian

by

Calvin D. Ahlbrandt
Mathematics Department
University of Missouri
Columbia, Missouri 65211

Don B. Hinton
Mathematics Department
University of Tennessee
Knoxville, Tennessee 37916

Roger T. Lewis[1]
Mathematics Department
Univ. of Ala. in Birmingham
Birmingham, Alabama 35294

Abstract. For a solution u of $\Delta^n u = 0$, where Δ is the Laplacian in m-space and n is a positive integer, we establish a transformation through inversion in the unit sphere which transforms u into a solution v of $\Delta^n v = 0$. For n = 1 this is the classical Kelvin transformation and for m = n = 2 this is the Michell transformation. Applications are given to spectral theory and oscillation theory.

A result of classical analysis is that if $u = u(x_1, \cdots, x_m)$ is a solution of Laplace's equation, i.e., $\Delta u = 0$, then $v = v(x_1, \cdots, x_m)$ defined by

$$v = \frac{1}{r^{m-2}} u\left(\frac{x_1}{r^2}, \cdots, \frac{x_m}{r^2}\right), \; r^2 = x_1^2 + \cdots + x_m^2 \tag{1}$$

is another such solution [6, p. 243] or [10, p. 84]. The above transformation of independent variable is the Kelvin transformation or inversion in the unit sphere. In this paper we show that an analogous result holds for powers of the Laplacian. Inversion in the unit sphere may be viewed as interchanging singular behavior at the origin and at infinity. For certain differential operators, radial symmetry, when present, is preserved.

M. Riesz [22] used a variable change which he termed the Kelvin transform to perform balayage for the Riesz kernels k_α. The case $\alpha = -2n$ gives rise to the operator $(-1)^n\Delta^n$. However, it does not appear the methods of Riesz can be applied to derive Theorem 1 below even in the case of m = 1. Our method directly uses properties of the Laplacian Δ. For m = 1, Theorem 1 below is a special case of Theorem 2.3 of [1]. Michell [19] used complex variable methods to

[1] Supported by NSF grant number MCS-8005811.

develop the inversion for the biharmonic in the plane. He also stated the correct result for the biharmonic in R^3 but did not provide a proof [19, p. 142]. Sternberg and Eubanks [24] presented conditions for $m = n = 2$ under which the change of independent variables must be a linear fractional (Möbius) transformation.

Let R^m denote the vector space of all real m-tuples, $R_+^m = R^m \setminus \{0\}$, and $C^{2n}(R_+^m)$ be the set of all real functions on R_+^m for which all partial derivatives of order $\le 2n$ exist and are continuous. For $x = (x_1, \cdots, x_m) \in R^m$, define $\|x\|^2 = x_1^2 + \cdots + x_m^2$. The inversion map f in R_+^m is given by $t = f(x)$ where

$$x = (x_1, \cdots, x_m),\ t = (t_1, \cdots, t_m) = \left(\frac{x_1}{\|x\|^2}, \cdots, \frac{x_m}{\|x\|^2}\right). \quad (2)$$

It will be convenient to define Laplacians Δ_x, Δ_t by:

$$\Delta_x = \sum_{i=1}^{m} \frac{\partial^2}{\partial x_i^2}, \quad \Delta_t = \sum_{i=1}^{m} \frac{\partial^2}{\partial t_i^2}.$$

Note that $\|t\| = \|x\|^{-1}$ in (2). We also make use of the identity for the Laplacian

$$\Delta(gh) = g\Delta h + h\Delta g + 2\nabla g \cdot \nabla h \quad (3)$$

where ∇ indicates the gradient.

A straightforward calculation yields the following.

Lemma 1. *If* $g(t) = \|t\|^2$ *for* $t \in R_+^m$, *then*

(i) $\Delta_t g = 2m$

(ii) $\Delta_t(g^{-n}) = -2n(m - 2n - 2)\|t\|^{-2n-2}$ *for* $n = 1, 2, \cdots$.

Lemma 2. *If* $n \in \{0, 1, 2, \cdots\}$ *and* $z = z(t) \in C^{2n+2}(R_+^m)$, *then*

$$\Delta_t^n\{\|t\|^2 \Delta_t z - 4n \sum_{i=1}^{m} t_i \frac{\partial z}{\partial t_i}\} = \|t\|^2 \Delta_t^{n+1} z + 2n(m-2n-2)\Delta_t^n z. \quad (4)$$

Proof. The proof is by induction. For $n = 0$, (4) is immediate. Assume Lemma 2 for $n = k$. Then a calculation using (3) and Lemma 1 (i) gives:

$$\Delta_t^{k+1}\{||t||^2 \Delta_t z - 4(k+1)\sum_{i=1}^{m} t_i \frac{\partial z}{\partial t_i}\}$$

$$= \Delta_t^k\{\Delta_t[||t||^2 \Delta_t z - 4(k+1) \sum_{i=1}^{m} t_i \frac{\partial z}{\partial t_i}]\}$$

$$= \Delta_t^k\{2m\Delta_t z + ||t||^2\Delta_t^2 z + 2 \sum_{i=1}^{m} 2t_i \frac{\partial}{\partial t_i}(\Delta_t z)$$

$$- 4(k+1) \sum_{i=1}^{m}[t_i\Delta_t(\frac{\partial z}{\partial t_i}) + 2 \frac{\partial^2 z}{\partial t_i^2}]\}$$

$$= \Delta_t^k\{2m \Delta_t z + ||t||^2 \Delta_t^2 z - 4k \sum_{i=1}^{m} t_i \frac{\partial}{\partial t_i} (\Delta_t z)$$

$$- 8(k+1) \Delta_t z\}$$

$$= \Delta_t^k\{||t||^2 \Delta_t(\Delta_t z) - 4k \sum_{i=1}^{m} t_i \frac{\partial}{\partial t_i} (\Delta_t z)$$

$$+ (2m - 8k - 8)\Delta_t z\}.$$

Applying the induction hypothesis to the first two terms on the right hand side of the above reduces it to

$$||t||^2 \Delta_t^{k+1}(\Delta_t z) + 2k(m-2k-2)\Delta_t^k(\Delta_t z) + (2m-8k-8)\Delta_t^{k+1} z$$

$$= ||t||^2 \Delta_t^{k+2} z + 2(k+1)(m-2k-4)\Delta_t^{k+1} z.$$

Thus (4) holds for $n = k + 1$ and the proof is complete.

Theorem 1. *If* $n \in \{1, 2, \cdots\}$ *and* $z = z(t) \in C^{2n}(R_+^m)$, *then*

$$\Delta_x^n y = ||x||^{-m-2n} \Delta_t^n z \tag{5}$$

where $y(x) = ||x||^{-m+2n} z(t)$ *and* $t = f(x)$ *is given by* (2).

Proof. We establish (5) by induction, For $n = 1$, this is the classical Kelvin transformation, which follows readily from Theorem 3.1 of [2]. Assume now (5) holds for $n = k$. Then by (5) for $n = 1$, (3), and Lemma 1 (ii), we have for $y(x) = ||x||^{-m+2k+2} z(t)$ that,

$$\Delta_x^{k+1} y = \Delta_x^k \Delta_x \{||x||^{-m+2k+2} z\} = \Delta_x^k \Delta_x \{||x||^{2-m} (||t||^{-2k} z)\}$$

$$= \Delta_x^k \{||x||^{-m-2} \Delta_t (||t||^{-2k} z)\}$$

$$= \Delta_x^k \{||x||^{-m-2} [-2k(m-2k-2)\ ||t||^{-2k-2} z + ||t||^{-2k} \Delta_t z$$

$$+ 2 \sum_{i=1}^m (-k)\ ||t||^{-2k-2} (2t_i) \frac{\partial z}{\partial t_i}]\}$$

$$= \Delta_x^k \{||x||^{2k-m} [-2k(m-2k-2)z + ||t||^2 \Delta_t z - 4k \sum_{i=1}^m t_i \frac{\partial z}{\partial t_i}]\}.$$

Hence by the induction hypothesis and Lemma 2,

$$\Delta_x^{k+1} y = ||x||^{-m-2k} \Delta_t^k [-2k(m-2k-2)z + ||t||^2 \Delta_t z - 4k \sum_{i=1}^m t_i \frac{\partial z}{\partial t_i}]$$

$$= ||x||^{-m-2k} [-2k(m-2k-2)\ \Delta_t^k z + ||t||^2 \Delta_t^{k+1} z + 2k(m-2k-2) \Delta_t^k z]$$

$$= ||x||^{-m-2k-2} \Delta_t^{k+1} z$$

and the proof is complete.

To consider applications of Theorem 1 let $q, k \in C(R_+^m)$ with $k(x) > 0$. Define τ by

$$(\tau y)(x) = \frac{1}{k(x)} \{\Delta_x^n y + q(x) y\},\ y = y(x) \in C^{2n}(R_+^m). \tag{6}$$

Hence by Theorem 1, with $t = f(x)$, $z(t) = ||x||^{m-2n} y(x)$,

$$(\tau y)(x) = \frac{1}{k(x)} \{||x||^{-m-2n} \Delta_t^n z + q(x)\ ||x||^{2n-m} z\} = ||x||^{2n-m} (Tz)(t) \tag{7}$$

where

$$(Tz)(t) \equiv \frac{1}{K(t)}\{\Delta_t^n z + Q(t)z\}$$

with $K(t) = k(x)\,\|x\|^{4n}$ and $Q(t) = q(x)\,\|x\|^{4n}$. If we define the transformation U on the complex Hilbert space $L^2(R_+^m;\, k) = \{y\!:\!y$ is Lebesgue measurable and $\int_{R_+^m} k|y|^2 < \infty\}$ by $U[y](t) = \|x\|^{m-2n}y(x)$, $t = f(x)$, then a calculation shows that U is a linear isometry from $L^2(R_+^m;\, k)$ onto $L^2(R_+^m;\, K)$. Note that the Jacobian of f is given by

$$|\det f'(x)| = \|x\|^{-2m} .$$

From (7) we see that for $z = Uy$, $\tau y = U^{-1}TUy$; hence τ and T restricted to $C^{2n}(R_+^m)$ are unitarily equivalent. Thus we may relate spectral properties of τ at $\infty(0)$ to those of T at $0(\infty)$.

One such area of application is that of singular boundary value problems and the question of whether boundary conditions are needed at a singular point. Such questions are usually phrased in terms of essential self-adjointness of some symmetric operator. The spectral theory of singular partial differential equations has a voluminous literature; for a few recent results we refer to [4, 7, 9, 12, 13, 14, 15, 16, 23] and their references. The question of essential self-adjointness with point singularities is dealt with in [7, 9, 13, 14, 15, 16]. In terms of the operators τ and T above, we see that τ restricted to $C_0^\infty(R_+^m)$ is essentially self-adjoint in the complex Hilbert space $L^2(R_+^m\,;\, k)$ if and only if T restricted to $C_0^\infty(R_+^m)$ is essentially self-adjoint in the complex Hilbert space $L^2(R_+^m;\, K)$. An example dealing with a more general second order operator than Δ is considered in [2].

A more direct result may be obtained in oscillation theory. The above calculations show that y satisfies

$$\Delta_x^n\, y + q(x)y = 0 \tag{8}$$

if and only if $z(t) = ||x||^{m-2n}y$, $t = f(x)$, satisfies

$$\Delta_t^n z + Q(t)z = 0, \; Q(t) = q(x)\,||x||^{4n}. \tag{9}$$

Thus zeros of y are transformed to zeros of z, and nodal domains of y are transformed to nodal domains of z. We will define (8) to be oscillatory at ∞ if for each R there is a non-trivial solution y which has a nodal domain in $R^m \backslash B_R$ where $B_R = \{x \in R^m : ||x|| \le R\}$. Otherwise (8) will be called nonoscillatory at ∞. Similar definitions apply to z at 0. A specific oscillation criterion (cf.[8,17]) is that $\Delta_x y + q(x)y = 0$ is oscillatory at ∞ if for some $\delta < 2 - m$ ($\delta \le 0$ if $m = 2$)

$$\lim_{R\to\infty} \int_{1\le ||x|| \le R} ||x||^{\delta}\, q(x)\,dx = \infty .$$

This yields that $\Delta_t z + Q(t)z = 0$ is oscillatory at 0 if for some $\delta < 2 - m$($\delta \le 0$ if $m = 2$)

$$\lim_{r\to 0} \int_{r\le ||t|| \le 1} ||t||^{4-2m-\delta}\, Q(t)\,dt = \infty.$$

A number of oscillation and nonoscillation criteria for higher order elliptic equations may be found in [3, 5, 20, 21] and their references. In particular, Allegretto [5] has established a Kneser theorem for (8), namely that (8) is nonoscillatory at ∞ if $2n \ge m$, m is even, and

$$\limsup_{||x||\to\infty} (-1)^{n+1}\, q(x)\,||x||^{2n}(\ln||x||)^{c} < K \tag{10}$$

where c and K are certain constants depending on m and n. This theorem gives that $\Delta_t^n z + Q(t)z = 0$ is nonoscillatory at 0 if

$$\limsup_{||t||\to 0} (-1)^{n+1}\, Q(t)\,||t||^{2n}(-\ln||t||)^{c} < K \tag{11}$$

where the constants in (11) are the same as those in (10). Note the transformation $t = f(x)$ preserves certain Euler-type equations, i.e, if $q(x) = k/||x||^{2n}$ in (8), then $Q(t) = k/||t||^{2n}$ in (9).

As another example consider the nonlinear equation

$$\Delta_t^n z + Cz^\mu = 0,\ \mu = \frac{m+2n}{m-2n},\ C = \text{constant}. \tag{12}$$

If $z = z(t) \in C^{2n}(R_+^m)$ is a nonnegative solution of (12) and $y(x) = ||x||^{-m+2n} z(t)$, $t = f(x)$, then by Theorem 1, y satisfies the same equation, i.e.,

$$\Delta_x^n y + C\, y^\mu = 0.$$

The case $n = 1$ of (12) occurs in the study of Riemannian metrics which have some invariance properties under conformal transformations [18].

In the ordinary differential equation case, i.e., $m = 1$, a transformation theory for general self-adjoint operators is developed in [1]. Also for $m = 1$, it has been shown by Hadass [11] that the form of the differential equation of a complex variable, $y^{(n)}(z) + p(z)y(z) = 0$, is invariant under a general Möbius transformation.

As a final area of application we mention the existence of spherically symmetric solutions of

$$\Delta_x^n\, y = f(x,y). \tag{13}$$

This problem has been considered in [25]. Under the transformation of Theorem 1, we see that (13) is transformed into

$$\Delta_t^n\, z = ||t||^{-m-2n} f(t/||t||^2,\ ||t||^{m-2n} z). \tag{14}$$

Thus the existence of solutions of (13) on R^m is equivalent to the existence of solutions of (14) on R_+^m together with certain continuity conditons at ∞.

REFERENCES

1. C.D. Ahlbrandt, D.B. Hinton, and R.T. Lewis, The effect of variable change on oscillation and disconjugacy criteria with applications to spectral theory and asymptotic theory, J. Math. Anal. Appl., 81 (1981), 234-277.
2. ___________, Transformations of second-order ordinary and partial differential operators, Proc. Royal Soc. Edinburgh, to appear.
3. W. Allegretto, Nonoscillation theory of elliptic equations of order 2n, Pacific J. Math. 64 (1976), 1-16.
4. ___________, Finiteness criteria for the negative spectrum and nonoscillation theory for a class of higher order elliptic operators, in "Spectral Theory of Differential Operators", pp. 9-12, I.W. Knowles and R.T. Lewis editors, North Holland, New York, 1981.
5. ___________, A Kneser theorem for higher order elliptic equations, Can. Math. Bull. 20 (1977), 1-8.
6. R. Courant, and D. Hilbert, Methods of Mathematical Physics, Vol. II, (Wiley-Interscience, New York, 1962).
7. A. Devinatz, Essential self-adjointness of Schrödinger-type operators, J. of Functional Anal. 25 (1977), 58-69.
8. G.J. Etgen, and R.T. Lewis, The oscillation of elliptic systems, Math.Nachr. 94 (1980), 43-50.
9. W.D. Evans, On the essential self-adjointness of powers of Schrödinger-type operators, Proc. Royal Soc. Edin. 79A (1977), 61-77.
10. P.R. Garabedian, Partial Differential Equations. (Wiley, New York, 1964).
11. R. Hadass, On the zeros of the solutions of the differential equation $y^{(n)}(z) + p(z)y(z)=0$. Pacific J. Math 31 (1969), 33-46.
12. H. Kalf, On the characterization of the Friedrichs extension of ordinary or elliptic differential operators with a strong singular potential, J. Functional Anal. 10 (1972), 230-250.
13. ___________, Self-adjointness for strongly singular potentials with $a - |x|^2$ fall-off at infinity, Math. Z. 133 (1973), 249-255.
14. H. Kalf and J. Walter, Strongly singular potentials and essential self-adjointness of singular elliptic operators in $C_0^\infty(R^n\setminus\{0\})$, J. Functional Anal. 10 (1972), 114-130.
15. I. Knowles, On essential self-adjointness for singular elliptic differential operators, Math. Ann. 227 (1977), 155-172.
16. ___________, On essential self-adjointness for Schrödinger operators with wildly oscillating potentials, J. Math. Anal. Appl. 66 (1978), 574-585.
17. K. Kreith and C. Travis, Oscillation criteria for self-adjoint elliptic equations, Pacific J. Math. 41 (1972), 743-753.
18. C. Loewner and L. Nirenberg, Partial differential equations invariant under conformal or projective transformation, from Contributions to Analysis (Academic Press, New York, 1974).
19. J.H. Michell, The inversion of plane stress, Proc. London Math. Soc. 34 (1901), 134-142.
20. E. Müller-Pfeiffer, Kriterien für die Oszillation von elliptischen Differentialgleichungen höherer Ordnung, Math. Nachr. to appear.
21. E.S.Noussair, Oscillation of elliptic equations in general domains, Can. J. Math. 27 (1975), 1239-1245.
22. M. Riesz, Intégrales de Riemann-Liouville et Potentiels, Acta-Szeged 9 (1938), 1-42.
23. M. Schechter, On the spectra of singular elliptic operators, Mathematika 23 (1976), 107-115.
24. E. Sternberg and R.A. Eubanks, On the method of inversion in the two-dimensional theory of elasticity, Quarterly of Applied Math. 8 (1951), 392-395.
25. W. Walter and H. Rhee, Entire solutions of $\Delta^p u = f(r,u)$. Proc. Royal Soc. Edinburgh 82A (1979), 189-192.

Nonlinear Second Order Boundary Value Problems:
An Existence-Uniqueness Theorem of S. N. Bernstein

John V. Baxley

1. Introduction.

In 1912, S. N. Bernstein [6] proved that the boundary value problem consisting of the second order ordinary differential equation

$$y'' = f(x,y,y'), \quad a \leq x \leq b \tag{1}$$

and the fixed end conditions $y(a) = A$, $y(b) = B$ has a unique solution if $f(x,y,z)$ satisfies

(i) for each fixed $z \in \mathbb{R}$, the functions f, $\frac{\partial f}{\partial x}$, $\frac{\partial f}{\partial y}$ are all continuous for $(x,y) \in [a,b] \times \mathbb{R}$;

(ii) there exists $k > 0$ for which $\frac{\partial f}{\partial y} > k$ on $[a,b] \times \mathbb{R}^2$;

(iii) there exist nonnegative functions $\alpha(x,y)$, $\beta(x,y)$ such that

$$|f(x,y,z)| \leq \alpha(x,y)z^2 + \beta(x,y), \text{ for } (x,y,z) \in [a,b] \times \mathbb{R}^2,$$

and $\alpha(x,y)$, $\beta(x,y)$ are bounded on each bounded subset of $[a,b] \times \mathbb{R}$.

This theorem, which is quoted in the calculus of variations books by Achiezer [1] and Gelfand-Fomin [7], seems to have been otherwise largely ignored or unknown. Although much attention has been given in recent years to existence and uniqueness theorems for such two-point boundary value problems (see e.g. [2], [4], [5], [9]) none of these works references the Bernstein theorem. Recent work has allowed much more general boundary conditions, even nonlinear and not separated, and has relaxed the smoothness requirements of (i) considerably; meanwhile some trade-off has occurred between (ii) and (iii). Recent results have tended to relax (ii) to require only that f be nondecreasing in y for fixed x, z, but have required more in (iii), for example that $\frac{\partial f}{\partial z}$ is bounded so that f is almost linear in z or that $\frac{\partial f}{\partial z}$ grows no worse than $\log z$ as $|z| \to \infty$. Very recently, Jackson and Palamides [8] have obtained results which allow the quadratic behavior of (iii) but make other rather complicated hypotheses through which the boundary conditions are restricted by the behavior of f.

The purpose of this paper is to apply the methods of [3] to prove a simply stated improvement of the Bernstein theorem. Thus we assume familiarity with the contents of [3]; indeed the present paper is rightly viewed as a continuation of that work.

2. Main Results.

The careful reader will see that our results could be stated in a somewhat stronger, albeit more awkward, form. We prefer the simplicity of the slightly weaker statements.

Our basic assumptions on $f(x,y,z)$ are

(a) $f(x,y,z)$ is continuous on $[a,b] \times \mathbb{R}^2$;

(b_1) $f(x,y,z)$ is strictly increasing in y for fixed $(x,z) \in [a,b] \times \mathbb{R}$, or $f(x,y,z)$ is nondecreasing in y for fixed $(x,z) \in [a,b] \times \mathbb{R}$ and satisfies a uniform Lipschitz condition on each compact subset of $[a,b] \times \mathbb{R}^2$ with respect to z.

(b_2) for each $s > 0$, there exist $z_1 > s$, $z_2 < -s$ and real y_1, y_2 such that

$$f(x,y_1,z_1) > 0, \quad f(x,y_2,z_2) < 0$$

for $a \leq x \leq b$;

(c) For each pair $c,d \in \mathbb{R}$, there exist $M > 0$, $s > 0$ such that for all $(x,y) \in [a,b] \times [c,d]$

$$f(x,y,z) \geq -Mz^2, \text{ for } z \geq s,$$
$$f(x,y,z) \leq Mz^2, \text{ for } z \leq -s.$$

Let $\mathcal{C}$ be a (continuous) curve in $\mathbb{R}^2$ with parametric representation $y = y(\gamma)$, $z = z(\gamma)$, for $\gamma \in \mathbb{R}$, such that

(d_1) $y(\gamma)$, $z(\gamma)$ are nondecreasing and

(d_2) $z(\gamma) \to +\infty$ as $\gamma \to +\infty$, $z(\gamma) \to -\infty$ as $\gamma \to -\infty$.

Our first theorem deals with global existence of solutions of the initial value problem

$$y'' = f(x,y,y'), \quad a \leq x \leq b, \tag{1}$$
$$(y(a),y'(a)) \in \mathcal{C}. \tag{2}$$

Theorem 1. Suppose f satisfies the conditions (a) through (c) and $\mathcal{C}$ satisfies (d_1), (d_2). Then the initial value problem (1), (2) has at least one solution which exists on the entire interval $a \leq x \leq b$.

Our other three theorems deal with the boundary value problem consisting of (1), (2) and the mixed boundary condition

$$h(y(a),y'(a),y(b),y'(b)) = 0. \tag{3}$$

About the boundary condition (3) we shall assume

(e_1) $h(y(\gamma),z(\gamma),u,v)$ is a continuous function of $(\gamma,u,v) \in \mathbb{R}^3$;

(e_2) for (u,v) fixed, $h(y(\gamma),z(\gamma),u,v)$ is a nondecreasing function of γ;

(e_3) There exists $N > 0$ so that for fixed γ and $|u| \geq N$, $|v| \geq N$, then $h(y(\gamma),z(\gamma),u,v)$ is nondecreasing function of each of the variables u, v.

Here are two existence theorems.

Theorem 2. Suppose f and ζ satisfy the hypotheses of theorem 1 and that h satisfies (e_1), (e_2), (e_3). Suppose further that for each fixed pair (γ,u), there exist v_1, v_2 so that $h(y(\gamma),z(\gamma),u,v_2) \geq 0$ and $h(y(\gamma),z(\gamma),u,v_1) \leq 0$. Then the boundary value problem (1), (2), (3) has at least one solution.

The simplest case of theorem 2 is that in which the boundary conditions have the form $y(a) = A$, $y'(b) = B$. On the other hand, the simplest case of the next theorem is that of the fixed end conditions $y(a) = A$, $y(b) = B$.

Theorem 3. Suppose that both $f(x,y,z)$ and $f(x,y,-z)$ satisfy the conditions (a) through (c) and that ζ satisfies (d_1), (d_2). Suppose further that h satisfies (e_1), (e_2), (e_3) and for each fixed γ, there exist (u_i,v_i), $i = 1, 2$ so that $h(y(\gamma), y'(\gamma),u_2,v_2) \geq 0$ and $h(y(\gamma),z(\gamma),u_1,v_1) \leq 0$. Then the boundary value problem (1), (2), (3) has at least one solution.

At the risk of some oversimplification, the following comments are designed to shed light on Theorem 3. Using the shooting method on the problem with fixed end conditions $y(a) = A$, $y(b) = B$, one begins with a solution with large initial slope satisfying $y(a) = A$, hoping thereby to overshoot $y(b) = B$. If this solution exists for $a \leq x \leq b$, the hypotheses on f guarantee that the slope $y'(x)$ will remain large throughout $a \leq x \leq b$ and consequently assures that the altitude $y(b)$ can be made larger than B. If, however, solutions with large initial slope fail to exist on the entire interval $a \leq x \leq b$, it turns out that with $f(x,y,z)$ satisfying hypotheses (a) through (c), one can only guarantee that $y'(b)$ can be made arbitrarily large using global solutions satisfying $y(a) = A$. If also $f(x,y,-z)$ satis- hypotheses (a) through (c), one can then change variables, replacing x by

$a + b - x$, to verify the "looking-back" lemma (cf. [3, Lemma 2.9]) and use this lemma to ensure that $y(b)$ can be made arbitrarily large using global solutions satisfying $y(a) = A$.

We pass to the uniqueness question.

Theorem 4. Suppose f and ζ satisfy the hypotheses of theorem 1 and suppose h satisfies (e_2) and also

(f_1) for each fixed γ, if u, v both increase, then $h(y(\gamma),z(\gamma),u,v)$ increases,

(f_2) either $z(\gamma)$ is a strictly increasing function of $\gamma \in \mathbb{R}$ or $h(y(\gamma),z(\gamma),u,v)$ is a strictly increasing function of u for fixed (γ,v).

Then the boundary value problem (1), (2), (3) has at most one solution.

The purpose of the hypothesis (f_2) in this last theorem is to rule out the case of two solutions which differ by a constant. If the differential equation $y'' = f(x,y,y')$, $a \leq x \leq b$, does not have two solutions differing by a constant, then hypothesis (f_2) is not required.

The Bernstein theorem is essentially a corollary of theorems 3 and 4. However, it should be observed that the theorems here require that f be jointly continuous in all three variables x, y, z whereas the Bernstein theorem required only joint continuity in x, y for fixed z. Of course, the theorem here removes the other smoothness requirements of Bernstein, weakens the requirement (ii), and allows a much bigger variety of boundary conditions.

3. Global Existence.

We apply the program of [3] to prove theorem 1. The following two lemmas are fundamental. The conditions (a) through (c) on $f(x,y,z)$ are assumed satisfied in both lemmas.

Lemma 1. Suppose $\phi_1(x)$, $\phi_2(x)$ have continuous second derivatives on an interval $[a_1,b_1) \subset [a,b]$ and satisfy

$$\phi_1''(x) \leq f(x,\phi_1(x),\phi_1'(x)),\ \phi_2''(x) \geq f(x,\phi_2(x),\phi_2'(x))$$

for $a_1 \leq x < b_1$. Suppose further that $\phi_1(a_1) \leq \phi_2(a_1)$, $\phi_1'(a_1) \leq \phi_2'(a_1)$ and $\phi_1(a_1) + \phi_1'(a_1) < \phi_2(a_1) + \phi_2'(a_1)$. Then $\phi_1'(x) \leq \phi_2'(x)$, $\phi_1(x) \leq \phi_2(x)$ for

$a_1 \leq x < b_1$.

Proof. If f is nondecreasing in y and satisfies the Lipschitz condition of condition (b_1), the proof is a straightforward application of the maximum principle for linear differential inequalities; the details are essentially the same as [3, Lemma 2.11]. In the alternative case of (b_1) that $f(x,y,z)$ is strictly increasing in y for fixed (x,z), we proceed as follows. Let $u(x) = \phi_2(x) - \phi_1(x)$; then $u''(x) = \phi_2''(x) - \phi_1''(x) \geq f(x,\phi_2(x),\phi_2'(x)) - f(x,\phi_1(x),\phi_1'(x))$. Note that $u(a_1) \geq 0$, $u'(a_1) \geq 0$ and $u(a_1) + u'(a_1) > 0$. If there exists $c \in (a_1,b_1)$ for which $u'(c) < 0$, then u must attain a positive maximum on $[a_1,c)$ at some point d. Then

$$u(d) = \phi_2(d) - \phi_1(d) > 0, \quad u'(d) = \phi_2'(d) - \phi_1'(d) = 0 ,$$

whence

$$u''(d) \geq f(d,\phi_2(d),\phi_2'(d)) - f(d,\phi_1(d),\phi_1'(d)) > 0 .$$

Hence, u has a minimum at d, a contradiction.

Lemma 2. Let $\phi_k(x)$, $k = 1, 2, \ldots$, be a solution of $y'' = f(x,y,y')$ on some interval $I_k \subset [a,b]$ and suppose there exists $x_k \in I_k$ such that $c = \inf\{\phi_k(x_k)\} > -\infty$; $\phi_k'(x_k) \to +\infty$ as $k \to \infty$. Then given $K \geq 0$, there exists n such that $\phi_n'(x) \geq K$ for $x \in I_n$, $x \geq x_n$.

Proof. By (b_2), there exist $z_1 > K$ and real y_1 so that $f(x,y_1,z_1) > 0$ for $a \leq x \leq b$. By (c) there exist $M > 0$, $s > 0$ so that for all $(x,y) \in [a,b] \times [c,y_1]$,

$$f(x,y,z) \geq -Mz^2 \quad \text{for } z \geq s.$$

By increasing s if necessary, we may assume that $s \geq z_1$. Choose n so large that (we assume, by (b_1), that $y_1 \geq c$)

$$\phi_n'(x_n) > s \exp [M(y_1 - c)]. \tag{4}$$

We complete the proof by showing that $\phi_n'(x) > z_1$ for $x \in I_n$, $x \geq x_n$. If our assertion is false, since $s \geq z_1$, we may let d be the first point to the right of x_n for which $\phi_n'(x) = s$.

Note that $\phi_n(x_n) \geq c$ and since $s > 0$, ϕ_n is increasing on $[x_n,d]$. We assert the existence of a point $\tilde{c} \in [x_n,d)$ for which $\phi_n(\tilde{c}) \geq y_1$. If no such point exists, then $(x,\phi_n(x)) \in [a,b] \times [c,y_1]$ and $\phi_n'(x) \geq s$ for $x \in [x_n,d)$. Thus $f(x,\phi_n(x),\phi_n'(x)) \geq -M[\phi_n'(x)]^2$ for $x \in [x_n,d)$, and $\phi_n(x)$ is a solution of

$$y'' \geq -M(y')^2$$
$$y(x_n) = \phi_n(x_n) , \quad y'(x_n) = \phi_n'(x_n)$$

for $x_n \leq x < d$.

The solution of the problem

$$y'' = -M(y')^2$$
$$y(x_n) = \phi_n(x_n) - \varepsilon, \quad y'(x_n) = \phi_n'(x_n)$$

(where $\varepsilon > 0$) is

$$\psi(x) = \phi_n(x_n) - \varepsilon + M^{-1} \log [M \phi_n'(x_n)(x-x_n) + 1].$$

By lemma 1,

$$\psi'(x) \leq \phi_n'(x) , \quad \psi(x) \leq \phi_n(x)$$

for $x_n \leq x < d$. It is easy to verify that the equation $\psi(x) = y_1$ has a unique solution $\bar{x} > x_n$ which (because $\phi_n(x_n) \geq c$) satisfies the inequality

$$M \phi_n'(x_n)(\bar{x} - x_n) + 1 \leq \exp [M(y_1 - c + \varepsilon)].$$

Since

$$\psi'(x) = \frac{\phi_n'(x_n)}{M \phi_n'(x_n)(x - x_n) + 1}$$

is decreasing for $x \geq x_n$, then for $x_n \leq x \leq \bar{x}$,

$$\psi'(x) \geq \psi'(\bar{x}) \geq \frac{\phi_n'(x_n)}{\exp[M(y_1 - c + \varepsilon)]} .$$

Using (4) and choosing $\varepsilon > 0$ sufficiently small we have

$$\psi'(x) > s, \quad \text{for } x_n \leq x \leq \bar{x} .$$

Since $\phi_n'(x) \geq \psi'(x)$ for $x_n \leq x < d$, and $\phi_n'(d) = s$, it follows that $\bar{x} < d$. But then $\phi_n(\bar{x}) \geq \psi(\bar{x}) = y_1$, contradicting our assumption that no such point exists.

Now we know there exists $\tilde{c} \in [x_n, d)$ for which $\phi_n(\tilde{c}) \geq y_1$ and $\phi_n'(x) > s \geq z_1$ for $x_n \leq x \leq \tilde{c}$. We now let $\bar{c}$ be the first point to the right of x_n for which $\phi_n'(x) = z_1$. Then $\bar{c} > \tilde{c}$ and $\phi_n(\bar{c}) > \phi_n(\tilde{c}) \geq y_1$. Let $u(x) = \phi_n(x) - z_1 x$ for $\tilde{c} \leq x \leq \bar{c}$. Then $u'(x) = \phi_n'(x) - z_1 > 0$ for $\tilde{c} \leq x < \bar{c}$ and thus u is increasing on $[\tilde{c},\bar{c}]$. However,

$$u''(\bar{c}) = \phi_n''(\bar{c}) = f(\bar{c}, \phi_n(\bar{c}), \phi_n'(\bar{c}))$$
$$\geq f(\bar{c}, y_1, z_1) > 0,$$

which implies that u has a minimum at $\bar{c}$, contradicting the fact that u is increasing on $(\tilde{c},\bar{c})$, and the proof of lemma 2 is complete.

Proof of Theorem 1. The proof proceeds almost verbatim as in [3, sections 2 and 3]. Lemma 1 gives the conclusion of [3, lemma 2.11] and lemma 2 gives the conclusion of [3, lemma 2.6]. With the exception of lemma 2.9 (the "looking-back" lemma) and lemma 2.10 (2), which are not needed to prove Theorem 1, the other lemmas of [3, section 2] after lemma 2.6 follow just as before. Lemma 2.9 was used only in the proof of lemma 2.10 (2), and lemma 2.10 was used at only one place in [3, section 3]; our present hypotheses allow the use of lemma 2.10 (1) at that point in the proof.

4. Boundary Value Problems.

The proofs of [3, section 4] can be adapted without any essential change to prove theorems 2-4. A close examination of [3, section 4] shows that lemma 2.10 (2) is not used and that lemma 2.9 is only used in the proof of the equation (4.2), which is not needed for our present theorem 2 or theorem 4. Under the hypotheses of our present theorem 3, lemma 2.9 of [3] is easily verified and consequently the equation (4.2) of [3] is valid.

Acknowledgement. The author is grateful to Professor W. N. Everitt for bringing the Bernstein theorem to his attention.

REFERENCES

1. N. I. Akhiezer, The Calculus of Variations, Blaisdell, New York, 1962.

2. P. Bailey, L. F. Shampine, and P. Waltman, Nonlinear Two Point Boundary Value Problems, Academic Press, New York, 1968.

3. J. V. Baxley and S. E. Brown, Existence and uniqueness for two-point boundary value problems, Proc. Roy. Soc. Edinburgh 88A (1981), 219-234.

4. J. Bebernes and R. Gaines, A generalized two-point boundary value problem, Proc. Amer. Math. Soc. 19 (1968), 749-754.

5. S. R. Bernfield and V. Lakshmikantham, An Introduction to Nonlinear Boundary Value Problems, Academic Press, New York, 1974.

6. S. N. Bernstein, Sur les equations du calcul des variations, Ann. Sci. Ecole Norm. Sup. 29 (1912), 431-485.

7. I. M. Gelfand and S. V. Fomin, Calculus of Variations, Prentice-Hall, Englewood Cliffs, N.J., 1963.

8. L. K. Jackson and P. K. Palamides, An existence theorem for a nonlinear two point boundary value problem, to appear.

9. H. B. Keller, Existence theory for two-point boundary value problems, Bull. Amer. Math. Soc. 72 (1966), 728-731.

A DEFINITENESS RESULT FOR DETERMINANTAL OPERATORS

Paul Binding* and Patrick J. Browne*

Dedicated to Professor F.V. Atkinson

1. Introduction. Let $H_1,\ldots,H_k$ be Hilbert spaces and $H = \overset{k}{\underset{r=1}{\otimes}} H_r$ their tensor product. For $1 \le i,j \le k$, let $V_{ij}\colon H_i \to H_i$ be bounded symmetric operators with associated quadratic forms

$$v_{ij}(x_i) = (V_{ij}x_i,x_i),\quad x_i \in H_i,\quad 1 \le i,j \le k.$$

We write $V(x)$ for the $k\times k$ matrix with (i,j)th element $v_{ij}(x_i)$ and $\delta_0(x)$ for the determinant of this matrix.

Each operator V_{ij} can be regarded as acting in H with its action on a decomposable tensor given by

$$V_{ij}(x_1 \otimes\ldots\otimes x_k) = x_1 \otimes\ldots\otimes V_{ij}x_i \otimes\ldots\otimes x_k$$

with this definition extended to all of H by linearity and continuity. We can then form the determinantal operator

$$\Delta_0 = \det(V_{ij}),$$

the determinant being expanded formally. This is a standard construction

---

* Research supported in part by grants from the NSERC of Canada.

in multiparameter spectral theory where definiteness properties of Δ_0 are of prime importance. When x is a decomposable tensor we have $(\Delta_0 x, x) = \delta_0(x)$. Suitable references are [4,8].

We write

$V_{ij} \geq 0$ if $v_{ij}(x_i) \geq 0$ for all $x_i \in H_i$,

$V_{ij} > 0$ if $v_{ij}(x_i) > 0$ for all $x_i \neq 0$, $x_i \in H_i$,

$V_{ij} \gg 0$ if $v_{ij}(u_i) \geq \alpha > 0$ for some constant $\alpha > 0$ and for all $u_i \in H_i$, $\|u_i\| = 1$.

Analogous definitions hold in H and also for the operator Δ_0. We also define two Minkowski conditions for the array of operators $[V_{ij}]$ as follows:

M_1: $-V_{ij} > 0$ whenever $i \neq j$, $1 \leq i,j \leq k$,

M_2: $\sum_{j=1}^{k} V_{ij} > 0$, $1 \leq i \leq k$.

Corresponding weak and strong versions of these conditions, denoted by M_i^w, M_i^s, $i = 1,2$, are obtained by using ≥ 0 and $\gg 0$ respectively in place of > 0.

We write M for the combination of M_1 and M_2 with M^w and M^s defined similarly. We aim to relate the Minkowski conditions to standard multiparameter conditions (denoted by RD_Δ and LD_Δ in [4]):

R: $\Delta_0 > 0$

L: For some $\underset{\sim}{\omega} \in \mathbb{R}^k$, $\|\underset{\sim}{\omega}\| = 1$, we have

$$\sum_{j=1}^{k} \omega_j \Delta_{0ij} > 0,\ 1 \leq i \leq k,$$

where Δ_{0ij} denotes the cofactor of V_{ij} in the determinantal expansion of Δ_0. Again there are weak and strong versions of these conditions.

As further notation, $\tilde{M}_1$, $\tilde{M}_2$ will denote M_1, M_2 with each V_{ij} replaced by

$$\tilde{V}_{ij} = \sum_{n=1}^{k} V_{in} a_{nj}$$

for some non-singular $k \times k$ matrix of constants $[a_{nj}]$. It will be unnecessary to display the dependence of $\tilde{V}_{ij}$ on $[a_{nj}]$.

Our principal result, to be established in the subsequent sections, is

THEOREM 1. *$\tilde{M}^s$ holds, if, and only if, R^s and L^s both hold.*

We shall also discuss the relationships between the weaker versions of the conditions.

We shall now relate some of our results to those in the literature of multiparameter spectral theory developed by Atkinson [1] and subsequent workers. The implication $M \Rightarrow R$, established below, is a development of Minkowski's result. The implication $R \Rightarrow \tilde{M}_1^w$ is contained in [1, Theorem 9.7.1] while $R^s \Rightarrow \tilde{M}_1^s$ was proved explicitly by us in [7, Theorem 1]. The connexions between M and L are related to the following result, contained in [2, Theorem (2.3), p. 134], although we do not use it directly.

PROPOSITION 1. If M_1^w holds, then the following are equivalent.

(i) *Each element of $V(x)^{-1}$ is nonnegative.*

(ii) *Each principal minor of $V(x)$ is positive.*

(iii) $\tilde{M}_2$ *holds with* $[a_{nj}]$ *diagonal.*

In particular, M implies these statements, as we shall see.

Condition M involves only Hermitian forms on the individual spaces H_i. Conditions R and L involve determinantal operators on the full tensor product space H. Checking R and L is thus a non-trivial task although the following is known for the strong case.

PROPOSITION 2. $\Delta_0 >> 0$ *on decomposable tensors* $\Rightarrow R^s$.

This is a result of Atkinson [1, Theorem 7.8.2] in finite dimensions, extended to separable Hilbert spaces by Binding [3].

The combination of R^s and L^s, termed "properness", was originally isolated by us in connexion with the "comparison cone"

$$C = \{V(x)^{-1}\underset{\sim}{q} \mid 0 \neq x_i \in H_i, \underset{\sim}{q} \in \mathbb{R}^k_-\},$$

$\mathbb{R}^k_-$ being the nonpositive orthant of $\mathbb{R}^k$. Under certain conditions, (cf. [6, 8]), the eigentuples $\underset{\sim}{\lambda}$ of a multi-parameter problem

$$(T_r + \sum_{s=1}^{k} \lambda_s V_{rs})x_r = 0 \neq x_r \in H_r, \; 1 \leq r \leq k,$$

may be compared with those of a similar system (distinguished by primes) via the relation $\underset{\sim}{\lambda} - \underset{\sim}{\lambda}' \in C$.

If M holds then by Propositions 3 and 4 below, $V(x)^{-1}$ has positive entries and thus $C \subseteq \mathbb{R}^k_-$ and our comparison is at least as strong as the component-wise partial order. All previous such results have been restricted to strong definiteness.

We remark that properness also leads to the richest spectral theory for multiparameter problems. See, for example, [5, 8] for abstract oscillation, comparison, completeness and expansion results based on variational methods. The other definiteness conditions have proved more difficult to use although there are expansion theorems based on Atkinson's "limiting eigenvectors", [1, Chapter 11], and M may prove useful there.

The matrix $A = [a_{ij}]$ used to define $\widetilde{M}$ corresponds to a non-singular linear transformation of the eigenvalue $(\underset{\sim}{\lambda})$ space and thus Proposition 5 below may be regarded as providing a canonical form, M_1^s, for L^s. Actually we shall see that $V_{ii} >> 0$ and $\Delta_{0ij} >> 0$ can be achieved in addition. Likewise
M^s is a canonical form for properness and again we can take each $\Delta_{0ij} >> 0$.

2. Consequences of the Minkowski conditions. We shall need the following easily established facts.

LEMMA 1. *If* A *is a self-adjoint operator on a Hilbert space* H_1 *and* $A^\dagger = A \otimes I$ *is the operator induced by* A *on* $H = H_1 \otimes H_2$, *then* $A >> 0 \Rightarrow A^\dagger >> 0$, $A > 0 \Rightarrow A^\dagger > 0$ *and* $A \geq 0 \Rightarrow A^\dagger \geq 0$. *If* A, B *are two self-adjoint operators on a Hilbert space* H, $AB = BA$ *and* $A \geq 0$, $B \geq 0$, *then* $AB \geq 0$ *with corresponding results holding with* ≥ 0 *replaced throughout by* > 0 *or* $>> 0$.

PROPOSITION 3. $M \Rightarrow R$, $M^s \Rightarrow R^s$, $M^w \Rightarrow R^w$.

Proof. The implication $M \Rightarrow R$ is obvious in the case $k = 1$. Suppose the result holds for arrays of operators $[V_{ij}]$ of sizes $1,2,\ldots,k$ and consider an array $[V_{ij}]$ of size $(k+1)\times(k+1)$ for which M holds.

We write

$$S_i = \sum_{j=1}^{k+1} V_{ij}, \quad D_i = V_{ii} - S_i, \quad 1 \le i \le k+1.$$

Then $S_i > 0$ by M_2 and $D_i > 0$ by M_1. Now

$$\Delta_0 = \begin{vmatrix} S_1+D_1 & V_{1,2} & \cdots & V_{1,k+1} \\ V_{2,1} & S_2+D_2 & \cdots & V_{2,k+1} \\ \vdots & \vdots & & \vdots \\ V_{k+1,1} & V_{k+1,2} & \cdots & S_{k+1}+D_{k+1} \end{vmatrix}$$

Note that operators from different rows of this determinant commute. The term independent of $S_1,\ldots,S_{k+1}$ is a determinant whose columns sum to the column of zero operators; thus this term is zero. The term involving only S_1 is

$$S_1 \begin{vmatrix} D_2 & \cdots & V_{2,k+1} \\ \vdots & & \vdots \\ V_{k+1,2} & \cdots & D_{k+1} \end{vmatrix}$$

and the determinant here is of size $k\times k$ and satisfies M. By our inductive hypothesis and Lemma 1, the term is positive as will be all the terms involving just one S_i. In like fashion, the terms involving progressively $2,3,\ldots,k$ of the S_i's have as "coefficients" operator determinants of size smaller than $k\times k$ and satisfying M. Hence all these terms are positive. The final term is $S_1\ S_2\ \ldots\ S_{k+1}$ which is positive by Lemma 1. Thus $\Delta_0 > 0$ and the result follows by induction.

The result $M^s \Rightarrow R^s$ is proved similarly except that inequalities > 0 in the above argument must be replaced by $\gg 0$. Likewise we obtain $M^w \Rightarrow R^w$.

A careful analysis of the proof shows that we actually have

$$M_1^w \text{ and } M_2 \Rightarrow R,$$
$$M_1^w \text{ and } M_2^s \Rightarrow R^s.$$

PROPOSITION 4. $M \Rightarrow L$, $M^s \Rightarrow L^s$, $M^w \Rightarrow L^w$.

Proof. We shall concentrate on the implication $M \Rightarrow L$; the other results follow from similar arguments with appropriate strengthening or weakening of the inequalities involved.

Let $[V_{ij}]$ be a $k \times k$ array of operators satisfying M where $k \geq 3$ – the result is vacuous should $k = 1$ and trivial should $k = 2$. We shall show that it is possible to use $\underset{\sim}{\omega} = (1,0,\ldots,0) \in \mathbb{R}^k$ to satisfy the definition of L.

When $i = 1$, $\sum_{j=1}^{k} \omega_j \Delta_{0ij}$ becomes Δ_{011}, and using Proposition 3 we see that $\Delta_{011} > 0$. For $i > 1$, we must investigate the determinant

$$(-1)^{i+1} \begin{vmatrix} V_{12} & \cdots & V_{1k} \\ \vdots & & \\ V_{k2} & \cdots & V_{kk} \end{vmatrix} \leftarrow \text{row } i \text{ deleted}$$

which, after permutation of columns, can be written as

$$(-1) \begin{vmatrix} V_{1,i} & V_{1,2} & \cdots & V_{1,i-1} & V_{1,i+1} & \cdots & V_{1,k} \\ \vdots & & & & & & \\ V_{k,i} & V_{k,2} & \cdots & V_{k,i-1} & V_{k,i+1} & \cdots & V_{k,k} \end{vmatrix} \leftarrow \text{row } i \text{ deleted}.$$

We now have to investigate, in general, a determinant of the form $\det(T_{ij})_{1\le i,j\le n}$, where

$$T_{ij} > 0, \quad 1 \le j \le n,$$

$$T_{ij} < 0, \quad i > 1,\ j \ne i, \quad 1 \le i,j \le n,$$

$$\sum_{j=1}^{n} T_{ij} > 0, \quad 1 \le i \le n.$$

It is trivial to see that such a determinant is positive if $n = 1$. Suppose all such determinants of sizes $1,2,\ldots,n$ are positive and consider one of size $(n+1)\times(n+1)$. As before we write

$$S_i = \sum_{j=1}^{n+1} T_{ij}, \quad D_i = T_{ii} - S_i, \quad 2 \le i \le n+1,$$

and our determinant becomes

$$\begin{vmatrix} T_{11} & T_{12} & \cdots & T_{1,n+1} \\ T_{21} & S_2+D_2 & \cdots & T_{2,n+1} \\ \vdots & \vdots & & \vdots \\ T_{n+1,1} & T_{n+1,2} & \cdots & S_{n+1}+D_{n+1} \end{vmatrix}$$

The "coefficients" of terms involving one or more of the S_i's are determinants of the type under study, and will be positive by our inductive hypothesis, while the remaining term is expressible as

$$\left(\sum_{j=1}^{n+1} T_{ij}\right) \begin{vmatrix} D_2 & \cdots & T_{2,n+1} \\ \vdots & & \vdots \\ T_{n+1,2} & \cdots & D_{n+1} \end{vmatrix}$$

which is positive since the determinant satisfies condition M. Thus, by induction, all determinants of this type are positive and so our result is established.

3. Consequences of conditions L and R

PROPOSITION 5. *L^s implies the existence of a non-singular matrix $A = [a_{ij}]$ of constants so that M_1^s and $\Delta_{0ij} >> 0$, $1 \le i,j \le k$, are satisfied for the array $[\tilde{V}_{ij}]$, $\tilde{V}_{ij} = \sum_{n=1}^{k} V_{in}a_{nj}$.*

Proof. By means of a preliminary rotation we may assume $\underset{\sim}{\omega} = (0,\ldots,0,1) \in \mathbb{R}^k$ in the definition of L^s. Let

$$\underset{\sim}{v}_i = \underset{\sim}{v}_i(u_i) = (v_{i1}(u_i),\ldots,v_{i,k-1}(u_i)) \in \mathbb{R}^{k-1}, \; u_i \in H_i,$$

$$C_i = \overline{co}\{\underset{\sim}{v}_i(u_i) \mid \|u_i\| = 1,\; u_i \in H_i\}; \quad 1 \le i \le k.$$

Thus C_i is essentially the projection onto $\underset{\sim}{\omega}^{\perp}$ of the closed convex hull of the ith row vectors of $V(u)$ for $\|u_i\| = 1$. For the time we shall work entirely within $\underset{\sim}{\omega}^{\perp}$ viewed as $\mathbb{R}^{k-1}$. Note that any collection of k-1 of the sets C_i is linearly independent for otherwise L^s is violated.

We define

$$\hat{C}_i = \overline{co} \bigcup_{j \ne i} C_i,$$

$$\underset{\sim}{c}_i = \text{the point of minimal norm in } \hat{C}_i,\; \underset{\sim}{c}_i \in \underset{\sim}{\omega}^{\perp},$$

$$D_i = \{\underset{\sim}{x} \in \mathbb{R}^{k-1} \mid \underset{\sim}{x}\,\underset{\sim}{c}_i \le \|\underset{\sim}{c}_i\|^2\},$$

$$D = \bigcap_{i=1}^{k} D_i .$$

Note that $\underset{\sim}{c}_i \neq \underset{\sim}{0}$ for otherwise L^s is contravened. This shows that $\underset{\sim}{0}$ is an interior point of each D_i and hence

$$\underset{\sim}{0} \in \text{int } D.$$

Next we claim that D is bounded. To see this, consider any particular choice of $u_i \in U_i$, $1 \le i \le k$. Then by [4, Lemma 3.4], applied to a problem with singleton row ranges, the $\underset{\sim}{v}_i$ cannot lie in an open half-space. It follows that

$$\underset{\sim}{0} \in \sum := co\{\underset{\sim}{v}_1(u_1),\ldots,\underset{\sim}{v}_k(u_k)\}.$$

Suppose $\underset{\sim}{d} \in \text{int } D \setminus \sum$. Then there is $t \ge 0$ such that $t\underset{\sim}{d} \in \text{int } D$ and

$$t\underset{\sim}{d} = \sum_{i \neq j} \beta_i \underset{\sim}{v}_i(u_k),\ \beta_i \ge 0,\ \sum_{i \neq j} \beta_i = 1.$$ Since $\underset{\sim}{v}_i(u_i)\underset{\sim}{c}_j \ge \|\underset{\sim}{c}_j\|^2$, we have $t\underset{\sim}{d}\underset{\sim}{c}_j \ge \|\underset{\sim}{c}_j\|^2$ which contradicts $t\underset{\sim}{d} \in \text{int } D$. Hence int $D \subseteq \sum$ which establishes our claim.

Now we have

$$D = (co\{\underset{\sim}{d}_1,\ldots,\underset{\sim}{d}_k\})^{\text{o}},\quad \underset{\sim}{d}_i = \underset{\sim}{c}_i/\|\underset{\sim}{c}_i\|^2,$$

$$D^{\text{o}} = co\{\underset{\sim}{d}_1,\ldots,\underset{\sim}{d}_k\},$$

and also, $\underset{\sim}{0} \in \text{int } D^{\text{o}}$ (see [9, Cor. 14.5.1, p. 125]. Thus there are positive numbers $\alpha_1,\ldots,\alpha_k$ such that

$$\sum_{i=1}^{k} \alpha_i \underset{\sim}{c}_i = \underset{\sim}{0}\ .$$

Now define $\underset{\sim}{e}_i$, $1 \le i \le k$, to be the vertices of D and $\underset{\sim}{f}_i = (\varepsilon\underset{\sim}{e}_i, 1) \in \mathbb{R}^k$, where $\varepsilon > 0$ will be chosen below. It is clear that the $\underset{\sim}{f}_i$ are linearly independent since $\det[f_{ij}]_{1\le i,j\le k}$ is equal to

$$\varepsilon^k \det[e_{ij}-e_{kj}]_{1\le i,j\le k-1},\ \underset{\sim}{e}_i = (e_{i1},\ldots,e_{i,k-1}),$$

which is proportional to the volume of εD. Without loss, we may assume in addition that $\det\,[f_{ij}] > 0$. Note that D has positive volume since $\underset{\sim}{0} \in \text{int } D$.

The transformation matrix $[a_{ij}]$ to give $\tilde{V}_{ij}$ will be the inverse of $[f_{ij}]$ and this leads to the transformed cofactors [4; Lemma 2.3, Theorem 2.4]

$$\begin{aligned}\tilde{\Delta}_{0ij} &= \det[f_{ij}] \sum_{n=1}^{k} \Delta_{0in} f_{jn} \\ &= \det[f_{ij}]\left(\sum_{j=1}^{k-1} \Delta_{0in}\varepsilon e_{jn} + \Delta_{0ik}\right).\end{aligned}$$

Our first condition on ε is that it be small enough to guarantee that these quantities are $\gg 0$ - note that $\Delta_{0ik} \gg 0$ by hypothesis.

It remains to give further restrictions on ε to ensure M_1^s. First note that $\underset{\sim}{e}_i\underset{\sim}{c}_j = \|\underset{\sim}{c}_j\|^2$ for any $j \neq i$ and since

$$\sum_{j=1}^{k} \alpha_j \underset{\sim}{c}_j \underset{\sim}{e}_j = 0\ ,$$

we have

$$\underset{\sim}{c}_i\underset{\sim}{e}_i = -\,\alpha_i^{-1} \sum_{j\neq i} \alpha_j \underset{\sim}{c}_j \underset{\sim}{e}_i = -\,\alpha_i^{-1} \sum_{j\neq i} \alpha_j \|\underset{\sim}{c}_j\|^2 < 0.$$

Further, if we put $\underset{\sim}{\nu}_i = (-\underset{\sim}{c}_i, \varepsilon\|\underset{\sim}{c}_i\|^2)$, $1 \le i \le k$, then from above

$$\underset{\sim}{\nu}_i \underset{\sim}{f}_i = -\,\underset{\sim}{c}_i\underset{\sim}{e}_i + \varepsilon\|\underset{\sim}{c}_i\|^2 = q_i > 0$$

and

$$\underset{\sim}{v}_i \underset{\sim}{f}_j = -\underset{\sim}{c}_i \varepsilon \underset{\sim}{e}_j + \varepsilon\|c_j\|^2 = 0, \; j \neq i \;.$$

Thus the inverse of $[f_{ij}]$ has as i^{th} column $\underset{\sim}{v}_i/q_i$. We now calculate, for a given choice of $u_i \in U_i$,

$$\begin{aligned}\tilde{v}_{ij}(u_i) &= -\sum_{n=1}^{k-1} v_{in}(u_i)\frac{1}{q_j}c_{jn} + v_{ik}(u_i)\varepsilon\|\underset{\sim}{c}_j\|^2/q_j \\ &\leq \; -\|\underset{\sim}{c}_j\|^2/q_j + O(\varepsilon), \text{ for } i \neq j \\ &< 0 \;, \text{ for sufficiently small } \varepsilon.\end{aligned}$$

This gives the remaining conditions on ε so as to give the desired result.

REMARK. A similar condition on ε will ensure that $\tilde{v}_{ii}(u_i) > 0$ in addition, as mentioned in the introduction. The converse of Proposition 5 fails, as easy examples show with all $V_{ij} \ll 0$.

PROPOSITION 6. *L^s and R^s imply $\tilde{M}^s$.*

Proof. By Proposition 5 we may assume M_1^s and $\Delta_{0ij} \gg 0$, $1 \leq i,j \leq k$, after a non-singular linear transformation. Since $\delta_0(x)$ and $\delta_{0ij}(x)$ are all positive for non zero x, $V(x)^{-1}$ has positive entries. By R^s and [7, Theorem 1], there exists $\underset{\sim}{p} \in \mathbb{R}^k$ so that

$$\sum_{j=1}^{k} p_j V_{\ell j} \gg 0, \; 1 \leq \ell \leq k. \tag{*}$$

We then obtain

$$p_i = \sum_{j=1}^{k} \sum_{\ell=1}^{k} (V(x))^{-1}_{i\ell} v_{\ell j}(x) p_j > 0.$$

The extra transformation required is given by the non-singular diagonal matrix $D = \operatorname{diag}(p_1,\ldots,p_k)$. It is easily seen that M_1^s and $\Delta_{0ij} \gg 0$ persist after applying D while M_2^s corresponds to (*).

EXAMPLE. The following example shows that

$$(L \text{ and } R) \not\Rightarrow \widetilde{M}^w,$$

so the weaker versions of Proposition 6 both fail.

Let H be a Hilbert space with orthonormal basis $e_1, e_2, \ldots$ and put $Te_n = -e_n/n$, $Se_n = e_n/n$, thereby defining bounded operators T and S. On $H \otimes H$ consider the 2×2 array $[V_{ij}]$

$$\begin{bmatrix} I & T \\ I & S \end{bmatrix}.$$

V_1, V_2, the cones generated by the points $(v_{i1}(u_i), v_{i2}(u_i)) \in \mathbb{R}^2$, $i = 1,2$, are as shown in the accompanying manutract.

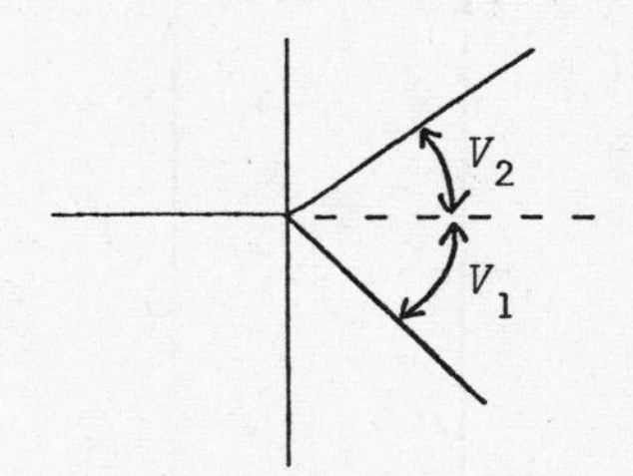

It is easy to check that this array satisfies L and R but cannot be transformed to satisfy $\widetilde{M}^w$.

References

[1] F.V. Atkinson, Multiparameter Spectral Theory, vol. 1, Academic Press, 1972.

[2] A. Berman, R.J. Plemmons, Nonnegative Matrices in the Mathematical Sciences, Academic Press, 1979.

[3] P.A. Binding, Another positivity result for determinantal operators, Proc. Roy. Soc. Edin., 86A (1980), 333-337.

[4] P.A. Binding, Multiparameter definiteness conditions, Proc. Roy. Soc. Edin., 89A (1981), 319-332.

[5] P.A. Binding, Multiparameter variational principles, SIAM J. Math. Anal., to appear.

[6] P.A. Binding, P.J. Browne, A variational approach to multiparameter eigenvalue problems in Hilbert space, SIAM J. Math. Anal., 9 (1978), 1054-1067.

[7] P.A. Binding, P.J. Browne, Positivity results for determinantal operators, Proc. Roy. Soc. Edin., 81A (1978), 267-271.

[8] P.A. Binding, P.J. Browne, Comparison cones for multiparameter eigenvalue problems, J. Math. Anal. Appl. 77 (1980), 132-149.

[9] R.T. Rockafellar, Convex Analysis, Princeton U. Press, 1970.

On second-order left-definite boundary value problems

C. Bennewitz and W.N. Everitt

Christer Bennewitz
Uppsala University
Department of Mathematics
Thunbergsvägen 3
S - 752 38 UPPSALA, Sweden

1. Introduction

This paper is concerned with eigenvalue problems associated with second-order linear differential equations which, at least formally, can be written

$$M[y] = \lambda S[y] \quad \text{on} \quad [a,b] .$$

Here M and S are formally symmetric quasi-differential expressions of the second and at most first order respectively, λ is a complex-valued parameter and $[a,b]$ a compact interval. More explicitly, we will consider the eigenvalue problem

$$M[y] \equiv -(p(y'-ry))' - \overline{r}p(y'-ry) + qy =$$
$$= \lambda((\rho y)' - \overline{\rho}y' + wy) \equiv \lambda S[y] \quad \text{in} \quad [a,b]$$
$$y(a)\cos\gamma - (p(y'-ry) + \lambda\rho y)(a)\sin\gamma = 0$$
$$y(b)\cos\delta + (p(y'-ry) + \lambda\rho y)(b)\sin\delta = 0$$

where γ, δ are fixed numbers in $(-\frac{\pi}{2}, \frac{\pi}{2}]$, p, q and w real-valued and r, ρ complex-valued functions. Note that for $r \equiv \rho \equiv 0$, then (1.1) reduces to a standard Sturm-Liouville eigenvalue problem with separated boundary conditions. The need for including (when $\rho \not\equiv 0$) the eigenvalue parameter λ in the boundary conditions will be clarified presently.

We will, in a certain sense, require minimal smoothness of the coefficients p, q, r, ρ and w (the exact conditions are given in Section 2), so the problem (1.1) will have to be interpreted as a first order system eigenvalue problem, namely

$$\begin{pmatrix} y^{[0]} \\ y^{[1]} \end{pmatrix}' - \begin{pmatrix} r & 1/p \\ q & -\bar{r} \end{pmatrix}\begin{pmatrix} y^{[0]} \\ y^{[1]} \end{pmatrix} = \lambda\left\{\begin{pmatrix} 0 & 0 \\ \bar{\rho} & 0 \end{pmatrix}\begin{pmatrix} y^{[0]} \\ y^{[1]} \end{pmatrix}' - \begin{pmatrix} \rho/p & 0 \\ w-\bar{r}\rho & 0 \end{pmatrix}\begin{pmatrix} y^{[0]} \\ y^{[1]} \end{pmatrix}\right\}$$

$$\begin{aligned} & y^{[0]}(a)\cos\gamma - y^{[1]}(a)\sin\gamma = 0 \\ & y^{[0]}(b)\cos\delta - y^{[1]}(b)\sin\delta = 0\,. \end{aligned} \qquad (1.2)$$

For sufficiently smooth coefficients this system becomes equivalent with (1.1) on setting $y^{[0]} = y$, $y^{[1]} = p(y' - ry) + \lambda\rho y$. Since, in the case of (1.2) it is natural to impose the boundary conditions on $y^{[0]}$ and $y^{[1]}$ this is a first explanation of the occurence of λ in the boundary conditions of (1.1).

To correctly interprete the system (1.2) we must solve for $\begin{pmatrix} y^{[0]} \\ y^{[1]} \end{pmatrix}'$. It will also be necessary to consider the inhomogeneous system corresponding to

$$M[y] = \lambda S[y] + S[f]\,, \quad f \text{ given.} \qquad (1.3)$$

Here f satisfies conditions stated in Section 2. Thus the final form of our boundary value problem is

$$\begin{pmatrix} y^{[0]} \\ y^{[1]} \end{pmatrix}' - \begin{pmatrix} r-\lambda\rho/p & 1/p \\ q-R & -\bar{r}+\lambda\bar{\rho}/p \end{pmatrix}\begin{pmatrix} y^{[0]} \\ y^{[1]} \end{pmatrix} = \begin{pmatrix} 0 & 0 \\ \bar{\rho} & 0 \end{pmatrix}\begin{pmatrix} f^{[0]} \\ f^{[1]} \end{pmatrix}' - \begin{pmatrix} \rho/p & 0 \\ w-\bar{r}\rho+\lambda|\rho|^2/p & 0 \end{pmatrix}\begin{pmatrix} f^{[0]} \\ f^{[1]} \end{pmatrix} \qquad (1.4)$$

where $R = \lambda(w - \bar{r}\rho - r\bar{\rho}) + \lambda^2|\rho|^2/p$

$$\begin{aligned} & y^{[0]}(a)\cos\gamma - y^{[1]}(a)\sin\gamma = 0 \\ & y^{[0]}(b)\cos\delta + y^{[1]}(b)\sin\delta = 0\,. \end{aligned} \qquad (1.5)$$

Setting $y^{[0]} = y$, $y^{[1]} = p(y' - ry) + \rho(\lambda y + f)$, $f^{[0]} = f$ and $f^{[1]}$

arbitrary the first order system (1.4) becomes, for smooth coefficients, equivalent to (1.3). The essential conclusion to draw at this stage is that for inhomogeneous boundary value problems of the form (1.3) the appropriate boundary conditions are imposed not on y only but involve also the function $\lambda y + f$ to the right of the equality sign. Note that, again, when $\rho \equiv 0$ so that S is multiplication by w the boundary conditions reduce to the usual ones.

It is, in fact, a feature of all equations such as (1.3) that to obtain proper symmetry of the eigenvalue problem the boundary conditions must in general contain the parameter λ or, which is the same thing, the boundary conditions must be imposed on the pair of functions occuring to the left and right hand side of the equality. References to the relevant literature on this are given later in this section.

To obtain orthogonal expansions in eigenfunctions a properly chosen scalar product must be introduced. The natural choice for this is a "Dirichlet integral" belonging to either M or S (see however, Weyl [27] and Kaper, Lekkerkerker and Zettl [16] where a scalar product is used which in a sense is "dual" to a Dirichlet integral of M). Since a scalar product must be positive we speak about left-definite and right-definite problems respectively. In this paper we will mainly discuss the left-definite case which differs more from the classical case than the right-definite case does. The latter is discussed in its general form in Everitt and Zettl [14], Everitt [12] and, essentially, in Naimark [17].

The main goal of this paper is to show how eigenfunction expansions may be derived in the general left-definite case using the methods of Titchmarsh [26] but we also bring out the proper operator theoretical setting of the problem. There are particular difficulties in the case when the metric is degenerate, i.e. only semi-positive definite. The methods with which these problems are handled are related to those of Pleijel [20], where M is the Laplacian and S multiplication by a non-definite function, and Bennewitz [3] which deals with a more general situation.

The paper is an offspring of Bennewitz and Pleijel [2], Pleijel [21, 22, 23], Bennewitz [3, 4, 5] and Everitt [11, 12]. The results in [12] are essentially aimed at extending the function-theoretic methods of Titchmarsh [26], which were devised to study a special case of (1.1) as a right-definite boundary value problem. However, the methods adopted in [12] are only partially successful in considering the left-definite case. The main difficulties encountered in [12] are connected with the proper choice of the boundary conditions for the inhomogeneous problem (1.3). These difficulties are resolved by combining the methods of [12] with those of [3, 4, 5].

There is a growing literature on left-definite boundary value problems. In particular, already Kamke [15] gave expansion theorems for regular boundary value problems. Whereas Pleijel [21, 22, 23] and Bennewitz [3, 5] concentrate on singular problems, particularly Pleijel [21] gives much information on the proper choices of boundary conditions in our case. Coddington and de Snoo [6, 7] have discussed extensions of the method in Bennewitz [3] to certain types of regular non-scalar equations. A more general theory for first order systems was given by Schäfke and Schneider [24]. This was extended to singular cases in a number of papers, for example Niessen [18], Niessen and Schneider [19] and, using different methods, Bennewitz [5]. All these papers require more smoothness of the coefficients than this account but deal of course mainly with higher order equations. More details about some of these contributions may be found in the survey paper by Schneider [25] where further references are given. See also Section 8.

The contents of the paper are as follows.

In Section 2 we state our assumptions regarding the coefficients of the expressions M and S and indicate why they are in a certain sense minimal for the eigenvalue problem to make sense. We also derive a "Dirichlet formula" which is basic for the rest of the paper. Section 3 is devoted to a discussion of the Hilbert space which gives the framework for the spectral theory. In Section 4 we discuss the choice of boundary conditions and identify the eigenvalues as the zeros of a certain integral (entire) function.

Section 5 introduces the Green's kernel of the boundary value problem and gives the expansion theorem for the case of a non-degenerate metric, whereas the case of a degenerate metric is treated in Section 6. Section 7 gives the operator-theoretical interpretation of the problem and finally Section 8 contains some general remarks concerning various generalizations. In particular, the case when some of the coefficients are measures is discussed.

2. Basic assumptions

We consider the formal differential expressions

$$M[y] \equiv -(p(y' - ry))' - \overline{r}\, p(y' - ry) + qy$$
$$S[y] \equiv (\rho y)' - \overline{\rho} y' + wy ,$$

which are of the most general symmetric form for second and first order expressions respectively, in an interval I of the real line. In order that we may interprete $M - \lambda S$ as a quasi-differential expression for any complex number λ, symmetric when λ is real, we assume

(i) $p: I \to \mathbb{R}$ and $1/p \in L_{loc}(I)$

(ii) $q: I \to \mathbb{R}$ and $q \in L_{loc}(I)$

(iii) $r: I \to \mathbb{C}$ and $r \in L_{loc}(I)$

(iv) $\rho: I \to \mathbb{C}$ measurable and $|\rho|^2/p \in L_{loc}(I)$

(v) $w: I \to \mathbb{R}$ and $w - \overline{r}\rho - r\overline{\rho} \in L_{loc}(I)$.

The formal symmetry of M and S requires that p, q and w are real-valued. The integrability conditions are needed in order that an existence theorem be valid for the equation $(M - \lambda S)[y] = 0$. The equation (1.3) will always be interpreted as the first order system (1.4) in $y^{[0]} = y$ and $y^{[1]} = p(y' - ry) + \rho(\lambda y + f)$. For the standard existence theorem to be valid for (1.4) when $f \equiv 0$, it is required that the matrix $\begin{pmatrix} r-\lambda\rho/p & 1/p \\ q-R & -\overline{r}+\lambda\overline{\rho}/p \end{pmatrix}$ is locally integrable for any λ which gives the integrability conditions in (i) - (v). (Recall that $R = \lambda(w - \overline{r}\rho - r\overline{\rho}) + \lambda^2|\rho|^2/p)$. The integrability of ρ/p follows from (i) and (iv) since $\rho/p = \rho/\sqrt{|p|} \cdot \sqrt{|p|}/p$ where both factors are in L^2_{loc}. To ensure that $S \not\equiv 0$ we also assume

(vi) ρ is <u>not</u> a real-valued integral of $-w$ throughout I.

If $u \in AC_{loc}(I)$ then wherever $\rho \in AC_{loc}$ we obtain

$$Su = (\rho - \overline{\rho})u' + (\rho' + w)u$$

which explains (vi).

When dealing with the left-definite case we will also make the (preliminary) assumption

(vii) $p \geq 0$

and correspondingly we assume in the right-definite case

(vii)' $\rho \equiv 0$ and $w \geq 0$.

That w does not vanish a.e. then follows from (vi). The assumption (vii)' could be relaxed in the right-definite case; compare the last remark of Section 8.

In the right-definite case we need to solve (1.3) for f such that $\sqrt{w}\, f \in L^2_{loc}(I)$. In view of (v) and (vii)' this implies that $wf \in L_{loc}$ so the existence theorem is valid for (1.4). The solution y, being in $AC_{loc}(I)$, then also satisfies $\sqrt{w}\, y \in L^2_{loc}(I)$.

When dealing with the left-definite case we will instead need to solve (1.3) for $f \in AC_{loc}(I)$ such that $\sqrt{p}(f' - rf) \in L^2_{loc}(I)$. The existence theorem is valid for (1.4) if $\rho/p\, f$ and $\overline{\rho}f' - (w - \overline{r}\rho + \lambda|\rho|^2/p)f$ are in $L_{loc}(I)$. The first holds since $\rho/p \in L_{loc}(I)$ and the second follows from

$$\overline{\rho}f' - (w - \overline{r}\rho + \lambda|\rho|^2/p)f = \overline{\rho}/\sqrt{p} \cdot \sqrt{p}(f' - rf) - (w - \overline{r}\rho - r\overline{\rho} + \lambda|\rho|^2/p)f$$

where on the right hand side the factors of the first term are in $L^2_{loc}(I)$ and the second term is in $L_{loc}(I)$ by (iv), (v). The solution y is also in $AC_{loc}(I)$ and since this is true also for $y^{[1]} = p(y' - ry) + \rho(\lambda y + f)$ it follows that $\sqrt{p}(y' - ry) = \frac{1}{\sqrt{p}} y^{[1]} - \rho/\sqrt{p}(\lambda y + f)$ is in $L^2_{loc}(I)$.

It is convenient to denote by $D(M,S)$ the set of functions y which satisfy an equation (1.3) with f satisfying the conditions mentioned above for the left- or right-definite case, as the case may be. It is then nearly obvious, in either case, that $D(M,S)$ is independent of λ.

Now let $[\alpha,\beta]$ be a compact subinterval of I and put

$$\overset{\beta}{\underset{\alpha}{(u,v)_M}} = \int_\alpha^\beta \{p(u' - ru)\overline{(v' - rv)} + q\, u\,\overline{v}\}$$

$$\overset{\beta}{\underset{\alpha}{(u,v)_S}} = \int_\alpha^\beta \{-\rho\, u\,\overline{v}' - \overline{\rho}\, u\,\overline{v}' + w\, u\,\overline{v}\}.$$

The integrals $\overset{\beta}{\underset{\alpha}{(\cdot,\cdot)}}_M$ and $\overset{\beta}{\underset{\alpha}{(\cdot,\cdot)}}_S$ are the most obvious choices of Dirichlet integrals belonging to M and S respectively.

If $u \in D(M,S)$ and $M[y] = S[f]$, then for any $v \in AC_{loc}(I)$ which in the left-definite case also satisfies $\sqrt{p}(v' - rv) \in L^2_{loc}(I)$ the Dirichlet formula

$$\overset{\beta}{\underset{\alpha}{(y,v)}}_M = \overset{\beta}{\underset{\alpha}{(f,v)}}_S + [y^{[1]}\overline{v}]_\alpha^\beta \tag{2.1}$$

holds for $\alpha, \beta \in I$. The existence of the integrals follows by a reasoning similar to that above concerning existence of solutions. $[\ldots]_\alpha^\beta$ denotes an outintegrated part, $y^{[1]} = p(y' - ry) + \rho f$ and the formula follows on differentiating with respect to β.

In this account we will only deal with regular boundary value problems so that we assume that $I = [a,b]$ is compact. Thus (i) - (v) hold with $L_{loc}(I)$ replaced by $L(I)$. In the sequel we will write $(\cdot,\cdot)_M$ and $(\cdot,\cdot)_S$ for $\overset{b}{\underset{a}{(\cdot,\cdot)}}_M$ and $\overset{b}{\underset{a}{(\cdot,\cdot)}}_S$ respectively.

Remark. The term "Dirichlet integral belonging to M" applies to any hermitean form $\overset{\beta}{\underset{\alpha}{(\cdot,\cdot)}}_M$ for which the difference $\overset{\beta}{\underset{\alpha}{(u,v)}}_M - \int_\alpha^\beta Mu\,\overline{v}$ can be written as an outintegrated part. Thus our Dirichlet integrals $(\cdot,\cdot)_M$ and $(\cdot,\cdot)_S$ are certainly not uniquely determined by M and S; in the next section we will in fact use a modified Dirichlet integral belonging to M.

3. Hilbert spaces

In the right-definite case $(u,v)_S$ reduces to $\int_a^b w\,u\,\overline{v}$ and this is the scalar product to be used. The corresponding Hilbert space is then the wheighted L^2-space $L^2_w(I)$.

In the left-definite case it would be possible, with proper positivity assumptions, to use $(u,v)_M$ as a scalar product. However, since we wish to use the boundary conditions (1.5) symmetry requirements will in general force the use of another Dirichlet integral belonging to M. See Section 4, but reference is also made to Everitt [11] and particularly Pleijel [21] for a thorough discussion of this. Consequently we set

$$(u,v)_{\gamma\delta} = (u,v)_M + \cot\gamma\,|u(a)|^2 + \cot\delta\,|u(b)|^2 \tag{3.1}$$

where the $\cot\gamma$ term is dropped if $\gamma = 0$ and similarly with $\cot\delta$.

It is appropriate to remark at this point that adding similar terms to $(\cdot,\cdot)_S$ in the right-definite case makes it possible to study right-definite problems with λ-dependent boundary conditions. This is well known and will also be evident from the discussion in Section 4.

Returning to the left-definite problem we will from now on assume (i) - (vii) of Section 2. Now define

$$H = \{u \in AC(I) \mid \sqrt{p}(u' - ru) \in L^2(I);\ u(a) = 0 \text{ if } \gamma = 0;\ u(b) = 0 \text{ if } \delta = 0\}.$$

It follows that $(\cdot,\cdot)_{\gamma\delta}$, $(\cdot,\cdot)_M$ and $(\cdot,\cdot)_S$ are well defined on H. To be able to use $(\cdot,\cdot)_{\gamma\delta}$ as our scalar product we now assume

(viii) $(u,u)_{\gamma\delta} \geq 0$ for $u \in H$.

Assuming (viii) the condition (vii) is in fact redundant. (Cf. Dunford-Schwarz [10, Lemma XIII.7.29] and Bennewitz [4, Lemma 2.1]).

Lemma 3.1. If (viii) holds, then $p \geq 0$ a.e. in I.

Proof. Let φ be a bounded measurable function and put

$$u(x) = \exp\{\int_a^x r\} \int_a^x e^{i\lambda t} \exp\{-\int_a^t r\}\, \varphi(t)/p(t)\,dt .$$

Then by the Riemann-Lebesgue lemma $u \to 0$ uniformly as $\lambda \to \pm\infty$. Furthermore $p|u' - ru|^2 = |\varphi|^2/p$. It follows that $(u,u)_{\gamma\delta} \to \int_a^b |\varphi|^2/p$. Positivity of $(\cdot,\cdot)_{\gamma\delta}$ thus implies $1/p \geq 0$ a.e. in I. Unless $\delta = 0$ it is clear that $u \in H$. In the exceptional case we modify u by subtracting

$$v(x) = u(b) \exp\{-\int_x^b r\} \int_a^x 1/p \{\int_a^b 1/p\}^{-1}$$

which tends uniformly to zero as $\lambda \to \pm\infty$. Furthermore

$$p|v' - rv|^2 = \frac{1}{p} |u(b)|^2 \exp\{2\int_x^b -\mathrm{Re}\; r\}\{\int_a^b 1/p\}^{-1}$$

which $\to 0$ in $L(I)$ as $\lambda \to \pm\infty$. Finally $u - v \in H$ so the claim of the lemma is generally valid.

The assumption (viii) does not exclude the possibility that $(u,u)_{\gamma\delta} = 0$ for some non-zero $u \in H$. The situation is clarified by the following lemma.

Lemma 3.2. Assuming (i) - (iii) and (viii), then $(u,u)_{\gamma\delta} = 0$ at most on a 1-dimensional subspace of H which consists of the eigenfunctions to $\lambda = 0$ of (1.5), (1.6). These functions have no zeros in (a,b).

Proof. From $(u,u)_{\gamma\delta} = 0$ follows by Cauchy-Schwarz´ inequality that $(u,v)_{\gamma\delta} = 0$ for all $v \in H$, in particular those for which $v(a) = v(b) = 0$. Let $g(x) = \exp\{-\int_a^x r\}$. Integrating by parts we obtain

$$0 = (u,v)_{\gamma\delta} = \int_a^b p\,F\,\overline{(gv)'}$$

where $F(x) = (u'(x) - r(x)u(x))/\overline{g(x)} - \frac{1}{p(x)}\int_a^x qu/\overline{g}$. It follows that $\sqrt{p}\,F \in L^2(I)$ and $F \in L(I)$. Since $\int_a^b (gv)' = 0$ we obtain

$$\int_a^b p(F - C/p)\overline{(gv)'} = 0 \tag{3.2}$$

for any constant C. Put $C = \int_a^b F / \int_a^b 1/p$ and $v(x) = \frac{1}{g(x)}\int_a^x (F - C/p)$.

One immediately verifies that $v \in AC(I)$, $\sqrt{p}(v' - rv) \in L^2(I)$ and $v(a) = v(b) = 0$. For these choices of C and v the equation (3.2) becomes $\int_a^b p|F - C/p|^2 = 0$ so that pF is constant. Differentiation of pF shows that $u \in D(M,S)$ and $Mu = 0$. Finally (2.1) gives, for $v \in H$,

$$0 = (u,v)_{\gamma\delta} = [u^{[1]}v]_a^b + \cot \gamma\, u(a)\overline{v(a)} + \cot \delta\, u(b)\overline{v(b)}$$

with the usual modifications if γ or $\delta = 0$. This shows that u must also satisfy (1.5). We will show in Section 4 that the eigenvalues are simple so it only remains to show that u has no interior zeros in I. Because of the existence theorem u can not vanish on a subinterval of I. Assuming c to be an interior zero we may set $v(x) = u(x)$ for $x \leq c$ and $v(x) \equiv 0$ for $x > c$. Then v defines an element of H which is linearly independent of u and for which $(v,v)_{\gamma\delta} = 0$ as an easy consequence of (2.1). This contradiction completes the proof.

The above proof is clearly a simple adaptation of the methods of the calculus of variation. For a related, but more elaborate, result see Lemma 5.1 of Bennewitz [3].

If, in addition to the necessary condition (vii), $g \geq 0$ and γ, δ are both in $[0, \frac{\pi}{2}]$ it is clear that (viii) holds. However, (viii) may be valid even if all these conditions are violated. Nevertheless, H is always complete with respect to the (semi-)norm induced by $(\cdot,\cdot)_{\gamma\delta}$. This fact is essential only in Section 7, but it seems appropriate to prove it at this point.

Theorem 3.3. H is a Hilbert space with scalar product $(\cdot,\cdot)_{\gamma\delta}$ except in the degenerate case when $(\cdot,\cdot)_{\gamma\delta}$ is only semi-definite. In this case, any subspace with finite codimension in H on which $(\cdot,\cdot)_{\gamma\delta}$ is non-degenerate is a Hilbert space with scalar product $(\cdot,\cdot)_{\gamma\delta}$.

To prove the theorem we need several lemmata.

Lemma 3.4. H is a Hilbert space with norm-square

$$|u(a)|^2 + \int_a^b p|u' - ru|^2 . \tag{3.3}$$

Proof. Let $u_1, u_2, \ldots$ be a Cauchy sequence in H and put $v_j = \sqrt{p}(u_j' - ru_j)$ so that

$$u_j(x) = \exp\{\int_a^x r\}\{u_j(a) + \int_a^x v_j(t) \exp\{-\int_a^t r\}/\sqrt{p(t)}\, dt\} .$$

Then v_j is a Cauchy sequence in $L^2(I)$ so that u_j converges uniformly to $u(x) = \exp\{\int_a^x r\}\{u(a) + \int_a^x v(t)\exp\{-\int_a^t r\}/\sqrt{p(t)}\, dt\}$ where $u(a) = \lim_j u_j(a)$ and v is the L^2-limit of v_j. Since $u \in AC(I)$, $\sqrt{p}(u' - ru) = v \in L^2(I)$ and u obviously is the limit of u_j in the norm-square (3.3) this completes the proof.

The next lemma is related to the classical Poincaré inequality.

Lemma 3.5. Suppose (i) - (iii) and (vii). Then there exists a finite number of points $c_1, \ldots, c_n$ in I and constants $A_1, \ldots, A_n$ such that for $u \in H$

$$\int_I |q||u|^2 \leq \frac{1}{2} \int_I p|u' - ru|^2 + \Sigma A_j |u(c_j)|^2 . \tag{3.4}$$

All linear forms $u \to u(c)$, c fixed, are uniformly bounded in H, i.e. convergence in H implies uniform convergence.

Proof. For x and $y \in J \subset I$ one has

$$u(x) = \{u(y) + \int_y^x (u' - ru) \exp(\int_t^y r) dt\} \exp(\int_y^x r)$$

so that Cauchy-Schwarz' inequality gives

$$|u(x)|^2 \leq A\{|u(c)|^2 + \int_J 1/p \int_J p|u' - ru|^2\} . \tag{3.5}$$

Here c is some fixed point in J and $A = 2\exp 2\int_a^b |\mathrm{Re}\ r|$. Taking $J = I$ and $c = a$ this shows the second statement of the theorem.

Multiplying (3.5) by $|q(x)|$ and integrating over J gives

$$\int_J |q||u|^2 \leq A \int_J |q|\{|u(c)|^2 + \int_J 1/p \int_J p|u' - ru|^2\}.$$

We may now divide I into finitely many subintervals J such that $A\int_J |q| \int_J 1/p \leq 1/2$ on each subinterval. Adding the various inequalities obtained in this way gives (3.4) so the proof is complete.

It is clear that in the same way one may prove the compactness of the form $\int_I q u \overline{v}$ in H. This compactness appears to be the fundamental reason that Theorem 3.3 is true.

The final lemma needed is

Lemma 3.6. Let H be a Banach space with norm $\|\cdot\|$ and let $|\cdot|$ be a bounded seminorm on H such that on some subspace with finite codimension $\|\cdot\|$ and $|\cdot|$ are equivalent norms. Then they are equivalent norms on any subspace of H on which $|\cdot|$ is a norm.

This is a standard fact but can be found as Lemma 2.7 of [3] where a proof is given.

Proof of Theorem 3.3. Consider the subspace H_0 of H obtained by setting $u(a) = u(b) = u(c_1) = \ldots = u(c_n) = 0$ where $c_1, \ldots, c_n$ are the points in (3.4). Using (3.4) we then obtain, for $u \in H_0$, that

$$\frac{1}{2}\int_I p|u' - ru|^2 \leq \int_I \{p|u' - ru|^2 + q|u|^2\} \leq \frac{3}{2}\int_I p|u' - ru|^2 .$$

It follows that $|u(a)|^2 + \int_I p|u' - ru|^2$ and $(u,u)_{\gamma\delta}$ are equivalent normsquares on H_0 and so a reference to Lemma 3.4 and Lemma 3.6 completes the proof.

4. Boundary conditions and eigenvalues

The Dirichlet formula (2.1) may be written

$$(u,v)_{\gamma\delta} = (f,v)_S + [u_{\gamma\delta}^{[1]}\,\overline{v}]_a^b \tag{4.1}$$

where $u \in D(M,S)$, $Mu = Sf$, $v \in AC(I)$ and belongs to the Hilbert space (L_w^2 or H). Furthermore we have set $u_{\gamma\delta}^{[1]}(a) = u^{[1]}(a) - u(a)\cot\gamma$ and $u_{\gamma\delta}^{[1]}(b) = u^{[1]}(b) + u(b)\cot\delta$ with the usual modifications if γ or δ equals 0.

Now suppose that u and v are in $D(M,S)$ and $Mu = Sf$, $Mv = Sg$. Then (4.1) gives

$$(f,v)_S - (u,g)_S = [u\,\overline{v_{\gamma\delta}^{[1]}} - u_{\gamma\delta}^{[1]}\,\overline{v}]_a^b \tag{4.2}$$

$$(f,v)_{\gamma\delta} - (u,g)_{\gamma\delta} = [f\,\overline{v_{\gamma\delta}^{[1]}} - u_{\gamma\delta}^{[1]}\,\overline{g}]_a^b \tag{4.3}$$

It is natural that any set of linear homogeneous conditions on $D(M,S)$ or, to be exact, pairs (u,f) where $Mu = Sf$, which ensures that the outintegrated parts in (4.2) or (4.3) vanish is called a symmetric boundary condition for the eigenvalue problem $Mu = \lambda Su$ in the right- and left-definite cases respectively. It is clear, by rank considerations, that two linearly independent conditions suffice in both cases, and such a symmetric boundary condition is called <u>selfadjoint</u>.

Clearly any set of conditions that make the outintegrated part in (4.1) vanish will be a symmetric boundary condition for both the right- and left-definite cases. Such a boundary condition could be called <u>strong</u>. The term <u>natural</u> has also been used. Confining ourselves to <u>separated</u> conditions, i.e. conditions that do not simultaneously involve a and b, there are only four minimal possibilities for the outintegrated part in (4.1) to vanish, namely

I $\quad v(a) = v(b) = 0$

II $\quad v(a) = 0\,,\ u^{[1]}_{\gamma\delta}(b) = 0$

III $\quad u^{[1]}_{\gamma\delta}(a) = 0\,,\ v(b) = 0$

IV $\quad u^{[1]}_{\gamma\delta}(a) = u^{[1]}_{\gamma\delta}(b) = 0$

(There is also the non-separated possibility $v(b) = cv(a)$, $u^{[1]}_{\gamma\delta}(b) = \bar{c}\, u^{[1]}_{\gamma\delta}(a)$ where c is a non-zero complex number.) Of these, only IV is genuinely a selfadjoint boundary condition for both the right- and left-definite cases. Comparing with (4.2), (4.3), the other conditions involve, if $M[u] = S[f]$, in the right-definite case conditions on u but in the left-definite case conditions on f. For eigenfunctions u satisfying $M[u] = \lambda S[u]$ this is of course the same thing except for $\lambda = 0$. This flaw can be remedied by, in the left-definite case, letting the conditions on v in I - III be built into the Hilbert space itself. This is the course taken here (cf. the definition of H).

The boundary conditions (1.5) that we impose are thus strong, selfadjoint boundary conditions related to the forms $(\cdot,\cdot)_{\gamma\delta}$ and $(\cdot,\cdot)_S$. It is clear that if we wish this to hold we have essentially no option in the choice of the Dirichlet integrals $(\cdot,\cdot)_{\gamma\delta}$ and $(\cdot,\cdot)_S$. Any eigenvalue of (1.4) with the (separated) boundary conditions (1.5) must be simple. This follows since by the existence theorem there is only one linearly independent solution satisfying the boundary condition at one endpoint. It is also clear from similar considerations that (1.4), (1.5) has a unique solution for any $f \in H$ as soon as λ is not an eigenvalue.

If the boundary conditions (1.5) together with the equation $M[u] = S[f]$ are to determine an operator, mapping u on f, there must be no solution of $S[f] = 0$ satisfying the boundary conditions, i.e. no f so that (1.4), (1.5) is satisfied for $y \equiv 0$. We must therefore investigate the existence of such solutions.

In the right-definite case $S[f] = 0$ means simply $wf = 0$ a.e. in I so that f is the zero-element of the Hilbert space; thus no complications arise in this case.

In the left-definite case the equation is

$$\begin{aligned} &(\rho f)' - \overline{\rho}\, f' + wf = 0 \\ &(\rho f)(a) = 0 \qquad (f(a) = 0 \ \text{ if } \ \gamma = 0) \\ &(\rho f)(b) = 0 \qquad (f(b) = 0 \ \text{ if } \ \delta = 0) \end{aligned} \tag{4.4}$$

where f and ρf are in $AC(I)$. Multiplying the equation by $\overline{f}$ and taking the imaginary part we obtain $(\operatorname{Im} \rho |f|^2)' = 0$ so that $\operatorname{Im} \rho |f|^2$ is constant. The boundary conditions imply that the constant is zero. If $f(x) \neq 0$ for some $x \in I$, then $f \neq 0$ in a neighbourhood of x. In this neighbourhood ρ must therefore be real and, because f and ρf are absolutely continuous, in AC_{loc}. Thus in this neighbourhood the equation reduces to $(\rho' + w)f = 0$ so that necessarily $\rho' + w = 0$ there. Define

$$\begin{aligned} H_\infty &= \{f \in H \,|\, f \text{ and } \rho f \in AC(I) \text{ and satisfy } (4.4)\} \\ \widetilde{H} &= \{f \in H \,|\, (f, H_\infty)_{\gamma\delta} = 0\}. \end{aligned}$$

The notation H_∞ indicates that this space may be thought of as the eigenspace belonging to $\lambda = \infty$ of the boundary value problem (1.4), (1.5). Let Ω be the largest (relatively) open subset of I such that ρ is a real-valued integral function of $-w$ in each component of Ω. The previous discussion then gives the following concrete description of the spaces H_∞ and $\widetilde{H}$.

<u>Theorem 4.1</u>. If $\Omega = \emptyset$, then $\dim H_\infty = 0$. Otherwise, $\dim H_\infty = \infty$ and H_∞ consists exactly of those $f \in H$ which vanish outside Ω and for which $(\rho f)(a) = (\rho f)(b) = 0$. $\widetilde{H}$ consists of those $u \in H$ which

on any component of Ω solves $M[u] = 0$. Furthermore, if $a \in \Omega$, $\rho(a) = 0$ and $\gamma \neq 0$ u has to satisfy $u^{[1]}_{\gamma\delta}(a) = 0$. A similar condition holds at b .

Proof. Only the statements about $\widetilde{H}$ remain to prove. If $M[u] = 0$ on Ω and $f \in H_\infty$, then (4.1) gives

$$(u,f) = [\, u^{[1]}_{\gamma\delta} \, \overline{f}\,]_a^b$$

which shows that $\widetilde{H}$ contains the elements claimed. Conversely, using the method of proof of Lemma 3.1 it follows that all elements of $\widetilde{H}$ are of this kind.

Corollary 4.2. 1. $\widetilde{H} \cap H_\infty = \{0\}$.
2. Eigenfunctions to different eigenvalues can not coincide.

Proof. 1. In the case of a degenerate metric $(\cdot,\cdot)_{\gamma\delta}$ $\widetilde{H} \cap H_\infty$ could concievably contain the degenerate space. However, by (vi) of Section 2 and Theorem 4.1 an element of H_∞ must have at least one zero in (a,b) which by Lemma 3.2 excludes the possibility that it is a non-zero element of the degenerate space.

2. If $M[y] = S[f]$, (1.5) is satisfied and $u \in H_\infty$, then according to (4.3),

$$(y,u)_{\gamma\delta} = (y,u)_{\gamma\delta} - (f,0)_{\gamma\delta} = 0$$

so that $y \in \widetilde{H}$. If $M[y] = \lambda S[y]$ and (1.5) is satisfied for two different values of λ it follows that $y \in \widetilde{H} \cap H_\infty$ so 2. follows from 1.

Assuming u and v to be eigenfunctions to the eigenvalues λ and μ respectively (4.1) gives

$$\lambda(u,v)_S = (u,v)_{\gamma\delta} = \overline{\mu}(u,v)_S \, .$$

It follows that $(u,v)_S = (u,v)_{\gamma\delta} = 0$ unless $\lambda = \overline{\mu}$. In particular a non-real eigenvalue λ is possible only if $(u,u)_S = (u,u)_{\gamma\delta} = 0$. This is feasible only in the degenerate left-definite case, but then Lemma 3.2 and Corollary 4.2 imply $\lambda = 0$. Thus we obtain

Theorem 4.3. The eigenvalues are real in both the right- and left-definite cases. Eigenfunctions to different eigenvalues are orthogonal with respect to both $(\cdot,\cdot)_S$ and $(\cdot,\cdot)_{\gamma\delta}$. For an eigenfunction u to a non-zero eigenvalue $(u,u)_{\gamma\delta}$ and $(u,u)_S \neq 0$.

Now define

$$D_{\gamma\delta} = \{y \in D(M,S) \mid (1.4),\ (1.5) \text{ holds for some } f \in H\}.$$

Lemma 4.4. $D_{\gamma\delta}$ is dense in $\tilde{H}$ and H_∞ is closed.

Proof. Choose λ so that it is not an eigenvalue of (1.4), (1.5), e.g. non-real, and assume $(f, D_{\gamma\delta})_{\gamma\delta} = 0$. One can then determine (uniquely) u and v in $D_{\gamma\delta}$ so that $M[u] = \lambda S[u] + S[f]$, $M[v] = \bar{\lambda} S[v] + S[u]$ and the boundary conditions are satisfied. Then

$$(u,u)_{\gamma\delta} = (u,u)_{\gamma\delta} - (f,v)_{\gamma\delta} = (u, \bar{\lambda} v + u)_{\gamma\delta} - (\lambda u + f,\ v)_{\gamma\delta} = 0$$

where the last equality follows from (4.3) and the boundary conditions. Hence $u = 0$ or, in the degenerate case, u is an eigenfunction to $\lambda = 0$. In the first case it follows that $f \in H_\infty$, in the second that f is in the same equivalence class as an element of H_∞, namely $\lambda u + f$. The proof is complete.

Remark. In essentially the same way one may prove that $D_{\gamma\delta}$ is dense in $L_w^2(I)$ in the right-definite case.

The basis for the Titchmarsh method of proving eigenfunction expansions is a characterization of the eigenvalues as the zeros of a certain integral (entire) function. With this in mind we define

$\emptyset(x,\lambda)$ and $\chi(x,\lambda)$ as the solutions of $M[y] = \lambda S[y]$ satisfying the initial conditions

$$\begin{aligned} \emptyset(a,\lambda) &= \sin\gamma & \chi(b,\lambda) &= -\sin\delta \\ \emptyset^{[1]}(a,\lambda) &= \cos\gamma & \chi^{[1]}(b,\lambda) &= \cos\delta \end{aligned} \tag{4.5}$$

so that the boundary condition (1.5) imposed at a is satisfied by $\emptyset$ and that at b by χ. As functions of λ, the functions $\emptyset$, $\emptyset^{[1]}$, χ and $\chi^{[1]}$ are then integral functions, uniformly in x. This follows from the standard theorem on dependence on parameters applied to (1.4). Note that this means e.g. that $\sqrt{p}(\emptyset' - r\emptyset) = \frac{1}{\sqrt{p}}\emptyset^{[1]} - \lambda\rho/\sqrt{p}\,\emptyset$ is integral as an $L^2(I)$-valued function so that an expression such as $(\chi(\cdot,\lambda), \emptyset(\cdot,\overline{\lambda}))_S{}_\alpha^\beta$ or $(\emptyset(\cdot,\lambda), f)_{\gamma\delta}{}_\alpha^\beta$ is an entire function, uniformly with respect to α and β.

It is clear that λ is an eigenvalue precisely when $\emptyset$ and χ are linearly dependent, i.e. when the wronskian

$$\chi(x,\lambda)\emptyset^{[1]}(x,\lambda) - \chi^{[1]}(x,\lambda)\emptyset(x,\lambda) \tag{4.6}$$

vanishes (for one and hence for all x). Now, if $M[u] = \lambda S[u]$ and $M[v] = \overline{\lambda}\, S[v]$, then according to (2.1)

$$\lambda\,(u,v)_S{}_a^x + [u^{[1]}\overline{v}]_a^x = (u,v)_M{}_a^x = \lambda\,(u,v)_S{}_a^x + [u\,\overline{v^{[1]}}]_a^x$$

so that

$$u(x)\overline{v^{[1]}(x)} - u^{[1]}(x)\overline{v(x)} \text{ is constant.} \tag{4.7}$$

Hence, setting

$$W(\lambda) = \chi(x,\lambda)\overline{\emptyset^{[1]}(x,\overline{\lambda})} - \chi^{[1]}(x,\lambda)\overline{\emptyset(x,\overline{\lambda})}\,,$$

$W(\lambda)$ is actually independent of x and an integral function of λ. For $x = a$ the wronskian (4.6) coincides with $W(\lambda)$ so that the eigenvalues are the zeros of $W(\lambda)$. Incidentally, for $x = b$ the wronskian at $\overline{\lambda}$ coincides with $\overline{W(\lambda)}$. This shows that, in the non-definite case when non-real eigenvalues may occur, λ is an eigenvalue precisely if $\overline{\lambda}$ is (it may be shown that they also have the same algebraic multiplicity). See also Atkinson, Everitt and Ong [1].

In the right- and left-definite cases there are no non-real eigenvalues so $W(\lambda)$ is not identically zero. Thus there are, in this case, at most denumerably many eigenvalues without finite cluster points. For future reference we note also that

$$\begin{aligned} \emptyset(x,\lambda)\overline{\emptyset^{[1]}(x,\overline{\lambda})} - \emptyset^{[1]}(x,\lambda)\overline{\emptyset(x,\overline{\lambda})} &= 0 \\ \chi(x,\lambda)\overline{\chi^{[1]}(x,\overline{\lambda})} - \chi^{[1]}(x,\lambda)\overline{\chi(x,\overline{\lambda})} &= 0 . \end{aligned} \tag{4.8}$$

By (4.7) these expressions are namely independent of x, and at $x = a$ and $x = b$ respectively they vanish.

Lemma 4.5. $W'(\lambda) = -(\chi(\cdot,\lambda), \emptyset(\cdot,\overline{\lambda}))_S$. $\quad$ (4.9)

Proof. By (2.1) we have

$$(\chi(\cdot,\mu),\emptyset(\cdot,\overline{\lambda}))_M = \mu(\chi(\cdot,\mu),\emptyset(\cdot,\overline{\lambda}))_S + [\chi^{[1]}(x,\mu)\overline{\emptyset(x,\overline{\lambda})}]_a^b$$

$$(\chi(\cdot,\mu),\emptyset(\cdot,\overline{\lambda}))_M = \lambda(\chi(\cdot,\mu),\emptyset(\cdot,\overline{\lambda}))_S + [\chi(x,\mu)\overline{\emptyset^{[1]}(x,\overline{\lambda})}]_a^b .$$

Taking the difference we obtain

$$(\mu-\lambda)(\chi(\cdot,\mu),\emptyset(\cdot,\overline{\lambda}))_S = [\chi(x,\mu)\overline{\emptyset^{[1]}(x,\overline{\lambda})} - \chi^{[1]}(x,\mu)\overline{\emptyset(x,\overline{\lambda})}]_a^b = W(\lambda) - W(\mu)$$

since the initial conditions for $\emptyset$ at a and for χ at b are independent of $\overline{\lambda}$ and μ respectively. Dividing by $\lambda - \mu$ and letting $\mu \to \lambda$ we obtain (4.9) completing the proof.

5. Eigenfunction expansion

With the functions $\emptyset$, χ and W of Section 4 we now define, for $f \in H$ ($f \in L^2_w(I)$ in the right-definite case)

$$\Phi(x,\lambda;f) = \frac{\chi(x,\lambda)}{W(\lambda)} \int_a^x (f,\emptyset(\cdot,\overline{\lambda}))_S + \frac{\emptyset(x,\lambda)}{\overline{W(\overline{\lambda})}} (f,\chi(\cdot,\overline{\lambda}))_S \ . \tag{5.1}$$

The occurence of $\overline{W(\overline{\lambda})}$ in the second term is best understood if $\frac{\chi}{W}$ is kept together in both terms; however for computations the form of (5.1) is convenient as it stands. It is clear that Φ, for fixed f, is a meromorphic function of λ with possible poles at the zeros of $W(\lambda)$, which, as we know, coincide with the zeros of $W(\overline{\lambda})$. The interest of the function Φ is the following.

Theorem 5.1. $\Phi(x,\lambda;f)$ is the unique solution of the boundary value problem (1.4), (1.5) when λ is not an eigenvalue.

Proof. The proof is by direct computation and we omit most of the details. In the computations one uses that

$$\frac{\chi(x,\lambda)\overline{\emptyset^{[1]}(x,\overline{\lambda})}}{W(\lambda)} - \frac{\emptyset(x,\lambda)\overline{\chi^{[1]}(x,\overline{\lambda})}}{\overline{W(\overline{\lambda})}} = 1$$

$$\frac{\chi(x,\lambda)\overline{\emptyset(x,\overline{\lambda})}}{W(\lambda)} = \frac{\emptyset(x,\lambda)\overline{\chi(x,\overline{\lambda})}}{\overline{W(\overline{\lambda})}}$$

$$\frac{\chi^{[1]}(x,\lambda)\overline{\emptyset^{[1]}(x,\overline{\lambda})}}{W(\lambda)} = \frac{\emptyset^{[1]}(x,\lambda)\overline{\chi^{[1]}(x,\overline{\lambda})}}{\overline{W(\overline{\lambda})}}$$

all of which follow from (4.8). Setting $u(x) = \Phi(x,\lambda;f)$ and $u^{[1]} = p(u' - ru) + \rho(\lambda u + f)$ one then finds that

$$u^{[1]}(x) = \frac{\chi^{[1]}(x,\lambda)}{W(\lambda)} \int_a^x (f,\emptyset(\cdot,\overline{\lambda}))_S + \frac{\emptyset^{[1]}(x,\lambda)}{\overline{W(\overline{\lambda})}} \int_x^b (f,\chi(\cdot,\overline{\lambda}))_S \ . \tag{5.2}$$

From this and (5.1) it is clear that (1.5) is satisfied. Further differentiation shows that $\binom{u}{u^{[1]}}$ satisfies (1.4).

To prove a Parseval formula with the methods of Titchmarsh [26, Chapter II] we need two lemmata.

Lemma 5.2. Assume λ is not an eigenvalue of (1.4), (1.5).

1. If $f \in H$ and $\Phi = \Phi(\cdot,\lambda;f)$, then

$$(\Phi,\Phi)_{\gamma\delta} \leq |\mathrm{Im}\ \lambda|^{-2}(f,f)_{\gamma\delta} .$$

2. If $f \in H$, $M[u] = S[f]$ and (1.5) holds, then

$$\Phi(x,\lambda;f) = u(x) + \lambda\,\Phi(x,\lambda;u) .$$

Proof. 1. By Theorem 5.1 Φ solves $M[\Phi] = \lambda S[\Phi] + S[f]$ and the boundary conditions (1.5) are satisfied. Hence (4.3) gives

$$0 = (\lambda\Phi + f,\ \Phi)_{\gamma\delta} - (\Phi,\ \lambda\Phi + f)_{\gamma\delta} = (\lambda - \overline{\lambda})(\Phi,\Phi)_{\gamma\delta} + (f,\Phi)_{\gamma\delta} - (\Phi,f)_{\gamma\delta} .$$

Thus $(\Phi,\Phi)_{\gamma\delta} = \dfrac{\mathrm{Im}\ (\Phi,f)_{\gamma\delta}}{\mathrm{Im}\ \lambda}$ so 1. follows from the Cauchy-Schwarz inequality.

2. With $v(x) = u(x) + \lambda\Phi(x,\lambda;u)$ we obtain

$$M[v] = S[f] + \lambda^2 S[\Phi] + \lambda S[u] = \lambda S[v] + S[f] .$$

Additionally the boundary conditions (1.5) are satisfied for this equation since they are linear and satisfied for the equations $M[u] = S[f]$ and $M[\Phi] = \lambda S[\Phi] + S[u]$. Hence v must coincide with $\Phi(\cdot,\lambda;f)$.

Lemma 5.3. 1. The function $\lambda \to (\Phi(\cdot,\lambda;f),\ g)_{\gamma\delta}$ where f, g are fixed in H, is meromorphic and regular except possibly at the eigenvalues of (1.4), (1,5).

2. If λ is a non-zero eigenvalue of (1.4), (1.5) and ψ is the corresponding eigenfunction, normalized so that $(\psi,\psi)_{\gamma\delta} = 1$, then the residue of $z \to (\Phi(\cdot,z;f),g)_{\gamma\delta}$ at $z = \lambda$ is $-\hat{f}\overline{\hat{g}}$ where $\hat{f} = (f,\psi)_{\gamma\delta}$, $\hat{g} = (g,\psi)_{\gamma\delta}$.

Proof. 1. From the properties of $\emptyset$ and χ and (5.1), (5.2) it follows that $\Phi(x,\lambda;f)$ and $\Phi^{[1]}(x,\lambda;f)$ are meromorphic in λ, uniformly with respect to x in I, with poles at most at the zeros of $W(\lambda)$. Hence

$$\sqrt{p}(\Phi' - r\Phi) = \frac{1}{\sqrt{p}}\Phi^{[1]} - \lambda\rho/\sqrt{p}\,\Phi - \rho)\sqrt{p}\,f$$

is analytic as an $L^2(I)$-valued function except at the eigenvalues of (1.4), (1.5) so that 1. follows.

2. Being an eigenvalue λ is real so that $\emptyset(\cdot,\lambda) = \emptyset(\cdot,\overline{\lambda}) = c\,\psi$ and $\chi(\cdot,\lambda) = \chi(\cdot,\overline{\lambda}) = d\,\psi$ for some finite, non-zero constants c and d. According to Lemma 4.4 we therefore have

$$W'(\lambda) = -(\chi(\cdot,\lambda),\ \emptyset(\cdot,\overline{\lambda}))_S = -\overline{c}\,d(\psi,\psi)_S\,.$$

By (4.1) $1 = (\psi,\psi)_{\gamma\delta} = \lambda(\psi,\psi)_S$ so that $W'(\lambda) = -\overline{c}\,d/\lambda \neq 0$. The residue at λ is thus $-\lambda(\psi(f,\psi)_S,g)_{\gamma\delta} = -\lambda(f,\psi)_S\,\hat{g}$. Again from (4.1) follows $(f,\psi)_{\gamma\delta} = \lambda(f,\psi)_S$ which completes the proof.

In the case when $(\cdot,\cdot)_{\gamma\delta}$ is non-degenerate we now have all necessary ingredients for an expansion theorem. We order the eigenvalues increasingly, labeling negative eigenvalues with negative and positive eigenvalues with positive integers. For $n \neq 0$ we then let ψ_n be the eigenfunction belonging to λ_n, normalized so that $(\psi_n,\psi_n)_{\gamma\delta} = 1$. Put $\hat{f}_n = (f,\psi_n)_{\gamma\delta}$ for $f \in H$, $n \neq 0$.

Theorem 5.4. Suppose $(\cdot,\cdot)_{\gamma\delta}$ is non-degenerate.

1. For f, $g \in H$ and at least one in $\tilde{H}$, holds the Parseval formula

$$(f,g)_{\gamma\delta} = \Sigma\,\hat{f}_n\overline{\hat{g}_n}\,.$$

2. If $f \in \tilde{H}$, then the Fourier series $\Sigma\,\hat{f}_n\psi_n(x)$ converges to f in the norm of H and uniformly on I.

Proof. 1. The proof can be made as the proof of Titchmarsh [26, Theorem 2.7(ii)], first proving the formula for $f \in D_{\gamma\delta}$ and then by approximation for $f \in \tilde{H}$. Note that $D_{\gamma\delta}$ is dense in $\tilde{H}$ according to Lemma 4.4.

2. The convergence in norm is a standard consequence of the Parseval identity. According to Lemma 3.5 convergence in norm implies uniform convergence so the proof is complete.

Note that it follows that there are infinitely many eigenvalues exactly if $\dim \tilde{H} = \infty$ but that the latter is not necessarily true. Referring to the set Ω of Theorem 4.1 it is clear that $\dim \tilde{H} < \infty$ exactly if the number of points in $I \setminus \Omega$ is finite and that this number then equals $\dim \tilde{H}$ and also the dimension of a maximal subspace of H on which $(\cdot,\cdot)_S$ is non-degenerate. Clearly this case must be regarded as exceptional. In the same way it is clear that there are infinitely many positive eigenvalues precisely if there is an infinite-dimensional subspace of H on which $(\cdot,\cdot)_S$ is positive definite. A similar statement holds for negative eigenvalues.

6. The degenerate case

We now assume that $\lambda = 0$ is an eigenvalue to (1.4), (1.5) and let ψ_0 be a corresponding eigenfunction. In this section it is therefore assumed that $(\psi_0,\psi_0)_{\gamma\delta} = 0$. If $(\psi_0,\psi_0)_S \neq 0$ then according to Lemma 4.5 $W'(0) \neq 0$ so we may calculate the residue of $(\Phi(\cdot,\lambda;f),g)_{\gamma\delta}$ at $\lambda = 0$ as in Lemma 5.3. The value is $-(f,\psi_0)_S(\psi_0,g)_{\gamma\delta}/(\psi_0,\psi_0)_S = 0$ so that there is actually no singularity at 0. Setting $\hat{f}_0 = (f,\psi_0)_S/(\psi_0,\psi_0)_S$ we therefore obtain the following expansion theorem.

<u>Theorem 6.1</u>. Suppose $(\psi_0,\psi_0)_{\gamma\delta} = 0$ but $(\psi_0,\psi_0)_S \neq 0$.

1. For $f, g \in H$ and at least one in $\tilde{H}$ holds the Parseval formula

$$(f,g)_{\gamma\delta} = \sum_{n\neq 0} \hat{f}_n \overline{\hat{g}}_n .$$

2. If $f \in \tilde{H}$, the series $\Sigma \hat{f}_n \psi_n(x)$ converges to f in the semi-norm of H and uniformly on I.

<u>Proof</u>. The proof of the Parseval formula and the semi-norm convergence is identical to the corresponding part of the proof of Theorem 5.4. To prove the uniform convergence, first consider the space $H_1 = \{f \in \tilde{H} | (f,\psi_0)_S = 0\}$. Since eigenfunctions are orthogonal in $(\cdot,\cdot)_S$ clearly $\psi_n \in H_1$ for $n \neq 0$ but $\psi_0 \notin H_1$. Consequently, H_1 is a Hilbert space with scalar product $(\cdot,\cdot)_{\gamma\delta}$ according to Theorem 3.3.

Hence the result on uniform convergence follows as before for elements of H_1. For general $f \in \tilde{H}$ set $g = f - \hat{f}_0\psi_0$. It is clear that $\hat{g}_n = \hat{f}_n$ for $n \neq 0$ and $g \in H_1$. Hence $\sum_{n\neq 0} \hat{f}_n\psi_n(x)$ converges uniformly to $f(x) - \hat{f}_0\psi_0(x)$ which completes the proof.

All of the above assumed that $(\psi_0,\psi_0)_S \neq 0$ which may of course not be true; if not the analysis becomes more involved. The reason is that 0 is now in general a double pole of $\Phi(x,\lambda;f)$, another aspect of which is the appearance of an algebraic eigenspace at 0.

<u>Lemma 6.2</u>. Assume $(\psi_0,\psi_0)_{\gamma\delta} = (\psi_0,\psi_0)_S = 0$. Then there exists $\dot{\psi}_0 \in \tilde{H}$ so that $M[\dot{\psi}_0] = S[\psi_0]$ and (1.5) holds. In addition, $(\dot{\psi}_0,\psi_n)_{\gamma\delta} = (\dot{\psi}_0,\psi_n)_S = 0$, but $(\dot{\psi}_0,\dot{\psi}_0)_{\gamma\delta} = (\psi_0,\dot{\psi}_0)_S \neq 0$. Furthermore, for $f \in H$ satisfying $(f,\psi_0)_S = (f,\dot{\psi}_0)_S = 0$ there is a unique $y \in \tilde{H}$, also satisfying $(y,\psi_0)_S = (y,\dot{\psi}_0)_S = 0$ and for which $M[y] = S[f]$ and (1.5) holds.

<u>Proof</u>. Let u be the solution of $M[u] = S[f]$ with vanishing data at a. Then (4.2) gives

$$(f,\psi_0)_S = (f,\psi_0)_S - (u,0)_S = u(b)\,\overline{\psi_{0\gamma\delta}^{[1]}}(b) - u_{\gamma\delta}^{[1]}(b)\,\overline{\psi_o(b)}\,.$$

Since ψ_0 satisfies the boundary condition at b but does not have vanishing data there it follows that the boundary conditions at b for $M[u] = S[f]$ are satisfied precisely if $(f,\psi_0)_S = 0$. We may thus take $\dot{\psi}_0$ as the solution of $M[\dot{\psi}_0] = S[\psi_0]$ with vanishing data at a. and then (1.5) will be satisfied. By (4.1) we also obtain

$$\lambda_n(\dot{\psi}_0,\psi_n)_S = (\dot{\psi}_0,\psi_n)_{\gamma\delta} = (\psi_0,\psi_n)_S = 0\,.$$

The last equality follows from the orthogonality of eigenfunctions in $(\cdot,\cdot)_S$. Also from (4.1) follows $(\psi_0,\dot{\psi}_0)_S = (\dot{\psi}_0,\dot{\psi}_0)_{\gamma\delta}$ and this must be non-zero since otherwise $\psi_0 \in H_\infty$ by Lemma 3.2 which contradicts Corollary 4.2.1.

Now, with u as above, put $c = (u,\dot{\psi}_0)_S/(\dot{\psi}_0,\psi_0)_S$ and set $y =$ $= u - c\,\psi_0$. It is clear that $M[y] = S[f]$ and (1.5) is still satisfied. Additionally we obtain $(y,\dot{\psi}_0)_S = 0$. An application of (4.2) then gives

$$(y,\psi_0)_S = (y,\psi_0)_S - (f,\dot{\psi}_0)_S = 0$$

so the existence of y is proved. On the other hand, the difference of two such solutions must be a multiple of ψ_0, orthogonal with respect to $(\cdot,\cdot)_S$ to $\dot{\psi}_0$, so necessarily zero. The proof is complete.

Since we have assumed $(\psi_0,\psi_0)_S = 0$ it is clear from Lemma 4.4 that $W'(0) = 0$ so the calculation of the residue at 0 of $(\Phi(\cdot,\lambda;f),g)_{\gamma\delta}$ becomes more complicated. We circumvent this difficulty with the aid of the next lemma.

<u>Lemma 6.3</u>. Suppose $(\psi_0,\psi_0)_{\gamma\delta} = (\psi_0,\psi_0)_S = 0$. Then if f, g are in H and $(f,\psi_0)_S = (f,\dot{\psi}_0)_S = 0$ the function $\lambda \mapsto (\Phi(\cdot,\lambda;f)g)_{\gamma\delta}$ is regular at $\lambda = 0$.

Proof. For N sufficiently large $\lambda^N/W(\lambda)$ has a removable singularity at 0. According to Lemma 6.2 we can find u_0 so that $M[u_0] = S[f]$, the boundary conditions (1.5) are satisfied and u_0 has the properties of f. The procedure can thus be repeated indefinitely so we obtain a sequence $u_0, u_1, \ldots$ such that $M[u_{j+1}] = S[u_j]$ and (1.5) is satisfied for $j = 0, 1, \ldots$. According to Lemma 5.2.2 we then have

$$\Phi(x,\lambda;f) = u_0(x) + \lambda\Phi(x,\lambda;u_0) = \ldots = \sum_{j=0}^{N-1} \lambda^j u_j + \lambda^N \Phi(x,\lambda;u_{N-1})$$

where the last term is regular at 0 as an element of H for sufficiently large N. This completes the proof.

Dividing, if necessary, ψ_0 and $\dot{\psi}_0$ by the same constant, we may assume that $(\dot{\psi}_0,\dot{\psi}_0)_{\gamma\delta} = 1$. Define

$$\dot{f}_0 = (f,\dot{\psi}_0)_{\gamma\delta} = (f,\psi_0)_S \quad \text{for } f \in H$$

$$\hat{f}_0 = (f,\dot{\psi}_0)_S - \dot{f}_0(\dot{\psi}_0,\dot{\psi}_0)_S \quad \text{for } f \in H$$

$$H_1 = \{f \in \tilde{H} \mid (f,\psi_0)_S = (f,\dot{\psi}_0)_S = 0\}$$

so that H_1 consists exactly of those $f \in \tilde{H}$ for which $\dot{f}_0 = \hat{f}_0 = 0$. It follows from Lemma 6.2 that all eigenfunctions to non-zero eigenvalues are in H_1 but that ψ_0 and $\dot{\psi}_0$ are not. The expansion theorem in the present case reads as follows.

Theorem 6.4. Suppose $(\psi_0,\psi_0)_{\gamma\delta} = (\psi_0,\psi_0)_S = 0$.

1. For $f, g \in H$ and at least one in $\tilde{H}$ holds the Parseval formula

$$(f,g)_{\gamma\delta} = \dot{f}_0\overline{\dot{g}_0} + \sum_{n\neq 0} \hat{f}_n \overline{\hat{g}_n}\ .$$

2. For $f \in \tilde{H}$ the series $\dot{f}_0\dot{\psi}_0(x) + \Sigma\, \hat{f}_n\psi_n(x)$ converges fo f with respect to $(\cdot,\cdot)_{\gamma\delta}$ and uniformly in I.

Proof. Put $h = f - \dot{f}_0\dot{\psi}_0 - \hat{f}_0\psi_0$. It is then clear that $\hat{h}_n = \hat{f}_n$ for $n \neq 0$ and $h \in H_1$. However, it follows from Lemma 6.3 that for $f \in H_1$ the proof is exactly as the proof of Theorem 6.1. Hence we have

$$(f,g)_{\gamma\delta} = (h,g)_{\gamma\delta} + \dot{f}_0(\mathring{\psi}_0,g)_{\gamma\delta} + \hat{f}_0(\psi_0,g)_{\gamma\delta} = \sum_{n\neq 0} \hat{f}_n\overline{\hat{g}_n} + \dot{f}_0\overline{\dot{g}_0}$$

which proves 1. Furthermore, the series $\sum_{n\neq 0}\hat{f}_n\psi_n(x)$ converges in $(\cdot,\cdot)_{\gamma\delta}$ and uniformly on I to $f(x) - f_0\dot{\psi}_0(x) - \hat{f}_0\psi_0(x)$ which completes the proof.

Remark. When $(\psi_0,\psi_0)_S = 0$ it may be seen that $W''(0)$ equals some non-zero constant times $(\dot{\psi}_0,\dot{\psi}_0)_{\gamma\delta}$ so that the pole at 0 of $(\Phi(\cdot,\lambda;f),g)_{\gamma\delta}$ is at most double. It is then possible to calculate the residue at 0 to be $-\dot{f}_0\overline{\dot{g}_0}$. It does not, however, appear quite simple to give strict proofs of these facts directly from the expression for $\Phi(x,\lambda;f)$. Lemma 6.3 circumvents the difficulty and once the expansion theorem is known the residues are easily calculable. In retrospect this shows that $N = 2$ is sufficient in the proof of Lemma 6.3.

7. Operator theory

It is obvious the the mapping $f \mapsto \Phi(\cdot,\lambda;f)$ should be interpreted as the resolvent operator of an operator representing the boundary value problem (1.4), (1.5). To carry out this interpretation we therefore define

$$H_1 = \begin{cases} \tilde{H} \text{ if } (\cdot,\cdot)_{\gamma\delta} \text{ is non-degenerate} \\ \{f \in \tilde{H} \mid (f,\psi_0)_S = 0\} \text{ if } (\psi_0,\psi_0)_S \neq 0 \\ \{f \in \tilde{H} \mid (f,\psi_0)_S = (f,\dot{\psi}_0)_S = 0\} \text{ if } (\psi_0,\psi_0)_S = 0, \end{cases}$$

using the notation of the earlier sections. By Theorem 3.3 it is clear that H_1 is a Hilbert space with scalar product $(\cdot,\cdot)_{\gamma\delta}$. Now put

$$R_\lambda f = \Phi(\cdot,\lambda;f) \quad \text{for } f \in H_1 .$$

R_λ is then well-defined as soon as λ is not an eigenvalue of (1.4), (1.5). The basic properties of the operator R_λ are given in the next lemma.

Lemma 7.1. (i) $R_\lambda: H_1 \to H_1$ and $\|R_\lambda\| \leq 1/|\mathrm{Im}\,\lambda|$.

(ii) $R_\lambda f = 0$ only if $f = 0$.

(iii) $R_\lambda - R_\mu = (\lambda - \mu) R_\lambda R_\mu$.

(iv) $(R_\lambda)^* = R_{\overline{\lambda}}$.

Proof. For $f \in H$ certainly $\Phi(\cdot, \lambda; f) \in \widetilde{H}$ so the range of R_λ is certainly in $\widetilde{H}$. $R_\lambda f$ solves $M[y] = \lambda S[y] + S[f]$ with the conditions (1.5) so in the degenerate case (4.2) gives

$$0 = (\lambda R_\lambda f + f, \psi_0)_S - (R_\lambda f, 0)_S = \lambda (R_\lambda f, \psi_0)_S + (f, \psi_0)_S .$$

Hence $(R_\lambda f, \psi_0)_S = 0$ if $f \in H_1$. If $(\psi_0, \psi_0)_S = 0$ we obtain in the same way

$$0 = (\lambda R_\lambda f + f, \dot{\psi}_0)_S - (R_\lambda f, \psi_0)_S = \lambda (R_\lambda f, \dot{\psi}_0)_S$$

if $f \in H_1$ which together with Lemma 5.2.1 proves (i).

To prove (ii), note that $R_\lambda f = 0$ means that $y = 0$ solves (1.4), (1.5) so that $f \in H_\infty$. By Corollary 4.2 this implies $f = 0$. The equation $M[y] = \lambda S[y] + S[(\lambda - \mu) R_\mu f]$ is satisfied by $R_\lambda f - R_\mu f$ together with the boundary conditions (1.5) so the uniqueness of this solution proves (iii).

Finally, the equations and boundary conditions satisfied by $R_\lambda f$ and $R_{\overline{\lambda}} g$ together with (4.3) show that

$$(R_\lambda f, g)_{\gamma\delta} - (f, R_{\overline{\lambda}} g)_{\gamma\delta} = (R_\lambda f, \overline{\lambda} R_{\overline{\lambda}} g + g)_{\gamma\delta} - (\lambda R_\lambda f + f, R_{\overline{\lambda}} g)_{\gamma\delta} = 0$$

for all $f, g \in H_1$ which proves (iv).

It is a well-known fact that from (i) - (iv) it follows that the operator $T = \lambda E + R_\lambda^{-1}$, where E is the identity, is independent of λ, selfadjoint and has R_λ as its resolvent operator. The independence of λ follows in fact directly on multiplying (iii) from the left and right by R_λ^{-1} and R_μ^{-1} respectively, and then (iv) gives

$$T^* = \overline{\lambda} E + R_{\overline{\lambda}}^{-1} = T .$$

<u>Theorem 7.2</u>. The operator $T: H_1 \to H_1$ has domain $D_{\gamma\delta} \cap H_1$ and maps $u \in D_{\gamma\delta} \cap H_1$ onto the unique $f \in H_1$ such that $u, u^{[1]} = p(u' - ru) + \rho f$ satisfies (1.4), (1.5).

<u>Proof</u>. $Tu = f$ means $(T - \lambda E)u = f - \lambda u$, i.e. $u = R_\lambda(f - \lambda u)$. Hence u satisfies $M[u] = \lambda S[u] + S[f - \lambda u] = S[f]$ and the boundary conditions as stated. The uniqueness of f follows from Corollary 4.2.1.

To obtain the Parseval formula and hence the expansion theorems of Sections 5 and 6 we must show that T has a discrete spectrum without finite cluster points, i.e. that R_λ is a compact operator. There are a number of ways of proving this. A simple means is provided by the following lemma.

<u>Lemma 7.3</u>. If $g \in AC(I)$ and $\sqrt{p}(g' - rg) \in L^2(I)$, then $H_1 \ni f \to \overset{\beta}{\underset{\alpha}{(f,g)}}_S$ are linear forms on H_1, uniformly bounded with respect to $\alpha, \beta \in I$.

<u>Proof</u>. We have

$$\overset{\beta}{\underset{\alpha}{(f,g)}}_S = \int_\alpha^\beta \{ -\rho/\sqrt{p}\, f\, \overline{\sqrt{p}(g' - rg)} - \sqrt{p}(f' - rf)\, \overline{\rho/\sqrt{p}\, g} + (w - \overline{r}\rho - r\overline{\rho}) f \overline{g} \} .$$

The first term may be estimated by $\{ \int_I |\rho|^2/p|f|^2 \int_I p|g' - rg|^2 \}^{1/2}$, the second term similarly and the last term by $\{ \int_a^b |w - \overline{r}\rho - r\overline{\rho}||f|^2 \}^{1/2}$ times a similar factor involving g. The integrals containing g are finite and those containing f may be estimated by $(f,f)_{\gamma\delta}$ with the aid of Lemma 3.5. The proof is complete.

<u>Theorem 7.4</u>. R_λ is a compact operator.

<u>Proof</u>. We need to prove that if $f_j \to 0$ weakly in H_1, then

$\Phi_j = R_\lambda f_j \to 0$ strongly. Now according to the previous lemma, $(f_j,\emptyset(\cdot,\bar{\lambda}))_S$ over $[a,x]$ and $(f_j,\chi(\cdot,\bar{\lambda}))_S$ over $[x,b]$ are uniformly bounded with respect to x and tend pointwise to zero as $j\to\infty$. Hence from (5.1), (5.2) the same is true for $\Phi_j(x)$ and $\Phi_j^{[1]}(x)$. By Lemma 3.5 $f_j(x)$, $j = 1, 2, \ldots$ are also uniformly bounded with respect to x and tend pointwise to zero. Hence

$$p|\Phi_j' - r\,\Phi_j|^2 + q|\Phi_j|^2 =$$

$$= \frac{1}{p}\,|\Phi_j^{[1]}|^2 - 2\,\mathrm{Re}\,\{\rho/p(\lambda\Phi_j + f_j)\overline{\Phi_j^{[1]}}\} + |\rho|^2/p\,|\lambda\,\Phi_j + f_j|^2 + q|\Phi_j|^2$$

tends pointwise to zero and is bounded by a function in $L(I)$. Thus by Lebesgue's theorem on majorized convergence the theorem follows.

8. Final remarks

To handle a singular case of (1.4) by the methods of Titchmarsh [26] calls for the introduction of an $m(\lambda)$-function as a first step. There are certain difficulties here since in the singular case the function $\Phi(x,\lambda;f)$ could only be written in a form analogous to (5.1) if the, unwarranted, assumption is made that $(\cdot,\cdot)_S$ is a bounded form on H. It is our intention to return to these questions shortly.

There are several ways of embedding the left-definite boundary value problem (1.4), (1.5) in an operator theoretical setting alternative to that of Section 7. One way is to prove the compactness of the form $(\cdot,\cdot)_S$ with respect to $(\cdot,\cdot)_{\gamma\delta}$ on the lines of the proof of Lemma 3.5. In the degenerate case one must then work in the space H_1 of Section 7. This gives rize to an invertible, compact and selfadjoint operator G on H_1 such that $(u,v)_S = (Gu,v)_{\gamma\delta}$ for $u, v \in H_1$. It remains to show that G is the inverse of the operator defined by

(1.4), (1.5) in H_1 or at least that the eigenfunctions of G are exactly those of (1.4), (1.5). This can be done by a regularity theorem for weak solutions of (1.4), proved on the lines of Lemma 3.2. This approach is not so promising in the case of a singular problem since there is then no reason to assume that $(\cdot,\cdot)_S$ is a bounded form on H_1.

Another approach would be to consider the problem in the light of the general theory of [3], which makes use of the abstract theory of symmetric relations on a Hilbert space. In [3] are introduced minimal and maximal relations associated with a pair of differential expressions in a given Hilbert space and the possible selfadjoint boundary conditions are then characterized. To carry out this in our case creates, because of the low degree of smoothness required from the coefficients, certain difficulties. If these difficulties could be overcome, this approach would certainly work as well for singular cases, the expansion theorem being obtained on the lines of Section 6 of [3].

The approach of Pleijel [22], although not designed for coefficients with the present low degree of smoothness, could presumably be made to work in the present case. It would, however, require some stricter positivity condition than ours in the singular case. This theory also lacks an expansion theorem in terms of explicit integral transforms in the singular case.

Our problem could be considered a special case of an S-hermitean system according to Schäfke, Schneider, Niessen and others [18, 19, 24]. Again our smoothness conditions are less restrictive than those of these authors. We do not know whether this is a serious obstacle, however. In any case, a singular left-definite case can only be handled by the theory of these authors in case $\rho \equiv 0$ and the $(\cdot,\cdot)_S$-form is bounded in the Hilbert space.

Finally, our problem can be handled by the methods scetched in [5], but also this theory so far lacks an expansion theorem in terms of explicit integral transforms in the singular case.

Some comments on the non-definite case where neither $(\cdot,\cdot)_{\gamma\delta}$ nor $(\cdot,\cdot)_S$ is positive may be in order. Assuming (i) - (vii) of Section 2 it will be easy to show, using Lemma 3.5, that there can only be finitely many non-real eigenvalues, each associated with a finite dimensional algebraic eigenspace. Also, there will be only finitely many eigenvalues for which $(\psi_n,\psi_n)_{\gamma\delta} \leq 0$ for the corresponding eigenfunctions. It then appears feasible to use the methods of Section 6 to obtain an expansion theorem. There are, however, many unsettled problems in this area. In particular, the only technique which has been successfully applied to, so far rather special, singular problems is the rather complicated one of Daho and Langer [8, 9].

A final comment concerns the question of whether our smoothness assumptions (i) - (v) can be further relaxed. There have been recent discussion (see e.g. Birkeland [28] and further references there) of the (right-definite) Sturm-Liouville equation

$$-(pu')' + \nu u = \lambda \mu u \tag{8.1}$$

where ν and μ are measures. This case was also discussed already in Atkinson´s book [29, Chaper 1]. It is somewhat surprising that this equation is actually a special case of our equation (1.4). To see this, put

$$r(x) = \frac{1}{p(x)} \int_a^x d\nu\,, \quad q = -p\,r^2 \quad \text{and} \quad \rho(x) = \int_a^x d\mu\,, \quad w \equiv 0\,.$$

It is then a simple matter to check that formally the equation (1.1) reduces to (8.1). Choosing appropriate Dirichlet integrals belonging to M and S one may also obtain the (simplest choices of) Dirichlet integrals $\int p|u'|^2 + \int |u|^2 d\nu$ and $\int |u|^2 d\mu$ associated with (8.1). It is clear that r, q, ρ and w defined as above have the properties (i) - (v) of Section 2 so that (8.1) is apparently less general than our equation (1.4). An interesting fact to notice is then that an operator of the form S may have a positive definite Dirichlet integral, corresponding to positivity of the measure μ. It should perhaps be remarked, though, that such an operator can hardly be called a first order operator, being multiplication by ρ' !

References

[1] F.V. Atkinson, W.N. Everitt and K.S. Ong, On the m-coefficient of Weyl for a differential equation with an indefinite weight function. Proc. London Math. Soc. (3) 29 (1974), 368-384.

[2] C. Bennewitz and Å. Pleijel, Selfadjoint extensions of ordinary differential operators. Proc. of the Coll. on Math. Analysis, Jyväskylä, Finland 1970, Lecture Notes in Mathematics 419 (1974), 42-52, (Springer-Verlag; Heidelberg).

[3] C. Bennewitz, Spectral theory for pairs of differential operators. Ark. Mat. 15 (1977), 33-61.

[4] C. Bennewitz, A generalization of Niessen's limit-circle criterion. Proc. Royal Soc. Edinburgh (A) 78 (1977), 81-90.

[5] C. Bennewitz, Spectral theory for hermitean differential systems. Proc. of the Int. Conf. on Spectral Theory of Differential Operators, University of Alabama Birmingham, USA, 1981. Mathematical Studies 55, 61-68 (North Holland; Amsterdam 1981).

[6] E.A. Coddington and H.S.V. de Snoo, Differential subspaces associated with pairs of ordinary differential expressions. J. Differential Equations 35 (1980), 129-182.

[7] E.A. Coddington and H.S.V. de Snoo, Regular boundary value problems associated with pairs of ordinary differential expressions. Lecture Notes in Mathematics 858 (1981) (Springer-Verlag; Heidelberg).

[8] K. Daho and H. Langer, Some remarks on a paper of W.N. Everitt. Proc. Royal Soc. Edinburgh (A) 78 (1977), 71-79.

[9] K. Daho and H. Langer, Sturm-Liouville operators with an indefinite weight function. Proc. Royal Soc. Edinburgh (A) 78 (1977), 161-191.

[10] N. Dunford and J.T. Schwarz, Linear operators II. (Interscience; New York 1963).

[11] W.N. Everitt, Some remarks on a differential expression with an indefinite weight function. Mathematical Studies 13 (1974), 13-28. (North Holland; Amsterdam).

[12] W.N. Everitt, On certain regular ordinary differential expressions and related differential operators. Proc. of the Int. Conf. on Spectral Theory of Differential Operators, University of Alabama Birmingham, USA, 1981. Mathematical Studies 55 (1981), 115-167, (North-Holland; Amsterdam).

[13] W.N. Everitt and D. Race, On necessary and sufficient conditions for the existence of Carathéodory solutions of ordinary differential equations. Questiones Mathematicae 2 (1978), 507-512.

[14] W.N. Everitt and A. Zettl, Generalized symmetric ordinary differential expressions I: the general theory. Nieuw Archief voor Wiskunde (3) XXVII (1979), 363-397.

[15] E. Kamke, Zum Entwicklungssatz bei polaren Eigenwertaufgaben. Math. Z. 45 (1939), 706-718.

[16] H. Kaper, Lekkerkerker and A. Zettl, Transport problems and spectral theory for non-definite ordinary differential equations. To appear.

[17] M.A. Naimark, Linear differential operators. Part II. (Ungar; New York 1968).

[18] H.D. Niessen, Singuläre S-hermitesche Rand-Eigenwertprobleme. Manuscripta Math. 3 (1970), 35-68.

[19] H.D. Niessen and A. Schneider, Spectral theory for left-definite systems of differential equations: I and II. Mathematical Studies 13 (1974), 29-44 and 45-56, (North-Holland; Amsterdam 1974).

[20] Å. Pleijel, Le problème spectral de certaines équations aux dérivées partielles. Ark. Mat. Astr. Fys. 30 A no. 21 (1944), 1-47.

[21] Å. Pleijel, Generalized Weyl circles. Lecture Notes in Mathematics 415 (1974), 211-226 (Springer-Verlag; Heidelberg).

[22] Å. Pleijel, A positive symmetric ordinary differential operator combined with one of lower order. Mathematical Studies 13 (1974), 1-12. (North-Holland; Amsterdam).

[23] Å. Pleijel, Symmetric boundary conditions for Sturm-Liouville equations. Dept. of Mathematics, Uppsala Univ. Sweden; Report No. 1977:1.

[24] F.W. Schäfke and A. Schneider, S-hermitesche Rand-Eigenwertprobleme. I. Math. Ann. 162 (1965), 9-26.

[25] A. Schneider, On spectral theory for the linear selfadjoint equation $Fy = \lambda Gy$. Lecture Notes in Mathematics 846 (1981), 306-332, (Springer-Verlag; Heidelberg).

[26] E.C. Titchmarsh, Eigenfunction expansions. Part I. (Oxford University Press, 1962).

[27] H. Weyl, Über gewöhnliche lineare Differentialgleichungen mit singulären Stellen und ihre Eigenfunktionen (2. Note). Nachr. Königl. Gesellschaft der Wissenschaften zu Göttingen. Math.-physik. Klasse 442-467 (1910).

[28] B. Birkeland, A singular Sturm-Liouville problem treated by non-standard analysis. Math. Scand. 47 (1980), 275-294.

[29] F.V. Atkinson, Discrete and continuous boundary value problems. (Academic Press; New York 1964).

A VON NEUMANN FACTORIZATION OF SOME SELFADJOINT EXTENSIONS OF POSITIVE SYMMETRIC DIFFERENTIAL OPERATORS AND ITS APPLICATION TO INEQUALITIES

Richard C. Brown

1. <u>Introduction</u> Let H and H' be Hilbert spaces and $T: H \to H'$ a closed densely defined operator. Then a well known theorem of Von Neumann (cf. Kato [8],p. 275) states that T^*T is selfadjoint, and that the domain of (T^*T) is a core of T (in other words the closure of the restriction of the graph of T to the domain of T^*T is the graph of T). Furthermore, a straightforward application of the first representation theorem for quadratic forms (see [8], ch. 6), yields the additional facts:

(i) The form $\|Ty\|^2$ is closed and bounded below by the least element in the spectrum of T^*T.

(ii) The domain of T is the domain of the square root of T^*T.

In this paper we are going to apply these ideas to factor a class of selfadjoint extensions generated by symmetric differential expressions with positive coefficients and with one singular endpoint. As an application of our factorization we are able to give simple new proofs (based on (i)) of Dirichlet and several related inequalities.

Specifically suppose [a,b) is an interval with $-\infty < a < b \leq \infty$ and let p_i, $i = 1,\ldots,n$, w be real locally lebesgue integrable functions on [a,b). Assume further that p_0, w are positive and locally uniformly bounded away from zero; also suppose that $p_i \geq 0$ a.e., $i = 1,\ldots,n-1$, and that $p_n \geq -\gamma w$ a.e. for some positive γ. Let $M[y]) := w^{-1}y^{[2n]}$ where $y^{[2n]}$ is the quasi-derivative of order 2n. Then in the terminology of Naimark the endpoint a is regular and b is singular. Under the above conditions it is known that M determines closed densely defined minimal and maximal operators $T_o(M)$, $T(M)$ in the Hilbert space $L^2_w(a,b)$ of complex valued functions y with norm $\left(\int_a^b w|y|^2\right)^{1/2}$. Further $T_o(M)$ is a restriction of $T(M)$ and $T_o^*(M) = T(M)$ so that $T_o^*(M)$ is symmetric. If we also assume that $p_n > 0$ (this can always be arranged by translating M by a number $d: = \gamma + \varepsilon$, $\varepsilon > 0$) $T_o(M)$ and its selfadjoint extensions are positive. For further details, standard definitions, proofs, etc. relating to the theory of symmetric differential operators we refer the reader to Naimark [12] Ch. V, (see also Kauffman, Read, and Zettle [10]).[1]

As stated above one of the purposes of this paper will be to factor certain selfadjoint extensions of positive definite translations of $T_o(M)$ in the Von Neumann form T^*T and to study the structure of T and T*. To this end in section 3 we define certain 'minimal' and 'maximal' operators L_o and L and construct a family of intermediate operators $L_o \subset L_s \subset L$. The properties of these operators and their adjoints are derived in Lemmas 3-7. Then it is shown (Theorem 1) that $L_s^*L_s$ is the factorization we want. That is, $L_s^*L_s$ is a selfadjoint extension of $T_o(M)$ and restriction of $T(M)$. It is worth remarking that L_o, L do not lie in $L^2_w(a,b)$; in fact their ranges lie in an n + 1 fold cross product of $L^2(a,b)$. Further L and its restrictions are 1-1 normally solvable operators and $||L(y)||^2$ is a translations of the Dirichlet form $\sum_{i=0}^{n}\int_a^b p_i|y^{(n-i)}|^2$. Thus our factorization is different from other recent factorizations which write M as a power or product of standard differential expressions of lower degree (e.g. [10], Ch. V, or [14]). It is rather striking however that the theory of L_o and L will be found to parallel the theory of ordinary differential operators in many formal respects.

1Although both references assume w = 1, the general case is entirely similar.

Once the factorization theory is complete it is easy using the fact that $\|Ly\|^2$ is bounded below (equivalently that L^{-1} is bounded) to prove the Dirichlet inequality

$$(1.1) \qquad \sum_{i=0}^{n} \int_a^b p_i |y^{(n-i)}|^2 \geq \mu_0 \int_a^b w|y|^2$$

where μ_0 is the least element in the spectrum of L^*L, $y \in L^2_w(a,b)$, and $y^{(n-1)}$ is absolutely continuous. A precise statement and proof (Theorem 2) of this inequality as well as certain corollaries is given in section 4. As a short digression this section also contains a sketch of the application of the L, L_o idea to the concept of the Dirichlet index. The last section is application of our machinery to a class of inequalities (Theorem 3) which we label "dual Dirichlet inequalities". These inequalities are multivariable in form and thus are quite different in appearance from (1.1). However, in our setting they are just as easy to prove as the ordinary Dirichlet inequality. They arise by an estimation of the norm of the "generalized inverse" of L^* (just as (1.1) arises by estimating the norm of the inverse of L).

§2. Notation and Preliminary Lemmas We adopt the following notations $L^i(a,b)$, $i = 1, 2$, is the classical lebesgue space of complex valued functions on $[a,b)$, while $L^1_{loc}[a,b)$ denotes the locally lebesgue integrable functions in $[a,b)$. If H is any Hilbert space $\|\cdot\|$, $[\cdot,\cdot]$ denote its norm and inner product. (We usually rely on the context to distinguish between norms or inner products of different spaces; however, for emphasis we often write $\|\cdot\|_w, [\cdot,\cdot]_w$ to distinguish the norm and inner product on $L^2_w(a,b)$.) Let $\{H_i\}$ be a collection of Hilbert spaces $\mathbb{C}^n$ complex n dimensional Euclidean space. If

$$\tilde{\beta}_k := (\beta_1, \ldots, \beta_k) \in \mathop{X}_{i=1}^{k} H_i,$$

we define

$$\|\tilde{\beta}_k\| := \left(\sum_{i=1}^{k} \|\beta_i\|^2\right)^{1/2}$$

$$[\tilde{\beta}_k, \tilde{\beta}'_k] := \sum_{i=1}^{k} [\beta_i, \beta'_i].$$

If $T: H \to H'$ is an operator where H and H' are Hilbert spaces $D(T)$, $N(T)$, $R(T)$, and $G(T)$ stand for its domain, null space, range, and graph respectively. T is said to be normally solvable if T has both closed graph and closed range. $\sigma(T)$, $\sigma_p(T)$, $\sigma_{ap}(T)$, $\sigma_r(T)$ and $\rho(T)$ denote the spectrum, point spectrum, approximate point spectrum, spectral radius, and resolvent of T.

If $b < \infty$ and p_0^{-1}, $p_i, i = 1, \ldots, n$, $w \in L^1(a,b)$, we say that b is regular (otherwise,as mentioned above b is singular). If b is regular, $AC^{(j)}$ denotes the functions y having absolutely continuous j^{th} derivative on $[a,b]$ (we write AC^o as AC); in the singular case the notation refers to local absolute continuity. Finally if $y \in AC^{(n-1)}$, $\hat{y}(a)$, signifies $(y(a), y'(a), \ldots, y^{(n-1)}(a))$ considered as a column vector.

The following result may be proved by the methods of Naimark [12], Ch. V or Goldberg, [6], Ch. 6. However, for completeness we include a proof.

LEMMA 1. Let $p^{-1} \in L^1_{loc}(a,b)$. Define the operator $L^2_w(a,b) \to L^2(a,b)$ by $L(y) := p^{1/2}\, y^{(n)}$ on

$$D := \{y \in L^2_w(a,b) \cap AC^{(n-1)}: \; p^{1/2} y^{(n)} \in L^2(a,b)\}.$$

Then L is a densely defined closed operator with adjoint $L_o^+ y := (-1)^n\, w^{-1} (p^{1/2} z)^{(n)}$ on

$$D_o^+ := \{z \in L^2(a,b): (p^{1/2}z)^{(n-1)} \in AC;\ w^{-1}(p^{1/2}z)^{(n)} \in L^2(a,b);$$

$$(p^{1/2}z)^{(i)}(a) = 0,\ i = 0,\ldots,n-1;\ \{y,z\}(b^-) = 0, \forall y \in D\},$$

where

$$\{y,z\}(t) := \sum_{i=1}^{n} (-1)^{(i-1)} y(t)^{(n-i)} (p^{1/2}z)(t)^{(i-1)}.$$

Proof: We first introduce some additional operators. Let L^+ be given by $(-1)^n w^{-1}(P^{1/2}z)^{(n)}$ on

$$D^+ := \{z \in L^2(a,b): (p^{1/2}z) \in AC^{(n-1)};\ w^{-1}(p^{1/2}z)^{(n)} \in L_w^2(a,b)\ .$$

Let $L_o^{+\prime}$ be the restriction of L^+ to functions of compact support in (a,b). With these definitions it is evident that $L_o^{+\prime} \subset L_o^+ \subset L^+$. Similarly, L_o' is the restriction of L to functions of compact support in (a,b) and L_o is the restriction of L to

$$D_o := \{y \in D: \hat{y}(a) = 0,\ \{y,z\}(b^-) = 0,\ z \in D^+\}.$$

Thus $L_o' \subset L_o \subset L$. When b is a regular endpoint, it is also immediate from these definitions that

$$D_o^+ = \{z \in D^+: (p\,z)^{(i)}(a) = (p\,z)^{(i)}(b) = 0,\ i = 0,\ldots,n-1\}$$

and

$$D_o = \{y \in D: \hat{y}(a) = \hat{y}(b) = 0\}.$$

Intergration by parts furthermore gives Green's formula:

$$[Ly, z] = [y, L^+z]_w = \{y, z\}(b^-) - \{y, z\}(a).$$

To see that the operators L_o', L_o, and L are densely defined it suffices to show that $L_o'^* = L^+$. We first consider the regular case: If $p^{-1} \varepsilon L^1(a, b)$ application of the Cauchy-Schwartz ineqaulity shows

$$y = \int_a^t \frac{(t - s)^{(n-1)}}{(n - 1)!} p^{-1/2} f \, ds$$

exists; since it is continuous, $y \varepsilon L_w^2(a, b)$. This shows that $p^{1/2}y^{(n)} = f$ is solvable for all f in $L^2(a,b)$ and thus L is onto. Similarly if $(-1)^n w^{-1}(pz)^{(n)} = f$ where $f \varepsilon L_w^2(a, b)$,

$$(p^{1/2}z)(t) = \int_a^t \frac{(t - s)^{(n-1)}}{(n - 1)!} (-1)^n w^{-1}(p^{1/2}z)^{(n)} \mathbf{w} \, ds$$

$$\leq C||f||_w.$$

By Cauchy-Schwartz again z is seen to be in $L^2(a,b)$. Next we claim $R(L_o)^\perp = N(L^+)$ and $R(L_o^+)^\perp = N(L)$. That $R(L_o)^\perp \subset N(L^+)$ and $R(L_o^+)^\perp \subset N(L)$ follow from a suitably generalized fundamental lemma of the calculus of variations. Since however we have not found a statement that quite matches our setting, we sketch a proof here. For i = 0, ..., n - 1 define $\eta_i := p^{-1/2}x^i$. Let ξ be a function such that $\int_a^b p^{1/2}|\xi|^2 < \infty$ and $[p^{1/2} \xi, \eta_i] = 0$ for all η_i. ξ is easily seen to be integrable. Now $[p^{1/2}\xi, \eta_o] = \int_a^b \xi$. Integration of each $[p^{1/2}\xi, u_i]$ by parts yields successively:

$$\int_a^b \int_a^t \xi = 0$$

$$\int_a^b \frac{(b - s)^{i-1}}{(i - 1)!} \xi \, ds = 0$$

$$\int_a^b \frac{(b - x)^{n-1}}{(n - 1)!} \xi \, ds = 0.$$

Further if $\psi := \int_a^t \frac{(t - s)^{n-1}}{(n-1)!} \xi \, ds$, $\psi^{(n-1)} \varepsilon AC$, $\psi^{(n)} = \xi$, $\hat{\psi}(a) = 0$, and (from (2.1)) $\hat{\psi}(b) = 0$. Hence $\psi \varepsilon D_o$ and $\xi \varepsilon R(L_o)$. Conversely if $\xi \varepsilon R(L_o)$, it is easy to check that $[\xi, n_i] = 0$ for all i. Suppose $f \perp R(L_o)$. Using a generalized lemma of the calculus of variation in Akhiezer [1] p. 198, we conclude that f is a linear combination of the η_i's. Another argument yielding the same conclusion is to observe that the intersections of the kernels of the functionals $\psi_i : R(L) \to ₵$ defined by $\psi_i(z) = [z, \eta_i]$ is by the above argument exactly $R(L_o)$. We can now use the linear dependence principle,e.g. Rudin [17] p. 62, to see that $f = \Sigma d_i \eta_i$. At any rate $p^{1/2}$ f is a.e. a polynomial of degree η - 1, proving that $R(L_o)^{\perp} \subset N(L^+)$. The argument that $R(L_o)^{\perp} \subset N(L)$ is similar. The reverse inclusions are consequences of Green's formulas. Since L^+ is onto, if $(\alpha, \beta) \varepsilon G(L_o^*)$ there exists α' in D^+ such that $(\alpha', \beta) \varepsilon G(L^+)$. By Greens formula $L^+ \subset L_o^*$ so that

$$\alpha - \alpha' \varepsilon N(L^+) = R(L_o)^{\perp} = N(L_o^*).$$

These facts imply that $(\alpha, \beta) \varepsilon G(L^+)$ and thus that $L^+ = L_o^*$. The proof that $L_o^{+*} = L$ follows the same logic interchanging L_o^+ for L_o and $N(L)$ for $N(L^+)$. From this we conclude that L and L^+ are closed operators and that L_o^+ is densely defined. Since $L_o \subset L$ and

$L_o^+ \subset L^*$, $L^* \subset L^+$ and $L^{+*} \subset L$. Applying Green's formula we see that L^* and L^{+*} satisfy exactly the boundary conditions characterizing L_o^+ and L_o respectively.

If b is a singular endpoint, recall first that the operators L_o', $L_o^{+}{}'$ are the restrictions of L and L^+ to functions of compact support. Let $\Delta \subset [a,b)$ be an arbitrary compact subinterval. Let $L_{o,\Delta}'$, $L_{o,\Delta}^+$ be respectively L_o and L_o^+ defined on Δ. Let L_Δ, L_Δ^+ be the restrictions of L and L^+ to Δ. The argument given above implies that $L_{o,\Delta}^* = L_\Delta^+$ and $L_{o,\Delta}^{+*} = L_\Delta$. Since the interval Δ is arbitrary we conclude that $L_o'^* \subset L^+$ and $L_o^{+}{}'^* \subset L$. The reverse inclusions are given by Green's relation. Hence L^+ and L are closed operators and L^+ is densely defined since it is an extension of $L_o^{+}{}'$ whose adjoint is an operator. Finally since $L_o^+ \supset L_o^{+}{}'$, $L_o^{+*} \subset L$. However by Green's relation $L \subset L_o^{+*}$; thus $L_o^{+*} = L$. A similar argument shows that $L_o^* = L^+$. Since $L^* \subset L^+$ Green's relation forces the boundary conditions defining L_o^+. Summarizing

$$L_o'^* = L_o^* = L^+ \;,$$

$$L_o^{+*} = L \;,\; L_o^{+*} = L \;,$$

$$L^* = L_o^+ \;,\; L^{+*} = Lo,$$

$$L_o = \overline{L_o'} \;,\; L_o^+ = \overline{L_o^{+}{}'}.$$

The last preliminary result of this section is a refinement of some standard facts of operator theory required in section 4 for the demonstration of our inequalities.

LEMMA 2. *Let H, H' be Hilbert spaces and let* $L: H \to H'$ *be a closed densely defined 1 - 1 operator with closed range in H'. Define* L^*L *as the restriction of* L^* *to* $R(L)$. *Then*

$$\|Ly\|^2 \geq \mu_0 \|y\|^2, \tag{2.1}$$

and

$$\|L^*y\|^2 \geq \mu_0 \operatorname{dist}(y, N(L^*))^2, \tag{2.2}$$

where μ_0 *is strictly positive and equals* $\inf\{\sigma(L^*L)\}$ *and where* $\operatorname{dist}(y,N(L^*)) := \inf\|y - g\|, g \in N(L^*)$. *If* $\mu_0 \in \sigma_p(L^*L)$ *equality in* (2.1) *holds for* $\psi \in D(L)$ *if and only if* ψ *is an eigenfunction corresponding to* μ_0. *Equality in* (2.2) *holds for* ψ *in* $D(L^*)$ *if and only if* $\psi \equiv Lg \bmod N(L^*)$ *where* g *is an eigenfunction corresponding to* μ_0. If $\mu_0 \notin \sigma_p(L^*L)$ *equality in* (2.1) *or* (2.2) *holds if and only if* $y = 0$. *However there are sequences* $\langle\psi_n\rangle$ *in* $D(L)$ *and* $\langle\psi_n'\rangle$ *in* $R(L)$ *such that* $\|L\psi_n - \mu_0\psi_n\| \to 0$ and $\|L^*\psi_n' - \mu_0\psi_n'\| \to 0$. *Finally* $D(L) = D\sqrt{(L^*L}$ *and* $D(L^*) = D\sqrt{LL^*}$.

Proof: As noted in the introduction by Von Neuman's theorem and the first representation theorem L^*L and LL^* are densely defined, self-adjoint, and $D(L) = D\sqrt{L^*L}$, $D(L^*) = D\sqrt{LL^*}$. It is clear also that L^*L and LL^* are 1 - 1 and have real positive spectra. The inequality (2.1) is also true by the first representation theorem. (see [8],Example 2.13,p. 326). However we find it convenient to give a different argument focusing on the fact that since L is 1 - 1 and has closed range, there exists a bounded inverse $A: R(L) \to D(L)$. Let P be orthogonal projection on $R(L)$. Clearly

$$\mu_0^{-1/2} = ||A|| = ||AP|| = ||(AP)^*|| = ||AP(AP)^*||^{1/2} =$$

$$= (\sigma_r(AP(AP^*)))^{1/2}.$$

AP is often called the Hilbert space pseudoinverse of L (cf. [11], ch. 6). Since $(AP)^* = (APP)^* = P(AP)^*$, $(AP)^*$ maps into R(L). Hence the operator $S: = AP(AP)^*$ is well defined. Since $(AP)^*$ also maps into $D(L^*)$ -(in fact, $(AP)^*(z)$ is the least L^2 norm solution in $D(L^*)$ of the equation $L^*y = z$) - and S is inverse to the operator L^*L, $\sigma(S) = \sigma(L^*L)^{-1}$ (Kato, [8], p. 177). Since S is also positive and self-adjoint we conclude that

$$||S|| = \sigma_r(S) = \mu_0^{-1} = \inf\{\sigma(L^*L)\})^{-1}.$$

We turn now to the statements concerning equality in (2.1), (2.2). Suppose $\mu_0^{-1} \varepsilon \sigma_p(S)$ and let ξ be an eigenfunction of S corresponding to μ_0^{-1}. Then $L^*L\xi = \mu_0\xi \Longleftrightarrow ||L\xi||^2 = \mu_0||\xi||^2$. Conversely suppose $\xi \varepsilon D(L)$ gives equality in (2.1). Without loss of generality assume $||L\xi|| = 1$. Set $\eta = L\xi$. Then

$$\begin{aligned}[AP\eta, AP\eta] &= \mu_0^{-1} \\ &= [(AP)^* (AP)\eta, \eta] \\ &\leq ||(AP)^*|| \; ||AP|| \\ &\leq \mu_0^{-1}.\end{aligned}$$

However, equality in Cauchy's inequality is possible if and only if $(AP)^* AP\eta = \mu_0^{-1}\eta$. Applying AP to both sides and using the definition of η, we obtain $AP(AP)^*\xi = \mu_0^{-1}\xi$. Equivalently $L^*L\xi = \mu_0\xi$. Since S is bounded and self-adjoint $\sigma(S) = \sigma_{ap}(S)$ (cf. [11], p. 234) so that if $\mu_0 \notin \sigma_p(S)$, then $\mu_0 \varepsilon \sigma_{ap}(S)$ and the next two statements follow from the definition of "approximate point spectrum". The inequality

(2.2) given the properties mentioned above of $(AP)^*$ expresses the fact that $\|(AP)^*\| = \mu_0^{-1/2}$. The case of equality in (2.2) is similar to the case of equality in (2.1). For suppose $\psi - Lg = \eta$ where $\eta \in N(L^*)$ etc. Then

$$\begin{aligned} \|L^*\psi\|^2 = \|L^*Lg\|^2 &= [L^*Lg, L^*Lg] \\ &= \mu_0 \|Lg\|^2 \\ &= \mu_0 \ \mathrm{dist}(\psi, N(L^*))^2. \end{aligned}$$

Suppose now

$$\|L^*z\|^2 = \mu_0 \ \inf\|z - N(L^*)\|^2 .$$

Set $z = Lg + \eta$ where $\eta \in N(L^*)$. Then $\|L^*Lg\|^2 = \mu_0\|Lg\|^2 \iff \|v\|^2 = \mu_0\|(AP)^*v\|^2$ where v is defined by $g = AP(AP)^*v$. By Cauchy's inequality and the fact that $\|AP\| = \|(AP)^*\| = \mu_0^{-1/2}$, $[(AP)(AP)^*v, v] \leq \mu_0^{-1}\|v\|^2$. So we have equality in Cauchy's inequality which implies $(AP)(AP)^*v = \mu_0^{-1}v$ or $g = \mu_0^{-1}v$. But $v = \mu_0^{-1} L^*Lv$ so $L^*Lg = \mu_0 g$. If $\mu_0 \notin \sigma_p(L^*L)$, $\mu_0 \notin \sigma_p(LL^*)$, since $LL^*g = \mu_0 g \Rightarrow LL^*LL^*g = \mu_0 LL^*g$ $\Rightarrow L(L^*L(L^*g) - \mu_0 L^*g) = 0$ which is impossible since L is 1 - 1. Therefore, $\mu_0 \in \sigma_{ap}(LL^*)$; and the argument is similar to that given above for the inequality $\|Ly\|^2 \geq \mu_0\|y\|^2$.

§3 The "Maximal" and "Minimal" operators L, L_o, their adjoints, and the factorization of certain selfadjoint extensions of $T_o(M_d)$

We set $\tilde{p}_n := p_n + dw$ where $d = \gamma + \varepsilon$ and introduce the following operators. Let L: $L^2_w(a,b) \to H := \overset{n}{\underset{i=0}{X}} L^2(a,b)$ be given by

$$Ly := \begin{pmatrix} p_0^{1/2} y^{(n)} \\ p_1^{1/2} y^{(n-1)} \\ \cdot \\ \cdot \\ \cdot \\ \tilde{p}_n^{1/2} y \end{pmatrix}$$

on the domain

$$\mathcal{D} := \{y \in L_w^2(a, b) \cap AC^{(n-1)},\ p_i^{1/2} y^{(n-i)} \in L^2(a, b),\ i = 0, \dots, n\}.$$

Let L_o' be the restriction of L to the domain $\mathcal{D}_o'$ of functions in $\mathcal{D}$ with compact support in [a, b). We shall show that the closure of $G(L_o')$ in $L_w^2(a, b) \times H$ yields a "minimal operator" L_o analogous to the minimal operator in the differential operator case.

LEMMA 3. L is a normally solvable operator.

Proof: Since $\|y\|_w \leq d^{-1/2} \|Ly\|$, it follows that $G(L)$ is closed if and only if $R(L)$ is closed. Also $\mathcal{D}$ is dense because the local integrability of the p_i implies that $\mathcal{D} \supset D(L_o')$ which is dense by Lemma 1. (Hence $\mathcal{D}_o'$ is also dense.) Now suppose $\langle y_\ell \rangle \to \alpha$ in $L_w^2(a, b)$ and $\langle Ly_\ell \rangle \to \langle \beta_1, \dots, \beta_{n+1} \rangle$ in H. Applying Lemma 1, $\alpha \in D(L)$ and $\beta_1 = L\alpha$. Let Δ be a compact subinterval of [a,b). Then $\tilde{p}_n \geq \varepsilon_\Delta > 0$ where ε_Δ depends on Δ and so $\tilde{p}_n^{-1/2}$ is essentially bounded on Δ. Hence on Δ $\langle \tilde{p}_n^{-1/2} (\tilde{p}_n^{1/2} y_\ell - \beta_{n+1}) \rangle \to 0$ so that $\beta_{n+1} = \tilde{p}_n^{1/2} \alpha$ on Δ. Since Δ is arbitrary, $\beta_{n+1} = \tilde{p}_n^{1/2} \alpha$ on all [a, b). Also on Δ $y_\ell^{(n)} \to \alpha^{(n)}$ since $p_0^{1/2}|_\Delta$ is bounded uniformly away from zero. These facts imply by well known a priori estimates (cf. [6] Lemma VI 6.1, p. 157) that the intermediate derivatives $y_\ell^{(n-i)} \to \alpha^{(n-i)}$ on Δ in the

L^2 norm. Hence $\|p_i^{1/2}y_\ell^{(n-i)} - p_i^{1/2}\alpha^{(n-i)}\|_\Delta \to 0$ because of the local integrability of the p_i, and we conclude $\beta_i = p_i^{1/2}\alpha^{(n-i)}$ on all $[a, b)$.

Now set $\tilde{z}_j := (z_1, \ldots, z_j) \in \overset{j}{\underset{i=1}{X}} L^2(a, b)$ and define recursively the expressions

$$\ell_1^+(\tilde{z}_1) = p_0^{1/2} z_1,$$
$$\ell_2^+(\tilde{z}_2) = -\ell_1^+(\tilde{z}_1)' + p_1^{1/2} z_2$$

$$\ell_{n+1}^+(\tilde{z}_{n+1}) = -\ell_n^+(\tilde{z}_n)' + p_n^{1/2} z_{n+1}.$$

We assume that the $\ell_i^+(\tilde{z}_i)$, $1 \leq i \leq n$, are absolutely continuous and define $L^+: H \to L^2(a, b)$ by $L^+(\tilde{z}_{n+1}) = \ell_{n+1}^+(\tilde{z}_{n+1})$. We denote the domain of L^+ by $\mathcal{D}^+$. The following Lemma may be proved by a calculation:

LEMMA 4. (<u>a Green's formula</u>). <u>For all</u> y <u>in</u> $\mathcal{D}$ <u>and</u> $\tilde{z}_{n+1} \in \mathcal{D}^+$

$$[Ly, \tilde{z}_{n+1}] - [y, L^+(\tilde{z}_{n+1})] = \{y, \tilde{z}_n\}(b^-) - \{y, \tilde{z}_n\}(a^+).$$

<u>where</u>

$$\{y, \tilde{z}_n\}(s) = \sum_{j=1}^{n} y^{(n-j)}(s)\, \overline{\ell_j^+(\tilde{z}_j)}\,(s)\ .$$

Let L_o^+ be the restriction of L^+ to the domain $\mathcal{D}_o^+$ such that

$$\{y, \tilde{z}_n\}(b^-) = \{y, \tilde{z}_n\}(a^+) = 0,\ \forall\, y \in \mathcal{D}$$

and L_o be the restriction of L to the domain $\mathcal{D}_o$ such that

$$\{y, \tilde{z}_n\}(b^-) = \{y, \tilde{z}_n\}(a^+) = 0,\ \forall\, \tilde{z}_{n+1} \in \mathcal{D}^+.$$

Note that $y^{(i-1)}(a)$ and $\ell^+(\tilde{z}_i)(a)$, $j = 1, \ldots, n$, vanish for $y \in \mathcal{D}_o$ and $\tilde{z}_{n+1} \in \mathcal{D}_o^+$ and that the same is true in the regular case $(b < \infty)$ at b.

LEMMA 5. $L_o'^{*} = L^{+}$; $L_o^{*} = L^{+}$; $L^{*} = L_o^{+}$; $L_o^{+*} = L$; $L^{+*} = L_o$. <u>Thus L^{+}, L_o^{+} are onto normally solvable operators</u>.

<u>Proof</u>: Once the required adjoint relations have been shown the second statement follows from the fact that L and L_o are 1-1 and densely defined (hence in particular L^{+} and L_o^{+} are operators) and from the closed range theorem.

We first consider the regular case. By the previous lemma and by standard theory $N(L^{+}) \subset R(L_o)^{\perp} = N(L^{*})$. Suppose $\tilde{z}_{n+1} \in R(L_o)^{\perp}$. Then

$$\sum_{i=0}^{n-1} \int_a^b p_i^{1/2}\, y^{(n-i)} \overline{z}_{i+1} + \int_a^b \tilde{p}_n\, y\, \overline{z}_{n+1} = 0.$$

We integrate this equation by parts, differentiating y and integrating the $p_i^{1/2} z_{i+1}$. The endpoint terms vanish since $y \in \mathcal{D}_o$. We obtain

$$\int_a^b y^{(n)} \left(\sum_{i=0}^{n} (-1)^{(i)}\, I^{(i)}(p_i^{1/2}\, \overline{z}_{i+1}) \right) = 0,^{2}$$

where $I^{(i)}$ signifies the i fold iterated integral.

$$\int_a^{t_i} \cdots \int_a^{t_1} p_i^{1/2}\, \overline{z}_{i+1}\, dt_1 \ldots dt_i \quad (t_i \equiv t).$$

From the fundamental lemma of the calculus of variations (see the discussion in the proof of Lemma 1) the term in parentheses is a polynomial of degree n-1; a computation then shows that $\tilde{z}_{n+1} \in N(L^{+})$. Thus $R(L_o)^{\perp} = N(L^{+})$. Now $R(L^{+}) \supset R(L^{+})$ so that L^{+} is onto. Suppose next $(\tilde{z}_{n+1}, \psi) \in G(L_o^{*})$. Because L^{+} is onto $(\tilde{z}'_{n+1}, \psi) \in G(L^{+})$ for some $\tilde{z}'_{n+1}$. As in the case of Lemma 1 we conclude successively that

---

2 Here p_n is understood to be $\tilde{p}_n$.

$\tilde{z}'_{n+1} - \tilde{z}_{n+1} \in N(L^+) \Rightarrow \tilde{z}'_{n+1} \in \mathcal{D}^+ \Rightarrow L_o^* \subset L^+$. From Green's formula $L_o^* = L^+$. We consider L^*. Clearly $L^* \subset L^+$. If $(\tilde{z}_{n+1}\ \psi) \in G(L^*)$, $[Ly, \tilde{z}_{n+1}] - [y, \psi] = 0$ and Green's formula implies $\{y, z_n\}(b^-) - \{y, z_n\}(a^+) = 0$ for all y in $\mathcal{D}$. For a given $1 \leq j \leq n$ let $y^{(n-i)}(b) = \delta_{ij}$ and $y^{(n-i)}(a) = 0$, $1 \leq i \leq n$. This forces $\ell_j^+(\tilde{z}_j)(b^-) = 0$. Similarly with $y^{(n-i)}(b) = 0$ and $y^{(n-i)}(a) = \delta_{ij}$, $\ell_j^+(\tilde{z}_i)(a^+) = 0$. Since this will hold for all j $(\tilde{z}_{n+1}, \psi) \in G(L_o^+)$. Consequently $L^* = L_o^+$, and because L is closed $L_o^{+*} = L$.

The singular theory follows the reasoning for L and L* in Lemma 1.

We now consider a family of restrictions of L characterized by the property $\mathcal{D}_s = \{y \in \mathcal{D}: \hat{y}(a) \in S\}$ where S is some subspace of $\mathbb{C}^u$. If $S = \mathbb{C}^u$ we identify L_s with L but note that if S is trivial, the resulting operator is not in general L_o since L_o is also characterized by boundary conditions at b.[3]

LEMMA 6. L_s^* <u>is the restriction of</u> L^+ <u>with domain</u>

$$\mathcal{D}_s^+ = \{\tilde{z}_{n+1} \in \mathcal{D}^+: \check{\ell}^+_{n-i}(\tilde{z}_{n-i})(a) \in S^\perp;$$

$$\{y, \tilde{z}_n\}(b^-) = 0, \forall\ y \in \mathcal{D}_s\}$$

where $\check{\ell}^+_{n-i}(\tilde{z}_{n-i})(a) := (\ell_n^+(\tilde{z}_n), \ldots \ell_1^+(z_1))(a)$.

PROOF. This follows from Lemma 4 and the fact that near b $\mathcal{D}_s^+$ must satisfy the boundary conditions of $\mathcal{D}_o^+$; and at a

$$\{y, \tilde{z}_n\}(a) = \sum_{i=1}^{n} y^{(n-i)}(s)\ \ell_i^\perp(\tilde{z}_i)(a) = 0.$$

LEMMA 7. $L^+Ly = y^{[2n]}$.

---

3 See the discussion of the "limit point" case in section 4 after Corollary 2.

Proof: Clearly $\ell_1^+(p_0^{1/2}y^{(n)}) = p_0 y^{(n)} = y^{[n]}$.

Assume that

$$\ell_j^+\left(p_0^{1/2}y^{(n)}, \ldots, p_{j-1}^{1/2}\, y^{(n-(j-1))}\right) = y^{[n+j-1]}.$$

Then

$$\ell_{j+1}^+\left(p_0^{1/2}y^{(n)}, \ldots, p_j^{1/2}y^{(j)}\right) = - y^{[n+j-1]'} + p_0^{1/2}(p_0^{1/2}y^j)$$
$$:= y^{[n+j]}.$$

In particular $\ell_{n+1}^+(Ly) = y^{[2n]}$.

Green's formula for symmetric ordinary differential expressions states

$$\int_a^b (\bar{g}\, M[f] - \bar{f}\, M[g]) = [f, g](b^-) - [f, g](a^+)$$

where

$$[f, g](s) := D(f, \bar{g})(s) - D(g, \bar{f})(s),$$

and

$$D(f, \bar{g})(s) := \sum_{i=0}^{n-1} f^{[i]}(s)\; \bar{g}(s)^{[2n-i-1]}.$$

Following [7] we call $D(f, \bar{g})$ the <u>Dirichlet form</u>.

LEMMA 7. <u>If</u> $\tilde{z}_{n+1} = Lg$ <u>for some</u> $g \in \mathcal{D}$, <u>then</u> $\{y, \tilde{z}_n\}(s) = D(y, \bar{g})(s)$.

Proof: By Green's formula (Lemma 4)

$$[Ly, Lg] - [y, L^+Lg] = \{y, \tilde{z}_n\}(b^-) - \{y, \tilde{z}_n\}(a^+).$$

However the Dirichlet formula (cf. [7], p. 285) tells us

$$(3.1) \quad \int_a^b \sum_{i=0}^{n} p_{n-i}\, \bar{g}^{(i)}y^{(i)} - \int_a^b y\, \overline{M[g]} = D(y, \bar{g})(b^-) - D(y, \bar{g})(a^+)$$

Since the left sides of these two equations are the same, the lemma follows by modifying g or y so that it vanishes near a.

THEOREM 1. The operators $T_1 := L^+ L_o$, $T_2 := L_o^+ L$ are selfadjoint 1-1 and onto. The domains of T_1 and T_2 are cores respectively of L_o and L and $T_o(M) \subset T_1, T_2 \subset T(M)$. Functions y in $D(T_1)$ or $D(T_2)$ satisfy respectively the boundary conditions $y^{[i]}(a) = 0$, $0 \le i \le n-1$, $D(f, \bar{y})(b^-) = 0$, $\forall f \in \mathcal{D}_o$ and $\{y, \tilde{z}_n\}(b^-) = 0$, $\forall \tilde{z}_{n+1} \in \mathcal{D}^+$, or $y^{[i]}(a) = 0$, $n \le i \le 2n-1$, and $D(f, \bar{y})(b^-) = 0$, $\forall f \in \mathcal{D}$.

Proof: That T_1 and T_2 are selfadjoint and have the core property follows from Von Neuman's Theorem. Since $L^+ = L_o^*$ and $L_o^+ = L^*$, the fact that T_1 and T_2 is 1-1 follows from the Fredholm Alternatives $N(L^+) \perp R(L_o)$, $N(L_o^+) \perp R(L)$, and the 1-1 ness of L_o and L. Since self-adjoint restrictions of T(M) have the same continuous spectrum as T(M), the closure of the ranges of T_1, T_2 and therefore their ontoness will follow once we know that T_1, T_2 are restrictions of T(M). But this is clear from Lemma 6. (Another way to prove that T_1 and T_2 have closed ranges is to note that they are closed operators and by Lemma 2 have bounded inverses.)

We turn now to the boundary conditions determining $D(T_1)$ and $D(T_2)$. The boundary conditions at a follow from the definitions of L_o and L_o^+. Suppose $g \in D(T_1)$ then g satisfies the boundary conditions of L_o at b. By Green's formula and since $L^+ = L_o^*$ for all $f \in \mathcal{D}_o$,

$$\begin{aligned}[L_o f, L_o g] - [f, L^+ L_o g] &= 0 \\ &= \{f, \tilde{z}_n\}(b^-),\ \tilde{z}_{n+1} = L_o g \\ &= D(f, \bar{g})(b^-).\end{aligned}$$

If however $g \in D(T(M)) \cap \mathcal{D}$ and satisfies the stated boundary conditions $g \in \mathcal{D}_o$. Further the Dirichlet formula (3.1) implies

$$[L_o f, L_o g] - [f, M[g]] = 0$$

so that $M[g] = L^+ L_o g$. The boundary conditions for T_2 follow in a similar manner.

COROLLARY 1. The statements of Theorem 1 are true with respect to $T_s := L_s^* L_s$. T_s has the boundary conditions of T_2 at b while at a $\hat{y}(a) \in S$ and $(y^{[2n-1]}(a), \ldots, y^{[n]}(a)) \in S^{\perp}$.

§4. Applications to Dirichlet Inequalities

THEOREM 2. Let $p_0, \ldots, p_n$ satisfy the hypotheses of Section 1. Then for all $y \in \mathcal{D}_o$ or $\mathcal{D}_s$ (1.1) is true where $\mu_0 = \inf\{\sigma(T_1 - d)\}$ or $\inf\{\sigma(L_s^* L_s - d)\}$. Further $\mu_0 > -\infty$; and if $\mu_0 \in \sigma_p(T_1)$ or $\sigma_p(L_s^* L_s)$ equality in (1.1) occurs if and only if ψ is an eigenfuction corresponding to μ_0. If $\mu_0 \notin \sigma_p(T_1)$ or $\sigma_p(L_s^* L_s)$ equality occurs if and only if $y = 0$. However, there is a sequence of functions $\langle\psi_\ell\rangle$ in $\mathcal{D}_o$ or $\mathcal{D}_s$ such that

$$\int_a^b \sum_{i=0}^{n} p_i |\psi_\ell^{(n-i)}|^2 - \mu_0 \int_a^b w|\psi_\ell|^2 \to 0.$$

Finally $\mathcal{D}_o = D\sqrt{T_1}$ and $\mathcal{D}_s = D\sqrt{L_s^* L_s}$.

Proof: If $p_n > 0$ the inequality and subsequent statements are immediate from Lemma 2 (inequality (2.2)) and the properties of L demonstrated in the last section. If $p_n < 0$ the inequality, etc. is true relative to $\tilde{p}_n$. But then

$$\int_a^b \sum_{i=0}^{n} p_i |y^{(n-i)}|^2 \geq (\mu_0' - d) \int_a^b w|y|^2,$$

where $\mu_0' = \inf\{\sigma(T_1)$ or $\inf\{\sigma(L_s^* L_s)$. The spectral mapping theorem then implies that $\mu_0 = \mu_0' - d$.

COROLLARY 1. For all y in $D(L_s^* L_s)$

$$||L_s^* L_s y||^2 \geq \mu_0 ||L_s y||^2 .$$

A similar inequality holds on $D(T_1)$.

Proof: We have

$$\|L_s y\|^2 = [L_s y, L_s y] = [L_s^* L_s y, y] \leq \|L_s^* L_s y\|^2 \|y\| / \|L_s^* L_s y\|$$
$$\leq \|L_s^* L_s y\|^2 \sup \|y\| / \|L_s^* L_s y\|$$
$$\leq \|L_s^* L_s y\|^2 \mu_0^{-1}.$$

The argument is the same on $D(T_1)$.

COROLLARY 2. *Suppose* $T_o(M)$ *is limit-n. Then* $L_s^* L_s$ *is the restriction of* $T(M)$ *to the domain determined by the boundary conditions* $\hat{y}(a) \in S$, $(\overset{[2n-1]}{y}(a), \ldots, \overset{[n]}{y}(a)) \in S^{\perp}$.

Proof: If $T_o(M)$ is limit-n, the restriction T of T(M) defined by the stated boundary conditions is known to be self-adjoint (cf. [13], p. 80). Since $L_s^* L_s$ is a self-adjoint restriction of T, the two operators are equal.

Remark: Writing (4.1) out when n = 1, $L_s = L$, and for $T_o(M)$ limit-point at b we obtain

$$\int_a^b |-(p_0 y')' + p_1 y| \geq \mu_0 \left(\int_a^b p_0 |y'|^2 + p_1 |y|^2 \right)$$

for all y in D(T(M)) satisfying $p_0 y'(a) = 0$.

If $z \in D(T(M))$, z is said to be M-Dirichlet if $p_i^{1/2} z^{(n-i)} \in L^2(a,b)$, $i = 0,\ldots,n$. The Dirichlet index of M is the dimension of the subspace spanned by the M-Dirichlet solutions of $M[f] = 0$. Provided that the coefficients p_i are sufficiently smooth (say in $C^{(n-i)}(a,b)$) then it is known (see [9]) that the Dirichlet index $\geq n$. It seems to be the case that the Dirichlet index is nearly always n if the p_i satisfy certain regularity conditions. Read [16] has recently shown that the index is 2 in the 4th order case when $p_0 = 1$. However the general question is still open. For further details see [9].

In our setting it is clear that z is an M-Dirichlet solution if and only if $L_o^* L z = L^+ L z = 0$, since $L_o^* L$ is the restriction of T(M) to M-Dirichlet functions in its domain. Further

$$R(L) = R(L_o) \oplus (R(L) \cap N(L^+)).$$

This follows since

$$\begin{aligned} R(L) &= R(L) \cap \mathop{X}_{i=0}^{n} L^2(a,b) \\ &= R(L) \cap (R(L_o) \oplus N(L_o^*)) \\ &= (R(L) \cap R(L_o)) \oplus (R(L) \cap N(L_o^*)) \\ &= R(L_o) \oplus (R(L) \cap N(L^+)). \end{aligned}$$

Hence the Dirichlet index of M is $\dim(G(L)/G(L_o))$; moreover the claim that the Dirichlet index is n is equivalent to the requirement that L_o be "limit-n". As for symmetric ordinary differential operators n-dimensional extension of L_o can be constructed using functions in $\mathcal{D}_o'$ satisfying linearly independent boundary conditions at a. We conclude therefore with the observation M has Dirichlet index n if and only if $\{y,\tilde{z}_n\}(b^-) = 0$, for all y in $\mathcal{D}$ and $\tilde{z}_{n+1}$ in $\mathcal{D}^+$. This result is in agreement with the results of Bradley, Hinton and Kauffman [4], Cor. 2.7. For a certain class of differential operators, they show that if the Dirichlet index is minimal, then their Dirichlet form D (corresponding to our $\{.,.\}$) vanishes at b for all Dirichlet functions in the domain of the maximal operator (cf. Lemma 7).

The equivalence between the minimality of the Dirichlet index and the vanishing of $\{y,z_n\}(b^-)$ in turn allows us to state:

COROLLARY 3. *If the Dirichlet index of* M *is* n, $D(L_s^* L_s) =$

$$\{f \in D(T(M)) \cap \mathcal{D}: \hat{y}(a) \in S;\ (y^{[2n-1]}(a), \ldots, y^{[n]}(a)) \in S^{\perp}\}.$$

§5. *"Dual" Dirichlet Inequalities* While Theorem 2 can be regarded as a new proof - under minimal conditions - of a familiar inequality studied by many authors (see e.g. [2]-[5],[14]). The inequalities in this section may be new in form. They are "dual" to (1.1) and the corollaries to Theorem 2 in that they depend on (2.2) instead of (2.1) of Lemma 2 and hold on the domain of L_s^* instead of L_s; equivalently while the previous inequalities are statements about the norm of the inverse AP for L_s, the ones given here are about $\|(AP)^*\|$.

Although everything in this section goes over to the case $p_n \geq -\gamma w$ we assume for technical simplicity that $p_n > 0$. Also to avoid notational complications we explicitly state the inequalities only for L^+ noting, however, that they carry over verbatim for L_s^*.

THEOREM 3. *Let* $z_{n+1} \in \mathcal{D}^+$ *then*

$$\|\ell_{n+1}^+(\tilde{z}_{n+1})\|^2 \geq \mu_0 (\sup|[\tilde{z}_{n+1}, L_0 y]|)^2,$$
$$\|L_0 y\| = 1.$$

Equality holds if and only if $\tilde{z}_{n+1} = Lg$ *and* g *is an eigenfunction of* L^*L *corresponding to* μ_0. *If* $\mu_0 \notin \sigma_p(L^*L)$ *there is a sequence* ψ_ℓ *in* R(L) *such that*

$$\left| \|\ell_{n+1}^+(\psi_\ell)\|^2 - \mu_0 \sup|[\psi_\ell, Ly]| \right| \to 0,$$
$$\|Ly\| = 1.$$

Proof: The only part of this theorem which is not obvious from Lemma 2 and the material in section 3 is the well-known duality lemma.

$$\operatorname{dist}(\tilde{z}_{n+1}, N(L^*)) = \sup |[\tilde{z}_{n+1}, \phi]|, \quad \phi \in N(L^*)^{\perp}, \|\phi\| = 1.$$

For a proof see [12] §5.8.

COROLLARY 4. $\|L_s L_s^*(\tilde{z}_{n+1})\| \geq \mu_0 \sup |[\tilde{z}_{n+1}, L_s y]|, \quad \|Ly\| = 1.$

Equality holds if and only if $\tilde{z}_{n+1} \equiv Lg \bmod N(L^*)$ *where* g *is an eigenfunction of* L^*L *corresponding to* μ_0. *However there is a sequence* $\langle\psi_\ell\rangle$ *in* $D(LL^*)$ *such that*

$$\lim_{\ell\to\infty} \big| \|LL^*\psi_\ell\| - \mu_0 \sup |[\psi_\ell, Ly]| \big| \to 0, \quad \|Ly\| = 1.$$

Proof:

$$\frac{\sup |[\tilde{z}_{n+1}, Ly]|}{\|LL^*(\tilde{z}_{n+1})\|} = \frac{\inf \|\tilde{z}_{n+1} - N(L^*)\|}{\|LL^*(\tilde{z}_{n+1})\|}, \quad \|Ly\| = 1$$

$$\leq \sup \frac{\|L^*(z_{n+1})\|}{\|LL^*(\tilde{z}_{n+1})\|} \cdot \frac{\inf \|\tilde{z}_{n+1} - N(L^*)\|}{\|L^*(\tilde{z}_{n+1})\|}$$

$$\leq \mu_0^{1/2} \cdot \mu_0^{1/2}.$$

The statements concerning equality are an exercise involving Lemma 2 whose proof we omit.

To illustrate Theorem 3 and Corollary 4 assume $n = 1$, $w = 1$. Then we obtain the inequalities:

$$(5.1)\quad \int_a^b |-(p_0^{1/2}z_1)' + p_1^{1/2}z_2|^2 \geq \mu_0 \sup\left(\left|\int_a^b z_1 p_0^{1/2}\bar{y}' + z_2 p_1^{1/2}\bar{y}\right|^2,\right.$$

$$\int_a^b p_0|y'|^2 + p_1|y|^2 = 1,$$

$$y(a) = 0,$$

and

$$(5.2)\quad \int_a^b p_0|(-(p_0^{1/2}z_1)' + p_1^{1/2}z_2)'|^2 + p_1|-p^{1/2}z_1)' + p^{1/2}z_2|^2 \geq$$

$$\mu_0^2 \sup\left(\left|\int_a^b z_1 p_0^{1/2}\bar{y}' + z_2 p_1^{1/2}\bar{y}\right|^2\right.$$

$$\int_a^b p_0|y'|^2 + p_1|y|^2 = 1,$$

$$p_0^{1/2} z_1(a) = 0\,.$$

In (5.1) $y \in \mathcal{D}_o$ and in (5.2) $y \in \mathcal{D}$ (so that $(z_1, z_2) \in \mathcal{D}_o^+$). In either case since $T_o(M)$ is limit point it has Dirichlet index 1, implying in view of the discussion following Corollary 2 that the boundary conditions at b for L_o and L_o^* are absent. In (5.1) μ_0 is the least element in the spectrum of the self-adjoint restriction of $-(p_0y')' + p_1y$ such that $y(a) = 0$ and in (5.2) the restriction is determined by $(p_0y')(a) = 0$.

REFERENCES

1. N.I. Akhiezer. The Calculus of Variations, (Blaisdell, New York, 1952).

2. J.S. Bradley and W.N. Everitt. Inequalities associated with regular and singular problems in the calculus of variations, Trans. Amer. Math. Soc. 182(1973), 303-321.

3. J.S. Bradley and W.N. Everitt. A singular inequality on a bounded interval, Proc. Amer. Math. Soc. 61(1978), 29-35.

4. J.S. Bradley, D.B. Hinton, and R.M. Kauffman. On the minimization of singular quadratic functionals, Proc. Roy. Soc. Edinburg, Sect. A 87(1981), 193-208.

5. W.N. Everitt and S.D. Wray. A singular spectral identity and equality involving the Dirichlet functional, to appear, Czech. Math. J.

6. S. Goldberg. Unbounded Linear Operators: Theory and Applications. (McGraw-Hill, New York, 1966).

7. D.B. Hinton. On the eigenfunction expansions of singular ordinary differential equations. J. Differential Equations, 24(1977), 282-308.

8. T. Kato. Perturbation Theory for Linear Operators, (Springer-Verlag, New York, 1966).

9. R.M. Kauffman. The number of Dirichlet solutions to a class of linear ordinary differential equations, J. Differential Equations, 31(1979), 117-129.

10. R.M. Kauffman, T.T. Read, and A. Zettl. The Deficiency Index Problem for Powers of Ordinary Differential Expressions, (Lecture Notes in Mathematics #621, Ed. A. Dold and B. Eckmann, Springer-Verlag, Berlin, Heidelberg, and New York, 1977).

11. A.M. Krall. Linear Methods of Applied Analysis, (Addison-Wesley, Reading, Mass., 1973).

12. D.G. Luenberger. Optimization by Vector Space Methods, (John Wiley, New York, 1969).

13. M.A. Naimark. Linear Differential Operators Part II, (Ungar, New York, 1968).

14. C.R. Putnam. An application of spectral theory to a singular calculus of variations problem, Amer. J. Math. 70(1948), 780-803.

15. T.T. Read. Factorization and discrete spectra for second order differential expressions. J. Differential Equations, 35(1980), 388-406.

16. T.T. Read. Dirichlet solutions of fourth order differential operators, Spectral Theory of Differential Operators, Ed. I. Knowles and R. Lewis, (Math. Study Series, North Holland Publishing Co., Amsterdam, 1981).

17. W. Rudin. Functional Analysis. (McGraw-Hill, New York, 1973).

Addendum

If M[y] is the two term expression $w^{-1}((-1)^n(p_0y^{(n)})^n + p_ny)$, the hypotheses on p_0 and w can be weakened to yield a minimal condi- setting. Concerning p_0 and w we require only that they be nonnegative and that p_0^{-1}, w be locally integrable. The proof that L is normally solvable (Lemma 3) will still go through with a slight alteration in the argument: Following the application of Lemma 1 the estimate

$$d\|y_\ell - \tilde{p}_n^{-1/2}\beta_{n+1}\|_w \leq \|\tilde{p}_n^{1/2}y_\ell - \tilde{p}_n^{1/2}\tilde{p}_n^{-1/2}\beta_{n+1}\|$$

together with the facts that $y_\ell \to \alpha$ in $L_w^2(a,b)$ and $\tilde{p}_n^{1/2}y_\ell \to \beta_{n+1}$ in $L^2(a,b)$ allows us to conclude that $\beta_{n+1} = \tilde{p}_n^{1/2}\alpha$ and that therefore L is closed. In this case it is unnecessary to use the compact subinterval Δ.

Thus for two term operators our results are consistent with those of Everitt and Wray [5].

"Inclusion theorems for solutions of differential equations with aid of pointwise or vector monotonicity"

L. Collatz, Hamburg

Summary This survey describes briefly some methods which use ideas of monotonicity for numerical solution of differential equations. In many cases the approximation methods combined with principles of monotonicity are the only ones which give lower and upper bounds for the wanted solutions. Numerical examples are given for linear and nonlinear partial differential equations and integral equations mostly connected with applications in science. The idea of vector-monotonicity can be used for calculating also derivatives of wanted solutions in not too complicated cases.

1. Introduction.

In this survey paper we deal with applications of the theory of differential equations to the numerical methods of calculation ot the solutions of ordinary (ODE) and of partial differential equations (PDE). There are many different methods of computing solutions. The most frequently used methods are perhaps the methods of finite differences, finite elements and variational methods. These methods give numerical approximate values for the solution, but it is for these methods difficult or impossible to give lower and upper bounds for the solutions (so called inclusions theorems) which one can guarantee and with which one can see, how many of the digits the computer has printed out are right.

This guarantee can be given in many (not too complicated) examples by using fixed point theorems of the functional analysis, approximation and optimization procedures, monotonicity properties of the solution and monotonicity theorems for iteration procedures. This can be illustrated by many examples of linear and nonlinear O.D.E. and P.D.E and integral equations, the most examples coming from applications in sciences. But in many applications there occur singularities as for instance reentering corners, singularities of the coefficients a.o.; it is important to take care of the type of singularity otherwise the numerical results would be unsatisfactory. Also for free boundary value problems it was possible in not too complicated problems to calculate inclusions for the free boundary one can guarantee. A new theory deals with vector valued operators of monotonic type which gives in simple cases inclusion theorems also for derivatives of solutions which may be in some cases even more important then the values of the solution itself.

There are several kinds of application of monotonicity:

I. Use of immediate comparison theorems

II. Use of operators of monotonic type

IIa) Using scalor monotonicity

IIb) Using vector monotonicity

III. Use of monotonic iterations.

2. Immediate comparison of neighoured problems

Given a problem P, the solution u of which is wanted. One is replacing P by a neighboured solvable Problem P^* with a solution u^*. We require for the Problem P^* the following properties:

1) It is easy to get a problem P^* and to construct it.
2) One can solve the problem P^* explicitly so that the solution u^* is known.
3) One can control the influence of the replacement in the problem; this means more precisely: One has possibilities for measuring the magnitude ρ of the replacement, f.i. a distance between original and changed coefficients in a differential equation. We write $||P^* - P|| \leq \rho$ and ρ is a computable value. Furthermore one has an idea of a distance between solutions, f.i. by using norms:

$$||u^* - u||$$

and we ask for a bound ε for this norm

$$||u^* - u|| \leq \varepsilon$$

The requirement 3) means, that one can calculate a bound for the error ε, if a bound ρ is computed.

This simple idea has been used very often in analysis, we give some examples.

3. Neighboured problems in differential equations.

A. The classical idea of lower- and upper functions; one can find this idea in many different areas of differential equations. An example:

The solution of the initial value problem for a function $y(x)$ of a real variable x (in $x \geq 0$) (Problem P)

$$(3.1) \qquad y'(x) = f(x) + g(x) \cdot (y(x))^2, \quad y(0) = y_o$$

with a given value $y_o > 0$ and given continuous functions $f(x) \geq 0$, $g(x) \geq 0$, $g(x) > 0$ for $x > 0$ has for a finite value p of x a "first" singularity; that means $y(x)$ is regular for $0 < x < p$.

The neighboured problem P^*

$$(3.2) \qquad y^{*\prime}(x) = f^*(x) + g^*(x) \cdot (y^*(x))^2, \quad y^*(0) = y_o^*$$

has for $y_o^* \geq y_o$, $f^*(x) \geq f(x)$, $g^*(x) \geq g(x)$ a first singularity p^* with $p^* \leq p$. Similarly one can get an upper bound $\hat{p} \geq p$ and one has an inclusion of the solution $y(x)$.

Example

$$(3.3) \qquad y'(x) = x^2 + xy^2, \quad y(0) = 1.$$

Lower function y^*: $y^{*\prime} = (b+cx)\, y^{*2}$, $y^*(0) = 1$ has for constants b,c the solution $y^* = (1-bx-\frac{c}{2}x^2)^{-1}$ with the pole at p^*, the zero of $1-bx-\frac{c}{2}x^2=0$. For $b=0$, $c=1$ one gets $p^*=\sqrt{2}\approx 1.414\ldots$. We have $p \leq p^*$, Fig. 1.

For an upper bound for $y(x)$ we can consider $\hat{y}'(x)=(a^2+\hat{y}^2)x$, $\hat{y}(0)=1$ with the solution $\hat{y}(x)=a \tan(\arctan(a^{-1})+\frac{a}{2}x^2)$. Now $x^2 \leq x\sqrt{2}$ holds for $0 \leq x \leq \sqrt{2}$ and for $a=\sqrt[4]{2}$ we get $\hat{p}=\frac{2}{a}(\frac{\pi}{2}-\arctan a^{-1})^{1/2} \approx 1.211$.

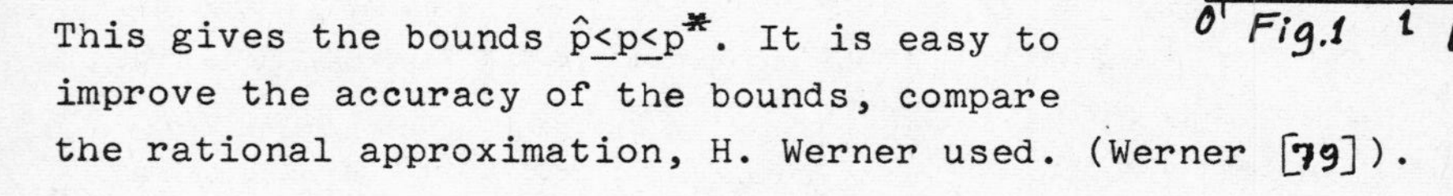

Fig.1

This gives the bounds $\hat{p} \leq p \leq p^*$. It is easy to improve the accuracy of the bounds, compare the rational approximation, H. Werner used. (Werner [79]).

Comparison theorems are used in theoretical and numerical analysis very often, f.i. as sub- and superharmonic functions, as lower and upper solutions (J. Schröder [80],p.199) a.o.

4. Scalar monotonicity-principle

There a very simple idea to get lower and upper bounds for wanted solutions, if one has monotonicity. The monotonicity is often the mathematical equivalent for a physical principle: Often enlargement of the influence (of the imput) causes enlargement of the observed effect (of the output). Of course this is not true in all cases, but if one can proove this principle then usually one get in this way an inclusion theorem easily to compute. Let us illustrate this principle by considering a simple situation in mechanics.

Displacement of a beam

We look on a solution y(x) of the boundary value problem with the O.D.E.

$$(4.1) \qquad Ly = \frac{d^2}{dx^2}\left(\alpha(x)\frac{d^2y}{dx^2}\right) = r(x) \text{ in the interval } J\text{: } a<x<b.$$

and the homogeneous boundary conditions

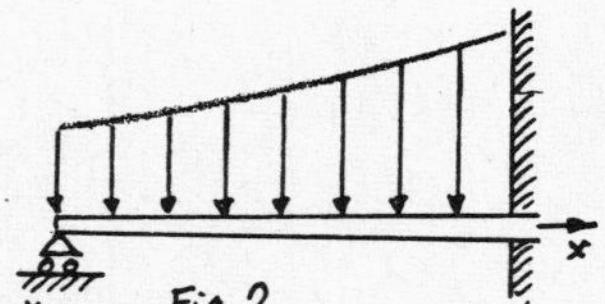

$$(4.2) \qquad U_\mu[y] = 0 \ (\mu=1,2,3,4), \text{ for instance}$$

$$(4.3) \qquad y(a) = y''(a) = y(b) = y'(b) = 0.$$

x=a Fig.2 x=b

y(x) may be interpreted as displacement of a beam with variable cross section under the influence of a load with density r(x) (with usual notations); the beam is supported at x=a and clamped at x=b, Fig. 3; r(x), α(x) are given functions, r(x) continuous in J, α(x) twice continuously differentiable in J and α(x)>0 in [a,b].

One has many numerical methods for calculating ẏ(x), and it is easy to compute y(x) with high accuracy. The simple example should only show how easy it is to get bounds on can guarantee. If one uses discretizations it is usually difficult or even impossible to get such bounds. But the monotonicity principle gives bounds with very few computation.

The idea is as follows: We are considering only "admissable" functions w(x), that means functions in a linear space R[J], which have continuous derivatives up to the order 4 and which satisfy the boundary conditions (4.2). Then we have a one-to-one correspondence between admissable displacements $w(x)$ and load densities $r(x)$, and enlargement of $r(x)$ causes enlargement of $w(x)$ what is easy to prove (the Green's function is non-negative). If we use a guess $\bar{w}(x)$ for $w(x)$ and if the corresponding density $\bar{r}(x)$ is greater then the given density $r(x)$, then holds: $\bar{w}(x)$ is greater (or equal) to the displacement $y(x)$ which belongs to $r(x)$. Analogeously one can get an lower bound $\underline{w}(x)$ and the inclusion (Collatz |52||67|)

$$(4.4) \qquad \underline{w}(x) \leq y(x) \leq \bar{w}(x)$$

For getting good numerical results we let $\underline{w}(x)$ and $\bar{w}(x)$ depend on some arbitrary parameters $\underline{a}_\nu$, resp. $\bar{a}_\nu$, $(\nu=1,2,\ldots p)$; we determine these parameters from the semi-infinite optimization problem

$$(4.5) \qquad \begin{array}{l} L\underline{w}(x,\underline{a}_\nu) \leq r(x) \leq L\bar{w}(x,\bar{a}_\nu) \\ \bar{w}(x,\bar{a}_\nu) - \underline{w}(x,\underline{a}_\nu) \leq \delta^* \\ \delta^* = \text{Min.} \end{array} \qquad \text{for all } x \in J$$

One can use an optimization problem with less parameters (which may give a little worse result, by calculating lower and upper bounds separatedly) for the upper bound

$$(4.6) \qquad \begin{array}{l} 0 \leq L\bar{w}(x,\bar{a}_\nu) - r(x) \leq \bar{\delta} \\ \bar{\delta} = \text{Min} \end{array} \qquad \text{for all } x \in J$$

and analogeously for the upper bound (compare f.i. Collatz |66| p.390)

Numerical example.

We take in (4.1)(4.3)

$$\alpha(x) = 2+x, \quad r(x) = \frac{2+3x}{2+x}, \qquad a = 0,\ b = 1$$

and as approximate solution

$$y \approx w(x) = (4x-4x^2+x^3)\left[a_1(1+x)+a_2x^2+a_3x^3+\ldots+a_px^p\right].$$

Then $w(x)$ satisfies all boundary conditions (4.3).

The values $\underline{a}_\nu$, $\bar{a}_\nu$ are determined by (4.5); it is not necessary, to solve the optimization problem accurately; for practical cases one discretizes usually the domain J and has a finite optimization problem , for which computer-programs are available. One gets in this way approximate values $\underline{a}_\nu^*$, $\bar{a}_\nu^*$ instead of the solution $\underline{a}_\nu$, $\bar{a}_\nu$ of (4.5), and corresponding functions $\underline{w}^*(x)$, $\bar{w}^*$; if then $L\underline{w}^* \leq r(x) \leq L\bar{w}^*$ is satisfied, one has the inclusion $\underline{w}^* \leq y \leq \bar{w}^*$ one can guarantee. In our example one gets (only with 2 parameters the error bound $|\bar{w}-y| \leq$ for $w = \frac{1}{2}(\underline{w}+\bar{w})$. Using more parameters one can improve this error bound.

Singularities.

If the problem has singularities one has to take care on the type of singularity otherwise every numerical method would give unsatisfactory results. Many problems with different types of singularities have been calculated on computers (compare f.i. Collatz [ ][ ], Whiteman [79] a.o.).

We give an example for illutrating the procedure. A model of the condensator-problem may be described by the boundary value problem for a function u(x,y), fig.3:

$$(4.7)\quad \begin{cases} \Delta u = 0 \text{ in } B \text{ (for all } x,y \text{ except the set } S = \quad B) \\ u = y \text{ on } S.\ (|x| \leq 1,\ y = \pm 1) \\ u = 0 \text{ at "infinity", fig.3.} \end{cases}$$

Of course one can consider only the quatrant Q: $x \geq 0$, $y \geq 0$ and the problem

$$(4.8)\quad \begin{cases} \Delta u = 0 \text{ for } (x,y) \in Q-S \\ u = 1 \text{ for } 0 \leq x \leq 1,\ y=1 \text{ (on } \Gamma_1) \\ u = 0 \text{ for } y = 0 \text{ and at "infinity"} \\ \frac{\partial u}{\partial x} = 0 \text{ for } x=0,\ y \geq 0 \quad \text{(on } \Gamma_2) \end{cases}$$

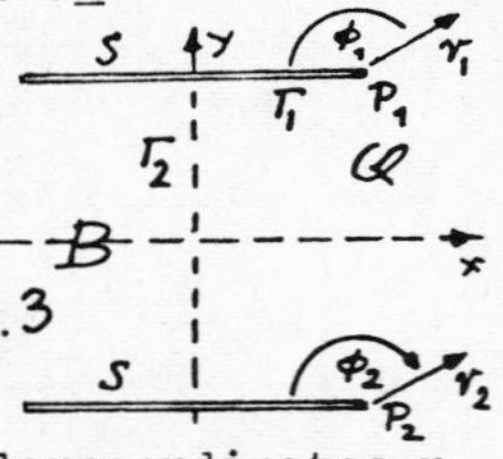

Fig.3

The points $x=\pm 1$, $y=\pm 1$ are singular, we introduce polarcoordinates r_i, ϕ_j at the points P_j (j=1,2) with $P_1(1,1)$, $P_2(1,-1)$ as in fig.3; then the function

$$w_{sg} = r_1^{1/2} \sin\left(\frac{\phi_1}{2}\right) - r_2^{1/2} \sin\left(\frac{\phi_2}{2}\right)$$

is vanishing for y = 0 and at "infinity" and has no singularity in Q-S and satisfies $\Delta w_{sg} = 0$. The same is true for

$$w_1 = \frac{y+1}{(x+1)^2+(y+1)^2} + \frac{y-1}{(x+1)^2+(y-1)^2}$$

and
$$w_2 = (x+1)\left\{\frac{y+1}{((x+1)^2+(y+1)^2)^2} + \frac{y-1}{((x+1)^2+(y-1)^2)^2}\right\}.$$

We choose as approximate function w for u

$$w = a_o\, w_{sg} + a_1\, w_1 + a_2\, w_2$$

and determine the parameters a_ν so that the conditions at Γ_1 and Γ_2 are satisfied as good as possible, but with the right signum of the error. If $w-u \geq 0$ on Γ_1 and $\frac{\partial w}{\partial x} \leq \frac{\partial u}{\partial x} = 0$ on Γ_2, then follows $w \geq u$ in Q-S and we have an upper bound for u; analogeously we can get a lower bound for u. I thank Dr. Christian Maas, Hamburg for numerical calculation on a computer; he has got an approximate solution w with $|w-u| \leq 0.0266$. (Higher accuracy is obtainable by using terms with more parameters)

These methods have been applied to many other problem; we mention only the crack problem of Whiteman-Babuska[80]. A simplified model asks for a function u(x,y) in a rectangle $R = \{(x,y),\ |x|<a, |y|<b\}$, from which the straight lines $S = \{(x,y),\ c\leq|x|\leq a,\ y=0\}$ are taken off.

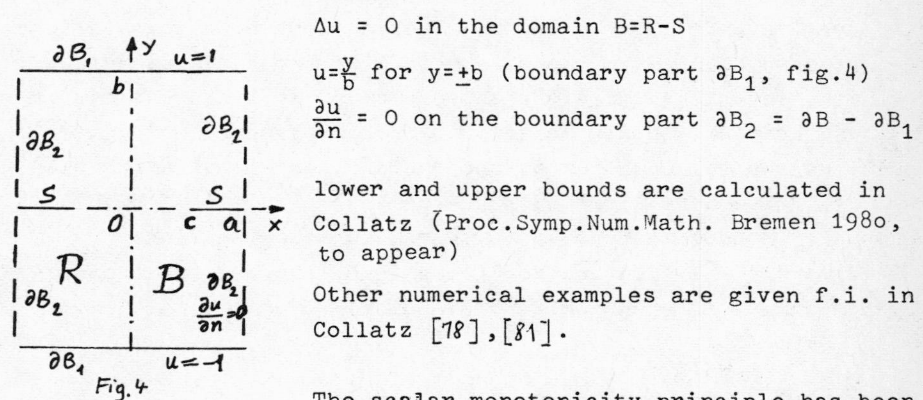

Fig.4

We have

$\Delta u = 0$ in the domain B=R-S

$u=\frac{y}{b}$ for $y=\pm b$ (boundary part ∂B_1, fig.4)

$\frac{\partial u}{\partial n} = 0$ on the boundary part $\partial B_2 = \partial B - \partial B_1$

lower and upper bounds are calculated in Collatz (Proc.Symp.Num.Math. Bremen 198o, to appear)

Other numerical examples are given f.i. in Collatz [78],[81].

The scalar monotonicity principle has been applied to numereous other types of problems, eigenvalue problems (J. Albrecht, Goerisch [81]) bifurcation problems, free boundary value problems (Hoffmann [78],[8o], Collatz [81]) a.o.
About numerical procedures compare f.i. Bredendiek [76], Grothkopf [81] and many others.

5. Operators of monotonic type

The example in Nr. 4 is a special case of "operators of monotonic type" T with the property (for functions v,w in the domain D of the operator T).

(5.1) From $Tv \leq Tw$ in B follows $v \leq w$ in B

T may be the vector, the components are a differential operator L in the a domain B (for instance an open connected bounded subset of a n-dimensional real point space R^n) and one or moere boundary operators $U_1,\ldots,U_k$ on parts of the boundary ∂B.

The property of an operator to be of monotonic type was prooved for rather general linear and nonlinear ordinary and partial differential operators of second order of elliptic and parabolic type and for special classes with hyperbolic P.D.E. compare Redheffer [67] Nickel [58], Walter [70], Wetterling [77], Hofmann [73], [77], Bandle [81] a.o.

As example we mention the first boundary value problem for real linear elliptic operators

$$(5.2) \qquad Lu(x_j) = - \sum_{j,k=1}^{n} a_{jk}(x_\ell)\frac{\partial^2 u}{\partial x_j \partial x_k} - \sum_{j=1}^{n} b_j(x_\ell)\frac{\partial u}{\partial x_j} + cu.$$

The real coefficients a_{jk}, b_j, c are given as continous bounded functions in an open bounded connected domain B of the n-dimensional space R^n of point $\{x_1,\ldots,x_n\}$. We suppose $c \geq 0$ in B and the matrix $A = (a_{jk})$ as uniformly positive definit in B. The operator $U_1(u(x_\ell)) = u(x_\ell \varepsilon \partial B)$ may be the operator of the boundary values of a function $u(x_k)$ along te boundary ∂B of B, then the vector $Tu = (Lu, U_1(u))$ is an operator of monotonie type, that means: for two functions $v,w \in C^2(B) \cap C(B+\partial B)$ (v,w are continous in $B + \partial B$ and twice continuously partially differentiable in B) holds:

$$(5,3) \qquad \left\{ \begin{array}{l} Lv \leq Lw \text{ in } B \\ v \leq w \text{ on } \partial B \end{array} \right\} \text{ has the consequence } v \leq w \text{ in } B.$$

The inequalities mean pointwise inequalities for all $x \in B$, resp. $x \in \partial B$ in the classical sense of inequalities for reals numbers.

This can be generalized to nonlinear P.D.E, for more general boundary conditions a.o. (Collatz [66], Bohl [74], Schröder [80], Glashoff-Werner [79] a.o.)

6. Monotonic Iterations.

Many problems with differential- and integral equations can be formulated as fixed-point equations for operators T, that means as equations for an element u with

$$Tu = u$$

where T is a given (linear or nonlinear) operator, which maps a given domain D of a certain partially ordered. Banach space R into this Banach space. At first one had considered iteration procedures for calculation of u by determining a sequence u_n of approximate solution by

$$(6.1) \qquad u_{n+1} = Tu_n \qquad (n = 0,1,\dots)$$

starting with an element $u_o \in D$. The applicability of the theory for the iteration (6.1) was generalized to a great extend by J. Schröder [62], [80] by introducing the theory of monotonically decomposible operators T. These operators are defined as sums of syntone operators T_1 and antitone operators T_2:

$$(6.2) \qquad T = T_1+T_2.$$

An operator T is called syntone (resp. antitone), if

$$(6.3) \qquad v \leq w \text{ implies } Tv \leq Tw \text{ (resp. } Tv \geq Tw) \text{ for all } v,w \in D$$

For instance a nonlinear Hammerstein integral operator

$$(6.4) \qquad Tu(x_1,\dots,x_n) = \int_B K(x_1,\dots,x_n,t_1,\dots,t_n)\phi(u(t_1,\dots,t_n))dt_1\dots dt_n$$

is monotonically decomposible, if K is integrable and real in BxB and if the function $\phi(z)$ is a realvalued function of bounded variation for all z in the range of $u(t_1,\dots,t_n)$. Here B is supposed as a bounded closed connected subset of the n-dimensional space R^n of points $(x_1,\dots,x_n)$.

Then one is working with initial elements $v_o,w_o \in D$ and calculating

$$(6.5) \qquad \begin{cases} v_1 = T_1v_o + T_2w_o \\ w_1 = T_1w_o + T_2v_o \end{cases} .$$

If the initial conditions

$$(6.6) \qquad v_o \leq v_1 \leq w_1 \leq w_o$$

are satisfied and if T_1,T_2 are compact operators, then one can apply Schauders fixed point theorem (Schauder [30] Schröder [80] Collatz [66] a.o.).

7. Vector-monotonicity

There are many problems in applications, for which it is more important to calculate the derivatives of unknown functions then the values of the functions itself. For these cases it can be useful to apply a vector-monotonicity instead the hitherto in Nr. 4 considered monotonicity for values of functions. Furthermore many cases of the monotonicity of Nr. 4 can be considered as special cases of the more general vector-monotonicity.

We describe the vector-monotonicity at first in the linear case, and consider a simple case of the biharmonic equation for a function $u(x,y)$ in an open connected bounded domain B of the x-y-plane. For a function $u \in C^4(B) \cap C^2(\partial B)$ one can proove easily:

Lemma:

$$(7.1) \begin{cases} \text{From } \Delta\Delta u \geq 0 \text{ in } B,\ u \geq 0 \text{ on } \partial B,\ -\Delta u \geq 0 \text{ on } \partial B \\ \quad \text{follows } -\Delta u \geq 0 \text{ in } B,\ u \geq 0 \text{ in } B \end{cases}$$

The proof is obvious: We put $-\Delta u = v$; then we have $-\Delta v \geq 0$ in B, $v \geq 0$ on ∂B and this has the consequence $v \geq 0$ in B. Therefore we have $v = -\Delta u \geq 0$ in B, $u \geq$ on ∂B, and this gives $u \geq 0$ in B.

We can include other cases too by generalizing (7.1) by admitting inequalities and equations together. We look on functions $u(x)=u(x_1,\ldots,x_n)$ in a certain open bounded connected domain B of the n-dimensional point-space R^n, which are sufficiently often continuously differentiable and which satisfy on certain subsets B_j, C_j, D_j of B the conditions

$$(7.2) \qquad T_j u \geq 0 \text{ on } B_j \quad (j = 1,\ldots,k)$$

$$(7.3) \qquad M_j u = 0 \text{ on } C_j \quad (j = 1,\ldots,p).$$

We suppose that (7.2) (7.3) have the consequence

$$(7.4) \qquad N_j u \geq 0 \text{ on } D_j \quad (j = 1,\ldots,q).$$

The operators T_j, M_j, N_j may be given, and may contain derivatives. We include the case, that we have no conditions of the form (7.3); we speak of "vector monotonicity" in the case $k + p \geq 2$.

We give an example with occurence of an equation (7.3) at the boundary (the following lemma holds for a domain B as above with boundary ∂B

and ν as inner normal).

Lemma:

$$(7.5)\qquad -\Delta v \geq -\Delta w \text{ in } B$$
$$(7.6)\qquad v = w \text{ on } \partial B$$

has the consequence

$$(7.7)\qquad v \geq w \text{ in } B, \ \frac{\partial v}{\partial \nu} \geq \frac{\partial w}{\partial \nu} \text{ on } \partial B.$$

$v(x)$ and $w(x)$ are supposed as sufficiently smooth.

Proof: For $v-w = \varepsilon$ holds $-\Delta\varepsilon \geq 0$ in B, $\varepsilon = 0$ on ∂B and therefore $\varepsilon \geq 0$ in B; and $\varepsilon = 0$ on ∂B, $\varepsilon \geq 0$ in B has the consequence $\frac{\partial \varepsilon}{\partial \nu} \geq 0$ on ∂B.

Now we consider nonlinear problems; T_j, M_j, N_j may now be nonlinear, v and w may be functions which are sufficiently smooth.

We assume (analogeously to (7.2)(7.3)(7.4))

$$(7.8)\qquad T_j v \geq T_j w \text{ on } B_j \qquad (j = 1,\dots,k)$$
$$(7.9)\qquad M_j v = M_j w \text{ on } C_j \qquad (j = 1,\dots,p)$$

and we suppose, that (7.8)(7.9) have the consequence

$$(7.1o)\qquad N_j v \geq N_j w \text{ on } D_j \qquad (j = 1,\dots,q).$$

Again we speak of vector-monotonicity in the case $k + p \geq 2$.

8. Examples for calculating derivatives with aid of vector-monotonicity

We suppose, that in (7.8)(7.1o) we have $M_1v=v$, $N_1v=v$ for $p=q=1$. We consider the boundary value problem for a function $u(x)$:

$$(8.1)\qquad Tu(x) = r(x) \text{ in } B$$
$$(8.2)\qquad u(x) = s(x) \text{ on } \partial B;$$

Let $v(x)$ be a function, satisfying the boundary condition and the inequality:

(8.3) $\quad Tv(x) \leq r(x)$ in B

(8.4) $\quad v(x) = s(x)$ on ∂B;

If T, M_1 satisfies the vector monotonicity, we have $v(x) \leq u(x)$ in B. We have for the difference $\varepsilon(x) = v(x) - u(x)$

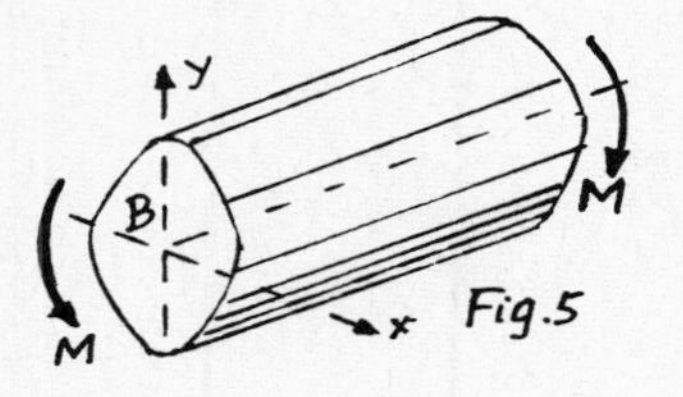

Fig.5

(8.5) $\quad \varepsilon(x) \leq 0$ in B, $\varepsilon(x) = 0$ on ∂B

and therefore

$$\frac{\partial \varepsilon}{\partial \nu} \leq 0 \text{ on } \partial B$$

or

(8.6) $$\frac{\partial v}{\partial \nu} \geq \frac{\partial u}{\partial \nu} \text{ on } \partial B.$$

Analogeously we can get a lower bound for $\frac{\partial u}{\partial \nu}$, and we have an inclusion for the derivative $\frac{\partial u}{\partial \nu}$ along the boundary. Other bounds for derivatives are given by J. Schröder [80] p.203, p.290.

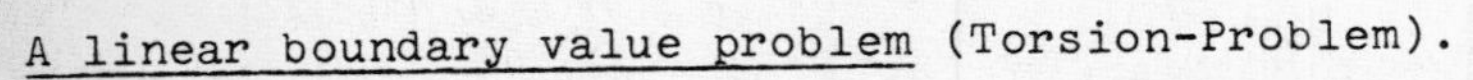

A linear boundary value problem (Torsion-Problem).

$B = \{(x,y),\ \phi(x,y) > 0\}$ with $\phi = 4-2x^2-2y^2-x^2y^2$ may be the cross-section of a beam under the influence of a torque, fig. 5,6. We wich to calculate a function $u(x,y)$ wich

$$-\Delta u(x,y) = 1 \text{ in } B,\ u = 0 \text{ on } \partial B.$$

We are especially interested on $\frac{\partial u}{\partial \nu}$ along the boundary $\phi = 0$: from these values one can calculate tensions along ∂B.

We choose as approximate solution w for u

$$w(x,y) = \Phi \cdot \sum_{j=1}^{p} a_j w_j(x,y)$$

Fig.6

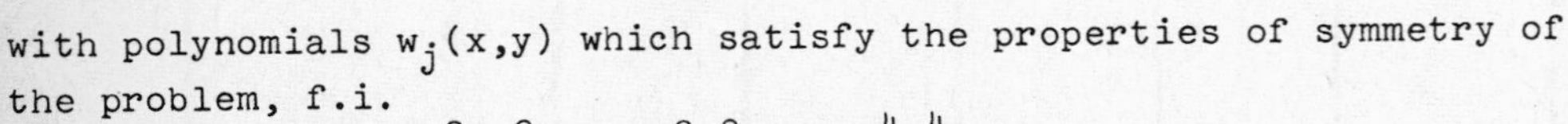

with polynomials $w_j(x,y)$ which satisfy the properties of symmetry of the problem, f.i.

$$w_1=1,\ w_2=x^2+y^2,\ w_3=x^2y^2,\ w_4=x^4y^4, \ldots$$

and determine parameters $\underline{a}_j$ (resp. $\bar{a}_j$) so that for the corresponding function $\underline{w}$ (resp. $\bar{w}$) holds

$$-\Delta \underline{w} \leq 1 \leq -\Delta \bar{w} \text{ in } B.$$

then we have for the error $\underline{\varepsilon} = \underline{w}-u$ (resp. $\bar{\varepsilon} = \bar{w}-u$)

$\underline{\varepsilon} \leq 0 \leq \bar{\varepsilon}$ in B, $\frac{\partial \underline{\varepsilon}}{\partial \nu} \geq \frac{\partial u}{\partial n} \geq \frac{\partial \bar{\varepsilon}}{\partial n}$, $\frac{\partial \underline{w}}{\partial \nu} \leq \frac{\partial u}{\partial \nu} \leq \frac{\partial \bar{w}}{\partial \nu}$

The table gives the results for p=1,2,3

p	$\lvert\frac{\partial \bar{w}}{\partial \nu} - \frac{\partial \underline{w}}{\partial \nu}\rvert \leq$
1	0.30339
2	0.035729
3	0.002492

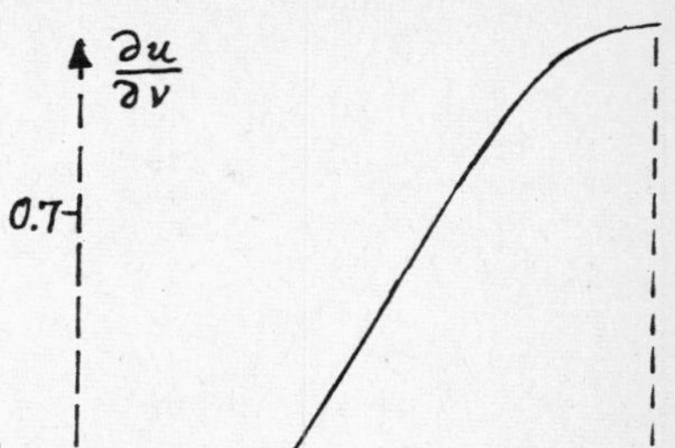

For p=3 are the values of the parameters

$\underline{a}_1$ = 0.112 174	$\bar{a}_1$ = 0.112 500
$\underline{a}_2$ = 0.006 297	$\bar{a}_2$ =-0.006 250
$\underline{a}_3$ = 0.002 189	$\bar{a}_3$ = 0.002 098

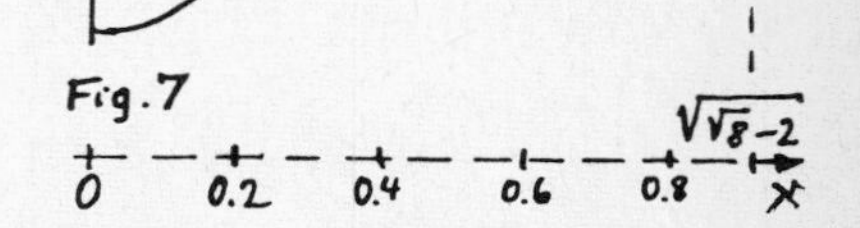

Fig.7

Fig. 7 schows the graph of $\frac{\partial u}{\partial \nu}$ along the curve $\phi = 0$ as function of y.

A linear problem (bending moment)

We consider the displacement y(x) of a loaded beam as in (4.1) (4.2), but with the boundary conditions of clamped ends: $U_\mu y=0$ (μ=1,2,3,4) or

$$y(a) = y'(a) = y(b) = y'(b) = 0$$

instead of (4.3), Fig. 8. Here we will look on the bending moments at the ends, and for simplicity (only for illustrating the method, let us take $\alpha(x) = 1$ (dimensionsless written) and suppose symetry:

$$a = -1,\ b = 1,\ r(x) = r(-x).$$

Let be v(x), w(x) two functions with continuous forth derivatives in the interval J = [-1,1]; then we have: From

$$\begin{cases} Lv \leq Lw \text{ in } J \\ U_\mu v = U_\mu w \text{ for } x=\pm 1 (\mu=1,2,3,4) \end{cases}$$

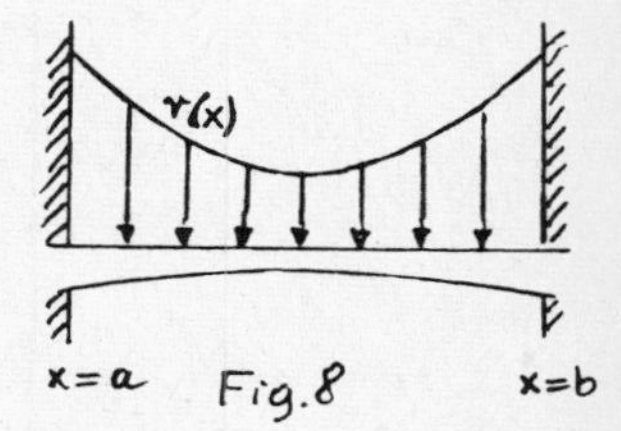

Fig.8

it follows

$$\begin{cases} v \leq w \text{ in } J \\ v''(1) \leq w''(1) \\ v''(0) \geq w''(0) \end{cases}$$

This gives the possibility for inclusion of y"(-1)=y"(1) and of y"(0).

Numerical example: $r(x)=\exp(x^2)$:

We choose as approximate solution

$$w(x) = \sum_{j=1}^{p} a_j w_j(x)$$

with $w_j(x) = j - (j+1)x^2 + x^{2(j+1)}$ $(j = 1,2,\ldots,p)$,
especially $w_1 = 1-2x^2 + x^4$, $w_2 = 2-3x^2 + x^6,\ldots$

We determine parameters $\underline{a}_j$ for a lower bound $\underline{w}$ and other values of the parameters $\overline{a}_j$ for an upper bound $\bar{w}$ for y. The table gives the error bound $\delta = \bar{w}''(1) - \underline{w}''(1)$ in dependence of the number of parameters

p	$\bar{w}''(1)-\underline{w}''(1)$
1	0.572 7
2	0.040 75
3	0.002 728
4	0.000 152

For p=4 are the values of the parameters and of the approximate values for $w''(1)$

$\underline{a}_1 = 0.041\ 661\ 5$	$\overline{a}_1 = 0.041\ 666\ 7$
$\underline{a}_2 = 0.002\ 787\ 3$	$\overline{a}_2 = 0.002\ 795\ 9$
$\underline{a}_3 = 0.000\ 280\ 5$	$\overline{a}_3 = 0.000\ 271\ 2$
$\underline{a}_4 = 0.000\ 046\ 4$	$\overline{a}_4 = 0.000\ 050\ 8$
$\underline{w}''(1) = 0.417\ 365$	$\bar{w}''(1) = 0.417\ 517$

Further results for inclusions of bending moments in more complicated cases will be given, Collatz [82].

A Nonlinear boundary value problem.

We take for illutration a very simple problem: We consider the classical equation for an unknown function u(x,y)

(8.7) $$Tu = -\Delta u - e^u = 0 \text{ in } B = \{(x,y),\ r^2 = x^2+y^2 < 1\}$$

the boundary condition $\quad u = 0$ on $\partial B = \{(x,y),\ r = 1\}$
and ask for an inclusion for the normal derivative along the boundary

(8.8) $$k = \left(\frac{\partial u}{\partial r}\right)_{r=1}.$$

We take with respect to the symetry of the problem as approximate solution

$$u \approx w(r) = (1-r^2)\sum_{\nu=0}^{p} a_\nu r^{2\nu}.$$

We have for p=0

$$w(r) = a_o(1-r^2); Tw = 4a_o - e^{a_o(1-r^2)} \quad ; k=-2a_o$$

Fig. 9 shows Tw for some values of a_o; one sees:

$$Tw \le 0;\ w \le u \text{ for } a_o = \tfrac{1}{4}$$
$$Tw \ge 0;\ w \ge u \text{ for } a_o \approx 0.3574,$$

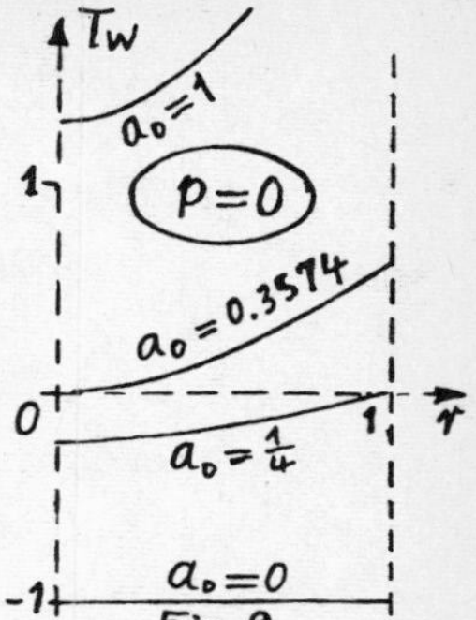

Fig. 9

and therefore

$$-0.72 \le k = \left(\frac{\partial u}{\partial r}\right)_{r=1} \le -0.5$$

This shows in principle the possibility of inclusion for the derivative.

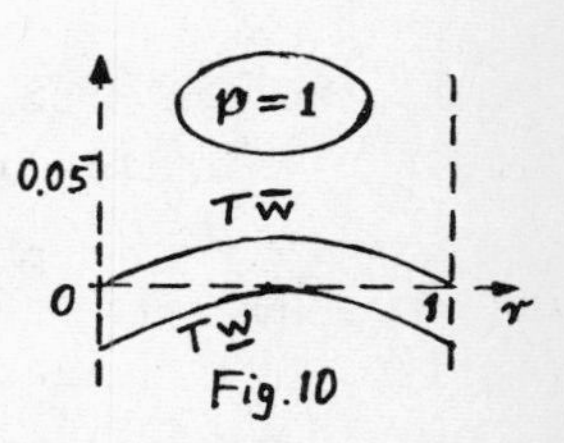

Fig. 10

We get better values for p=1, Fig. 1o

	coefficients		bounds for k $=-2(a_o+a_1)$
For lower bound $\underline{w}$	$\underline{a}_o = 0.3137$	$\underline{a}_1 = -0.0229$	$\underline{k} = -0.5816$
for upper bound $\overline{w}$	$\overline{a}_o = 0.3205$	$\overline{a}_1 = -0.0235$	$\overline{k} = -0.594$

therefore $\quad -0.594 \le k = \left(\frac{\partial u}{\partial r}\right)_{r=1} \le -0.5816$

Of course one can improve these bounds by using more parameters a_ν.

I thank Dipl. Math. Uwe Grothkopf for numerical calculations on a computer.

Prof. Dr. Lothar Collatz
Institut für Angewandte Mathematik
Bundesstraße 55
2ooo Hamburg 13
Germany

References

Albrecht, J. [61] Monotone Iterationsfolgen Num.Math. 3(1961) 345-358

Bandle, C. [81] Abschätzung der Randwerte bei nichtlinearen elliptischen Gleichungen aus der Plasmaphysik, Internat.Ser.Num.Math. Vol 56 (1981) 1-17.

Bohl, E. [74] Monotonie, Lösbarkeit und Numerik bei Operatorgleichungen. Springer, 1974, 255 S.

Bredendiek E. - L. Collatz [76] Simultan Approximation bei Randwertaufgaben, Internat.Ser.Num.Math. 3o (1976), 147-174.

Collatz, L. [52] Aufgaben monotoner Art, Arch.Math.Anal.Mech. 3 (1952), 366-376.

Collatz, L. [66] Functional Analysis and Numerical Mathematics, Academic Press, 1966, 473p.

Collatz, L. [67] Monotonie bei gewöhnlichen Differentialgleichungen 4. Ordnung, Internat. Ser.Num.Math. 7 (1967) 171-178.

Collatz, L. [78] Numerical treatment of some singular boundary value problems, Lect.Notes in Math., Bd.6o3, Springer 1978, 41-5o.

Collatz, L. [81] Anwendung von Monotoniesätzen zur Einschließung der Lösungen von Gleichungen, Jahrbuch Überblicke Mathematik 1981, 189-225.

Collatz, L. [82] Monotonie und Vektormonotonie, Vortrag GAMM-Tagung, Budapest, 1982, to appear

Glashoff, K. - B. Werner [79] Inverse Monotonicity of Monotone L-Operators with Applications..., J.Math.Anal.Appl. 72 (1979) 89-1o5.

Goerisch, F. [81] Über die Anwendung einer Verallgemeinrung des Lehmann-Maehly-Verfahrens zur Berechnung von Eigenwertschranken. Internat.Ser.Num.Math, Vol. 56 (1981) 58-72

Grothkopf, U. [81] Anwendungen der nichtlinearen Optimierung auf Randwertaufgaben bei partiellen Differentialgleichungen, Internat. Ser. Numer. Math. 56 (1981) 73-82.

Hoffmann K.H. - H.J. Kornstaedt [8o] Zum inversen Stefan Problem Internat. Ser. Math. Vol. 53 (198o) 115-143

Hoffmann K.H. [78] Monotonie bei nichtlinearen Stefan-Problemen, Internat.Ser.Num.Math. Vol 39 (1978) 162-19o.

Hofmann, W. [73] Monotoniesätze für hyperbolische Anfangswertaufgaben und Einschließung von Lösungen, Num.Math. 24 (1973) 137-149

Hofmann, W. [77] Generalization of a maximum priciple for the wave operator..., Bull da Sociendale Brasiliera de Matem. 1977

Nickel, K. [58] Einige Eigenschaften von Lösungen der Prandtlschen Grenzschichtdifferentialgleichung, Arch.rat.Mech.Anal.4 (1958)1-31.

Redheffer, R.M. [67] Differentialungleichungen unter schwachen Voraussetzungen, Abhandl.Math.Sem.Univ.Hamburg 31 (1967) 33-5o.

Schauder, J. [3o] Der Fixpunktsatz in Funktionenräumen, Studia Math.2 (193o), 171-182.

Schröder, J. [62] Invers-monotone Operatoren, Arch.Rat.Mech.Anal. 1o (1962), 276-295.

Walter, W. [7o] Differential and Integral Inequalities, Springer 197o, 352 S.

Werner, H. [79] Extrapolationsmethoden zur Bestimmung der beweglichen Singularitäten von Lösungen gewöhnlicher Differentialgleichungen, Internat.Ser.Num.Math. Vol. 49, 1979, 159-176.

Wetterling, W. [77] Quotienteneinschließung bei Eigenwertaufgaben mit partieller Differentialgleichung, Int.Ser.Num.Math.38 (1977)213-218.

Whiteman, J.R. [79] Two dimensional biharmonic problems, Proc.Manchester/Liverpool Sommer school on Numerical Solut. of P.D.E's. Oxford Univ. Press 1979, to appear.

Whiteman, J.R. - Babuska [8o], A Crack problem as test problem for numerical methods, posed at the Symposium on Finite Element Methods for nonlinear and singular problems (Whiteman, Mitchell, Morton) at Durham June 198o.

THE LIOUVILLE-GREEN ASYMPTOTIC THEORY FOR SECOND-ORDER DIFFERENTIAL EQUATIONS: A NEW APPROACH AND SOME EXTENSIONS

M. S. P. EASTHAM

1. In the asymptotic theory of the differential equation

$$\{p(x)y'(x)\}' + q(x)y(x) = 0 \quad (X \leqslant x < \infty), \tag{1.1}$$

the Liouville-Green transformation

$$y(x) = (p(x)|q(x)|)^{-\frac{1}{4}}z(t), \qquad t = \int_X^x \{|q(u)|/p(u)\}^{\frac{1}{2}}\,du \tag{1.2}$$

is used to derive the transformed equation

$$d^2z/dt^2 + \{\pm 1 + R(t)\}z(t) = 0, \tag{1.3}$$

where $\quad R(t) = p^{\frac{1}{4}}(x)|q(x)|^{-\frac{3}{4}}\frac{d}{dx}p(x)\frac{d}{dx}\{p(x)|q(x)|\}^{-\frac{1}{4}}$

and $\pm = \text{sgn}\, q$. The standard conditions on p and q are

(i) p and q are real-valued in $[X,\infty)$,

(ii) $p > 0$ and q nowhere zero in $[X,\infty)$,

(iii) p and q have continuous second derivatives in $[X,\infty)$.

If it is further assumed that

$$\int_X^\infty \{|q(u)|/p(u)\}^{\frac{1}{2}}\,du = \infty, \tag{1.4}$$

then $t \to \infty$ as $x \to \infty$ by (1.2), and therefore the t-interval in (1.3) is $[0,\infty)$. The final conditions which are usually imposed are

$$\int_X^\infty p(p|q|)^{-5/2}\{(p|q|)'\}^2\,dx < \infty \tag{1.5}$$

and

$$\int_X^\infty (p|q|)^{-3/2}|\{p(p|q|)'\}'|\,dx < \infty, \tag{1.6}$$

these implying that $\displaystyle\int_0^\infty |R(t)|\,dt < \infty$.

Then, by a standard asymptotic result, (1.3) has solutions

$$z(t) \sim e^{\pm it} \quad \text{or} \quad z(t) \sim e^{\pm t}$$

depending on whether $q > 0$ or $q < 0$. Correspondingly, by (1.2), (1.1) has solutions with the familiar Liouville-Green asymptotic forms

$$y \sim (p|q|)^{-\frac{1}{4}} \exp\left(\pm i \int_X^x (|q|/p)^{\frac{1}{2}}\, du\right) \quad (q > 0) \tag{1.7}$$

or

$$y \sim (p|q|)^{-\frac{1}{4}} \exp\left(\pm \int_X^x (|q|/p)^{\frac{1}{2}}\, du\right) \quad (q < 0). \tag{1.8}$$

We refer to (2, pp.118-122; 3, pp.60-63; 9, pp.190-202) for detailed accounts of the asymptotic theory outlined here. It was also shown in (2, pp.122) that, when $p = 1$, (1.6) implies (1.4) and (1.5), and therefore only (1.6) need be stated in this case. Refinements of (1.7)-(1.8) can be made in various directions and we add (1) and (11) to the references already given.

The conditions (1.4)-(1.6) represent a major restriction on the applicability of the transformation (1.2) to the asymptotic theory of (1.1). This restriction is illustrated by the familiar example $p(x) = x^{\alpha}$, $q(x) = \pm x^{\beta}$ with real α and β. Here (1.4)-(1.6) hold if $\alpha - \beta < 2$. The opposite condition $\alpha - \beta > 2$ can however be covered by a different method (16, Theorem 9.1), and we discuss this case further in §6(a). The borderline condition $\alpha - \beta = 2$ gives an Euler equation which is soluble explicitly.

In this paper, we avoid the use of (1.2) and we give an alternative and more systematic approach to the asymptotic theory of (1.1). Our method provides a new proof of (1.7) and (1.8) but, more important, it also covers the case where (1.4)-(1.6) are not satisfied. Even the Euler equation appears merely as an example of a general borderline case which again our method will cover. Other advantages of our method are:

(a) p and q are allowed to be complex-valued;

(b) it is not necessary in all cases to assume that p and q are twice differentiable, the condition (1.6) then being eliminated;

(c) extensions to higher-order equations are possible.

We conclude these introductory remarks with a further comment about complex-valued coefficients. Although (1.7) and (1.8) are normally established for real-valued p and q (2, 3, 5, 9, loc. cit.), it is possible to extend (1.2) to complex-valued coefficients by replacing $|q|$ by q in (1.2). The variable t is then complex and (1.3) can be treated as an equation in the complex domain (6, pp.178-185, Theorem 5.6.2) and in terms of the associated integral equation (12, § 5.8). In our method, however, complex-valued p and q are covered without special consideration.

2. In (1.1) we assume that p and q are complex-valued and nowhere zero in $[X,\infty)$. We begin by writing (1.1) as a first-order system

$$Y' = AY, \tag{2.1}$$

where

$$A = \begin{pmatrix} 0 & 1/p \\ -q & 0 \end{pmatrix}, \quad Y = \begin{pmatrix} y \\ py' \end{pmatrix}.$$

The next step is to diagonalize A by writing

$$A = T^{-1}DT,$$

where

$$D = \begin{pmatrix} i(q/p)^{\frac{1}{2}} & 0 \\ 0 & -i(q/p)^{\frac{1}{2}} \end{pmatrix}, \quad T = \begin{pmatrix} 1 & 1 \\ i(pq)^{\frac{1}{2}} & -i(pq)^{\frac{1}{2}} \end{pmatrix}.$$

The substitution

$$Y = TZ \tag{2.2}$$

in (2.1) then gives

$$Z' = (D - T^{-1}T')Z = A_1 Z, \tag{2.3}$$

where

$$A_1 = D - \tfrac{1}{4}\left(\frac{p'}{p} + \frac{q'}{q}\right)\begin{pmatrix} 1 & -1 \\ -1 & 1 \end{pmatrix}. \tag{2.4}$$

Up to this point we have assumed that p and q have continuous derivatives. If we now further assume that the second derivatives of p and q exist and are continuous, then we can repeat the diagonalization procedure which led from (2.1) to (2.3). The point of this process is that, under suitable conditions, (2.3) is then transformed into the standard first-order system

$$U' = (\Lambda + R_1)U \tag{2.5}$$

with Λ diagonal and R_1 being $L(X,\infty)$. Thus, we write

$$A_1 = T_1^{-1}D_1T_1,$$

where

$$D_1 = \begin{pmatrix} \sqrt{(\mu^2 + P^2)} - P & 0 \\ 0 & -\sqrt{(\mu^2 + P^2)} - P \end{pmatrix} \tag{2.6}$$

and

$$T_1 = \begin{pmatrix} 1 & -Q \\ Q & 1 \end{pmatrix}, \tag{2.7}$$

with the notation

$$\mu = i(q/p)^{\frac{1}{2}}, \quad P = \tfrac{1}{4}\left(\frac{p'}{p} + \frac{q'}{q}\right), \tag{2.8}$$

$$Q = P/\{\mu + \sqrt{(\mu^2 + P^2)}\}. \tag{2.9}$$

The inverse matrix T_1^{-1} exists if $Q^2 + 1 \neq 0$, that is, if

$$\mu^2 + P^2 \neq 0 \tag{2.10}$$

throughout $[X,\infty)$, and this is now a further condition on p and q.

In (2.3) we make the second substitution

$$Z = T_1U \tag{2.11}$$

to obtain

$$U' = (D_1 - T_1^{-1}T_1')U. \tag{2.12}$$

By (2.7) we have

$$T_1^{-1}T_1' = \frac{Q'}{1+Q^2}\begin{pmatrix} Q & -1 \\ 1 & Q \end{pmatrix} \tag{2.13}$$

and, after a calculation using (2.9), we obtain

$$\frac{Q'}{1+Q^2} = \frac{\mu^2}{\mu^2 + P^2}\left(\frac{P}{\mu}\right)'. \tag{2.14}$$

If we therefore add the assumption that

$$\frac{\mu^2}{\mu^2 + P^2}\left(\frac{P}{\mu}\right)' \quad \text{is} \quad L(X,\infty), \tag{2.15}$$

(2.12) takes the form (2.5) with

$$\Lambda = \mathrm{dg}\left(\sqrt{(\mu^2 + P^2)} - P - \frac{QQ'}{1+Q^2},\ -\sqrt{(\mu^2 + P^2)} - P - \frac{QQ'}{1+Q^2}\right) \tag{2.16}$$

$$= \mathrm{dg}(\mu_1, \mu_2),$$

say. Provided that $\mathrm{Re}(\mu_1 - \mu_2)$ does not change sign in (X,∞), the standard asymptotic theorem of (2, p.88) can be applied to (2.5). Thus, if

$$\mathrm{Re}\ \sqrt{(\mu^2 + P^2)} \text{ does not change sign in } (X,\infty), \tag{2.17}$$

(2.5) has solutions

$$U_j = \{e_j + o(1)\}\exp\left(\int_X^x \mu_j(t)\,dt\right) \tag{2.18}$$

for $j = 1, 2$, where $e_1 = \begin{pmatrix}1\\0\end{pmatrix}$ and $e_2 = \begin{pmatrix}0\\1\end{pmatrix}$. On transforming back to (1.1) via (2.11) and (2.2), we obtain corresponding solutions

$$y_1 = (1+Q^2)^{-\frac{1}{2}}\{1 + Q + o(Q) + o(1)\}\exp\left(\int_X^x \{\sqrt{(\mu^2 + P^2)} - P\}\,dt\right),$$

$$y_2 = (1+Q^2)^{-\frac{1}{2}}\{1 - Q + o(Q) + o(1)\}\exp\left(\int_X^x \{-\sqrt{(\mu^2 + P^2)} - P\}\,dt\right),$$

(2.19 - 2.20)

on substituting for μ_1 and μ_2 from (2.16).

3. We now start to draw our results together, assuming that p and q are twice continuously differentiable. We divide our analysis of (2.19-2.20) into the following three cases, which correspond to $P = o(\mu)$, $P \sim (\text{const.})\mu$ and $\mu = o(P)$ as $x \to \infty$.

I $(pq)'/pq = o\{(q/p)^{\frac{1}{2}}\}$ $(x \to \infty)$.

II $(pq)'/pq \sim \sigma(q/p)^{\frac{1}{2}}$ $(x \to \infty)$, where σ is a non-zero constant and $\sigma^2 \neq 16$.

III $(q/p)^{\frac{1}{2}} = o\{(pq)'/pq\}$ $(x \to \infty)$.

Other conditions will be imposed on p and q in each case as necessary.

Case I. Let

$$(pq)'/pq = o\{(q/p)^{\frac{1}{2}}\} \quad (x \to \infty) \tag{3.1}$$

and let (1.5) and (1.6) hold. Also, let (2.17) hold.

Here we have $P = o(\mu)$ and therefore (2.10) holds. In addition, our other condition (2.15) is implied by (1.5) and (1.6). By (2.9) we have $Q \to 0$ and, in (2.19-2.20), we have

$$\surd(\mu^2 + P^2) = \mu + O(P^2/\mu).$$

Now P^2/μ is $L(X,\infty)$ by (2.8) and (1.5). Hence, after an adjustment of constant multiples, (2.19-2.20) give

$$\left.\begin{aligned} y_1 &\sim (pq)^{-\frac{1}{4}}\exp\left(\int_X^x i(q/p)^{\frac{1}{2}}\,dt\right) \\ y_2 &\sim (pq)^{-\frac{1}{4}}\exp\left(-\int_X^x i(q/p)^{\frac{1}{2}}\,dt\right). \end{aligned}\right\} \tag{3.2}$$

This is our proof of (1.7)-(1.8) for complex-valued p and q and, in our approach, (3.1) replaces (1.4). We note that (2.17) is certainly satisfied in the case of real-valued coefficients.

4. Case II. Let

$$(pq)'/pq \sim \sigma(q/p)^{\frac{1}{2}} \qquad (x \to \infty) \tag{4.1}$$

where σ is a non-zero constant and $\sigma^2 \neq 16$, and let (1.5) and (1.6) hold. Also, let (2.7) hold.

Here we have $P \sim -\frac{1}{4}i\sigma\mu$ and therefore (2.10) holds when $\sigma^2 \neq 16$. By (2.9) we have

$$Q \to -\tfrac{1}{4}i\sigma/\{1 + \sqrt{(1 - \sigma^2/16)}\} \neq \pm 1, \tag{4.2}$$

and we substitute this into (2.19-2.20). To deal with $\sqrt{(\mu^2 + P^2)}$, we write

$$(pq)'/pq = \sigma(q/p)^{\frac{1}{2}}(1 + \phi), \tag{4.3}$$

by (4.1), where $\phi = o(1)$ and, by (1.6), ϕ' is $L(X,\infty)$. Then

$$\sqrt{(\mu^2 + P^2)} = \frac{1}{4}\,\frac{(pq)'}{pq}\sqrt{\{1 - 16\sigma^{-2}(1 + \phi)^{-2}\}}$$

$$= \frac{1}{4}\,\frac{(pq)'}{pq}\{(1 - 16\sigma^{-2})^{\frac{1}{2}} + 16\sigma^{-2}(1 - 16\sigma^{-2})^{-\frac{1}{2}}\phi + O(\phi^2)\},$$

on terminating the binomial expansion at the ϕ^2 term for example. Hence, assuming that

$$\phi^2(pq)'/pq \quad \text{is} \quad L(X,\infty) \tag{4.4}$$

and adjusting constant multiples in y_1 and y_2, we obtain from (4.2) and (2.19-2.20)

$$y_1 \sim (pq)^{-\frac{1}{4}+\frac{1}{4}\sqrt{(1-16\sigma^{-2})}}\exp I(x), \tag{4.5}$$

$$y_2 \sim (pq)^{-\frac{1}{4}-\frac{1}{4}\sqrt{(1-16\sigma^{-2})}}\exp\{-I(x)\}, \tag{4.6}$$

where

$$I(x) = 4(1 - 16\sigma^{-2})^{-\frac{1}{2}}\sigma^{-2}\int_X^x \phi(pq)'/pq\, dt. \tag{4.7}$$

If, further to (4.4),

$$\phi(pq)'/pq \quad \text{is} \quad L(X,\infty),$$

the exponential factors in (4.5-4.6) can be replaced by unity.

Referring again to the familiar coefficients

$$p(x) = x^{\alpha}, \quad q(x) = \pm x^{\beta}, \tag{4.8}$$

(4.1) holds when $\alpha - \beta = 2$ and then (1.1) is an Euler equation. The function ϕ in (4.3) is now zero and the right-hand sides of (4.5) and (4.6) reduce to the exact solutions of the Euler equation.

5. Case III. Let $(pq)'$ be nowhere zero in $[X,\infty)$ and let

$$(q/p)^{\frac{1}{2}} = o\{(pq)'/pq\} \quad (x \to \infty). \tag{5.1}$$

Let

$$(\mu/P)' \text{ be } L(X,\infty). \tag{5.2}$$

Also, let (2.17) hold.

Here (5.2) replaces (1.5) and (1.6) and, since $\mu = o(P)$ now, it is (5.2) that implies (2.15), Also, $Q \to 1$ as $x \to \infty$ in (2.9). Hence (2.19-2.20) give

$$y_1 = \{2 + o(1)\}\exp\left(\int_X^x \{\tfrac{1}{2}\mu^2/P + O(\mu^4/P^3)\}\, dt\right), \tag{5.3}$$

$$y_2 = o\{(pq)^{-\frac{1}{2}}\}\exp\left(-\int_X^x \{\tfrac{1}{2}\mu^2/P + O(\mu^4/P^3)\}\, dt\right) \tag{5.4}$$

when the definition of P in (2.8) is used.

If, in addition,

$$\mu^2/P \text{ is } L(X,\infty), \tag{5.5}$$

(5.3-5.4) can be written as

$$y_1 \sim 1, \tag{5.6}$$

$$y_2 = o\{(pq)^{-\frac{1}{2}}\} \tag{5.7}$$

after adjustment of a constant multiple.

The feature of (5.4) and (5.7) is of course the appearance of an o-estimate rather than a true asymptotic formula. The o-term derives from (2.18) and it is possible to improve the precision of this "error bound" by re-examining the proof of

(2.18). This procedure is carried out in (11; see also the references in 11) in the usual Liouville-Green situation corresponding to our Case I.

In Case III, however, it is more convenient to use the formula

$$y_2(x) = y_1(x) \int_X^x \{p(t)y_1^2(t)\}^{-1}\,dt$$

to obtain an improved asymptotic form for a a second solution y_2 of (1.1) in terms of y_1, as given by (5.3) or (5.6). Thus, in the case of (5.6-5.7), we have

$$y_2 = \{1 + o(1)\} \int_X^x \frac{1}{p(t)}\{1 + o(1)\}\,dt. \tag{5.8}$$

If, further, $1/p$ is $L(X,\infty)$ we can replace the integral in (5.8) by an infinite integral and write, for real-valued p,

$$y_2 \sim \int_x^\infty 1/p(t)\,dt. \tag{5.9}$$

The improvement of (5.8) over (5.7) is indicated if we note that (5.1) gives $1/p = o(\{(pq)^{-\frac{1}{2}}\}')$.

Referring back to the coefficients (4.8), the conditions (5.1) and (5.2) both give

$$\alpha - \beta > 2. \tag{5.10}$$

Also, (5.5) holds and hence (5.6), (5.8) and (5.9) give

$$y_1 \sim 1, \tag{5.11}$$

$$y_2 \sim \begin{cases} x^{1-\alpha} & (\alpha \neq 1) \\ \log x & (\alpha = 1) \end{cases}$$

again after adjusting constant multiples. We note that (5.11) confirms the limit-point nature of the equation

$$(x^\alpha y')' + x^\beta y = 0$$

when (5.10) holds, a fact which has been noted elsewhere

(8, p.38) using less precise information about the solutions.

Finally in this section, we note that (5.5) is not implied by (5.1) and (5.2), as the example

$$p(x) = \exp x^{\alpha}, \quad q(x) = x^{\beta} \exp x^{\alpha}$$

shows when $\alpha > 0$ and $\alpha - 2 \leq \beta < 2(\alpha - 1)$.

6. We conclude with a number of comments on various aspects of our method.

(a) Alternative conditions for (5.6) and (5.8), (5.9).

In (16, Theorem 9.1, pp.379-381), it is shown that a solution satisfying (5.6) exists if

(i) p is real-valued

and (ii) $\int_X^{\infty} \left(\int_J \frac{1}{p(t)} \, dt \right) |q(x)| \, dx$ converges,

where J is either (X,x) or (x,∞).

Here (ii) reduces to (5.10) when the coefficients are (4.8). There are no differentiability conditions in (16) such as we have in (5.1) and (5.2). In our approach, however, (5.6) arises merely as a special case of (5.3), and the appearance of P in (5.3) means that some differentiability conditions are necessarily involved in (5.3). It is not clear whether (5.3) can be obtained using the methods of (16, Theorem 9.1).

(b) The condition (2.17).

It is not difficult to formulate explicit properties of p and q which guarantee that (2.17) is satisfied, and we give a list of three separate such properties indicating to which of the three cases in §§ 3-5 they relate.

A. p and q are real-valued (I, II, III).

B. $\omega < \arg(q/p) < 2\pi - \omega$ for a fixed $\omega > 0$ (I).

C. c_1p and c_2q are real-valued, where c_1 and c_2 are constants with c_1/c_2 non-real (II, III).

(c) Coefficients differentiable once only.

The second diagonalization in §2, and with it the assumption of twice differentiability on p and q, can be avoided if the system (2.3) can itself be investigated asymptotically. The two main asymptotic theorems for systems in general are those of Levinson and Hartman-Wintner, and we refer to (15) for a discussion of these theorems. Levinson's theorem, which we used to obtain (2.18), applies to (2.12) rather than (2.3). However, subject to the special condition

$$|\text{Im}(q/p)^{\frac{1}{2}}| \geq (\text{const.}) > 0 \tag{6.1}$$

in (X,∞), the Hartman-Wintner theorem (15, pp.6-7) applies directly to (2.3) and gives solutions

$$Z_j = \{e_j + o(1)\}\exp\left(-\int_X^x \{(-1)^j(q/p)^{\frac{1}{2}} + P\}\,dt\right)$$

for $j = 1, 2$, where P is as in (2.8). The immediate consequence for (1.1) is that there are solutions satisfying (3.2). In addition to (6.1), a more usual type of condition is required which describes the perturbation of (2.3) from a diagonal system. This condition is that P should be $L^2(X,\infty)$. We state the result of applying the Hartman-Wintner theorem as follows.

Case I'. Let (6.1) hold and let

$$(pq)'/pq \text{ be } L^2(X,\infty).$$

Then (1.1) has solutions satisfying (3.2).

The most obvious situation to apply (6.1) is where p and q are real-valued with $q < 0$. Referring back to (1.7) and (1.8), we note that it is not always recognised in the literature that a distinction can be drawn between the conditions (apart from the sign of q) under which these formulae can be established, in the sense that (1.8) does not require twice differentiability of p and q nor a condition such as (1.6). This point is not however overlooked in (2, loc cit) and (16, pp.319-320).

(d) Higher-order differential equations.

For higher-order equations there are transformations that are similar to (1.2) in that a change of independent variable in terms of the coefficients is involved. Again, some restriction to real-valued coefficients is required. We refer to (10, 13) for third-order equations, to (17) for fourth-order equations, and to (7) for a two-term higher-order equation. The method in this paper can be applied to higher-order equations in general, and the details have been worked out for the fourth-order case and for higher-order equations with coefficients which resemble powers of x. We refer to (4, 14) and the bibliographies of these two papers for the details. Further applications of the method of this paper to higher-order equations are under investigation.

References

1. F.V. Atkinson, Proc. Glasgow Math. Assoc. 3 (1956-8) 105-111.
2. W.A. Coppel, Stability and asymptotic behaviour of differential equations (Heath, Boston, 1965).

3. M.S.P. Eastham, Theory of ordinary differential equations (Van Nostrand Reinhold, 1970).

4. M.S.P. Eastham, Proc. Roy. Soc. London, to appear.

5. W.N. Everitt, Czechoslovak Math. J., to appear.

6. E. Hille, Ordinary differential equations in the complex domain (Wiley, New York, 1976).

7. D.B. Hinton, J. Diff. Equations 4 (1968) 590-596.

8. R.M. Kauffman, T.T. Read and A. Zettl, Lecture Notes in Mathematics 621 (Springer, 1977).

9. F.W.J. Olver, Asymptotics and special functions (Academic Press, 1974).

10. G.W. Pfeiffer, J. Diff. Equations 11 (1972) 145-155.

11. J.G. Taylor, J. Math. Anal. Appl. 85 (1982) 79-89.

12. E.C. Titchmarsh, Eigenfunction expansions, Part 1 (2nd ed., Oxford, 1962).

13. K. Unsworth, Quart. J. Math. (Oxford) 24 (1973) 177-188.

14. M.S.P. Eastham and C.G.M. Grudniewicz, J. London Math. Soc. 24 (1981) 255-271.

15. W.A. Harris and D.A. Lutz, J. Math. Anal. Appl. 48 (1974) 1-16.

16. P. Hartman, Ordinary differential equations (Wiley, 1964).

17. P.W. Walker, J. Diff. Equations 9 (1971) 108-132.

Department of Mathematics
Chelsea College (University of London)
London SW3 6LX.

NON-SELF-ADJOINT OPERATORS AND THEIR ESSENTIAL SPECTRA

W. D. Evans, R. T. Lewis* and A. Zettl

Dedicated to Professor F. V. Atkinson

Introduction

In view of the increasing amount of work being done on non-self-adjoint differential operators and their spectra, there is a need to examine some of the techniques and results which have proved their worth in self-adjoint problems and to determine how they can be extended or modified for the non-self-adjoint cases. That particular need motivated much of the work whose results appear in this article. The first part of the paper deals with abstract results, mainly concerning accretive operators and operators defined by sectorial forms, the emphasis partly being on a discussion of the effects relatively bounded and relatively compact perturbations have on these operators. Our prime concern is the study of the essential spectra of these operators. As the operators are not self-adjoint the various essential spectra which are to be found in the literature are in general different sets, ranging from the complement of the semi-Fredholm domain used by Kato in [9] to the complement of eigenvalues of finite multiplicity which are isolated points in the resolvent set as used by Browder and recently by Reed and Simon. The fact that these essential spectra are coincident in the self-adjoint problems which have dominated research on the spectral theory of differential operators to date, has meant that some of the interesting properties of these sets have remained undetected. Moreover, the definition used by Glazman in [5] in terms of singular sequences, which has been the most widely used in problems involving differential operators, possesses properties which are not shared by some of the other essential spectra. For example, if S is a closed extension of a closed operator T, Glazman's essential spectrum of S

---

*Partly supported by NSF Grant No. MCS-8005811

contains that of T but this is rather special to this essential spectrum. The main objective of Section 1 is to study such points. In the second part of the paper we apply the abstract results obtained in §1 to ordinary differential operators in $L^2(\Omega)$, where $\Omega = (-\infty,\infty)$ or $[0,\infty)$. Our concern here is with describing the techniques which can be used rather than with obtaining the most general results, although we do try to be as abstemious as possible with our assumptions. The methods used are also applicable in more general circumstances (see e.g. [1], [2], [3]) when the coefficients of the differential operators and forms are large and the Sobolov spaces used throughout this paper are replaced by weighted spaces.

We shall assume that the reader has some familiarity with the basic concepts and especially with the relevant chapters of [5], [9] and [19].

§1. Abstract theory

1.1. Accretive operators

The numerical range of a linear operator T with domain $\mathcal{D}(T)$ and range $\mathcal{R}(T)$ in a Hilbert space H is the set

$$\Theta(T) = \{(Tu,u): u \in \mathcal{D}(T), \|u\| = 1\}$$

in $\mathbb{C}$, when $(\cdot,\cdot)$ and $\|\cdot\|$ are the inner product and norm in H. The set $\Theta(T)$ is known to be convex and hence if $\Delta = \mathbb{C} \setminus \overline{\Theta(T)} \neq \emptyset$, it has either one or two connected components. If T is closed and $\lambda \in \Delta$ then $T - \lambda I$ is semi-Fredholm with $\mathrm{nul}(T - \lambda I) = 0$ and $\mathrm{def}(T - \lambda I)$ constant in each connected component of Δ. The constant $m = \mathrm{def}(T - \lambda I)$ is called the deficiency index of T if Δ is simply connected. Otherwise, T has a pair (m_1,m_2) of deficiency indices, $m_i = \mathrm{def}(T - \lambda I)$, $\lambda \in \Delta_i$, $i = 1,2$ for the connected components Δ_1, Δ_2 of Δ, these deficiency indices being unequal in general. If $\mathrm{def}(T - \lambda I) = 0$ for $\lambda \in \Delta$ $(\Delta_1$ or $\Delta_2)$ then Δ $(\Delta_1$ or $\Delta_2)$ is a subset of the resolvent set $\rho(T)$ of T and

$$\|(T - \lambda I)^{-1}\| \leq 1/\mathrm{dist}\left(\lambda, \overline{\Theta(T)}\right)$$

(see [9, Theorem V:3.2]).

An operator T is accretive if $\Theta(T)$ lies in the right half plane in $\mathbb{C}$, i.e.,

$$\mathrm{Re}\ (Tu,u) \geq 0 \qquad u \in \mathcal{D}(T).$$

T is said to be quasi-accretive if $T + c$ is accretive for some $c \in \mathbb{R}$. In view of the preceding remarks a closed accretive operator T is such that for $\mathrm{Re}\ \lambda < 0$, $T - \lambda I$ is semi-Fredholm with zero nullity and constant deficiency. Moreover, if $\mathrm{def}(T - \lambda I) = 0$ for $\mathrm{Re}\ \lambda < 0$ then

$$\{\lambda : \mathrm{Re}\ \lambda < 0\} \subseteq \rho(T)$$

and

$$\|(T - \lambda I)^{-1}\| \leq 1/|\mathrm{Re}\ \lambda| . \tag{1.1}$$

Such an operator is said to be m-accretive. It can be shown (see [9]) that an m-accretive operator is a densely defined accretive operator with no proper accretive extensions, hence the suffix m- which stands for maximal with respect to being accretive. Some useful instances of m-accretive operators are provided by the following theorem. Recall that an operator J defined on H is called a conjugation operator if for $x, y \in H$,

$$(Jx,Jy) = (y,x); \quad J^2x = x.$$

Such a map is conjugate linear on H. Furthermore a densely defined linear map T in H is said to be J-symmetric if

$$JTJ \subset T^* \tag{1.2}$$

and J-self-adjoint if $JTJ = T^*$. If T is a closed J-symmetric operator

$$\mathrm{nul}(T - \lambda I) \leq \mathrm{nul}(T^* - \bar{\lambda} I), \qquad \lambda \in \mathbb{C} \tag{1.3}$$

with equality if T is J-self-adjoint [5, §22]. Note that we follow Kato [9] and define defT to be the codimension of the range R(T) in H. Glazman, however, defines defT as nulT*. The two definitions are identical if R(T) is closed.

Theorem 1.1. (i) A closed symmetric operator T is m-accretive iff T is self-adjoint and $T \geq 0$.

(ii) A closed J-symmetric operator is m-accretive iff T is J-self-adjoint and accretive.

(iii) If $-iT = S^*$, where S is maximal symmetric then T is m-accretive; in particular iS is m-accretive if S is self-adjoint.

Proof. (i) If T is self-adjoint and $T \geq 0$ then the left-half-plane lies in the resolvent set of T and so $\mathrm{def}(T - \lambda I) = 0$ for $\mathrm{Re}\,\lambda < 0$. Conversely, if T is m-accretive and symmetric then $\mathrm{def}(T - \lambda I) = 0$ for $\mathrm{Re}\,\lambda < 0$ and hence for all $\lambda \in \Delta = \mathbb{C} \setminus [0,\infty)$; T is therefore self-adjoint.

(ii) The proof of this is similar on using the fact that a closed J-symmetric operator T is J-self-adjoint iff $\mathrm{def}(T - \lambda I) = 0$ for some, and hence all, $\lambda \in \Delta$ [5, §22].

(iii) This is proved in [10].

Let us define the following numbers

$$\ell(T) = \inf\{\mathrm{Re}\,\lambda : \lambda \in \sigma(T)\} \tag{1.4}$$

$$\theta(T) = \inf\{\mathrm{Re}\,\lambda : \lambda \in \Theta(T)\}. \tag{1.5}$$

If T is m-accretive then clearly, as $\sigma(T) \subseteq \Theta(T)$ we must have $\theta(T) \leq \ell(T)$. If T is self-adjoint then $\theta(T) = \ell(T)$ [9, §V3.10] but this is not the case in general. Indeed, if T is a 2×2 matrix with distinct eigenvalues α, β and normalised eigenvectors ϕ, ψ then $\Theta(T)$ is a closed elliptical disc with foci at α, β, minor axis $\gamma|\alpha - \beta|/\delta$ and major axis $|\alpha - \beta|/\delta$, where $\gamma = |(\phi,\psi)|$, $\delta = \sqrt{1 - \gamma^2}$ [8, p. 166].

1.2 Relative bounded perturbations

A linear operator P is said to be T-bounded if $\mathcal{D}(T) \subseteq \mathcal{D}(P)$ and there exist non-negative constants a, b such that

$$\|Px\|^2 \leq a\|Tx\|^2 + b\|x\|^2, \qquad x \in \mathcal{D}(T). \tag{1.6}$$

The infimum of the constants a is called the T-bound of P.

Theorem 1.2. Let T be a closed linear operator in H and suppose that for some $\alpha \in [0,2\pi)$ and $\gamma \in \mathbb{C}$, the set

$$\Lambda_{\alpha,\gamma} = \{z \in \mathbb{C}: \alpha < \arg(z - \gamma) < \pi + \alpha\}$$

lies in $\Delta(T) \cap \rho(T)$. Then, if P is T-bounded with T-bound < 1, $S = T + P$ is closed and there exists $R > 0$ such that $\lambda = re^{i\theta} \in \rho(S)$ for $r \geq R$ and $\theta = \alpha + \pi/2$.

In particular we have the following stability results for perturbations P of an operator T which are T-bounded with T-bound < 1:

(i) If T is maximal symmetric and P is symmetric, $T + P$ is maximal symmetric.

(ii) If T is self-adjoint and P is symmetric, $T + P$ is self-adjoint.

(iii) If T is J-self-adjoint, $\Lambda_{\alpha,\gamma} \subset \Delta(T)$ and P is J-symmetric then $T + P$ is J-self-adjoint.

(iv) If T is quasi-m-accretive and P is quasi-accretive then $T + P$ is quasi-m-accretive.

Proof. We can assume, without loss of generality, that $\gamma = 0$ since otherwise we can replace T by $T - \gamma$ throughout. Also, it is well known that S is closed.

If $\lambda = re^{i\theta}$, $\theta = \alpha + \pi/2$ then since $\Theta(T) \subset \mathbb{C} \setminus \Lambda_{\alpha,\gamma}$ we have for all $u \in \mathcal{D}(T)$

$$\pi/2 \leq \arg[\bar{\lambda}(Tu,u)] \leq 3\pi/2.$$

Hence $\operatorname{Re}[\bar{\lambda}(Tu,u)] \leq 0$ and

$$\|Tu\|^2 + |\lambda|^2 \|u\|^2 = \|(T - \lambda I)u\|^2 + 2\operatorname{Re}[\bar{\lambda}(Tu,u)] \leq \|(T - \lambda I)u\|^2.$$

Since P has T-bound < 1, there are numbers $a \in (0,1)$ and $b > 0$ such that

$$\begin{aligned} \|Pu\|^2 &\leq a^2\|Tu\|^2 + b^2\|u\|^2, \qquad u \in \mathcal{D}(T) \\ &\leq a^2\{\|Tu\|^2 + |\lambda|^2\|u\|^2\} \end{aligned}$$

if $|\lambda| \geq b/a$. By the above inequality

$$\|Pu\|^2 \leq a^2\|(T - \lambda I)u\|^2.$$

Since $\lambda = re^{i\theta} \in \rho(T)$, $(T - \lambda I)^{-1}$ is bounded on H and if $r \geq b/a$,

$$\|P(T - \lambda I)^{-1}\| \leq a < 1.$$

Hence from

$$S - \lambda I = [I + P(T - \lambda)^{-1}](T - \lambda)$$

it follows that $\lambda \in \rho(S)$ if $|\lambda| \geq b/a$.

If T is maximal symmetric then $\operatorname{def}(T - \lambda I) = 0$ for $\lambda \in \mathbb{C}_+$ (say, similarly for $\mathbb{C}_-$) and $\mathbb{C}_+ \subset \rho(T) \cap \Delta(T)$ since $\Theta(T) \subseteq \mathbb{R}$. From what we have just proved it follows that since $\Lambda_{o,o} = \mathbb{C}_+$, $\lambda = ik$ lies in $\rho(T + P)$ for k positive and large enough. Since $T + P$ is symmetric, $\Theta(T + P) \subseteq \mathbb{R}$ and consequently $\operatorname{def}(T + P - \lambda I) = 0$ for $\lambda \in \mathbb{C}_+$. This in turn implies that $T + P$ is maximal symmetric.

The proof of (ii) is similar to (i).

If T is J-self-adjoint, $\Delta(T) \subset \rho(T)$ since $\operatorname{def}(T - \lambda I) = \operatorname{nul}(T - \lambda I) = 0$ for $\lambda \in \Delta(T)$ and consequently, $\Lambda_{\alpha,\gamma} \subseteq \Delta(T) \cap \rho(T)$ as required. Also $T + P$ is J-symmetric with constant deficiency in each of the connected components of $\Delta(T + P)$; in fact in all of $\Delta(T + P)$ — see Zhikhar's result referred to in [5 , §22]. It follows that $\operatorname{def}(T + P - \lambda I) = 0$ for the λ values determined in the first part, and hence for all $\lambda \in \Delta(T+P)$. Therefore, $T + P$ is J-self-adjoint.

If T is quasi-m-accretive we can take $\Lambda_{\pi/2,\gamma}$ for some $\gamma \in \mathbb{R}$ in the first part of the theorem. $T + P$ is quasi-accretive with $\Theta(T + P) \leq \{\lambda\colon \operatorname{Re} \lambda > \beta\}$ for some $\beta > 0$. Since $T + P$ has constant deficiency for $\text{Re}\,\lambda < \beta$ it follows that this deficiency is zero and $T + P$ is quasi-m-accretive.

Corollary 1.3. Let T be a closable operator in H and P a closable T-bounded operator with T-bound < 1.

(i) If $\overline{T}$ is maximal symmetric and P is symmetric, $\overline{T + P} = \overline{T} + \overline{P}$ is maximal symmetric.

(ii) If T is essentially self-adjoint and P is symmetric, $T + P$ is essentially self-adjoint.

(iii) If $\overline{T}$ is J-self-adjoint, $\Lambda_{\alpha,\gamma} \subset \Delta(\overline{T})$ and P is J-symmetric, then $\overline{T + P}$ is J-self-adjoint.

(iv) If $\overline{T}$ is quasi-m-accretive and P is quasi-accretive, $\overline{T + P}$ is quasi-m-accretive.

Proof. Since P is closable with T-bound < 1, $\overline{P}$ has $\overline{T}$-bound < 1 and moreover, $\overline{T + P} = \overline{T} + \overline{P}$. The corollary then follows immediately from Theorem 1.2.

Theorem 1.4. Let T be m-accretive and let P be T-bounded with T-bound < 1. Then $S = T + P$ is closed and $\lambda \in \rho(S)$ for all large enough negative λ.

Proof. For $\lambda < 0$

$$\begin{aligned} \|(T - \lambda)\phi\|^2 &= \|T\phi\|^2 - 2\lambda\,\text{Re}[(T\phi,\phi)] + |\lambda|^2\|\phi\|^2 \\ &\geq \|T\phi\|^2 + |\lambda|^2\|\phi\|^2 \end{aligned} \tag{1.7}$$

since T is accretive. Also $\lambda \in \rho(T)$ since T is m-accretive. Since P has T-bound < 1

$$\begin{aligned} \|P\phi\|^2 &\leq \alpha\|T\phi\|^2 + \beta\|\phi\|^2, \qquad 0 < \alpha < 1, \qquad \beta > 0 \\ &\leq \alpha\,\|(T - \lambda)\phi\|^2 + \beta\|\phi\|^2 \end{aligned}$$

and so

$$\|P(T - \lambda I)^{-1}\phi\|^2 \le \alpha\|\phi\|^2 + \beta\|(T - \lambda)^{-1}\phi\|^2$$
$$\le (\alpha + \beta/|\lambda|^2)\|\phi\|^2$$

from (1.7). Hence $\|P(T - \lambda I)^{-1}\| < 1$ for $\lambda < 0$ and $|\lambda|$ large enough. The result therefore follows from the identity

$$S - \lambda I = [I + P(T - \lambda I)^{-1}](T - \lambda I).$$

1.3 Sectorial forms and m-sectorial operators

A sesquilinear form t with domain $\mathcal{D}(t)$ is said to be sectorial if its numerical range, i.e., the set

$$\Theta(t) = \{t[u]: u \in \mathcal{D}(t), \|u\| = 1\} \qquad (t[u] \equiv t[u,u]),$$

lies in a sector

$$S_{\theta,\gamma} = \{x + iy: x \ge \gamma, |y| \le \tan\theta(x - \gamma)\} \tag{1.8}$$

for some $\gamma \in \mathbb{R}$ and $\theta \in [0,\pi/2)$. An operator T is said to be m-sectorial if it is quasi-m-accretive and sectorial in the sense that its numerical range $\Theta(T)$ lies in a sector $S_{\theta,\gamma}$.

If t is a densely defined sectorial form which is closed in H (see [9, Ch. VI]) there is associated with t an m-sectorial operator T, this association being determined by

$$t[u,\phi] = (Tu,\phi), \qquad u \in \mathcal{D}(T) \subset \mathcal{D}(t), \quad \phi \in \mathcal{D}(T). \tag{1.9}$$

If we denote by h, the real part of t, i.e. $h = \frac{1}{2}(t + t^*)$, the closedness of t is equivalent to the property that for some $\alpha > -\gamma$

$$(h + \alpha)[u] \equiv h[u] + \alpha\|u\|^2 \tag{1.10}$$

determines a norm on $\mathcal{D}(t)$ with respect to which $\mathcal{D}(t)$ is complete and lies in H (or more precisely, can be identified in a one-one fashion with a subspace of H). Let $\mathcal{Q}(T)$ denote this Hilbert space; it is called the form domain of T and has $\mathcal{D}(T)$ as a dense subspace. The form h is of course symmetric (i.e. $h^*[u,v] \equiv \overline{h[v,u]} = h[u,v]$) and bounded below, and there is therefore associated with it a lower semi-bounded self-adjoint operator A (called the real part of T and written $\operatorname{Re} T$) with $\mathcal{D}(A) \subset \mathcal{Q}(T)$ and

$$h[u,\phi] = (Au,\phi); \qquad u \in \mathcal{D}(A), \quad \phi \in \mathcal{Q}(T).$$

If $\Theta(t)$ lies in the sector (1.8), $h \geq \gamma$ and $A \geq \gamma$. In fact $\mathcal{Q}(T) = \mathcal{D}([A-\gamma]^{1/2})$ and

$$(h-\gamma)[u,v] = ([A-\gamma]^{1/2}u, [A-\gamma]^{1/2}v)$$

(see [9, Theorem VI: 2.23]). The h norm in (1.10) is thus equivalent to the graph norm of $(A-\gamma)^{1/2}$ on its domain $\mathcal{Q}(T)$ and this is true for any $\alpha > -\gamma$ in (1.10).

As noted above the inclusion map $E: \mathcal{Q}(T) \to H$ is injective and continuous and consequently, for $h \in H$ and $\phi \in \mathcal{Q}(T)$

$$|(h,\phi)_H| \leq \|h\|_H \|\phi\|_{\mathcal{Q}}$$

where we have written $\mathcal{Q} \equiv \mathcal{Q}(T)$ and $\|\cdot\|_H$, $\|\cdot\|_{\mathcal{Q}}$ denote the norms in H and $\mathcal{Q}$. The map $F: h \to (h,\cdot)_H$ is therefore a linear injection of H into $\mathcal{Q}^{\#}$, the conjugate dual of $\mathcal{Q}$, and we have the triplet

$$\mathcal{Q} \xrightarrow{E} H \xrightarrow{F} \mathcal{Q}^{\#} \tag{1.11}$$

each of the maps E, F having dense range. From (1.8), for $x, y \in \mathcal{Q}$

$$\begin{aligned} |(t-\gamma)[x]| &\leq (h[x] - \gamma\|x\|^2) + |(\operatorname{Im} t)[x]| \\ &\leq (1+\tan\theta)[(h-\gamma)[x]] \\ &\leq (1+\tan\theta)\|x\|_{\mathcal{Q}}^2. \end{aligned}$$

It follows that [9, Problem I: 6.8 and I: 6.33]

$$|(t-\gamma)[x,y]| \leq \varepsilon(1+\tan\theta)\|x\|_{\mathcal{Q}}\|y\|_{\mathcal{Q}}$$

where $\varepsilon = 1$ if t is symmetric and $\varepsilon = 2$ otherwise. Hence $(t-\gamma)[\cdot,\cdot]$ is bounded on $\mathcal{Q} \times \mathcal{Q}$ and we may define a bounded linear operator $\hat{T}$ by

$$\hat{T}: \mathcal{Q} \to \mathcal{Q}^{\#}, \quad (\hat{T}x)(y) = t[x,y].$$

From [2 ,§7], it follows that $T = F^{-1}\hat{T}$ and for $\alpha > -\gamma$, $\hat{T} + \alpha F$ has range $\mathcal{Q}^{\#}$; $(\hat{T} + \alpha F)^{-1}$ therefore lies in $B(\mathcal{Q}^{\#},\mathcal{Q})$, the space of bounded linear operator mapping $\mathcal{Q}^{\#}$ into $\mathcal{Q}$.

1.4 Perturbations of sectorial forms

Let t be a closed, densely defined sectorial form in H and with the notation of §1.3, let p be a sesquilinear form defined on $\mathcal{Q} \times \mathcal{Q}$ and satisfying

$$|p[x]| \leq \alpha\| (A - \gamma)^{1/2}x\|^2 + \beta\| x \|^2, \qquad x \in \mathcal{Q} \tag{1.12}$$

for some positive numbers α, β. If $\alpha < 1$, then $s = t + p$ is closed and sectorial (see [9 , Theorem VI: 1.33]) and each of the forms s, t, p are bounded on $\mathcal{Q} \times \mathcal{Q}$. We therefore have

$$\hat{S} = \hat{T} + \hat{P} \tag{1.13}$$

in the notation of §1.3, where each of the maps in (1.13) is a bounded linear map of $\mathcal{Q}$ into $\mathcal{Q}^{\#}$. If we know from the outset that s and t are closed densely defined and sectorial, then $p = s - t$ automatically satisfies (1.13) since $\mathcal{Q}(s)$ and $\mathcal{Q}(t)$ are topologically isomorphic.

Since S and T, the operations associated with s and t, are m-sectorial, $\lambda \in \rho(S) \cap \rho(T)$ for $\mathrm{Re}\,\lambda$ large enough and negative. Moreover, $(\hat{S} - \lambda F)^{-1}$ and $(\hat{T} - \lambda F)^{-1}$ are in $B(\mathcal{Q}^{\#},\mathcal{Q})$ and from (1.13),

$$\begin{aligned} \hat{S} - \lambda F &= (\hat{T} - \lambda F) + \hat{P} \\ &= \{I^{\#} + \hat{P}(\hat{T} - \lambda F)^{-1}\}(\hat{T} - \lambda F) \end{aligned}$$

on $\mathcal{Q}$, where $I^{\#}$ is the identity on $\mathcal{Q}^{\#}$. Thus, on $\mathcal{Q}^{\#}$

$$(\hat{T} - \lambda F)^{-1} - (\hat{S} - \lambda F)^{-1} = (\hat{S} - \lambda F)^{-1}\,\hat{P}(\hat{T} - \lambda F)^{-1}$$

and since $(S - \lambda I)^{-1} = (\hat{S} - \lambda F)^{-1}F$,

$$(T - \lambda I)^{-1} - (S - \lambda I)^{-1} = (\hat{S} - \lambda F)^{-1}\,\hat{P}(T - \lambda I)^{-1} \tag{1.14}$$

on H. Note that in (1.14) the embedding map $\mathcal{Q} \to H$ is suppressed as the sense is clear.

1.5. Essential Spectra

Let T be a closed, densely defined linear operator in H and define the following subsets of $\mathbb{C}$:

$$\Phi^{+}(T) = \{\lambda: R(T - \lambda I) \text{ closed and } \mathrm{nul}(T - \lambda I) < \infty\}$$

$$\Phi^{-}(T) = \{\lambda: R(T - \lambda I) \text{ closed and } \mathrm{def}(T - \lambda I) < \infty\}$$

where $R(T-\lambda I)$ denotes the range of $T-\lambda I$,

$$\Phi_1(T) = \Phi^{+}(T) \cup \Phi^{-}(T)$$

$$\Phi_3(T) = \Phi^{+}(T) \cap \Phi^{-}(T)$$

$$\Phi_4(T) = \{\lambda: \lambda \in \Phi_3(T), \ \mathrm{nul}(T - \lambda I) = \mathrm{def}(T - \lambda I)\}$$

$$\Phi_5(T) = \{\lambda: \lambda \in \Phi_4(T) \text{ and a deleted neighbourhood of } \lambda \text{ lies in the resolvent set } \rho(T) \text{ of } T\}.$$

Let

$$\sigma_{ek}(T) = \mathbb{C} \setminus \Phi_k(T), \qquad k = 1, 3, 4, 5$$

$$\sigma_{e2}^{\pm}(T) = \mathbb{C} \setminus \Phi^{\pm}(T).$$

Each of these sets σ_{ek}, $\sigma_{e2}^{\pm}$ is used for the essential spectrum in the literature ([21, Ch. 11]; note that our notation differs from that in [12]. Schechter's set σ_{e1} is smaller than our σ_{e1} and our σ_{e2}^{+} is denoted by $\sigma_{e\alpha}$ by Schechter). $\Phi_1(T)$ is called the semi-Fredholm domain of T and $\Phi_3(T)$ the Fredholm domain of T [9]. The sets $\sigma_{ek}(T)$, $\sigma_{e2}^{\pm}$ are all closed subsets in $\mathbb{C}$ as their complements are open [9, Theorem IV: 5.31] and clearly

$$\sigma_{e1}(T) \subseteq \sigma_{e2}^{\pm}(T) \subseteq \sigma_{e3}(T) \subseteq \sigma_{e4}(T) \subseteq \sigma_{e5}(T).$$

In general the latter inclusions are strict (see [7]) but as we shall see in Theorem 1.5 below, the situation is somewhat more satisfactory for self-adjoint and J-self-adjoint operators.

Since $R(T)$ is closed if and only if $R(T^{*})$ is closed [9, Thm. IV: 5.13] and $\mathrm{nul}\, T^{*} = \mathrm{def}\, T$, $\mathrm{def}\, T^{*} = \mathrm{nul}\, T$, it follows that $\lambda \in \sigma_{e2}^{+}(T)$ iff $\bar{\lambda} \in \sigma_{e2}^{-}(T^{*})$. Also if $\lambda \in \sigma_{e4}(T) \setminus \sigma_{e2}^{+}(T) = \Phi^{+}(T) \setminus \Phi_4(T)$ then either λ is an eigenvalue of T (of finite geometric multiplicity) or $\lambda \in \sigma_r(T) = \{\lambda: \text{nul}(T - \lambda I) = 0, \text{def}(T-\lambda I) \neq 0\}$, the so-called residual spectru of T. Furthermore

$$\{\lambda: \lambda \in \sigma_r(T), R(T - \lambda I) \text{ closed}\} \subseteq \sigma_{e4}(T) \setminus \sigma_{e2}^{+}(T)$$

and

$$\sigma_r(T) \cap \Phi_3(T) \subseteq \sigma_{e4}(T) \setminus \sigma_{e3}(T).$$

Hence $\sigma_{e4}(T)$ will be different from $\sigma_{e3}(T)$ if $\sigma_r(T) \cap \Phi_3(T) \neq \emptyset$.

In each connected component $\Phi_1^{(n)}(T)$ say of the semi-Fredholm domain $\Phi_1(T)$, the index of $T - \lambda I$ is constant and moreover $\text{nul}(T - \lambda I)$ and $\text{def}(T - \lambda I)$ take constant values, $\nu_1^{(n)}$, $\mu_1^{(n)}$ say, except at an isolated set of values of λ [9, Ch. IV]. It follows from this that $\Phi_5(T)$ can be defined as the subset of $\Phi_4(T)$ with the property that each of its connected components intersects $\rho(T)$. For if $\Phi_1^{(n)}(T) \cap \Phi_4(T)$ intersects $\rho(T)$, $\text{nul}(T - \lambda I) = \text{def}(T - \lambda I) = 0$ and hence $\lambda \in \rho(T)$ for all $\lambda \in \Phi_1^{(n)}(T) \cap \Phi_4(T)$ except at isolated points where $0 \neq \text{nul}(T - \lambda I) = \text{def}(T - \lambda I) < \infty$. From this we see that $\lambda \in \sigma(T) \setminus \sigma_{e5}(T)$ if and only if λ is an isolated eigenvalue of T, $\overline{\lambda}$ is an isolated eigenvalue of T^* and their geometric multiplicities are equal. As usual $\sigma(T)$ represents the spectrum of T. It is possible for each $\lambda \in \sigma(T) \setminus \sigma_{ek}(T)$, $k = 1,2,3,4$ $(\sigma_{e2} \equiv \sigma_{e2}^{\pm})$ to be an eigenvalue of T, having finite geometric multiplicity if $k > 1$.

Theorem 1.5. (i) If T is self-adjoint the sets $\sigma_{ek}(T)$, $k = 1,3,4,5$, and $\sigma_{e2}^{\pm}(T)$ are identical and $\lambda \in \sigma(T) \setminus \sigma_{ek}(T)$ if and only if λ is an isolated eigenvalue of finite multiplicity.

(ii) If T is J-self-adjoint, the sets $\sigma_{ek}(T)$, $k = 1,3,4$ and $\sigma_{e2}^{\pm}(T)$ are identical.

Proof. (i) It suffices to show that $\Phi_1(T) \subseteq \Phi_5(T)$ or in view of the above remarks, that each connected component of $\Phi_1(T)$ intersects $\rho(T)$. But, since $\sigma(T) \subseteq \mathbb{R}$ every neighbourhood of every $\lambda \in \mathbb{C}$ intersects $\rho(T)$ and hence the result follows.

(ii) Since T is J-self-adjoint $\mathrm{nul}(T - \lambda I) = \mathrm{def}(T - \lambda I)$ for any $\lambda \in \mathbb{C}$. Hence $\Phi_1(T) \subseteq \Phi_4(T)$ and the result follows.

In applications it is often convenient to work with the following equivalent definitions of the sets σ_{ek}. Recall that a sequence $\{u_n\}$ is $\mathcal{D}(T)$ which is such that

$$u_n \rightharpoonup 0, \quad \|u_n\| = 1, \quad (T - \lambda I)u_n \to 0$$

is called a *singular sequence* of T corresponding to λ.

Theorem 1.6. Let T be a closed, densely defined linear operator in $\mathcal{H}$. Then

(i) $\lambda \in \sigma^+_{e2}(T)$ if and only if there exists a singular sequence of T corresponding to λ;

(ii) $\lambda \in \sigma^-_{e2}(T)$ if and only if there exists a singular sequence of T^* corresponding to $\bar{\lambda}$;

(iii) $\sigma_{e4}(T) = \bigcap_{P \text{ compact}} \sigma(T + P)$.

Proof. The result (i) is given in [23] and (ii) is an immediate consequence. The proof of (iii) may be found in [21, Theorem 1: 4.5].

If $T \subseteq S$, it follows from Theorem 1.6 that $\sigma^+_{e2}(T) \subseteq \sigma^+_{e2}(S)$ and $\sigma^-_{e2}(T) \supseteq \sigma^-_{e2}(S)$. We now investigate the relationship between the various essential spectra of S and T when S is a finite dimensional extension of T; recall that S is said to be an m-dimensional extension of T if the quotient space $\mathcal{D}(S)/\mathcal{D}(T)$ is of dimension m. This situation is particularly relevant for operators generated by ordinary differential expressions.

Theorem 1.7. Let S be a closed, m-dimensional extension of the closed densely defined operator T in $\mathcal{H}$. Then

(i) $\text{nul } T \leq \text{nul } S \leq \text{nul } T + m$,

(ii) $\text{def } S \leq \text{def } T \leq \text{def } S + m$,

(iii) and when T is a Fredholm operator, S is a Fredholm operator and $\text{index } S = \text{index } T + m$.

Proof. (i) Let $N(T)$, $N(S)$ denote the null spaces of T, S respectively. As $N(T) \subseteq N(S)$ it is immediate that $\text{nul } T \leq \text{nul } S$. Since $[N(S) \ominus N(T)] \cap \mathcal{D}(T) = \{0\}$ we have the direct sum

$$[N(S) \ominus N(T)] \dot{+} \mathcal{D}(T) \subseteq \mathcal{D}(S)$$

and hence

$$\begin{aligned}\dim N(S) - \dim N(T) &= \dim[N(S) \ominus N(T)] \\ &\leq m\end{aligned}$$

which completes the proof of (i).

(ii) Let $\mathcal{R}(T)$, $\mathcal{R}(S)$ denote the ranges of T, S respectively. Since $\mathcal{R}(S) \supseteq \mathcal{R}(T)$ it follows that $\text{def } S \leq \text{def } T$. Also, as S is an m-dimensional extension of T, there exists an m-dimensional subspace G of $\mathcal{D}(S)$ such that $\mathcal{D}(S) = \mathcal{D}(T) \dot{+} G$ and hence $\mathcal{R}(S) = \mathcal{R}(T) + SG$. If $\mathcal{R}$ denotes the complementary subspace of $\mathcal{R}(S)$ in $\mathcal{H}$ (so that $\dim \mathcal{R} = \text{def } S$) we therefore have

$$\begin{aligned}\mathcal{H} &= \mathcal{R}(S) + \mathcal{R} \\ &= \mathcal{R}(T) + (\mathcal{R} + SG)\end{aligned}$$

and consequently $\text{def } T \leq \dim(\mathcal{R} + SG)$. But $\mathcal{R} \cap SG \subseteq \mathcal{R} \cap \mathcal{R}(S) = \{0\}$ and so $\dim(\mathcal{R} + SG) = \dim \mathcal{R} + \dim(SG) \leq \text{def } S + m$. Thus (ii) is proved.

(iii) Let $\mathcal{H}_T$, $\mathcal{H}_S$ denote $\mathcal{D}(T)$, $\mathcal{D}(S)$ with the appropriate graph norms. Then $S \in B(\mathcal{H}_S, \mathcal{H})$, $T \in B(\mathcal{H}_T, \mathcal{H})$, where $B(X,Y)$ denotes the space of bounded linear maps of X into Y, and $\mathcal{H}_S = \mathcal{H}_T \dot{+} G$, $\dim G = m$. Let P be the projection of H_S onto H_T. Then $P \in B(H_S, H_T)$, nul $P = \dim G = m$ and def $P = 0$. Also

$$S = TP + S(I - P)$$

and $S(I - P)$ is a bounded finite dimensional operator of H_S into H. Therefore, $S(I - P)$ is compact as a map of H_S into H. Some well-known results on Fredholm operators can now be used. We have from [21, Theorems 3.1 and 3.2 of Ch. 1] that S is a Fredholm operator whenever T is Fredholm and furthermore

$$\text{index } S = \text{index } TP = \text{index } T + \text{index } P = \text{index } T + m.$$

The theorem is therefore proved.

Corollary 1.8. Let S be a closed, finite dimensional extension of the closed densely defined operator T in H. Then

$$\sigma_{e2}^{\pm}(T) = \sigma_{e2}^{\pm}(S) \quad \text{and} \quad \sigma_{ek}(T) = \sigma_{ek}(S), \quad k = 1,3. \tag{1.15}$$

Moreover,

$$\Phi_k(T) \cap \Phi_k(S) = \emptyset, \quad k = 4,5. \tag{1.16}$$

Proof. If $\lambda \in \Phi^+(T)$, $R(T - \lambda I)$ is closed and $\text{nul}(T - \lambda I) < \infty$. Since $R(S - \lambda I) = R(T - \lambda I) + (S - \lambda I)G$, for an m-dimensional subspace G, $R(S - \lambda I)$ is therefore closed (see [9 , Lemma III: 1.9]) and from Theorem 1.7(i) $\text{nul}(S - \lambda I) < \infty$. This implies that $\Phi^+(T) \subseteq \Phi^+(S)$ or $\sigma_{e2}^+(S) \subseteq \sigma_{e2}^+(T)$. Conversely, if $\lambda \in \sigma_{e2}^+(T)$, there exists a singular sequence of T corresponding to λ. As this is also a singular sequence of S corresponding to λ we have $\sigma_{e2}^+(T) \subseteq \sigma_{e2}^+(S)$ and hence $\sigma_{e2}^+(T) = \sigma_{e2}^+(S)$.

If $\lambda \in \Phi^-(T)$, $R(S - \lambda I)$ is closed as before and $\text{def}(S - \lambda I) < \infty$ from Theorem 1.7(ii); thus $\Phi^-(T) \subseteq \Phi^-(S)$. If $\lambda \in \Phi^-(S)$, $\text{def}(T - \lambda I) < \infty$ from Theorem 1.7(ii) and hence, on using [9 , Problem IV: 5.7 and Theorem IV: 5.2],

$\mathcal{R}(T - \lambda I)$ is closed. It follows that $\lambda \in \Phi^{-}(T)$ and consequently $\Phi^{-}(T) = \Phi^{-}(S)$; (1.15) is therefore proved.

The result (1.16) is an immediate consequence of Theorem 1.7(iii).

Note that an implication of (1.16) is that (1.15) does not hold for $k = 4,5$. If T is a closed symmetric operator with equal deficiency indices (n,n), $0 < n < \infty$, the upper and lower half-planes $\mathbb{C}_{\pm}$ lie in $\sigma_r(T) \cap \Phi_3(T)$ and hence in $\sigma_{e4}(T)$ and $\sigma_{e5}(T)$ but $\mathbb{C}_{\pm}$ lie in the resolvent set of every self-adjoint extension S of T. Moreover every such S is a finite dimensional extension of T.

Another useful result in applications is the following theorem from [21, ch. 11]. Recall that a linear operator P is said to be T-compact if $\mathcal{D}(P) \supseteq \mathcal{D}(T)$ and is compact as an operator from $\mathcal{D}(T)$, endowed with the graph norm of T, into $\mathcal{H}$ i.e., if $\{x_n\} \subset \mathcal{D}(T)$ and $\|x_n\| + \|Tx_n\| \leq K$ then $\{Px_n\}$ contains a subsequence which converges in $\mathcal{H}$. If P is T-compact and either T or P is closable then P has T-bound zero. [22, Theorem 9.7].

Theorem 1.9. If T is a closed densely defined operator in $\mathcal{H}$ and P is T-compact, then $S = T + P$ is closed. Moreover,

$$\sigma_{ek}(S) = \sigma_{ek}(T), \quad k = 1,3,4,$$

and

$$\sigma_{e2}^{\pm}(S) = \sigma_{e2}^{\pm}(T).$$

The latter result is not true for σ_{e5} (see [20, XIII.4 Ex. 2]).

If $\lambda \in \rho(T)$, $T - \lambda I$ is continuous and continuously invertible as a map from $\mathcal{D}(T)$ (with graph norm) into $\mathcal{H}$. It therefore follows that P is T-compact if and only if $P(T - \lambda)^{-1}$ is compact in $\mathcal{H}$ for some (and in fact, as a consequence, for all) $\lambda \in \rho(T)$. If $S = T + P$ and $\lambda \in \rho(T) \cap \rho(S)$,

$$S - \lambda I = T - \lambda I + P = \{I + P(T - \lambda I)^{-1}\}(T - \lambda I)$$

and hence

$$(T - \lambda I)^{-1} - (S - \lambda I)^{-1} = (S - \lambda I)^{-1} P(T - \lambda I)^{-1}.$$

It therefore follows that if P is T-compact, $(T - \lambda I)^{-1} - (S - \lambda I)^{-1}$ is compact. The latter property is also sufficient for two operators to have the same essential spectra σ_{ek}, $k \neq 5$. In fact, we have

Corollary 1.10. Let T, S be closed linear operators in H and suppose there exists an $\xi \in \rho(T) \cap \rho(S)$ such that $(T - \xi I)^{-1} - (S - \xi I)^{-1}$ is compact. Then $(T - \lambda I)^{-1} - (S - \lambda I)^{-1}$ is compact for all $\lambda \in \rho(T) \cap \rho(S)$ and

$$\sigma_{ek}(S) = \sigma_{ek}(T), \quad k = 1,3,4, \quad \sigma_{e2}^{\pm}(S) = \sigma_{e2}^{\pm}(T).$$

Proof. The first part is given in [20, XIII.4 Lemma 4]. The result $\sigma_{ek}(S) = \sigma_{ek}(T)$ is proved in [21, Theorem 1: 4.7] when $k = 4$ and the proof also applies verbatim to the case $k = 3$. The remaining results will follow if we prove that $\sigma_{e2}^{+}(S) = \sigma_{e2}^{+}(T)$.

From Theorem 1.9, $(T - \xi I)^{-1}$ and $(S - \xi I)^{-1}$ have the same essential spectrum σ_{e2}^{+}. It is therefore a question of proving that (without loss of generality we take $\xi = 0$) if $0 \in \rho(T)$ then $\lambda \in \sigma_{e2}^{+}(T)$ if and only if $1/\lambda \in \sigma_{e2}^{+}(T^{-1})$. To show this we use singular sequences and the identity

$$T - \lambda I = -\lambda(T^{-1} - \lambda^{-1}I)T. \tag{1.17}$$

Let $\{\phi_n\}$ be a singular sequence of T corresponding to λ and put $\psi_n = T\phi_n/\|T\phi_n\|$; notice that $\|T\phi_n\| \geq \delta > 0$ since $0 \in \rho(T)$. Then $\|\psi_n\| = 1$ and $\psi_n \rightharpoonup 0$ since $T\phi_n = (T - \lambda)\phi_n + \lambda\phi_n \rightharpoonup 0$. Moreover $(T^{-1} - \lambda^{-1}I)\psi_n \to 0$ from (1.17) so that $\{\psi_n\}$ is a singular sequence of T^{-1} corresponding to λ^{-1}.

Conversely, let $\{\psi_n\}$ be a singular sequence of T^{-1} corresponding to λ^{-1}. Then $T^{-1}\psi_n \not\to 0$ for otherwise we have the contradiction $\psi_n \to 0$. Let

$\{\psi_{nk}\}$ be a subsequence such that $\|T^{-1}\psi_{nk}\| \geq \delta > 0$ and put $\phi_{nk} = T^{-1}\psi_{nk}/\|T^{-1}\psi_{nk}\|$. It is easy to check that $\{\phi_{nk}\}$ is a singular sequence for T corresponding to λ. The corollary is therefore proved.

If T is self-adjoint, then perturbations P, other than ones which are T-compact, preserve the essential spectrum. (Recall that all of the essential spectra, $\sigma_{e2}^{\pm}$ and σ_{ek}, now coincide). An operator P is said to be T^2-compact if $\mathcal{D}(P) \supseteq \mathcal{D}(T)$ and $P(T^2 + I)^{-1}$ is compact. If P is symmetric and T^2-compact and if $S = T + P$ is self-adjoint, then S and T have the same essential spectra. For this result and a comprehensive treatment of the invariance of the essential spectrum of a self-adjoint operator see [7] and [20, ch. XIII.4].

Corollary 1.10 can be applied to (1.14) relating to m-sectorial operators S, T associated with closed densely defined sectorial forms $\mathfrak{s}$, $\mathfrak{t}$ in H. From (1.14) it follows that if $\hat{P}(T - \lambda I)^{-1}$ is compact as a map from H into $Q^{\#}$ for λ large and negative then $\sigma_{ek}(S) = \sigma_{ek}(T)$, $k = 1,3,4$ and $\sigma_{e2}^{\pm}(S) = \sigma_{e2}^{\pm}(T)$. This result is particularly useful when studying the essential spectra of differential operators as it is often expedient to define the operators in terms of sectorial forms (see [2]). A noteworthy consequence of (1.14) is the following result relating to the case when $T = \mathrm{Re}\, S$. The operator $\hat{P}$ is now determined by $\mathrm{Im}\, \mathfrak{s}$, where $\mathfrak{s}$ is the sectorial form which defines S. Let $(T - \lambda I)^{-1}$ be compact in H for some $\lambda \in \rho(T)$. Then, if $\{u_n\}$ is a bounded sequence in H, and $v_n = (T - \lambda I)u_n^{-1}$, $\{v_n\}$ is precompact in H. Moreover, we have (c.f (8.5) in [2])

$$\begin{aligned} \|\hat{P}v_n\|^2 &\leq K\|v_n\|_Q^2 \\ &\leq K(\mathfrak{t} + \alpha)[v_n]\,, \qquad \mathfrak{t} = \mathrm{Re}\, \mathfrak{s} \\ &= K([T+\alpha]v_n, v_n)_H \\ &\leq K\{|(u_n, v_n)_H| + \|v_n\|_H^2\} \\ &\leq K(\|v_n\|_H + \|v_n\|_H^2). \end{aligned}$$

It follows that $\{\hat{P}v_n\}$ is precompact in $\mathcal{Q}^{\#}$ and that $\hat{P}(T - \lambda I)^{-1}$ is therefore compact as a map from H into $\mathcal{Q}^{\#}$. Hence, from (1.14), $(S - \lambda I)^{-1}$ is also compact in H. Interchanging T and S, with $v_n = (S - \lambda I)^{-1}u_n$ now and $\{u_n\}$ bounded in H, we have

$$\begin{aligned} \|\hat{P}v_n\|^2 &\leq K(t + \alpha)[v_n] \\ &\leq K|(s + \alpha)[v_n]| \\ &= K|((S + \alpha)v_n, v_n)| \\ &\leq K(\|v_n\|_H + \|v_n\|_H^2) \end{aligned}$$

so that $\hat{P}(S - \lambda I)^{-1}: H \to \mathcal{Q}^{\#}$ is compact if $(S - \lambda I)^{-1}$ is compact in H. Consequently, $(T - \lambda I)^{-1}$ is compact from (1.14) with T and S interchanged and P replaced by -P. We have therefore proved that $(S - \lambda I)^{-1}$ is compact if and only if $(T - \lambda I)^{-1}$ is compact, a result proved by an alternative method in [9, Theorem VI: 3.3].

An attempt (unsuccessful) to generalise the preceding result led to the following considerations. First, define in analogy with (1.4) and (1.5)

$$\ell_{ek}(T) = \inf\{\mathrm{Re}\,\lambda : \lambda \in \sigma_{ek}(T)\}, \quad k = 1,3,4,5$$

and

$$\ell_{e2}^{\pm}(T) = \inf\{\mathrm{Re}\,\lambda : \lambda \in \sigma_{e2}^{\pm}(T)\}.$$

If T is self-adjoint the $\ell_{ek}(T)$ coincide and are denoted by $\ell_e(T)$.

Theorem 1.11. (i) Let S be a closed, densely defined linear operator in H and let t be a closed lower semibounded symmetric form. If $\mathcal{D}(S) \subseteq \mathcal{D}(t)$ and $\mathrm{Re}(Su,u) \geq t[u]$, $u \in \mathcal{D}(S)$, then $\ell_e(T) \leq \ell_{e2}^{+}(S)$, where T is the self-adjoint operator associated with t.
(ii) Let S be a closed densely defined linear operator in H and let t be a closed lower semi-bounded symmetric form. If $\mathcal{D}(S^*) \subseteq \mathcal{D}(t)$ and $\mathrm{Re}(S^*u,u) \geq t[u]$, $u \in \mathcal{D}(S^*)$, then $\ell_e(T) \leq \ell_{e2}^{-}(S)$.

(iii) If the hypotheses of (i) and (ii) hold, then $\ell_e(T) \leq \ell_{ek}(S)$, $k = 1,3,4$.

(iv) Let $\mathfrak{s}$ be a closed, densely defined sectorial form and

$$\operatorname{Re} \mathfrak{s}[u] \geq \mathfrak{t}[u], \quad u \in \mathcal{D}(\mathfrak{s}) \subseteq \mathcal{D}(\mathfrak{t})$$

where $\mathfrak{t}$ is a closed lower semi-bounded symmetric form. Then $\ell_e(T) \leq \ell^{\pm}_{e2}(S)$ and $\ell_e(T) \leq \ell_{ek}(S)$, $k = 1,3,4$, where S, T are the operators associated with $\mathfrak{s}$, $\mathfrak{t}$ respectively.

Proof. (i) This is proved in [13]. The following proof is also instructive. Let $\operatorname{Re} \lambda < \ell_e(T)$. Then there exists a compact symmetric operator P such that for some $\varepsilon > 0$, $(-\infty, \operatorname{Re} \lambda + \varepsilon) \cap \sigma(T + P) = \emptyset$ and as $T + P$ is self-adjoint

$$([T + P - \operatorname{Re} \lambda]u,u) \geq \varepsilon \|u\|^2, \quad u \in \mathcal{D}(T).$$

Since $\mathcal{D}(T)$ is a core of $\mathfrak{t}$, it follows that

$$\mathfrak{t}[u] + (Pu,u) - \operatorname{Re} \lambda \|u\|^2 \geq \varepsilon \|u\|^2, \quad u \in \mathcal{D}(\mathfrak{t})$$

and hence

$$\|(S + P - \lambda)u\| \|u\| \geq \operatorname{Re}([S + P - \lambda]u,u) \geq \varepsilon \|u\|^2, \quad u \in \mathcal{D}(S).$$

Hence, since $S + P$ is closed (as P is bounded), $(S + P - \lambda)^{-1}$ is both bounded and closed on $\mathcal{R}(S + P - \lambda)$ and consequently $\mathcal{R}(S + P - \lambda)$ is closed. As $\operatorname{nul}(S + P - \lambda) = 0$ we have $\lambda \in \Phi^+(S + P) = \Phi^+(S)$ from Theorem 1.9. This proves (i), and (ii) follows since $\lambda \in \sigma^+_{e2}(S^*)$ if and only if $\overline{\lambda} \in \sigma^-_{e2}(S)$.

(iii) This follows from (i) and (ii).

(iv) We have $\operatorname{Re}(Su,u) \geq \mathfrak{t}[u]$, $u \in \mathcal{D}(S) \subseteq \mathcal{D}(\mathfrak{s}) \subseteq \mathcal{D}(\mathfrak{t})$ and also $\operatorname{Re}(S^*u,u) \geq \mathfrak{t}[u]$, $u \in \mathcal{D}(S^*) \subseteq \mathcal{D}(\mathfrak{s})$, since S^* is the m-sectorial operator associated with $\mathfrak{s}^*$. The result therefore follows from (iii).

If S is an m-sectorial operator and $T = \operatorname{Re} S$ then from Theorem 1.11(iv),

$\ell_e(T) \le \ell_e(S) \equiv \min_{k=1,3,4}\left(\ell^{\pm}_{e2}(S), \ell_{ek}(S)\right)$. In view of the result that $(T - \lambda I)^{-1}$ and $(S - \lambda I)^{-1}$ are compact together we expected that $\ell_e(T) = \ell_e(S)$ but the truth or falsity of this result eludes us.

For a bounded operator A in $\mathcal{H}$, each of the essential spectra $\sigma^{\pm}_{e2}(A)$ and $\sigma_{ek}(A)$, $k = 1,3,4,5$ have the same radius [12,17]. Hence if $0 \in \rho(S)$ the numbers

$$r^{\pm}_e(S) = \min\{|\lambda| : \lambda \in \sigma^{\pm}_{e2}(S)\}$$

$$r_{ek}(S) = \min\{|\lambda| : \lambda \in \sigma_{ek}(S)\} \quad k = 1,3,4,5$$

coincide.

In applications it is frequently easier to establish the inequality required in the hypothesis of Theorem 1.11(iv) for a closable form $\mathfrak{s}$ with closure $\tilde{\mathfrak{s}}$. The next corollary shows that this is sufficient. It is possible for a form to be densely defined and sectorial, but not closable. However, a form $\mathfrak{s}$ which arises from a densely defined, sectorial operator S, i.e.

$$\mathfrak{s}[u.v] \equiv (Su,v) \quad \text{with} \quad \mathcal{D}(\mathfrak{s}) \equiv \mathcal{D}(S),$$

is closable [9, p. 318].

Corollary 1.12. Let $\mathfrak{s}$ be a closable, densely defined sectorial form for which

$$\operatorname{Re} \mathfrak{s}[u] \ge \mathfrak{t}[u], \quad u \in \mathcal{D}(\mathfrak{s}) \subseteq \mathcal{D}(\mathfrak{t})$$

where $\mathfrak{t}$ is a closed lower semi-bounded symmetric form. Then $\ell_e(T) \le \ell^{\pm}_{e2}(S)$ and $\ell_e(T) \le \ell_{ek}(S)$, $k = 1,3,4$, where S, T are the operators associated with $\tilde{\mathfrak{s}}$, $\mathfrak{t}$ respectively.

Proof. The proof follows from Theorem 1.11 above and Theorems VI: 1.12, VI: 1.17 of [9, pp. 314, 315].

For any nonzero complex number α, αT is defined by

$$(\alpha T)x = \alpha\cdot(Tx), \quad \mathcal{D}(\alpha T) = \mathcal{D}(T)$$

where T is a densely defined linear operator in a Hilbert space $\mathcal{H}$. Similarly, given a sesquilinear form s the form αs is defined by

$$(\alpha s)[u,v] = \alpha\cdot s[u,v] \quad \text{with} \quad \mathcal{D}(\alpha s) = \mathcal{D}(s).$$

Lemma 1.13. Let T be a closed, densely defined linear operator in the Hilbert space $\mathcal{H}$. If α is any nonzero, complex number then

$$\lambda \in \sigma_{e2}^{\pm}(T) \quad \text{if and only if} \quad \alpha\lambda \in \sigma_{e2}^{\pm}(\alpha T)$$

and

$$\lambda \in \sigma_{ek}(T) \quad \text{if and only if} \quad \alpha\lambda \in \sigma_{ek}(\alpha T), \quad k = 1,3,4,5.$$

Proof. First, note that $R(T - \lambda)$ is closed if and only if $R(\alpha T - \alpha\lambda)$ is closed, and $N(T - \lambda) = N(\alpha T - \alpha\lambda)$. Also, $y \in R(T - \lambda)^{\perp}$ if and only if $\alpha y \in R(\alpha T - \alpha\lambda)^{\perp}$. Hence, $\operatorname{def}(T - \lambda) = \operatorname{def}(\alpha T - \alpha\lambda)$. Finally, note that $\lambda \in \rho(T)$ if and only if $\alpha\lambda \in \rho(\alpha T)$. The conclusion now follows.

Corollary 1.14. Let s be a closable, densely defined form such that αs is sectorial for some nonzero complex number $\alpha = a - ib$. Let t be a closed lower semi-bounded symmetric form with $\mathcal{D}(t) \supseteq \mathcal{D}(s)$. Let αS and T be the operators associated with $\widetilde{\alpha s}$ and t. If

$$a \operatorname{Re} s[u] + b \operatorname{Im} s[u] = \operatorname{Re} \alpha s[u] \geq t[u], \quad u \in \mathcal{D}(s)$$

then $\lambda \in \sigma_{e2}^{\pm}(S)$ or $\lambda \in \sigma_{ek}(S)$, $k = 1,3,4$, implies that

$$a \operatorname{Re} \lambda + b \operatorname{Im} \lambda \geq \ell_e(T).$$

Proof. The proof follows from Corollary 1.12 and Lemma 1.13.

If S' is a closable, densely defined linear operator such that $e^{i\gamma} S' + z$ is sectorial for z complex and γ real, then for

$$s[u,v] \equiv (S'u,v) + e^{-i\gamma} z(u,v), \quad \mathcal{D}(s) = \mathcal{D}(S')$$

$e^{i\gamma}s$ is sectorial. The operator S associated with $e^{i\gamma}\tilde{s}$ is an extension of $S' + e^{-i\gamma}z$, c. f. [19, p. 282].

§2. Ordinary Differential Operators

2.1. A decomposition principle

Let T be a differential expression of the form

$$T = \sum_{j=0}^{n} a_j D^j \tag{2.1}$$

where each a_j is a complex valued function defined on some interval I and $D = d/dt$. Here we study operators generated by T in the space $L^2(I)$ for I an interval of the real line $\mathbb{R}$. Of particular interest are the two special cases $I = [0,\infty)$ and $I = \mathbb{R}$. In each case there are many such operators distinguished by their domains. Two basic ones are the so called minimal and maximal operators which we now define.

For each positive integer n let $AC_n(I)$ denote the set of complex-valued functions y on I such that $y^{(n-1)}$ is absolutely continuous on all compact subintervals of I.

Definition. The maximal operator $A_1 = A_1(T,I)$ is defined as follows:

$$\mathcal{D}(A_1) = \{y \in L^2(I) \cap AC_n(I) : Ty \in L^2(I)\}$$

$$A_1 y = Ty, \quad y \in \mathcal{D}(A_1).$$

Let A_c denote the restriction of A_1 to those $y \in \mathcal{D}(A_1)$ which have compact support in the interior of I. The minimal operator $A_o = A_o(T,I)$ is defined to be the closure $\tilde{A}_c$ of A_c. The fact that A_c is closable follows from the next lemma. We write $A_1(T)$ for $A_1(T,I)$ and $A_o(T)$ for $A_o(T,I)$ when I has been specified. When both I and T are understood we simply write A_1 and A_o.

Lemma 2.1. Let T be given by (2.1) and assume that

$$a_j \in C^j(I), \quad 0 \leq j \leq n \tag{2.2}$$

$$a_n(t) \neq 0, \quad t \in I. \tag{2.3}$$

Then

(1) A_o has dense domain and

$$A_o^*(T,I) = A_1(T^+,I) \tag{2.4}$$

$$A_1^*(T,I) = A_o(T^+,I) \tag{2.5}$$

where

$$T^+y = \sum_{j=0}^{n} (-1)^j(\bar{a}_j\ y)^{(j)} = \sum_{j=0}^{n} b_j y^{(j)} \tag{2.6}$$

with

$$b_j = \sum_{k=j}^{n} (-1)^k \binom{k}{j} \bar{a}_k^{(k-j)}, \quad 0 \leq j \leq n. \tag{2.7}$$

(2) The maximal operator $A_1(T,I)$ is a finite dimensional extension of the minimal operator $A_o(T,I)$ and $\dim A_1(T,I)/\dim A_o(T,I) \leq n$. For each complex number λ we have $\mathrm{nul}(A_o - \lambda) \leq n$ and $\mathrm{def}(A_o - \lambda) \leq n$.

(3) Suppose $I = [b,\infty)$. For each complex number λ we have $\mathrm{nul}(A_o - \lambda) = 0$.

(4) Let $I = [b,\infty)$. If any one of the four operators $A_o(T)$, $A_1(T)$, $A_o(T^+)$, $A_1(T^+)$ has closed range, then all four have closed range. In this case $A_1(T)$ and $A_1(T^+)$ are onto $L^2(0,\infty)$. In addition, $A_o(T)$ and $A_o(T^+)$ will have bounded inverses.

Proof. For a proof of the lemma see [6 , chapter VI].

Note that A_o and A_1 are closed since the adjoint of an operator is always closed.

The strong smoothness assumptions (2.2) can be weakened to merely local integrability provided the formal adjoint T^+ is defined appropriately in terms of quasi-derivatives. For details the interested reader is referred to Everitt and Zettl [4].

Lemma 2.2. Let T be given by (2.1) with coefficients a_j satisfying (2.2) and (2.3). Let $0 < b < \infty$. Then for any $\lambda \in \mathbb{C}$ we have

(i) nul $A_1(T-\lambda,[0,\infty)) = \text{nul } A_1(T-\lambda,[b,\infty))$.

(ii) The range of $A_1(T-\lambda,[0,\infty))$ is closed if and only if the range of $A_1(T-\lambda,[b,\infty))$ is closed.

Proof. The nullity of $A_1(T-\lambda,I)$ is the (maximum) number of linearly independent solutions of the equation

$$Ty = \lambda y \tag{2.8}$$

which lie in $L^2(I)$. If $y_1,\cdots,y_k$ are linearly independent solutions of (2.8) on $[0,\infty)$ which lie in $L^2[0,\infty)$ then their restrictions to $[b,\infty)$ are linearly independent $L^2[b,\infty)$-solutions of (2.8) on $[b,\infty)$. Hence nul $A_1(T-\lambda,[0,\infty)) \leq \text{nul } A_1(T-\lambda,[b,\infty))$. However, if $y_1,\cdots,y_k$ are linearly independent solutions of (2.8) on $[b,\infty)$ in $L^2[b,\infty)$ then $y_j^{(i)}(b)$, $i = 0,\cdots,n-1$, $j = 1,\cdots,k$ exist since b is a regular point of T — see Naimark [16, p. 63] — and each y_j can be extended to a solution of (2.8) on $[0,\infty)$. Clearly these extensions are linearly independent solutions of (2.8) in $L^2[0,\infty)$. It follows that nul $A_1(T-\lambda,[b,\infty)) \leq \text{nul } A_1(T-\lambda,[0,\infty))$ and the proof of (i) is complete.

To prove (ii) let $T = A_1(T-\lambda,[0,\infty))$ and $T_b = A_1(T-\lambda,[b,\infty))$. First suppose that $R(T)$ is closed. Let $f_k \in R(T_b)$ and assume that $f_k \to f$ in $L^2[b,\infty)$. Then $f_k = T_b g_k = (T-\lambda)g_k$ on $[b,\infty)$, $g_k \in D(T_b)$. Let y_k denote the solution of $(T-\lambda)y = 0$ on $[0,b]$ satisfying the initial condition $y_k^{(i)}(b) = g_k^{(i)}(b)$, $i = 0,\cdots,n-1$. Note that $g_k^{(i)}(b)$ exists since b is a regular point of T. Set $g_n^* = g_n$ on $[b,\infty)$ and $g_n^* = y_n$ on $[0,b]$, $f_n^* = f_n$ on (b,∞) and $f_n^* = 0$ on $[0,b]$. Then $g_n^* \in D(T)$ and $f_n^* = Tg_n^*$. Also $f_n^* \to f^*$ in $L^2[0,\infty)$ where $f^* = f$ on $[b,\infty)$ and $f^* = 0$ on $[0,b)$. Thus $f^* \in R(T)$ since $R(T)$ is closed. Let $f^* = Tg^* = (T-\lambda)g^*$ on $[0,\infty)$ with $g^* \in D(T)$. Then $f = f^* = (T-\lambda)g^*$ on $[b,\infty)$ with $g^* \in D(T_b)$.

Hence $f \in R(T_b)$ and we have shown that the range of T_b is closed.

To establish the converse implication, suppose that $R(T_b)$ is closed and $\{f_k\} \subseteq R(T)$ with $f_k \to f$ in $L^2[0,\infty)$. Let $f_k = Tg_k = (T - \lambda)g_k$ on $[0,\infty)$, $g_k \in D(T)$. Then $f_k = (T - \lambda)g_k$ on (b,∞) and $g_k\upharpoonright(b,\infty) \in D(T_b)$ with $f_k\upharpoonright(b,\infty) = T_b\big(g_k\upharpoonright(b,\infty)\big)$. Now note that $f_k \to f$ in $L^2[0,\infty)$ implies that $f_k \to f$ in $L^2(b,\infty)$. Since $R(T_b)$ is closed $f\upharpoonright(b,\infty) = T_b g = (T - \lambda)g$ on (b,∞) for some $g \in D(T_b)$. Again $g^{(i)}(b)$, $i = 0,\cdots,n-1$, exist since b is a regular point of T (and of $T - \lambda$). Let y be the solution of $Ty - \lambda y = f$ on $[0,b]$ determined by the initial conditions $y^{(i)}(b) = g^{(i)}(b)$, $i = 0,\cdots,n-1$. Such a solution exists since $T - \lambda$ is regular on $[0,b]$. Set $h = g$ on (b,∞) and $h = y$ on $[0,b]$. Then $f = (T - \lambda)h$ on $[0,\infty)$ and $h \in D(T)$. Hence $f \in R(T)$ and $R(T)$ is closed. This completes the proof of part (ii).

Lemma 2.3. Let I be an unbounded interval of the real line. Let T be given by (2.1) with coefficients satisfying (2.2) and (2.3). Suppose T is any closed operator satisfying $A_o(T,I) \subseteq T \subseteq A_1(T,I)$. Then $\sigma_{e1}(T) = \sigma^+_{e2}(T) = \sigma^-_{e2}(T) = \sigma_{e3}(T)$.

Proof. By Lemma 2.1 both the nullity and the deficiency of the minimal and maximal operators are finite. Hence both nul T and def T are finite. The remainder of the proof follows from the definitions of the various parts of the spectrum.

Lemma 2.4. Let T be given by (2.1) with coefficients satisfying (2.2), (2.3) with $I = [0,\infty)$ and let $0 < b < \infty$. Let T and T_b be any closed operators such that

$$A_o(T,[0,\infty)) \subseteq T \subseteq A_1(T,[0,\infty)),$$
$$A_o(T,[b,\infty)) \subseteq T_b \subseteq A_1(T,[b,\infty)).$$

Then

$$\sigma_{e3}(T) = \sigma_{e3}(T_b).$$

Proof. The case when T and T_b are the maximal operators of T on the intervals $[0,\infty)$ and $[b,\infty)$, respectively, follows from Lemma 2.3, part (ii) of Lemma 2.2 and the definition of σ_{e3}. The case in which T and T_b are the minimal operators of T on $[0,\infty)$ and $[b,\infty)$, respectively, follows from parts (1) and (4) of Lemma 2.1 and the case just considered in which T and T_b were the maximal operators of T on $[0,\infty)$ and $[b,\infty)$. Once the cases of the minimal and maximal operators have been established, then the remaining cases follow from parts (1) and (2) of Lemma 2.1 and Theorem 1.7.

Lemma 2.5. Let T, I and b be as in Lemma 2.4. Then

$$\sigma_{e4}(A_o(T,[0,\infty))) = \sigma_{e4}(A_o(T,[b,\infty))) \tag{2.9}$$

and

$$\sigma_{e4}(A_1(T,[0,\infty))) = \sigma_{e4}(A_1(T,[b,\infty))) \tag{2.10}$$

Proof. Let T, T_b denote the maximal operators on $[0,\infty)$, $[b,\infty)$, respectively. Let $R(T-\lambda)$, and hence $R(T_b-\lambda)$, be closed. From (i) of Lemma 2.2 we have

$$\text{nul}(T-\lambda) = \text{nul}(T_b-\lambda).$$

Also, $\text{def}(T-\lambda) = 0 = \text{def}(T_b-\lambda)$ by (3) of Lemma 2.1. Thus (2.10) follows. To prove (2.9) let T_o and T_{ob} denote the minimal operators of T on the intervals $[0,\infty)$ and $[b,\infty)$, respectively. Then $\text{nul}(T_o-\lambda) = 0 = \text{nul}(T_{ob}-\lambda)$ by Lemma 2.1 part (3). Using part (1) of Lemma 2.1 we have, if $\lambda \in \Phi_3(T_o) = \Phi_3(T_{ob})$, $\text{def}(T_o-\lambda) = \text{nul}(T_o^*-\bar{\lambda}) = \text{nul}\ A_1(T^+-\bar{\lambda},[0,\infty)) = \text{nul}\ A_1(T^+-\bar{\lambda},[b,\infty)) = \text{def}(T_{ob}-\lambda)$ and (2.9) follows. Here we also used part (i) of Lemma 2.2 (with $T-\lambda$ replaced by $T^+-\bar{\lambda}$).

2.2. Constant coefficient operators.

Throughout this section we assume that a_j is a real or complex constant and $a_N \neq 0$. Let

$$p(z) = \sum_{j=0}^{N} a_j z^j,$$

$$P = \{p(z) : \text{Re } z \geq 0\} \quad \text{and}$$

$$P_o = \{p(z) : \text{Re } z = 0\}.$$

Lemma 2.6. Let $I = \mathbb{R}$ or $\mathbb{R}^+$, $a_j = r_j \exp(i\theta_j)$, $0 \leq \theta_j < 2\pi$, $r_j \geq 0$. If $1 < N$ and N is odd assume that

$$\theta_{N-1} \neq -\theta_N + k\pi/2 \quad \text{for } k = \pm 1, \pm 2, \cdots, \quad \text{and } r_{N-1} > 0. \tag{2.11}$$

Let $T = A_o \exp(i\theta) + c$. Then, the following hold:

(i) T is accretive for some $\theta \in [0,2\pi]$ and $c > 0$. Furthermore, T is m-accretive if $I = \mathbb{R}$.

(ii) T is dissipative for some $\theta \in [0,2\pi]$ and $c < 0$. Furthermore, T is m-dissipative in case $I = \mathbb{R}$.

Proof. Let Fy denote the Fourier transform of y. If $y \in D(A_o)$ and $I = \mathbb{R}^+ = [0,\infty)$ extend y to $\mathbb{R}$ by defining it to be zero outside of $[0,\infty)$. Then

$$Fy(t) = (2\pi)^{-1/2} \int_{-\infty}^{\infty} y(x) \exp(ixt)dx$$

and for $y \in C_o^\infty$

$$Fy^{(j)}(t) = (it)^j \, Fy(t), \quad j = 0,1,\cdots.$$

For N even choose $\gamma = -\theta_N$ if $N = 4k$ and $-\theta_N + \pi$ if $N = 2k$. Let $\theta = \gamma$ if $0 \leq \gamma < 2\pi$ and let $\theta = \gamma + 2\pi$ if $-2\pi < \gamma < 0$. Then $\text{Re}\big(a_N \exp(i\theta)\big) = r_N > 0$ and $\text{Re}\big(\exp(i\theta)p(it)\big) = r_N t^N + \sum_{j=0}^{N-1} b_j t^j$ where b_j is real, $0 \leq j \leq N-1$. Hence there is a $c > 0$ such that

$$\text{Re}\big(\exp(i\theta)p(it)\big) + c > 0, \quad -\infty < t < \infty. \tag{2.12}$$

For N odd let $\gamma = -\theta_N$. Then

$\exp(i\gamma)p(it) = r_N(it)^N + r_{N-1}\exp\big(i(\theta_{N-1} - \theta_N)\big)(it)^{N-1} + \cdots$. Hence

$\mathrm{Re}\big[\exp(i\gamma)p(it)\big] = r_{N-1}\, i^{N-1}\cos(\theta_{N-1} - \theta_N)t^{N-1} + \sum_{j=0}^{N-2} b_j t^j$ with b_j real, $0 \leq j \leq N-2$. The coefficient b_{N-1}, of t^{N-1} is real and not zero by (2.11). If $b_{N-1} > 0$ let $\theta = \gamma + 2\pi$ and (2.12) holds. If $b_{N-1} < 0$ replace γ by $\gamma + \pi$ and proceed as above. Finally, let $\theta = \gamma + \pi$ if $\gamma + \pi \geq 0$ and let $\theta = \gamma + \pi + 2\pi$ if $\gamma + \pi < 0$, then (2.12) again holds.

Now for $y \in C_o^\infty$ with $(y,y) = 1$ we have

$$\begin{aligned}(e^{i\theta} A_o y, y) &= e^{i\theta}\sum_{j=0}^{N} a_j\big(y^{(j)}, y\big) = e^{i\theta}\sum_{j=0}^{N} a_j\big(Fy^{(j)}, Fy\big) \\ &= e^{i\theta}\sum_{j=0}^{N} a_j\big((it)^j\, Fy, Fy\big).\end{aligned}$$

Thus,

$$(Ty,y) = \big(\big(e^{i\theta}\, p(it) + c\big)Fy, Fy\big).$$

Hence by (2.12)

$$\mathrm{Re}(Ty,y) = \mathrm{Re}\int_{-\infty}^{\infty} [e^{i\theta}\, p(it) + c]\,|Fy(t)|^2\, dt \geq 0.$$

This completes the proof of the first part of (i). The first part of (ii) is established similarly. The second part follows from Theorem 2.7 below.

Theorem 2.7. Let $I = \mathbb{R}^+ = [0,\infty)$. Assume the hypothesis and notation of Lemma 2.6. Then

$$\sigma(A_o) = P, \tag{2.13}$$

$$\sigma_{e1}(A_o) = \sigma_{e2}^{\pm}(A_o) = \sigma_{e3}(A_o) = P_o, \tag{2.14}$$

$$\sigma_{e4}(A_o) = \sigma_{e5}(A_o) = P, \tag{2.15}$$

and

$$\sigma_r(A_o) \cap \Phi_3(A_o) = P \setminus P_o. \tag{2.16}$$

Proof. By Lemma 2.1 we have $\mathrm{nul}\, A_i = 0$ and $\mathrm{def}(A_i) \leq N$, $i = 0,1$. Hence $\Phi_1(A_i) = \Phi_3(A_i)$ and consequently $\sigma_{e1}(A_i) = \sigma_{e2}^{+}(A_i) = \sigma_{e2}^{-}(A_i) = \sigma_{e3}(A_i)$, $i = 0,1$.

Since $\text{nul}(A_o - \lambda) = 0$ when $I = \mathbb{R}^+$, then $\text{nul}(A_o - \lambda) = \text{def}(A_o - \lambda I)$ and $R(A_o - \lambda)$ being closed implies, by the closed graph theorem, that $\lambda \in \rho(A_o)$ and the resolvent set is open. Hence $\sigma_{e4}(A_o) = \sigma_{e5}(A_o)$.

Thus it only remains to show that $\sigma(A_o) = P$, $\sigma_{e3}(A_o) = P_o$, $\sigma_{e4}(A_o) = P$ and (2.16).

Let $z_j = z_j(\overline{\lambda})$ denote the distinct roots of $p^+(z) = (-1)^N \overline{a}_N z^N + (-1)^{N-1} \overline{a}_{N-1} z^{N-1} + \cdots + (-1)\overline{a}_1 z + \overline{a}_o = \overline{\lambda}$ and let m_j denote the multiplicity of z_j, $j = 1,\cdots,s$. Then the N functions

$$t^r \exp(z_j t), \quad r = 0,1,\cdots,m_j-1; \quad j = 1,\cdots,s \tag{2.17}$$

form a fundamental set of solutions of

$$T^+y = \overline{\lambda}y. \tag{2.18}$$

It is clear from (2.17) that if $\text{Re } z_j \geq 0$, $j = 1,\cdots,s$, then all nontrivial linear combinations of these functions lie outside of $L^2[0,\infty)$. It is also clear that $t^r \exp(z_j t) \in L^2[0,\infty)$ if $\text{Re } z_j < 0$. Thus (2.18) has a nontrivial solution in $L^2[0,\infty)$ if and only if $\text{Re } z_j < 0$ for some $j \in \{1,\cdots,s\}$.

To establish (2.16) suppose $R(A_o - \lambda)$ is closed and $0 \neq \text{def}(A_o - \lambda) = \text{nul}(A_o^* - \overline{\lambda}) = \text{nul}\left(A_1(T^+) - \overline{\lambda}\right)$. Then $\text{Re } z_j < 0$ for some j in $\{1,\cdots,s\}$. Now $p^+(z) = \overline{\lambda}$ implies $\lambda = \overline{p^+(z)} = p(-\overline{z})$. So $\text{Re } \overline{z}_j = \text{Re } z_j < 0$ and $\lambda \in P \setminus P_o$. On the other hand for any $\lambda \notin P_o$ there exists a $\delta > 0$ such that

$$|p(ix) - \lambda| \geq \delta, \quad x \in R. \tag{2.19}$$

Using (2.19) and the Fourier transform F we get

$$\|(A_o - \lambda)u\|^2 = \int_{-\infty}^{\infty} |p(ix) - \lambda|^2 |Fu(x)|^2 \, dx \geq \delta^2 \|u\|^2, \quad u \in D(A_o). \tag{2.20}$$

Here (2.20) is established first for $u \in C_o^\infty(R^+)$. The inequality then follows for all $u \in D(A_o)$ since A_o is closed. From (2.20) it follows that $(A_o - \lambda)^{-1}$ exists and is a bounded operator on $R(A_o - \lambda)$. Consequently $R(A_o - \lambda)$ is closed and $\lambda \in \Phi_3(A_o)$. To show that $\lambda \in \sigma_r(A_o)$ we must show that $\mathrm{def}(A_o - \lambda) = \mathrm{nul}\left(A_1(T^+) - \bar{\lambda}\right) \neq 0$, i.e., that $p^+(z) = \bar{\lambda}$ has a root z with $\mathrm{Re}\, z < 0$. Now $\lambda \in P \setminus P_o$ means that there exists a $z \in \mathbb{C}$ with $\mathrm{Re}\ z > 0$ such that $p(z) = \lambda$. But this implies that $p^+(-\bar{z}) = \bar{\lambda}$, and the proof of (2.16) is complete.

Next we show that

$$\sigma(A_o) \subseteq P \tag{2.21}$$

by showing that $\lambda \in \mathbb{C} - P$ implies $\lambda \in \rho(A_o)$. Suppose $p(z) = \lambda$ implies that $\mathrm{Re}\ z < 0$. Then $\bar{\lambda} = \overline{p(z)} = p^+(-z)$. Thus $\mathrm{Re}\ z_j > 0$, $j = 1,\cdots,s$, for all z_j in (2.17). Hence $\mathrm{nul}(A_o^* - \bar{\lambda}) = 0$. Since $A_o - \lambda$ is one-to-one we have $\mathrm{nul}(A_o - \lambda) = 0$. By (2.20), $\mathcal{R}(A_o - \lambda)$ is closed. Therefore $\lambda \in \rho(A_o)$.

To establish (2.14) we use Theorem VI: 7.2 of [6]. (Notice that, in the notation of [6], $\sigma_e(A_o) = \sigma_{e3}(A_o)$.) By Lemma 2.1, A_1 is a finite dimensional extension of A_o. Thus by Theorem 2.7 above and Theorem VI: 7.2 of [6] we have $\sigma_{e3}(A_o) = \sigma_{e3}(A_1) = P_o$. Formula (2.13) follows from (2.21), (2.16), and (2.14).

The proof of Theorem 2.7 is complete if we can show that $\sigma_{e4}(A_o) = P$. However, this follows from (2.13) and (2.16) since $\sigma_4(A_o) \supseteq \sigma_r(A_o) \cap \Phi_3(A_o)$.

The primary difference between the whole line case $I = R$ and the half line case $I = R^+$ is that the solutions (2.17) are not in $L^2(-\infty,\infty)$ for any z_j. Hence there is no residual spectrum in the case $I = R$.

Theorem 2.8. Let A_o be the minimal operator of T given by (2.1) with $a_j \in \mathbb{C}$, $a_N \neq 0$ on $\mathbb{R}$. Then, with $I = \mathbb{R}$,

$$\sigma(A_o) = \sigma_{e2}^{\pm}(A_o) = \sigma_{ek}(A_o) = P_o, \quad k = 1,3,4,5.$$

Proof. The proof is similar to the proof of Theorem 2.7.

2.3. Perturbations of bounded coefficient operators.

Let A and B be the maximal operators associated with

$$T = \sum_{k=0}^{N} a_k D^k$$

and

$$S = \sum_{k=0}^{N-1} b_k D^k$$

respectively.

Theorem 2.9. Let $I = [a,\infty)$, $-\infty < a < \infty$. Assume that a_N^{-1}, $a_k \in L^\infty(I)$, $0 \le k \le N$. Then, we have the following:

(i) B is A-bounded if and only if each b_k is in $L^2_{loc}(I)$ and

$$\sup_{0 \le s < \infty} \int_s^{s+1} |b_k|^2 < \infty, \quad 0 \le k \le N-1.$$

In this case, given $\varepsilon > 0$ there exists $K(\varepsilon) > 0$ such that for all $f \in \mathcal{D}(A)$

$$\|Bf\| \le \varepsilon\|Af\| + K(\varepsilon)\|f\|. \tag{2.22}$$

Furthermore, $A_1(T + S) = A_1(T) + A_1(S) = A + B$.

(ii) B is A-compact if and only if each b_k is in $L^2_{loc}(I)$ and

$$\lim_{s\to\infty} \int_s^{s+1} |b_k|^2 = 0, \quad 0 \le k \le N-1.$$

In this case

$$A_1(T + S) = A_1(T) + A_1(S) = A + B$$

and for $C = A_1(T + S)$

$$\sigma^{\pm}_{e2}(C) = \sigma^{\pm}_{e2}(A) \quad \text{and} \quad \sigma_{ek}(C) = \sigma_{ek}(A), \quad k = 1,3,4.$$

(Note that $k = 5$ is not included above.) Furthermore, in this case, $\text{ind}(A - \lambda) = \text{ind}(C - \lambda)$ for all $\lambda \in \Phi_3(A)$.

Proof. This follows directly from Theorem VI: 8.1 in [6 , pp. 166-167] and Theorem 1.9.

2.4. Location of the essential spectra.

In this section we indicate the variety of applications of Theorem 1.11 and its corollaries in the case of ordinary differential operators.

Let

$$T = \sum_{j=0}^{N} a_j(t)\, D^j, \quad a_N(t) \neq 0,$$

be defined on $C_o^N(0,\infty)$. If the coefficients are sufficiently differentiable then T can be written in the form

$$\ell_o\phi = (-1)^n\left(p_n(t)\phi^{(n)}\right)^{(n)} + \sum_{j=0}^{n-1}(-1)^j\left(p_j(t)\phi^{(j)} + q_j(t)\phi^{(j+1)}\right)^{(j)}$$

if N is even and

$$\ell_1\phi = \sum_{j=0}^{n}(-1)^j\left(p_j(t)\phi^{(j)} + q_j(t)\phi^{(j+1)}\right)^{(j)}$$

if N is odd. The corresponding forms are

$$(\ell_o\phi,\phi) = \int_o^\infty [p_n(t)|\phi^{(n)}|^2 + \sum_{j=0}^{n-1}\left((p_j(t) - q_j'(t)/2)|\phi^{(j)}|^2 + iq_j(t)\,\mathrm{Im}(\overline{\phi}^{(j)}\phi^{(j+1)})\right)]dt$$

and

$$(\ell_1\phi,\phi) = \sum_{j=0}^{n}\int_o^\infty\left((p_j(t) - q_j'(t)/2)|\phi^{(j)}|^2 + iq_j(t)\,\mathrm{Im}(\overline{\phi}^{(j)}\phi^{(j+1)})\right)dt.$$

Hence,

$$\mathrm{Re}(\ell_o\phi,\phi) = \int_o^\infty [\mathrm{Re}\; p_n(t)|\phi^{(n)}|^2 + \sum_{j=0}^{n-1}\left(\mathrm{Re}(p_j(t) - q_j'(t)/2)|\phi^{(j)}|^2 - \mathrm{Im}(q_j(t))\,\mathrm{Im}(\overline{\phi}^{(j)}\phi^{(j+1)})\right)]dt$$

and

$$\mathrm{Im}(\ell_o\phi,\phi) = \int_o^\infty [\mathrm{Im}\; p_n(t)|\phi^{(n)}|^2 + \sum_{j=0}^{n-1}\left(\mathrm{Im}(p_j(t) - q_j'(t)/2)|\phi^{(j)}|^2 + \mathrm{Re}(q_j(t))\,\mathrm{Im}(\overline{\phi}^{(j)}\phi^{(j+1)})\right)]dt.$$

The representations for $\mathrm{Re}(\ell_1\phi,\phi)$ and $\mathrm{Im}(\ell_1\phi,\phi)$ are similar.

One of the essential requirements needed in order to apply some of the results of section 1.5 is that a certain form be sectorial. A form $\mathfrak{s}[u,v]$ is sectorial if there are constants γ and $c > 0$ such that

$$|\mathrm{Im}\, s[u]| \le c\big(\mathrm{Re}\, s[u] - \gamma(u,u)\big), \qquad u \in \mathcal{D}(s).$$

Example 2.10. In the above expression for ℓ_o, let each $p_j(t)$ and each $q_j(t)$ be real-valued. Define $s[u,v] = (\ell_o u, v)$ with $\mathcal{D}(s) = C_o^{2n}(0,\infty)$. If there are real numbers γ, $c > 0$, and $\varepsilon_j > 0$, $j = 0,1,\cdots,n-1$, such that

$$\varepsilon_{n-1}|q_{n-1}(t)| \le 2\, c\, p_n(t),$$

$$\varepsilon_{j-1}|q_{j-1}(t)| + |q_j(t)|/\varepsilon_j \le c\big(2p_j(t) - q_j'(t)\big), \quad j = 1,\cdots,n-1,$$

and

$$|q_o(t)| \le 2\varepsilon_o\, c\big(p_o(t) - \gamma - q_o'(t)/2\big), \quad t \in [0,\infty)$$

then s is sectorial.

Proof. Since $2ab \le \varepsilon a^2 + \frac{1}{\varepsilon} b^2$ for $\varepsilon > 0$, then

$$
\begin{aligned}
|\mathrm{Im}\, s[\phi]| &= \sum_{j=0}^{n-1} \int_o^\infty q_j(t)\, \mathrm{Im}(\overline{\phi}^{(j)} \phi^{(j+1)})\,dt \\
&\le \int_o^\infty \varepsilon_{n-1}|q_{n-1}(t)||\phi^{(n)}|^2/2\, dt \\
&\quad + \sum_{j=1}^{n-1} \int_o^\infty \big(\varepsilon_{j-1}|q_{j-1}(t)| + |q_j(t)|/\varepsilon_j\big)|\phi^{(j)}|^2/2\, dt \\
&\quad + \int_o^\infty |q_o(t)||\phi|^2/2\varepsilon_o\, dt
\end{aligned}
$$

from which the proof follows.

Other examples of sectorial ordinary differential operators can be found in [9, p. 280] and recent work of Read [18]. Another weaker condition that arose in the results of section 1.5 is the requirement that αs be sectorial for some complex number $\alpha = a - ib$. The next example illustrates some conditions which insure that requirement.

Example 2.11. Consider the form

$$s[u,v] = \int_o^\infty [p_1 u'\overline{v}' + (p_o - q'/2)u\overline{v} + i\, q\, \mathrm{Im}(u'\overline{v})]dt,$$

$u, v \in C_o^1(0,\infty)$. With certain minimal conditions satisfied by the coefficients, we know that $\mathcal{s}$ is associated with a differential operator given by

$$\ell(\phi) = -(p_1\phi')' + q\phi' + p_o\phi.$$

Assume that there are constants $a_1 \geq 0$, $b_1 \geq 0$, $a_2 \geq 0$ and $b_2 < 0$ such that a_1 and a_2 are not both zero and

$$b_i \operatorname{Re}(q) - a_i \operatorname{Im}(q) \geq 0, \qquad t \in [0,\infty),$$

for $i = 1,2$.

Suppose that there are positive numbers ε_1 and ε_2 such that the forms

$$\int_o^\infty [a_i \operatorname{Re}(p_1) + b_i \operatorname{Im}(p_1) - \frac{\varepsilon_i}{2}\big(b_i \operatorname{Re}(q) - a_i \operatorname{Im}(q)\big)]|\phi'|^2$$
$$+ [a_i \operatorname{Re}(p_o - q'/2) + b_i \operatorname{Im}(p_o - q'/2) - \frac{1}{2\varepsilon_i}\big(b_i \operatorname{Re}(q) - a_i \operatorname{Im}(q)\big)]|\phi|^2 dt$$

are bounded below on $C_o^1(0,\infty)$ for $i = 1,2$. Then, there are numbers z complex and γ real such that

$$e^{i\gamma}(\mathcal{s}[u,v] + z[u,v]), \qquad u, v \in C_o^1(0,\infty)$$

is sectorial.

Proof. Since

$$\operatorname{Re} \mathcal{s}[u] = \int_o^\infty [\operatorname{Re}(p_1)|u'|^2 + \operatorname{Re}(p_o - q'/2)|u|^2 - \operatorname{Im}(q)\operatorname{Im}(\bar{u}\, u')]dt$$

and

$$\operatorname{Im} \mathcal{s}[u] = \int_o^\infty [\operatorname{Im}(p_1)|u'|^2 + \operatorname{Im}(p_o - q'/2)|u|^2 + \operatorname{Re}(q)\operatorname{Im}(\bar{u}\, u')]dt$$

then

$$a_i \operatorname{Re} \mathcal{s}[u] + b_i \operatorname{Im} \mathcal{s}[u] \geq \int_o^\infty [\big(a_i \operatorname{Re}(p_1) + b_i \operatorname{Im}(p_1)\big)|u'|^2$$
$$+ \big(a_i \operatorname{Re}(p_o - q'/2) + b_i \operatorname{Im}(p_o - q'/2)\big)|u|^2$$
$$- \big(b_i \operatorname{Re}(q) - a_i \operatorname{Im}(q)\big)|u||u'|]dt$$
$$\geq \int_o^\infty \{[a_i \operatorname{Re}(p_1) + b_i \operatorname{Im}(p_1) - \frac{\varepsilon_i}{2}\big(b_i \operatorname{Re}(q) - a_i \operatorname{Im}(q)\big)]|u'|^2$$
$$+ [a_i \operatorname{Re}(p_o - q'/2) + b_i \operatorname{Im}(p_o - q'/2)$$
$$- \frac{1}{2\varepsilon_i}\big(b_i \operatorname{Re}(q) - a_i \operatorname{Im}(q)\big)]|u|^2\}dt$$
$$\geq \beta_i \int_o^\infty |u|^2\, dt$$

for $i = 1,2$. This fact implies that the values of $\mathfrak{s}[u]$, $u \in C_o^1(0,\infty)$, lie in a sector in the complex plane whose angle is less than π.

Another requirement for the application of the results in section 1.5 is that a linear combination of the real and imaginary parts of a quadratic form be bounded below by a symmetric form—see Corollary 1.14. The next lemma indicates how this may be achieved.

Lemma 2.12. Let $\varepsilon_i > 0$ for $i = 0,1,2,\cdots,n-1$. For any $u \in C_o^{2n}(0,\infty)$

$$\begin{aligned}\operatorname{Re}(\ell_o\phi,\phi) \geq \int_0^\infty \big(\operatorname{Re}(p_n) - \varepsilon_{n-1}|\operatorname{Im} q_{n-1}|/2\big)|\phi^{(n)}|^2\, dt \\ + \sum_{k=1}^{n-1}\int_0^\infty \big(\operatorname{Re}(p_k - q_k'/2) - (2\varepsilon_k)^{-1}|\operatorname{Im} q_k| - \varepsilon_{k-1}|\operatorname{Im} q_{k-1}|/2\big)|\phi^{(k)}|^2 dt \\ + \int_0^\infty \big(\operatorname{Re}(p_o - q_o'/2) - |\operatorname{Im} q_o|/2\varepsilon_o\big)|\phi|^2\, dt.\end{aligned}$$

Proof. The proof follows from the inequality $2\alpha\beta \leq \varepsilon\alpha^2 + \beta^2/\varepsilon$, $\varepsilon > 0$, for numbers α and β, and the above expression for $\operatorname{Re}(\ell_o\phi,\phi)$.

A similar inequality holds for $\operatorname{Im}(\ell_o\phi,\phi)$. In the special case that q_n is real-valued, an inequality of this type can be derived for $(\ell_1\phi,\phi)$ in the same manner. For example, if

$$\ell_1\phi = (-1)^n\, \phi^{(2n+1)} + p(x)\phi, \qquad \phi \in C_o^{2n+1}(0,\infty)$$

then $\operatorname{Re}(\ell_1\phi,\phi) = \int_0^\infty \operatorname{Re}(p)|\phi|^2\, dt$.

Example 2.13. Let $p \in C[0,\infty)$ and $\mu = \liminf_{t\to\infty} \operatorname{Re}\big(p(t)\big)$. Let

$$\ell_1 = (-1)^n\, D^{2n+1} + p.$$

Then

$$\sigma_{e3}\big(A_o(\ell_1,[0,\infty))\big) \subseteq \{\lambda \in \mathbb{C}: \operatorname{Re}\lambda \geq \mu\}.$$

Proof. The proof follows from Lemma 2.4 and Theorem 1.11.

References

1. W. D. Evans, On the spectra of Schrödinger operators with a complex potential. Math. Ann. 255, 57-76 (1981).
2. _____, On the spectra of non-self-adjoint realisations of second-order elliptic operators. Proc. Roy. Soc. Edin. 90A, 71-105 (1981).
3. W. D. Evans, M. Kwong and A. Zettl, Lower bounds for the spectrum of ordinary differential operators. To appear in J. Diff. Eqns.
4. W. N. Everitt and A. Zettl, Generalized symmetric differential equations I. The basic theory. Niew Archief. voor Wiskunde (3) XXVII, 363-397 (1979).
5. I. M. Glazman, Direct Methods of Qualitative Spectral Analysis of Singular Differential Operators. (Israel Program of Scientific Translations: Jerusalem, 1965).
6. S. Goldberg, Unbounded Linear Operators. (McGraw-Hill: New York 1966).
7. K. Gustafson and J. Weidman, On the essential spectrum. J. Math. Anal. Appl. 25, 121-127 (1969).
8. P. Halmos, A Hilbert Space Problem Book. (Springer-Verlag: New York-Heidelberg 1974).
9. T. Kato, Perturbation Theory of Linear Operators. (Springer-Verlag: Berlin 1966).
10. M. Kwong and A. Zettl, Norm inequalities for dissipative operators on inner product spaces. Houston J. Math. 4, 543-557 (1979).
11. H. P. Kupcov, Kolmogorov estimates for derivatives in $L^2(0,\infty)$. Proc. Steklov Inst. Math. 138 (1975); Amer. Math. Soc. Transl.,101-125 (1977).
12. A. Lebow and M. Schechter, Semigroups of operators and measures of non-compactness. J. Func. An. 7, 1-26 (1971).
13. R. T. Lewis, Applications of a comparison theorem for quasi-accretive operators in a Hilbert space. To appear in Lecture Notes in Mathematics (Springer-Verlag: Proceedings, Dundee, Scotland 1982).

14. Su. I. Ljubič, On inequalities between the powers of a linear operator. Amer. Math. Soc. Transl. Ser. 2, 40, 39-84 (1964).

15. E. Müller-Pfeiffer, Spectral Theory of Ordinary Differential Operators. English Translation (Chichester: J. Wiley 1981)

16. M. A. Naimark, Linear Differential Operators, part II. (Ungar: New York 1968).

17. R. D. Nussbaum, Spectral mapping theorems and perturbation theorems for Browder's essential spectrum. Trans. Am. Math. Soc. 150, 445-455 (1970).

18. T. Read, Sectorial second order differential operators. Manuscript.

19. M. Reed and B. Simon, Methods of Modern Mathematical Physics I: Functional Analysis. (Academic Press: New York, 1972).

20. _____, Methods of Modern Mathematical Physics IV: Analysis of Operators. (Academic Press: New York 1978).

21. M. Schechter, Spectra of Partial Differential Operators. (London: N. Holland, 1971).

22. J. Weidmann, Linear operators in Hilbert spaces. English translation. (Berlin: Springer-Verlag, 1980).

23. F. Wolf, On the essential spectrum of partial differential boundary problems. Comm. Pure Appl. Math. 12, 211-228 (1959).

Acknowledgement: This work was done while the authors were participants in the Symposium on Ordinary Differential Equations and Operators held at The University of Dundee, Scotland, during the months of April, May, and June 1982. The authors were supported by the U. K. Science and Engineering Research Council, grant number GR/B/78083. The authors wish to express their gratitude for this support. In particular, we are especially grateful to Professor W. N. Everitt for extending the invitations and making the necessary arrangements.

A Concept of Adjointness and Symmetry of Differential Expressions Based on the Generalised Lagrange Identity and Green's Formula.

W.N. Everitt and F. Neuman

Notations

$A = A(x)$ — variable matrices, $n \times n$, $\varepsilon\ L(a,b) \equiv L$

$B = B(x)$

C — a constant matrix, $n \times n$

$*$ transposition of matrices

$\underline{y}, \underline{z}$, $\underline{a}, \underline{b}$ — column vectors, $n \times 1$ matrices

at the moment, everything is real; $' = d/dx$.

Consider two non-homogeneous systems

(1) $\underline{y}' = A(x)\underline{y} + \underline{a}$

(2) $\underline{z}' = B(x)\underline{z} + \underline{b}$

and C - a regular constant $n \times n$ matrix.

If $\underline{y}$ and $\underline{z}$ are solutions of (1) and (2), respectively, then

$$(\underline{y}^*C\underline{z})' = \underline{y}^{*\prime}C\underline{z} + \underline{y}^*C\underline{z}'$$

$$= (\underline{y}^*A^* + \underline{a}^*)C\underline{z} + \underline{y}^*C(B\underline{z} + \underline{b})$$

$$= \underline{y}^*(A^*C + CB)\underline{z} + \underline{a}^*C\underline{z} + \underline{y}^*C\underline{b}\ .$$

For the special case

(3) $B = -C^{-1}A^*C$

we have

$$(\underline{y}^*C\underline{z})' = \underline{a}^*C\underline{z} + \underline{y}^*C\underline{b}\ .$$

Define $L : = I.\frac{d}{dx} - A(x)$

$$(\text{i.e. } L\tilde{\underline{y}} = \tilde{\underline{y}}' - A(x)\tilde{\underline{y}}) ;$$

also $A^+ : = - C^{-1}A^*C$, see (3),

and L^+_C, or briefly only $L^+ : = I\frac{d}{dx} - A^*(x) = I\frac{d}{dx} + C^{-1}A^*C$,

$$\text{i.e., } L^+\tilde{\underline{z}} = \tilde{\underline{z}}' + C^{-1}A^*\tilde{\underline{z}} .$$

For our $\underline{y}$, $\underline{z}$ being solutions of (1) and (2) we have

$L\underline{y} = \underline{a}$, $L^+\underline{z} = \underline{b}$, hence

(generalised Lagrange identity for vector expressions L)

(4) $(\underline{y}^*C\underline{z})' = (L\underline{y})^*C\underline{z} + \underline{y}^*CL^+\underline{z}$.

The last relation holds for any $\underline{y}$ and $\underline{z}$ having $\underline{y}'$ and $\underline{z}'$, since given e.g. $\underline{y}$, we define $\underline{a} : = \underline{y}' - A(x)\underline{y}$, and $\underline{y}$ is uniquely determined as a solution of the system $\tilde{\underline{y}}' = A(x)\tilde{\underline{y}} + \underline{a}$ with given $A \in L$, given $\underline{a}$, given the same initial condition at some x_o as $\underline{y}$.

Note 1. C is an arbitrary constant regular matrix.

Note 2. If $Y(x)$ is a regular matrix solution of $LY = O$, O being the null matrix i.e.

$$Y' = A(x)Y , \tag{5}$$

then $- C^{-1}Y^{*-1}(x) = z(x)$ is a regular matrix solution of $L^+z = 0$, i.e.

$$z' = - C^{-1}A^*(x)C.z . \tag{6}$$

Let Z (i.e. Everitt-Zettl [4]) be the set of all $n \times n$ matrices $A(x) \in L$, whose elements $a_{i,i+1}(x) \neq 0$ for $i = 1, \ldots, n-1$, and $a_{i,j}(x) \equiv 0$ for $j \geqslant i+2$.

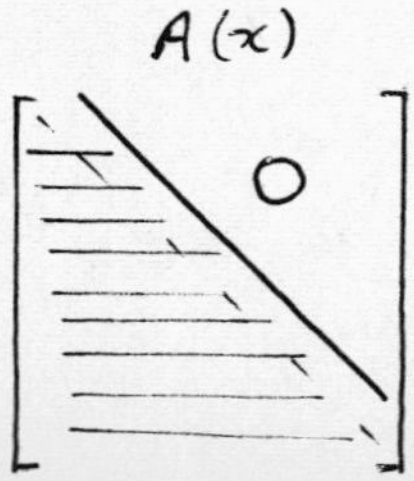

A(x)

O

Let ∇ denote the set of all constant regular matrices C satisfying

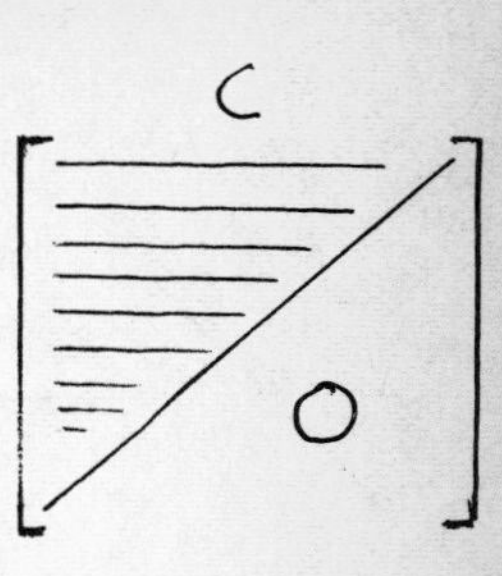

$$c_{ij} = 0 \quad \text{for} \quad i+j > n+1 .$$

We have

<u>Theorem</u>

If $A(x) \in Z$, <u>then</u>

$$- C^{-1}A^*(x)C \in Z ,$$

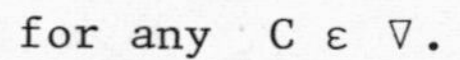

for any $C \in \nabla$.

<u>Note</u> In other words:

If A is of the Z form, then A^+ is again of the same form for any $C \in \nabla$ (remember, the definition of $+$ depends on C).

<u>Proof.</u>

If $C \in \nabla$, then it is regular and $\det C = c_{1n}c_{2,n-1} \ldots c_{n1} \neq 0$; hence $c_{i,n-i+1} \neq 0$ for all $i = 1, \ldots, n$.

C^{-1} has zero elements in the places (i, j) where $i+j \leqslant n$; its quazi-diagonal elements on places $(i, n+1-i)$ are $\frac{1}{c_{i,n+1-i}}$ $(\neq 0)$, $i = 1, \ldots, n$.

Hence $- C^{-1}A^*(x)C$ has $- \frac{1}{c_{n-i+1,i}} \cdot a_{i,i-1}(x) \cdot c_{i-1,n-i+2} \neq 0$ on the places $(n-i+1, n-i+2)$, $i = 2, \ldots, n$, (i.e., $(j, j+1)$ for $j = 1, \ldots, n-1$), and zero elements on (i, j) for $j \geqslant i+2$.

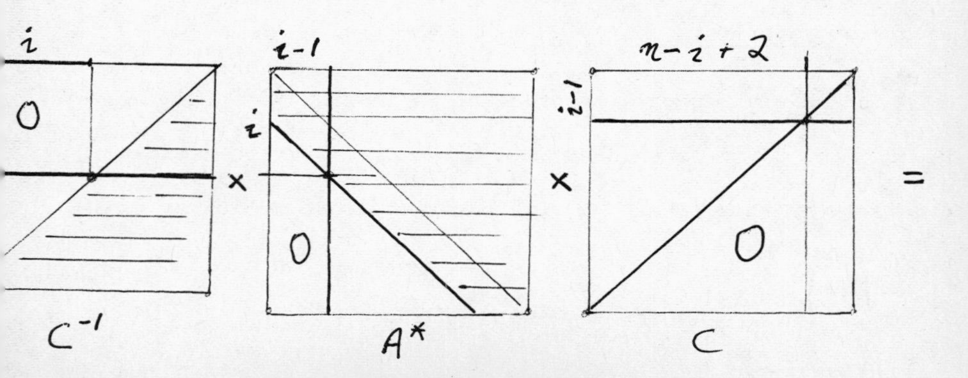

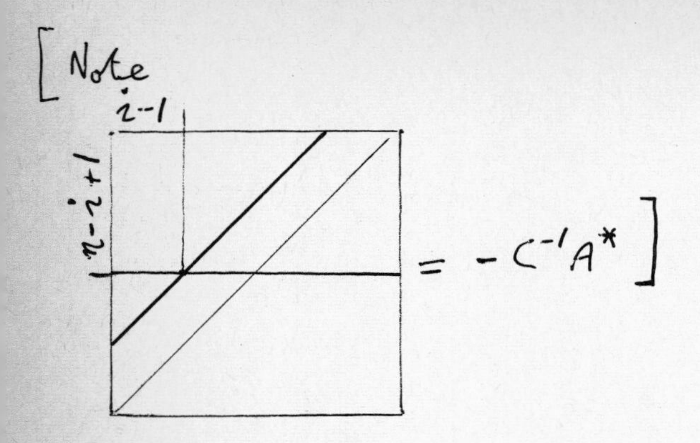

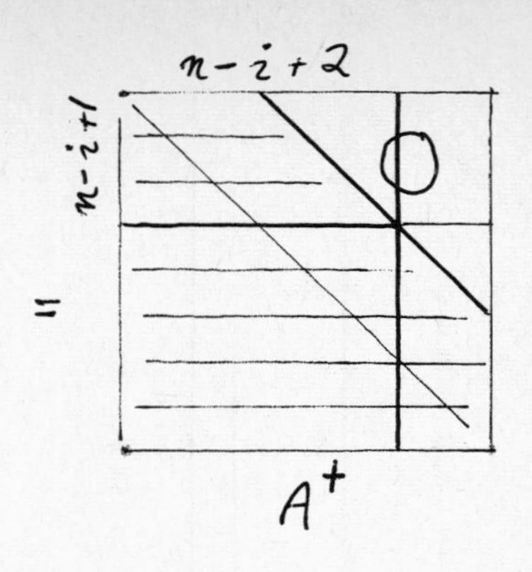

This completes the proof.

Suppose A is of the form Z, $C \in \nabla$. Then, due to our theorem, $A^+ \in Z$.

Choose $\underline{a} = \begin{pmatrix} 0 \\ 0 \\ a \end{pmatrix}$, $\underline{b} = \begin{pmatrix} 0 \\ 0 \\ b \end{pmatrix}$,

and

Define:

The scalar differential expression (associated with L) as

Ly_1, where $Ly_1 = a$ is the last relation of $L\underline{y} = \begin{pmatrix} 0 \\ 0 \\ a \end{pmatrix}$,

obtained by eliminating all coordinates of $\underline{y}$ from the system except of the first one, y_1.

The adjoint scalar differential expression (to L, with respect to C) is defined as L^+z_1, sometimes written more precisely as $L^+_C z_1$ where $L^+z_1 = b$ is the last relation of

$L^+\underline{z} = \begin{pmatrix} 0 \\ 0 \\ b \end{pmatrix}$, z_1 being the 1st coordinate of $\underline{z}$.

Note With respect to our theorem, the definitions have the following meaning. The operators are defined on those sets of y_1 (as z_1), where the formal expressions obtained by the elimination process as above (not only on those y_1, where y_1' is defined, since we require more, not only to have $\underline{a}$, but that this $\underline{a}$ is of the form $\begin{pmatrix} 0 \\ 0 \\ a \end{pmatrix}$).

Now Ly_1 and L^+z_1 are not vectors, but scalar expressions, and formula (4) reads

$$(7) \qquad (y^*Cz)' = (0, \ldots, 0, Ly_1) \cdot C \cdot \begin{pmatrix} z_1 \\ \vdots \\ z_n \end{pmatrix} +$$

$$+ (y_1, \ldots, y_n)^+ C \cdot \begin{pmatrix} 0 \\ 0 \\ L^+z_1 \end{pmatrix}$$

$$= c_{n1} z_1 \cdot Ly_1 + c_{1n} y_1 L^+ z_1 ;$$

remember $C \in \nabla$, hence $c_{ni} = c_{in} = 0$ for $i = z, \ldots, n$.

In accordance with [4], y_i is called $(i-1)$ quasi-derivative of y_1 ($y_i = y_1^{[i-1]}$) ; similarly for z_1. Remember, quasi-derivatives depend on the corresponding systems. The quasi-derivatives occur also on the left side of the identity. C occurs both on the left and on the right side in the definition of the adjoint expression.

Remarks

1. We still may define the adjoint expression for a larger class than ∇, if we require, that a particular $A(x) \in Z$ or a particular subset of Z should satisfy $- D^{-1}A^*(x)D \in Z$, and then define $A^+ := - D^{-1}A^*(x)D$ for this particular A or subset of Z.

2. We also need not be restricted to $A \varepsilon Z$; we may consider the class of those matrices $M(x)$ obtained from all $A(x) \varepsilon Z$ by taking all regular constant matrices N and forming $M(x) = N^{-1}A(x)N$, i.e. considering the set

$$\{N^{-1}A(x)N \,;\ N \text{ being regular, } A \varepsilon Z\}.$$

3. There is an open question, as to whether systems of the form $N^{-1}A(x)N$ are the most general systems to which we can construct or define a "reasonable" scalar differential expression. See the remarks at the end of the paper, especially [5].

If $C \varepsilon \nabla$ is taken so that

$$C = \begin{matrix} 0 & (-1)^{n-1} \\ 1^{-1} & \end{matrix}\,, \text{ then (7) reads}$$

(8) $$(y_n z_1 - y_{n-1} z_2 \pm \ldots +(-1)^{n-1} y_1 z_n)' = z_1 . Ly_1 - (-1)^n y_1 L^+ z_1 .$$

If $$C = \begin{bmatrix} 0 & 0 & . & . & . & \alpha_1 \\ 0 & & & & ⁄ & . \\ . & & & ⁄ & & 0 \\ \alpha_n & . & . & . & . & 0 \end{bmatrix}$$

then (7) becomes

(9) $$(\alpha_n y_n z_1 + \alpha_{n-1} y_{n-1} z_2 + \ldots + \alpha_1 y_1 z_n)' = \alpha_n z_1 Ly_1 + \alpha_1 y_1 L^+ z_1 .$$

Always L^+ is defined with respect to the corresponding C, i.e. L^+_C.

Now, let

(10) $$L^+ = -\frac{c_{n1}}{c_{1n}} L \,;$$

both c_{1n} and c_{n1} are not zero since C is regular.

Then (7) becomes

$$(11) \qquad (\underline{y}^*C\underline{y})' = 0$$

and we may <u>define</u> such an expression L satisfying (10) as <u>symmetric differential expression</u> (again, with respect to a particular C, that then occurs in (GENERALIZED GREEN'S FORMULA) (11); on the left side of (11) there are quasi-derivatives).

For the case in [4], $C = \begin{bmatrix} 0 & \cdot & \cdot & (-1)^{n-1} \\ \cdot & & \diagup & \cdot \\ \cdot & \diagup & & \cdot \\ 1 & \cdot & \cdot & 0 \end{bmatrix}$, $-\frac{c_{n1}}{c_{1n}}$ is $(-1)^n$. Hence, due to (10), let $L^+ = (-1)^n L$. Then (7) reads

$$(12) \quad y_n y_1 - y_{n-1} y_2 \pm \ldots + (-1)^{n-1} y_1 y_n = \text{constant} .$$

Let $C = \begin{bmatrix} 0 & 0 & \cdot & \cdot & \cdot & \alpha_1 \\ 0 & & & & \diagup & \cdot \\ \cdot & & \diagup & & & 0 \\ \alpha_n & \cdot & \cdot & \cdot & \cdot & 0 \end{bmatrix}$.

If $L^+ = -\frac{\alpha_n}{\alpha_1} L$, then (9) or (11) implies

$$\alpha_1 y_1 y_n + \alpha_2 y_2 y_{n-1} + \ldots + \alpha_n y_n y_1 = \text{constant} .$$

<u>Remark</u> If $c_{n1} = \pm c_{1n}$, then $L^{++} = L$, or more precisely, $(L_C^+)_C^+ = L$, ^+C means : with respect to C. However $(L_C^+)^+_{\pm C^*}$ is <u>always</u> L, important : $C \in \nabla$ implies $C^* \in \nabla$, that is not in general true for C^{-1}.

Coming back to systems,

$$A^+ = - C^{-1} A^* C .$$

Requiring $A^{++} = A$ gives $A^{++} = - C^{-1}(- C^{-1}A^*C)^*C =$

$$= C^{-1}C^*A(C^{-1}C^*)^{-1} = A .$$

Certainly,

if $C^{-1}C^{*} = I$ (unit matrix), i.e. $C^{*} = C$, C is symmetric and $A^{++} = A$ holds for all A; more precisely, $(A^{+}_{C})^{+}_{C} = A$.

However, $(A^{+}_{C})^{+}_{\pm C^{*}} = A$ always holds.

Remark: if $(A^{+}_{C})^{+}_{C} = A$ has to hold for a particular A, or for a particular subset of Z then C need no longer by symmetric.

We may also *define*

symmetric vector differential operators L requiring

$$L = L^{+},$$

i.e. $\underline{y}' - A(x)\underline{y} = \underline{y}' + C^{-1}A^{*}(x)C\,\underline{y}$ for all admissible $\underline{y}$.

By taking n vectors $\underline{y}$ forming a regular matrix Y, we can prove that

$$L = L^{+} \text{ is equivalent to } A = -C^{-1}A^{*}C = A^{+}.$$

We conclude with some remarks on the list of references given below. Early work on quasi-differential expressions is to be found in Halperin [6] (which contains reference to work before 1936) and Shin [7]. Both Atkinson [1] and Bor̊uka [2, pages 159-163] contain systematic accounts of generalized adjoint systems; in this context see also the account of Reid [8] and Coddington and Levinson [3, chapter 3]. This paper has been influenced by the work of Zettl; see [4] but also the earlier results in [9] which date back to 1965. For a very general systematic account of quasi-derivatives for scalar and system cases see the recent paper by Frentzen [5].

References

1. F.V. Atkinson. Discrete and continuous boundary value problems. (Academic Press, New York, 1964).

2. O. Borůvka. Differenciálne rovnice. (Slovenské Pedagogické nakadelství, Bratislava, 1964).

3. E.A. Coddington and N. Levison. Theory of ordinary differential equations. (McGraw-Hill, New York, 1955).

4. W.N. Everitt and A. Zettl. Generalized symmetric ordinary differential expression I : the general theory. Nieuw Archief voor Wiskunde (3) 27 (1979), 363-397.

5. Hilbert Frentzen. Equivalence, adjoints and symmetry of quasi-differential expressions with matrix-valued coefficients and polynomials in them. Proc. Royal Soc. Edinb.(A) 92(1982), 123-146.

6. I. Halperin. Closures and adjoints of linear differential operators. Ann. of Math. 38(1937), 880-919.

7. D. Shin. Existence theorems for the quasi-differential equation of the nth order. Dokl. Akad. Nauk. SSSR. 18(1938), 515-518.

8. W.T. Reid. Ordinary differential equations. (New York, Wiley, 1971).

9. A. Zettl. Formally self-adjoint quasi-differential operators. Rocky Mountain J. Math. 5(1975), 453-474.

ON QUADRATIC INTEGRAL INEQUALITIES ASSOCIATED WITH SECOND-ORDER SYMMETRIC DIFFERENTIAL EXPRESSIONS

W. N. Everitt and S. D. Wray

1. Introduction. Let I represent an arbitrary interval of the real line R; thus I may be open, half-open or closed, and bounded or unbounded. Let the end-points of I be denoted by a and b, so that $-\infty \le a < b \le \infty$. This paper is concerned with the integral inequality

$$\int_a^b \{p|f'|^2 + q|f|^2\} \ge \mu \int_a^b w|f|^2 \qquad (f \in D) \qquad (1.1)$$

where ' denotes classical differentiation on R, the coefficient functions p, q and w are real-valued on I, and the integrals are in the sense of Lebesgue. Here p and w are positive almost everywhere (Lebesgue measure) on I; $p^{-1} = 1/p$, q and w are locally integrable on I; D is a linear manifold of complex-valued functions chosen so that all three integrals in (1.1) are absolutely convergent; μ is a real number which depends only on the coefficients p, q and w.

In certain circumstances, and these are the concern of this paper, it is possible to characterize the best possible number μ in the inequality (1.1) in terms of the spectral properties of a differential operator T, where T is self-adjoint in the weighted Hilbert function space $L_w^2(I)$ and generated by the associated symmetric (*i.e.* formally self-adjoint) quasi-differential expression

$$w^{-1}(-(pf')' + qf) \text{ on } I. \qquad (1.2)$$

Inequalities of the form (1.1) have an interest in their own right but also have important applications in the calculus of variations and mathematical physics. The integral on the left-hand side of (1.1) is called the Dirichlet integral, sometimes the energy integral, of the differential expression (1.2).

For a connected account of integral inequalities and the calculus of variations see Hardy, Littlewood and Pólya [28, chapter 7]; for connections with mathematical physics see Hellwig [29]; and for connections between quadratic forms of the type (1.1) and differential equations see the recent book by Gregory [27].

For a brief account of work before 1970 on inequalities of the form (1.1) see the remarks in Bradley and Everitt [11, section 1], and Amos and Everitt [7, section 1]; particular attention is drawn to the results of Putnam [40] in 1948. Certain related results obtained before 1970 are discussed in the book of Mitrinović [37, section 2.23].

There has been new interest in quadratic inequalities of the form (1.1) since 1970; see Amos [3] and [4], Amos and Everitt [5], [6] and [7], Bradley and Everitt [11] and [12], Bradley, Hinton and Kauffman [13], Everitt, Kwong and Zettl [23], Everitt and Wray [24], Florkiewicz and Rybarski [26], Penning and Sauer [39], Sears and Wray [41], Wray [45] and [46].

It is the object of this paper to obtain a general theorem concerning the inequality (1.1) which includes most of the earlier results quoted above. The theorem is proved under the minimal conditions of local Lebesgue integrability on the coefficients p, q and w, but subject to requiring for the set D that all three integrals in (1.1), in particular the integral involving the coefficient q, are absolutely convergent. These conditions require for (1.2) the theory of linear quasi-differential expressions; see Naĭmark [38, chapter V], Everitt and Zettl [25], Walker [30].

Many of the results in this area, up to 1977, were considered and extended by Amos in his Ph.D. thesis [3], in which a connected account is to be found. Some of the results in [3] remain unpublished but reference to this work is made in this paper.

We may consider the inequality (1.1) as a problem in the calculus of variations, i.e. to minimize the integral on the left-hand side of (1.1) subject to the isoperimetric constraint $||f||_w = 1$, where the norm is taken in the space $L^2_w(I)$. This is a problem in the theory of weak variations, i.e. variations in which a control is exercised over both the function and its derivative in perturbing away from the extremal. However whilst the ideas of the calculus of variations lead to the introduction of the differential expression (1.2) to determine the extremal, if one exists, it soon becomes clear that the methods of the calculus of variations do not work effectively, in general, for this type of problem. The comments made by Hardy, Littlewood and Pólya [28, section 7.2] are particularly significant in this respect. Instead, a number of direct methods can be employed as considered below.

There are a number of different methods which may be used to link the inequality (1.1) with the spectral properties of the differential operator T, with domain D(T) in $L^2_w(I)$; these methods are interdependent but involve a number of different ideas. We list them below, with brief comments on, and references to each method:

1. Integral identities; a number of such identities are given in [28, chapter 7]; in the case of the inequality (1.1) but in the regular case only, i.e. when I = [a,b] is compact, there is a very elegant proof of the best possible result under minimal conditions on the coefficients, by this method; this is noted by Beesack [9] and [10, theorem 2]; it does not seem possible to extend this

method to the singular case.

2. Coercive quadratic forms; the general theory is given by Hildebrandt [30] with the required properties of compact embeddings given by Adams [1] in his book on Sobolev spaces; this method is applicable to regular and singular cases under minimal conditions, see [6] and [7], but is restricted to those cases of (1.1) for which the associated operator T has a discrete spectrum.

3. Spectral identities; this method has wide application to both regular and singular problems under minimal conditions; it requires a detailed knowledge of the spectral representation of the differential operator T; the singular case is considered in [24]; the method is also applicable when the Dirichlet integral in (1.1) is not absolutely convergent, see [41] and [45].

4. Sobolev spaces; in a number of cases it is possible to identify the domain D of (1.1) as a Sobolev-Hilbert function space embedded in L^2_w (I) and then to show that the domain D(T), of the self-adjoint operator T, is dense in D in the Sobolev norm; this method also has wide application to both regular and singular cases, see [3], [5], [11] and [12] although all of these references have some additional restrictions placed on the coefficients over those required in the minimal case; it is this method which is employed in this paper but now extended to the case when the coefficients p, q and w are only locally Lebesgue integrable.

5. Sesqui-linear forms; the representation theory of symmetric sesqui-linear forms in Hilbert space is considered in Kato [34, chapter 6, section 2] and Weidmann [44, chapter 5, section 5.5]; such representations are used in consideration of the inequality (1.1) in the last section of this paper; this method is closely linked with the Sobolev space method. as is made clear there.

6. Dirichlet index; this method has application not only to inequalities of the form (1.1) but to similar problems involving higher order derivatives; it has been applied to the case of continuous coefficients, see [13].

One of the difficulties in considering the inequality (1.1) lies in the fact that whilst the inequality is required on the maximal set D, the parameters of the inequality, i.e. the best possible number μ and the resulting cases of equality, may be determined by a differential operator T with domain D(T) in L_w^2 (I), and yet D(T)$\neq$D. In certain cases D(T) is a sub-set of D; if also it is possible to obtain an inequality of the form (1.1) valid on D(T), then it may prove possible to extend this result from D(T) to D and yet retain the same parameters for the inequality on D. Essentially this is the Sobolev space method adopted in this paper and previously employed in obtaining the results given in [3], [5], [11] and [12]. It is also of interest to note that the results given here are applicable not only to the regular case and the singular limit-point case, but also to certain singular limit-circle cases; see the results in sections 4 and 5 below. However, we are now able to give a proof with this method under the minimal local integrability conditions on the coefficients p, q and w; also the proof avoids the analytical complications of the direct construction method used originally in [11]. As in all previous studies of the inequality (1.1) certain additional constraints are required on the coefficients when the singular case holds; these constraints are essential to the determination of an operator T, self-adjoint in L_w^2 (I) and generated by the differential expression (1.2), so that the domain D(T) is contained in D, and T is bounded below in L_w^2 (I). Such additional constraints are not required in the regular case but are essential in the singular case, and an example is given below to illustrate this point; see section 5.18.

In stating the theorem in the next section it will be assumed that the given coefficients p, q and w are so chosen that D(T) is contained in D and that T is self-adjoint in L_w^2 (I). A similar, but not identical, approach was adopted in [5] but under the condition that the weight function w = 1 on I, and certain smoothness conditions on the coefficients p and q. The validity for such an approach rests upon the applicability of the resulting theorem in particular cases, and we shall see below in section 5 that nearly all known cases, and certain new cases, fall within the scope of the general result. Exceptions include the results obtained in [5], [12], [13] and the result given by Amos [3, theorems 3.1.1 and 3.1.3]; we comment on the difference between some of these results and the results obtained in this paper, in section 5.

One final point; throughout this paper we assume that the coefficients p and w are positive almost everywhere on the interval I. The need for these restrictions becomes clear in the analysis which follows; indeed if these restrictions are removed then the results given here may fail. If p changes sign essentially (Lebesgue) on I then the left-hand side of (1.1) is unbounded above and below on D even with the isoperimetric constraint $||f||_w = 1$. If p is positive almost everywhere and w is non-negative but vanishes on a sub-set of I of positive measure (but not identically on I) then the self-adjoint differential operator T in L_w^2 (I) can be defined; however even if T is bounded below then the inequality (1.1) may still not be valid on the maximal set D, <u>i.e.</u> it may happen that $\mu = -\infty$. For details of such results in the regular case see Everitt, Kwong and Zettl [23].

The contents of the paper are as follows. The general theorem is stated in section 2 and the proof given in section 3. In section 4 we state and prove a second theorem devoted to obtaining explicit conditions on the coefficients p, q and w in order to obtain a valid inequality, and we include a third theorem on the classification of the differential expression in one of the cases of the second theorem. Section 5 relates the results of this paper to earlier work in this area, and contains several new examples. In section 6 we employ the theory of symmetric sesqui-linear forms to obtain an alternative proof of the general theorem in section 2, and point out its relation to the first proof.

Notation. A symbol such as '$(x \in I)$' is to be read as 'for all elements of the set I.' The sets of integers $\{0,1,2,\ldots\}$ and $\{1,2,3,\ldots\}$ are denoted by N and N_1 respectively. Intervals of the real line R are represented by I, [a,b], [a,b), (a,b] and (a,b) where [and (indicate a closed and open end-point respectively; here the end-points a and b satisfy $-\infty \le a < b \le \infty$. The letter C denotes the complex field; if $z \in C$ then $\bar{z}$ denotes the complex conjugate of z.

The letters AC represent absolute continuity; a complex-valued function $f \in AC_{loc}(I)$, for an interval I of R, if f is absolutely continuous on all compact sub-intervals of I.

The letter L denotes Lebesgue integration ; L(I) and $L^2(I)$ represent the usual Lebesgue spaces; also $L_{loc}(I)$ and $L^2_{loc}(I)$ represent the collection of complex-valued functions which belong to the corresponding L-class on all compact sub-intervals of I. Note that, with this notation, $L_{loc}(I)$ is identical with L(I) if and only if the interval I is compact, i.e. I = [a,b].

The symbol $L^2_w(I)$, where w is a non-negative, Lebesgue measurable weight function on I, represents the collection of all complex-valued measurable functions f such that

$$\int_a^b w|f|^2 \equiv \int_a^b w(x)|f(x)|^2 dx < \infty.$$

The symbol H_o represents the Hilbert function space of all equivalence classes of functions in L_w^2 (I); in this space the inner-product and norm are represented by $(\cdot,\cdot)_o$ and $||\cdot||_o$ respectively, e.g.

$$(f,g)_o := \int_a^b wf\bar{g} \equiv \int_a^b w(x)f(x)\overline{g(x)}dx.$$

If w is positive almost everywhere on I then w^{-1} represents the reciprocal function, i.e. $w^{-1}(x) = \{w(x)\}^{-1}$ $(x \in I)$, and not an inverse function (which may not exist); similarly for p and p^{-1}.

Acknowledgements. The authors thank C. D. Ahlbrandt, R. J. Amos, P. R. Beesack, J. S. Bradley, W. D. Evans, M. K. Kwong, N. Sauer and A. Zettl for helpful discussions concerning the general inequality considered in this paper.

W. N. Everitt is indebted to his former student Roger Amos for his collaboration during the years 1974-77, and for his agreement to quote from the unpublished work of his Ph.D. thesis of 1977, in this paper.

S. D. Wray thanks the members of the Department of Mathematics, University of Dundee, for their hospitality during his visit in the summer of 1982 to participate in the Science and Engineering Research Council Symposium on Ordinary Differential Equations and Operators; he gratefully acknowledges an allowance derived from the Symposium grant, numbered GR/B/78083; this work was essentially finalised during the Symposium.

2. Statement of main theorem. Let I be any interval of R, bounded or unbounded, open or half-open or closed; let the end-points of I be a and b with $-\infty \le a < b \le \infty$.

The coefficient functions p, q and w involved in the inequality (1.1) are required to satisfy the following basic conditions:

$$\begin{aligned}
&\text{(i)} && p,\ q,\ w : I \to R \ \underline{\text{and are Lebesgue measurable on } I}; \\
&\text{(ii)} && p(x) \geq 0 \quad (x \in I) \quad \underline{\text{and}} \quad p^{-1} \in L_{loc}(I); \\
&\text{(iii)} && q \in L_{loc}(I); \\
&\text{(iv)} && w(x) > 0 \ (\text{almost all } x \in I) \quad \underline{\text{and}} \quad w \in L_{loc}(I).
\end{aligned} \tag{2.1}$$

Note that (ii) implies that $p(x)>0$ for almost all $x \in I$.

We define the following linear manifolds of the space $L_w^2(I)$:

$$\begin{aligned}
D := \{f : I \to C \ \underline{\text{such that}} \ &\text{(i)} \ f \in L_w^2(I), \\
&\text{(ii)} \ f \in AC_{loc}(I), \\
&\text{(iii)} \ p^{\frac{1}{2}} f' \ \underline{\text{and}} \ |q|^{\frac{1}{2}} f \in L^2(I)\},
\end{aligned} \tag{2.2}$$

$$\begin{aligned}
\Delta := \{f : I \to C \ \underline{\text{such that}} \ &\text{(i)} \ f \in L_w^2(I), \\
&\text{(ii)} \ f \ \underline{\text{and}} \ pf' \in AC_{loc}(I), \\
&\text{(iii)} \ w^{-1}(-(pf')'+qf) \in L_w^2(I)\}.
\end{aligned} \tag{2.3}$$

The linear operator T in the space H_o is defined as follows:

$$\begin{aligned}
D(T) := \{f \in \Delta \ \underline{\text{such that}} \ &\text{(i)} \lim_{a+} pf'g = 0 \ (g \in D) \\
\underline{\text{and}} \quad &\text{(ii)} \lim_{b-} pf'g = 0 \ (g \in D)\}
\end{aligned} \tag{2.4}$$

and $$Tf := w^{-1}(-(pf')'+qf) \quad (f \in D(T)). \tag{2.5}$$

The definition of D is consistent with previous work in this area; see in particular [11, section 2], [5, section 2], [7, section 2] and [24, section 2]. We note that within the framework of this paper D is the maximal linear manifold of $L_w^2(I)$ for which both terms in the Dirichlet integral on the left-hand side of (1.1) are absolutely convergent.

The linear manifold Δ is the domain of the maximal operator in H_o generated by the quasi-differential expression (1.2); for details see [25, section 6] or [30]; for the special case $w = 1$ on I see [38, section 17]; for the case with a weight function, but under smoothness conditions on all the coefficients, see [29, chapter 13].

The definition of the domain D(T) in (2.4) is unusual from the point of view of the determination of self-adjoint restrictions of the maximal operator, with domain Δ in H_o, by the application of normal boundary conditions at the end-points a and b of I; for the form of these normal boundary conditions see [38, section 18.1]. The end-point conditions in the definition (2.4) are adopted here for reasons particular to the discussion of the inequality (1.1); they are associated with, but not identical to, the strong limit-point property given below; they are also connected with results on the sesqui-linear form in H_o associated with T; for details see [24, (2.14) of section 2, and section 4]. In the theorem of section 4 below we show that this definition of D(T) is entirely suitable when it comes to considering applications of the general theorem of this section to special cases.

Given the domain D(T) the definition (2.5) of the differential operator T is in accord with the normal definition of such operators; see [38, section 17] and [25, section 6].

We recall the following results and definitions from the general theory of linear differential expressions:

(i) the differential expression (1.2) is said to be <u>regular</u> at the end-point b of I if for some $c \in (a,b)$ we have

$$\int_c^b (p^{-1} + |q| + w) + b < \infty ;$$

otherwise it is <u>singular</u> at b; a similar definition applies at the end-point a; see [38, section 15.1]; the interval I is taken to be closed at an end-point if and only if the differential expression is regular there; we speak of the <u>regular case</u> if and only if $I = [a,b]$ is compact;

(ii) the differential expression (1.2) is in the <u>limit-point</u> condition in $L^2_w(I)$ at the end-point b (similarly at the end-point a) if it is singular at b and for each $\lambda \in C$ there is a solution y of the differential equation

$$-(py')' + qy = \lambda wy \qquad \text{(on } I)$$

such that

$$\int^b w|y|^2 = \infty ;$$

it is well known that the differential expression is in the limit-point condition at b if and only if

$$\lim_{b-} (pf'g - fpg') = 0 \qquad (f,g \in \Delta), \qquad (2.6)$$

see the lemma in [38, p. 78];

(iii) the differential expression (1.2) is in the <u>strong limit-point</u> condition in $L^2_w(I)$ at the end-point b (similarly at the end-point a) if it is singular at b and

$$\lim_{b-} pf'g = 0 \qquad (f,g \in \Delta); \qquad (2.7)$$

note that this implies that the differential expression is in the limit-point condition at b;

(iv) the differential expression (1.2) is said to be in the <u>Dirichlet</u> condition in $L^2_w(I)$ if

$$\Delta \subset D; \qquad (2.8)$$

we also say that if I has the form $[a,b)$ or $(a,b]$ and (2.8) holds then the expression is in the Dirichlet condition at b or a respectively.

There are a number of known connections between the strong limit-point and Dirichlet conditions; for general accounts see

Everitt [15] and [16], and Kalf [33]. For a number of special results giving conditions, additional to (2.1), on the coefficients p, q and w which yield (2.7), and hence (2.6), and/or (2.8) see Amos and Everitt [7], Atkinson [8], Bradley and Everitt [11], Evans [14], Everitt [16], Everitt and Giertz [20], Everitt, Giertz and McLeod [21], Everitt, Giertz and Weidmann [22], Kwong [35] and [36]. Many of these results have significance for the determination of cases when the definition (2.4) of D(T) is appropriate for the application of the theorem given below.

If the differential expression (1.2) is not regular or limit-point at the end-point b then it is said to be in the limit-circle condition at b, see [38, section 17.5] or [29, section 13.3]; thus the differential expression is in the limit-circle condition at b if and only if for each $\lambda \in C$ every solution y of the differential equation

$$-(py')' + qy = \lambda wy \qquad \text{(on I)}$$

satisfies

$$\int^b w|y|^2 < \infty ;$$

similarly at the end-point a.

For the motivation of the definitions of limit-point and limit-circle see Titchmarsh [42, section 2.1, and sections 2.19-2.21], which contains references to the original work of Weyl.

With these remarks made and definitions given, we now state the general theorem concerning the inequality (1.1).

Theorem 1 *Let I be an arbitrary interval of the real line with end-points a and b, where $-\infty \le a < b \le \infty$; let the real-valued coefficients p, q and w satisfy the basic conditions (2.1) on I; let the notation of section 1 hold; let the linear manifolds D, Δ and D(T) of the Hilbert space H_0 be defined as in (2.2) to (2.4) above; let the opera-*

tor T on $D(T)$ in H_o be defined as in (2.5); suppose additionally that the coefficients p, q and w are chosen so that the following three conditions are satisfied:

1. $q = q_1 + q_2$ with $q_r \in L_{loc}(I)$ $(r=1,2)$ and

   (i) there exists $A \in [0,\infty)$ such that

$$q_1(x) + Aw(x) \geq 0 \quad (\text{almost all } x \in I) , \tag{2.9}$$

   (ii) there exist $k \in [0,1)$ and $K \in (0,\infty)$ such that

$$\int_a^b |q_2||f|^2 \leq k\int_a^b p|f'|^2 + K\int_a^b w|f|^2 \quad (f \in D); \tag{2.10}$$

2. $D(T) \subset D$; (2.11)

3. T is self-adjoint in H_o. (2.12)

Then it follows that:

   (i) T is bounded below in H_o, i.e. if

$$\mu := \inf\{(Tf,f)_o : f \in D(T) \text{ and } ||f||_o = 1\} \tag{2.13}$$

then

$$\mu > -\infty \text{ and } (Tf,f)_o \geq \mu(f,f)_o \quad (f \in D(T)) ; \tag{2.14}$$

   (ii) the following inequality holds

$$\int_a^b \{p|f'|^2 + q|f|^2\} \geq \mu \int_a^b w|f|^2 \quad (f \in D), \tag{2.15}$$

which is best possible in the sense that there exists a sequence $\{f_n : n \in N\}$ of H_o satisfying

$$f_n \in D(T) (\subset D) \text{ and } ||f_n||_o = 1 \quad (n \in N) \tag{2.16}$$

such that

$$\lim_{n\to\infty} \int_a^b \{p|f_n'|^2 + q|f_n|^2\} = \mu ; \tag{2.17}$$

   (iii) there is equality in (2.15) as follows: (α) when μ is in the (pure) continuous spectrum of T, if and only if f is null on I,

(β) <u>when</u> μ <u>is an eigenvalue of</u> T, <u>if and only if</u> f <u>is in the eigen-space of</u> T <u>at</u> μ.

<u>Proof</u> This is given in section 3 below.

<u>Remarks</u> 1. Note that with T self-adjoint in H_o it follows that $(Tf,f)_o \in R$ $(f \in D(T))$ so that the definition (2.13) is meaningful; also that

$$\mu = \inf\{\sigma(T)\} , \tag{2.18}$$

where $\sigma(T)$ denotes the spectrum of T; since $\sigma(T)$ is a closed set of R we have $\mu \in \sigma(T)$ so that (α) and (β) of (iii) of the above theorem cover all possibilities for μ; for the proof of (2.18) see [34, p. 278].

2. The cases (α) and (β) of (iii) of the theorem characterize completely the circumstances in which the isoperimetric problem in the calculus of variations represented by the inequality (1.1), has an unattained or attained solution respectively.

3. We show in section 5 below that the above theorem contains as special cases nearly all the previously known results on the inequality (1.1), and includes many new cases; the main exceptions are the results obtained in [5], [11] and the corresponding extensions given in [3], and [12].

4. We note that under the conditions of the theorem the end-point a or b may be regular, or singular limit-point or singular limit-circle; all possibilities are illustrated in the examples in section 5.

5. The conditions 1(ii), 2 and 3 of the theorem may be said to be indirect, <u>i.e.</u> not given directly on the coefficients p, q and w; the validity of this type of condition rests upon the applicability of the theorem; we show below in section 4 that the general theorem yields a variety of cases in which explicit condi-

tions can be given on the coefficients.

6. Condition 1(ii) is similar to a condition imposed by Kalf [33, p. 199, condition (iv)]; see also [22, p. 339], [7, p. 20], [17, p. 37 and references] and [32, condition (H_3), p. 293].

7. Condition 2 is similar to a condition imposed by Amos and Everitt [5, (2.9) of theorem 1].

8. Some restriction in the form of conditions (2.9) and (2.10) is essential if an inequality of the type (1.1) is to hold; these conditions allow the coefficient q to take negative values but essentially control the extent, with respect to the non-negative coefficients p and w, to which this can occur; an example is given in section 5.18 to show the necessity of this type of restriction.

9. Note that condition (2.11) of the theorem, i.e. $D(T) \subset D$, is less demanding than the Dirichlet condition (2.8), i.e. $\Delta \subset D$; this is one of the reasons why the general theorem above works in some limit-circle cases (recall that in certain general circumstances, e.g. $w = 1$ on I, the Dirichlet condition implies the strong limit-point condition in the singular case, see Everitt [16]); in his thesis, Amos has considered cases where D(T) is not a sub-set of D, see [3, theorem 3.1.1 and remarks on p. 85].

3. Proof of theorem 1

3.1. We begin by establishing that the operator T is bounded below in H_o, i.e. that (2.13) and (2.14) above hold.

Let $x, y \in (a,b)$, $x<y$, and let $f \in D(T)$; then on integration by parts we have

$$\int_x^y wTf\bar{f} = \int_x^y \{-(pf')' + qf\}\bar{f}$$

$$= -pf'\bar{f}\Big|_x^y + \int_x^y \{p|f'|^2 + q|f|^2\} .$$

Since $f \in D(T)$ and from condition 2 of the theorem $D(T) \subset D$, it follows from (2.4) that

$$\lim_{a+} pf'\bar{f} = \lim_{b-} pf'\bar{f} = 0.$$

Hence, letting $x \to a+$ and $y \to b-$ above, we obtain

$$(Tf,f)_o = \int_a^b wTf\bar{f} = \int_a^b \{p|f'|^2 + q|f|^2\} \quad (f \in D(T)). \tag{3.1}$$

Note that the integrals on the right of (3.1) are both absolutely convergent, since $D(T) \subset D$.

Now since $D(T) \subset D$ it follows that $|q|^{\frac{1}{2}} f \in L^2(I)$ $(f \in D(T))$, and from 1(ii) it follows that $|q_2|^{\frac{1}{2}} f \in L^2(I)$ $(f \in D(T))$; thus $|q_1|^{\frac{1}{2}} f \in L^2(I)$ $(f \in D(T))$. Hence from (3.1) we may write, on using 1(i) and (ii),

$$\begin{aligned}
(Tf,f)_o &= \int_a^b \{p|f'|^2 + q_1|f|^2 + q_2|f|^2\} \\
&= \int_a^b \{p|f'|^2 + (q_1 + Aw)|f|^2 + q_2|f|^2 - Aw|f|^2\} \\
&\geq \int_a^b \{p|f'|^2 - Aw|f|^2\} - \int_a^b |q_2||f|^2 \\
&\geq \int_a^b \{p|f'|^2 - Aw|f|^2\} - k\int_a^b p|f'|^2 - K\int_a^b w|f|^2 \\
&= (1-k)\int_a^b p|f'|^2 - (A + K)\int_a^b w|f|^2 \\
&\geq -(A + K)(f,f)_o \quad (f \in D(T)).
\end{aligned}$$

Thus T is bounded below in H_o.

3.2. We now take μ defined by (2.13) and note from the general theory of self-adjoint operators in Hilbert space, see Kato [34, p. 278], that μ may be characterized by (see remark 1 above)

$$\mu = \inf\{\sigma(T)\}, \tag{2.18}$$

that the inequality

$$(Tf,f)_o \geq \mu\,(f,f)_o \quad (f \in D(T)) \tag{2.14}$$

is best possible, and that there is equality if and only if f satisfies condition (α) or (β) at the end of the statement of the theorem.

From (2.14) and (3.1) we now have

$$\int_a^b \{p|f'|^2 + q|f|^2\} \geq \mu \int_a^b w|f|^2 \quad (f \in D(T)), \qquad (3.2)$$

which is best possible and with equality if and only if (α) or (β) is satisfied by f.

The problem now is to extend (3.2) to (1.1), i.e. to show that the inequality is valid on the larger set D and that in the process of this extension no additional cases of equality are introduced.

3.3. To carry out the extension from (3.2) to (1.1) we introduce a second Hilbert space H_1, where

$$H_1 := D \text{ endowed with the inner-product } (\cdot,\cdot)_1 ,$$

where

$$(f,g)_1 = \int_a^b \{pf'\bar{g}' + (q-\mu w)f\bar{g} + Bwf\bar{g}\} \quad (f,g \in D) \quad (3.3)$$

and

$$B = \max\{1, \mu + A + K + 1\} . \qquad (3.4)$$

Clearly (3.3) defines an inner-product provided that for all $f \in D$ we have $(f,f)_1 \geq 0$ and $||f||_1 = 0$ if and only if $f(x) = 0$ (almost all $x \in I$). These results may be deduced easily from the following analysis. By condition 1 we have

$$||f||_1^2 = \int_a^b \{p|f'|^2 + (q_1+Aw)|f|^2\} + (B-\mu-A)\int_a^b w|f|^2 + \int_a^b q_2|f|^2$$

$$\geq (1-k)\int_a^b p|f'|^2 + (B-\mu-A-K)\int_a^b w|f|^2$$

$$\geq (1-k)\int_a^b p|f'|^2 + \int_a^b w|f|^2 \geq \int_a^b w|f|^2 \quad (f \in D) , \qquad (3.5)$$

on using (3.4).

We note that H_1 is complete in the norm $||\cdot||_1$; to prove this result, we may easily adapt the proof of Lemma 2 of [24]: the closure of the quadratic form there is equivalent to the completeness of H_1. (We comment on this further in Section 6.2.) Thus, H_1 is indeed a Hilbert space.

Clearly, in the sense of set inclusion H_1 is contained in H_o; also from (3.5)

$$||f||_o \leq ||f||_1 \qquad (f \epsilon H_1)$$

and so in the terminology in the book of Adams [1, section 1.23] the space H_1 is continuously embedded in H_o.

3.4. From condition 2 of the theorem we see that the domain D(T) is a linear manifold not only of H_o but also of H_1. The next step is to show that D(T) is dense in the Hilbert space H_1. Suppose this is not the case. Then there exists $g \epsilon H_1$ with $||g||_1 = 1$ and $(f,g)_1 = 0$ for all $f \epsilon D(T)$, see Akhiezer and Glazman [2, section 17], and so, on integration by parts,

$$(f,g)_1 = \int_a^b \{pf'\bar{g}' + (q - \mu w)f\bar{g} + Bwf\bar{g}\}$$

$$= \lim_{b-} pf'\bar{g} - \lim_{a+} pf'\bar{g} + \int_a^b wTf\bar{g} - (\mu-B) \int_a^b wf\bar{g}$$

$$= (Tf,g)_o - (\mu-B)(f,g)_o \qquad (f \epsilon D(T))$$

on using the boundary conditions (i) and (ii) of (2.4). Thus

$$0 = (f,g)_1 = ((T - (\mu-B)E)f,g)_o \quad (f \epsilon D(T)) , \qquad (3.6)$$

where E is the identity operator in H_o. Now from the definition (3.4) we have $B \geq 1 > 0$, and since from (2.18) the real number μ is the lower bound of the spectrum of the self-adjoint operator T, it follows that $\mu - B \epsilon \rho(T)$, the resolvent set of T (see [2, section 44]). Thus

$$\{(T - (\mu-B)E)f : f\epsilon D(T)\} = H_o \tag{3.7}$$

and (3.6) and (3.7) imply that g is the null vector of H_o, i.e. $g(x) = 0$ (almost all $x\epsilon I$); this in turn implies that $||g||_1 = 0$. This is a contradiction, and so D(T) is dense in H_1.

3.5. As a consequence of the denseness of D(T) in H_1 we state the following result: given any $f\epsilon D$ and any $\varepsilon > 0$ there exists $g\epsilon D(T)$ (in general g will depend on both f and ε) such that

$$\int_a^b p|f' - g'|^2 < \varepsilon , \int_a^b |q||f - g|^2 < \varepsilon , \int_a^b w|f - g|^2 < \varepsilon . \tag{3.8}$$

This is the so-called approximation result from earlier work in [11, theorem 3] and [12, theorem 2]; see also [5, remark 2 of theorem 1].

To prove (3.8), under the conditions of the theorem in section 2 above, we note that from the result of the previous section given $f\epsilon D$ and $\varepsilon > 0$, and any $\ell \epsilon (0,1)$, we find $g\epsilon D(T)$ such that $||f - g||_1 < \ell\varepsilon$. Arguing as in the proof of (3.5) we obtain, from condition 1,

$$(1-k)\int_a^b p|f' - g'|^2 + \int_a^b (q_1 + Aw)|f - g|^2 + \int_a^b w|f - g|^2 < \ell\varepsilon .$$

All terms on the left are non-negative and so each one is less than $\ell\varepsilon$; from this and another application of the inequality 1(ii) it follows that (3.8) holds provided that the positive number ℓ is determined by (we omit details of the calculation)

$$\ell = \frac{1}{2} \min\{(1 - k), (A + 1)^{-1}, (k(1 - k)^{-1} + K)^{-1}\} .$$

Following the argument given in [5, section 6] it may now be shown that: given any $f\epsilon D$ and any $\eta > 0$ there exists $g\epsilon D(T)$ (in general g will depend on both f and η) such that

$$\left| \int_a^b p\{|f'|^2 - |g'|^2\}\right| < \eta , \left|\int_a^b q\{|f|^2 - |g|^2\}\right| < \eta$$

and

$$\left|\int_a^b w\{|f|^2 - |g|^2\}\right| < \eta. \tag{3.9}$$

3.6. With these results established we may now prove that the inequality (3.2) may be extended from D(T) to D, <u>i.e.</u> to give the inequality (1.1), with μ defined by (2.13) or, equivalently, (2.18). Again we follow the argument in [5, section 6]. Suppose, to the contrary, that there exists $f \in D$ such that for some number $\delta > 0$

$$\int_a^b \{p|f'|^2 + q|f|^2 - \mu w|f|^2\} = -\delta < 0.$$

If in (3.9) we take $\eta = \frac{1}{2}(2 + |\mu|)^{-1}\delta$ and then determine $g \in D(T)$ so that (3.9) holds for this number η, a calculation shows that

$$\int_a^b \{p|g'|^2 + q|g|^2 - \mu w|g|^2\} < -\tfrac{1}{2}\,\delta < 0$$

and this is a contradiction on (3.2) since $g \in D(T)$. Thus the required extension from D(T) to D is obtained.

Since we know that (3.2) is best possible on D(T) and also that $D(T) \subset D$ it follows that the inequality (1.1) is best possible on the maximal set D.

In fact to prove that (2.16) and (2.17) hold we note that if μ is an eigenvalue of T then there is an eigenvector $f \in D(T)$, with $||f||_o = 1$, such that $Tf = \mu f$; define $f_n := f$ $(n \in N)$ and the result follows. If μ is in the continuous spectrum of T then from the analysis in [38, section 14.9] we obtain the existence of a sequence $\{f_n : n \in N\}$ satisfying $f_n \in D(T)$ $(n \in N)$, $(f_n, f_n)_o = 1$ $(n \in N)$ and $(T - \mu E)f_n \to 0$ in H_o as $n \to \infty$; this implies that $((T - \mu E)f_n, f_n)_o \to 0$ in H_o as $n \to \infty$ and hence that

$$\int_a^b \{p|f_n'|^2 + (q - \mu w)|f_n|^2\} \to 0 \qquad \text{as } n \to \infty.$$

Thus $\{f_n : n \in N\}$ is the desired sequence.

3.7. To complete the proof of the theorem we have to discuss the cases of equality in (1.1) and here the methods of [5, sections 7 and 8] are suitable.

Define the functional $J : D \times D \to C$ by

$$J(f,g) := \int_a^b \{pf'\bar{g}' + (q - \mu w)f\bar{g}\} \qquad (f,g \in D) .$$

We have already shown that $J(f,f) \geq 0 \quad (f \in D)$; let $F \in D$ be chosen so that $J(F,F) = 0$. Let $\gamma \in C$; then $F + \gamma f \in D \quad (f \in D)$ and

$$\begin{aligned} 0 \leq J(F + \gamma f, F + \gamma f) &= J(F,F) + \gamma J(f,F) + \bar{\gamma} J(F,f) + |\gamma|^2 J(f,f) \\ &= \gamma J(f,F) + \bar{\gamma} J(F,f) + |\gamma|^2 J(f,f) \end{aligned}$$

again for all $f \in D$. If $J(f,F) = \bar{J}(F,f) \neq 0$ then, since γ is arbitrary, a contradiction may be obtained if γ is chosen appropriately, with $|\gamma|$ small. Thus

$$J(f,F) = 0 \qquad (f \in D) . \tag{3.10}$$

Since $D(T) \subset D$ it follows from (3.10) that

$$J(f,F) = 0 \qquad (f \in D(T)) ,$$

i.e.

$$\int_a^b \{pf\bar{F}' + (q - \mu w)f\bar{F}\} = 0 \qquad (f \in D(T)) .$$

On integration by parts, as in the proof of (3.6), it follows that

$$((T - \mu E)f, F)_o = 0 \qquad (f \in D(T))$$

and hence F is in the orthogonal complement of the range set

$$\{(T - \mu E)f : f \in D(T)\}$$

in the space H_o. Thus from [2, section 43] it follows that if μ is an eigenvalue of T then F belongs to the eigenspace of T at μ, and if μ belongs to the continuous spectrum of T then F is null on I.

This completes the proof of the theorem.

4. Two more theorems This section is primarily concerned with

obtaining explicit conditions on the coefficients p, q and w to enable a direct application of theorem 1 to be made; these are the subject of theorem 2 below.

In another result, theorem 3, we examine the classification of the differential expression (1.2) at the singular end-point in one of the cases of theorem 2; this case includes in particular some limit-circle examples.

Theorem 2. *Let* I *be an interval of the real line with end-points* a *and* b, *where* $-\infty\le a<b\le\infty$; *let the coefficients* p, q *and* w *satisfy the basic conditions* (2.1) *on* I; *then the conditions* 1, 2 *and* 3 *of theorem* 1 *on the coefficients* p, q *and* w *and the operator* T *are all satisfied in any one of the following three circumstances* (a), (b) *or* (c):

(a) *if* $I = [a,b)$ *with* $b\le\infty$ *and*

either (α) $b<\infty$ *and the conditions* (2.1) *are satisfied on the compact interval* $[a,b]$,

or (β) *there exists* $A\in(0,\infty)$ *such that*

$$|q(x)|\le Aw(x) \quad (\text{almost all } x\in[a,b)), \tag{4.1}$$

or (γ) *with* $q = q_1 + q_2$ (*as in theorem* 1)

(i) *there exists* $A\in[0,\infty)$ *such that*

$$q_1(x) + Aw(x) \ge 0 \quad (\text{almost all } x\in[a,b)), \tag{4.2}$$

(ii) *if the linear manifold* D_o *of* $L^2_w(I)$ *is defined by*

$$D_o := \{f\in L^2_w(I) \text{ such that } f\in AC_{loc}(I) \text{ and } p^{\frac{1}{2}}f'\in L^2_{loc}(I)\} \tag{4.3}$$

then there exist $k\in[0,1)$ *and* $K\in(0,\infty)$ *such that*

$$\int_a^x |q_2||f|^2 \le k\int_a^x p|f'|^2 + K\int_a^x w|f|^2 \quad (x\in[a,b) \text{ and } f\in D_o), \tag{4.4}$$

(iii) *the coefficients* p *and* w *satisfy one of*

$$w \notin L(I) , \qquad (4.5a)$$

or

$$\int_a^b w(x)\{ \int_a^x p^{-1}\}dx = \infty; \qquad (4.5b)$$

(b) *if* I = (a,b] *with* $-\infty \le a$ *and if the equivalent of* (β) *or* (γ) *of* (a) *holds at the end-point* a *of* I;

(c) *if* I = (a,b) *with* $-\infty \le a < b \le \infty$ *and if for some point* $c \in (a,b)$ *one of the conditions* (β) *or* (γ) *of* (a) *holds on* [c,b) *and one of the equivalent of* (β) *or* (γ) *of* (a) *holds on* (a,c].

Thus, if any one of (a), (b) *or* (c) *above holds then all the consequences of theorem 1 are valid*.

Proof. See below.

Remarks 1. Case (a)(α) of the theorem is the so-called regular case; see [6], [9], [10, theorem 2], [11, theorem 1], [31, theorem 2(i)], [41, theorem 2(ii)] and [45, theorem 4].

2. Case (a)(β) requires only the basic conditions (2.1) and the condition (4.1) for the inequality to hold; in particular, if q is null on I then the inequality is valid under (2.1) only.

3. Case (a)(γ) is related to results in [22] and [33]; condition (4.4) implies (2.10) since $D \subset D_0$, but the converse does not hold.

4. It will be shown in the next section that, with the exception of the results in [3, chapters 2 and 3], [5], [12] and [13], all the known previous cases of the inequality (1.1) are contained in theorem 2, *i.e.* those given in [3, section 5.4], [6], [7], [11], [24], [32], [39], [40] and [46].

Proof of theorem 2 We argue throughout the proof with real-valued functions; all the results extend trivially to general complex-valued functions if one considers their real and imaginary parts separately.

Consider case (a)(α), i.e. the regular case, in which $I = [a,b]$.

We deal first of all with the conditions (i) and (ii) in the definition (2.4) of the domain $D(T)$. Let $f \in D(T)$. Then choose $g \in D$ and $c \in I$ such that $g(c) \neq 0$, and define $h : I \to C$ by

$$h(x) = \begin{cases} g(c) , & a \leq x \leq c , \\ g(x) , & x > c . \end{cases}$$

Then $h \in D$ and $h(a) \neq 0$. Since (i) in (2.4) is satisfied by f, we have

$$(pf')(a) = 0 \tag{4.6}$$

on replacing g by the above function h in (i). A similar argument shows that

$$(pf')(b) = 0 . \tag{4.7}$$

Since obviously any $f \in \Delta$ satisfying (4.6) and (4.7) is contained in $D(T)$, we see that conditions (i) and (ii) in (2.4) are equivalent to the more usual Neumann boundary conditions (4.6) and (4.7).

The significance of the Neumann boundary conditions (in the regular case, and in singular cases with a regular end-point), in order to determine the correct operator T in the theory of the inequality (1.1), has been pointed out previously; see [5, theorem 2, remark 2], [7, sections 2 and 6.2] and [11, section 6].

In view of the fact that $D(T) = \{f \in \Delta : (pf')(a) = (pf')(b) = 0\}$, we may call upon the results of Everitt [17], which show that it is not necessary to make the assumptions 1-3 of theorem 1: these conditions are automatically satisfied. Specifically, (2.10) is satisfied with q replacing q_2, see [17, theorem 1]; we have

$$\int_a^b p|f'|^2 < \infty \qquad (f \in D(T)),$$

see [17, section 5], whence $D(T) \subset D$; and T is self-adjoint in H_o, see [17, theorem 2].

Now suppose that condition (a)(β) holds. The interval I is now $[a,b)$, with b a singular end-point, and the above discussion of the regular case shows that

$$D(T) = \{f\in\Delta : (pf')(a) = 0 \text{ and } \lim_{b-} pf'g = 0 \quad (g\in D)\}. \tag{4.8}$$

Clearly (4.1) implies condition 1 of theorem 1: take $q_1 = q$, $q_2 = 0$ to obtain (2.9). To establish that conditions 2 and 3 of theorem 1 are satisfied, we proceed as follows. The argument depends upon whether or not w is integrable over I.

If $w\notin L(I)$ then the result is a special case of that in case (a)(γ): again take $q_1 = q$, $q_2 = 0$; see the proof below in this case.

Suppose that $w\in L(I)$. The proof in this case employs a standard self-adjoint operator T_o associated with the differential expression (1.2), namely the Friedrichs extension of the minimal operator associated with it; this operator will be identified with the operator T defined in (2.4) - (2.5).

The operator T_o is defined as follows. Let

$$D(T_o) = \{f\in\Delta : (pf')(a) = \lim_{b-} pf' = 0\} \tag{4.9}$$

and then let

$$T_o f = w^{-1}(-(pf')' + qf) \qquad (f\in D(T_o)).$$

If the differential expression is in the limit-point condition at b, then by the lemma in [38, p. 78] we have

$$\lim_{b-} \{pf'g - pg'f\} = 0 \qquad (f,g\in\Delta). \tag{4.10}$$

Since $w\in L(I)$ and (4.1) holds, we have $q\in L(I)$ as well, and it is straightforward to check that $\phi\in\Delta\cap D$, where $\phi(x) = 1$ $(x\in I)$. Taking $g = \phi$ in (4.10), we see that

$$\lim_{b-} pf' = 0 \qquad (f\in\Delta),$$

which means that the condition at b in the definition (4.9) of $D(T_o)$ is superfluous in the limit-point case; this condition is needed in the limit-circle case, as we shall see. That both the limit-point and limit-circle cases may occur in these circumstances is made clear by theorem 3 below, with specific examples in section 5.15.

In the limit-point case we may again invoke [38, theorem 5', p. 80] to obtain the self-adjointness of T_o.

Now suppose that we have the limit-circle case at b; to show that T_o is self-adjoint we make use of the results of [38, Section 18.1].

Choose any two points x_1, $x_2 \epsilon (a,b)$ with $x_1 < x_2$, and define the functions ϕ_1, ϕ_2 by

$$\phi_1(x) = \begin{cases} 1 , & x \epsilon [a,x_1] , \\ \psi_1(x) , & x \epsilon [x_1,x_2] , \\ 0 , & x \epsilon [x_2,b) , \end{cases}$$

and

$$\phi_2(x) = \begin{cases} 0 , & x \epsilon [a,x_1] , \\ \psi_2(x) , & x \epsilon [x_1,x_2] , \\ 1 , & x \epsilon [x_2,b) , \end{cases}$$

where ψ_1 is a function satisfying $\psi_1(x_1) = 1$, $\psi_1(x_2)=(p\psi_1')(x_1) = (p\psi_1')(x_2) = 0$ and such that $\phi_1 \epsilon \Delta$, and ψ_2 is chosen similarly to make $\phi_2 \epsilon \Delta$; the existence of such functions ψ_1, ψ_2 is guaranteed by [25, theorem 5.2], given that (4.1) holds and $w \epsilon L(I)$. Now using [38, theorem 4, p. 75], we have that the domain of a self-adjoint operator derived from the differential expression consists of the set of all functions $f \epsilon \Delta$ which satisfy the conditions

$$\lim_{b-} (pf'\phi_1 - p\phi_1' f) - (pf'\phi_1 - p\phi_1' f)(a) = 0$$

and

$$\lim_{b-} (pf'\phi_2 - p\phi_2'f) - (pf'\phi_2 - p\phi_2'f)(a) = 0.$$

These conditions are easily seen to be equivalent to those in the definition of $D(T_o)$, see (4.9), and so the operator T_o is self-adjoint once more.

From now on, we deal with the limit-point and limit-circle cases simultaneously.

Since $D(T)$ is as in (4.8), and since the above function $\phi \in D$, we have

$$\lim_{b-} pf' = 0 \qquad (f \in D(T))$$

and so $\quad D(T) \subset D(T_o)$.

Now we show that $D(T_o) \subset D$. Let $f \in D(T_o)$; then f, $w^{-1}(-(pf')' + qf) \in L_w^2(I)$, and by integrating by parts we obtain the identity

$$\int_a^x ww^{-1}(-(pf')' + qf)f$$

$$= - pf'f\Big|_a^x + \int_a^x (pf'^2 + qf^2) \qquad (x \in I) . \tag{4.11}$$

Since $f \in L_w^2(I)$ and (4.1) holds, we have $|q|^{\frac{1}{2}} f \in L^2(I)$ and so, from (4.11),

$$\int_a^x pf'^2 = (pf')(x)f(x) + O(1) \qquad (x \to b-) . \tag{4.12}$$

If $p^{\frac{1}{2}}f' \notin L^2(I)$ then (4.12) shows that $(pf')(x)f(x) \to \infty$ as $x \to b-$; then there are $M > 0$ and $c \in I$ such that $(pf')(x)f(x) \geq M$ $(x \in [c,b))$, and we may assume that $(pf')(x) > 0$, $f(x) > 0$ $(x \in [c,b))$ (otherwise multiply f by -1); it follows that

$$|(pf')'(x)|f(x) \geq M|(pf')'(x)|/(pf')(x) \quad (\text{almost all } x \in [c,b))$$

and so

$$\int_c^x |(pf')'|f \geq M \int_c^x |(pf')'|/pf'$$

$$\geq M \left| \int_c^x (pf')'/pf' \right|$$

$$= M \left| \log(pf')(x) - \log(pf')(c) \right| \quad (x \in [c,b)). \quad (4.13)$$

Now, by (4.1), $w^{-1}qf \in L_w^2(I)$ and so $w^{-1}(pf')' \in L_w^2(I)$, since $f \in \Delta$; hence

$$\int_a^b |(pf')'|f = \int_a^b ww^{-1}|(pf')'|f < \infty$$

and so, from (4.13), we have that

$$0 < K_1 \leq (pf')(x) \leq K_2 \quad (x \in [c,b)) , \quad (4.14)$$

for some constants, K_1 and K_2. But $\lim_{b-} pf' = 0$, and this is a contradiction. Hence we must have $p^{\frac{1}{2}}f' \in L^2(I)$. Accordingly, $D(T_o) \subset D$.

The proof in this case is completed by showing that $D(T_o) \subset D(T)$, since this will identify T with T_o and hence show that T is self-adjoint. (We have already that $D(T) \subset D$, from what we have just shown.)

Let $f \in D(T_o)$ and let $g \in D$; since $f \in D$ and, by integration by parts,

$$\int_a^x ww^{-1}(-(pf')' + qf)g$$

$$= - pf'g\Big|_a^x + \int_a^x (pf'g' + qfg) , \quad (4.15)$$

it follows that $\ell = \lim_{b-} pf'g$ exists and is finite; if $\ell \neq 0$ then we may suppose it to be positive, and also that $(pf')(x) > 0$, $g(x) > 0$ and $(pf')(x)g(x) \geq \frac{1}{2}\ell$ on some interval $[c,b)$ (otherwise we multiply f or g by -1); then $|(pf')'(x)|g(x) \geq \frac{1}{2}\ell|(pf')'(x)|/(pf')(x)$ (almost all $x \in [c,b)$) and as before we obtain the contradictory result (4.14); hence $\ell = 0$, i.e. $\lim_{b-} pf'g = 0$, and so $f \in D(T)$.

This completes the proof of the theorem under condition (a)(β). Note that here Δ is not in general a sub-set of D; see case (iii)(a) of theorem 3 below.

We now consider the proof in the case where the conditions in (a)(γ) hold.

Since $D \subset D_o$, it is clear that condition 1 of theorem 1 is satisfied.

We show next that $\Delta \subset D$. Let $f \in \Delta$; then

$$\int_a^b w|w^{-1}(-(pf')' + qf)f| < \infty \tag{4.16}$$

and $f \in D_o$ since

$$\int_a^x pf'^2 = \int_a^b p^{-1}(pf')^2 \leq M \int_a^x p^{-1} \quad (x \in I),$$

where M is the supremum of $(pf')^2$ on $[a,x]$; since $f \in D_o$ we obtain from (4.2), (4.4) and (4.11)

$$(1-k) \int_a^x pf'^2 + \int_a^x (q_1 + Aw)f^2$$

$$\leq pf'f\Big|_a^x + (K+A)\int_a^x wf^2 + \int_a^x ww^{-1}(-(pf')' + qf) \quad (x \in I). \tag{4.17}$$

Suppose that either of the integrals on the left of the inequality in (4.17) does not have a finite limit as $x \to b$; then by (4.16) we have $(pf')(x)f(x) \to \infty$ as $x \to b-$, so that there is a constant $B > 0$ and a point $c \in I$ such that $f'(x)f(x) \geq B\, p^{-1}(x)$ (almost all $x \in [c,b)$); on integration we then obtain

$$f^2(x) \geq f^2(c) + 2B \int_c^x p^{-1} \quad (x \in [c,b)), \tag{4.18}$$

and also $f^2(x) > 0$ $(x \in [c,b))$; whichever of (4.5a) or (4.5b) holds, we obtain a contradiction since $f \in L^2_w(I)$ and in (4.18) $f(c) \neq 0$; thus $p^{\frac{1}{2}}f'$, $|q_1|^{\frac{1}{2}}f \in L^2(I)$, and we may now use (4.4) again to obtain $|q_2|^{\frac{1}{2}}f \in L^2(I)$; hence $f \in D$, and we have shown that $\Delta \subset D$, which implies condition 2 of theorem 1.

Now we show that

$$\lim_{b-} pf'g = 0 \quad (f\in\Delta, g\in D) , \tag{4.19}$$

which is stronger than the strong limit-point condition.

Let $f\in\Delta, g\in D$; then $f\in D$ and from (4.15) $\ell = \lim_{b-} pf'g$ exists and is finite; if $\ell \neq 0$ then as before we may suppose that $\ell > 0$ and that there is $c\in I$ such that $(pf')(x) > 0$, $g(x) > 0$ and $(pf')(x)g(x) \geq \frac{1}{2}\ell$ $(x\in[c,b))$; this gives $(pf')(x)|g'(x)| \geq \frac{1}{2}\ell|g'(x)|/g(x)$ (almost all $x\in[c,b)$) and on integration of this we obtain

$$\int_c^x pf'|g'| \geq \tfrac{1}{2}\ell\left|\log g(x) - \log g(c)\right| \quad (x\in[c,b)),$$

much as in the proof of (4.13); since $pf'g'\in L(I)$, this implies that

$$0 < K_1 \leq g(x) \leq K_2 \quad (x\in[c,b)),$$

for some constants, K_1 and K_2; but this is impossible if $w\notin L(I)$ (i.e. if (4.5a) holds) since $g\in L^2_w(I)$; if $w\in L(I)$ then from (4.5b) we have $p^{-1}\notin L(I)$ and as we have $\frac{1}{2}\ell p^{-\frac{1}{2}}(x) \leq p^{\frac{1}{2}}f'(x)g(x) \leq K_2p^{\frac{1}{2}}(x)f'(x)$ (almost all $x\in[c,b)$), whence $p^{-1}(x) \leq 4\, K_2^2p(x)(f'(x))^2/\ell^2$ (almost all $x\in[c,b)$), we again have a contradiction because $p^{\frac{1}{2}}f'\in L^2(I)$; thus $\ell = 0$, i.e. (4.19) holds.

In view of (4.8), (4.19) and the fact that $\Delta\subset D$ holds, we have

$$D(T) = \{f\in\Delta \ : \ (pf')(a) = 0\} ,$$

and so, by [38, theorem 5', p. 80], T is self-adjoint in H_o.

We have now established the desired results when the conditions in (a)(γ) are satisfied.

The proof of the theorem in case (b) is very similar to the above.

Finally, in case (c), the proof is a combination of the analysis required to deal with cases (a) and (b); we omit the details.

In the above proof, some information about the behaviour of the differential expression (1.2) at the end-point b in the case (a)(β) was obtained; the following theorem gives a complete classification for this case.

Theorem 3 Let $I = [a,b)$, where $-\infty<a<b\leq\infty$; let the coefficients p, q and w satisfy the basic conditions (2.1) on I; suppose also that condition (4.1) is satisfied by q and w. Then

(i) if $w\notin L(I)$ the differential expression (1.2) is in the strong limit-point and Dirichlet conditions at b, and in fact

$$\lim_{b-} pf'g = 0 \quad (f\in\Delta, g\in D); \tag{4.20}$$

(ii) if $w\in L(I)$ and $p^{-1}\in L(I)$ the differential expression is in the limit-circle and Dirichlet conditions at b;

(iii) if $w\in L(I)$ and $p^{-1}\notin L(I)$ then (a) if

$$\int_a^b w(x)\{\int_a^x p^{-1}\}^2 dx < \infty \tag{4.21}$$

the differential expression is in the limit-circle condition at b and is not in the Dirichlet condition, and (b) if

$$\int_a^b w(x)\{\int_a^x p^{-1}\}^2 dx = \infty \tag{4.22}$$

the differential expression is in the strong limit-point and Dirichlet conditions at b and, again, (4.20) holds.

Proof. (i) This information emerged in the proof of theorem 2; see the proof in the case (a)(γ).

(ii) Since in this case we have $q\in L(I)$, from (4.1), the results are a special case of those of Amos [4].

(iii)(a) Since $w\in L(I)$ and (4.21) holds we may choose $c\in(a,b)$ so that

$$3A^2 \int_c^b w(x)\{\int_c^x p^{-1}\}^2 dx \cdot \int_c^b w \le \tfrac{1}{2}\ , \qquad (4.23)$$

where A is as in (4.1).

Now let y be any solution of the differential equation

$$-(py')' + qy = 0 \quad \text{(on I)}$$

and let $y(c) = \beta_o$, $(py')(c) = \beta_1$; by integrating twice we obtain

$$y(x) = \beta_o + \beta_1 \int_c^x p^{-1} + \int_c^x p^{-1}(t)\{\int_c^t qy\}dt \quad (x\in[c,b))$$

and so

$$|y(x)|^2 \le 3|\beta_o|^2 + 3|\beta_1|^2\{\int_c^x p^{-1}\}^2 +$$

$$+ 3A^2\{\int_c^x p^{-1}\}^2 \cdot \int_c^x w \cdot \int_c^x w|y|^2 \quad (x\in[c,b)),$$

by the Cauchy-Schwarz inequality and the elementary inequality $(\alpha+\beta+\gamma)^2 \le 3(\alpha^2+\beta^2+\gamma^2)$ $(\alpha,\beta,\gamma\in R)$, again using (4.1). Thus we have

$$\int_c^x w|y|^2 \le 3|\beta_o|^2 \int_c^b w + 3|\beta_1|^2 \int_c^b w(t)\{\int_c^t p^{-1}\}^2\, dt +$$

$$+ 3A^2 \int_c^b w(t)\{\int_c^t p^{-1}\}^2 dt \cdot \int_c^b w \cdot \int_c^x w|y|^2 \quad (x\in[c,b))$$

and so, from (4.23), we obtain

$$\tfrac{1}{2}\int_c^x w|y|^2 \le 3|\beta_o|^2 \int_c^b w + 3|\beta_1|^2 \int_c^b w(t)\{\int_c^t p^{-1}\}^2 dt \quad (x\in[c,b)),$$

whence $y\in L_w^2(I)$. This establishes that the differential expression is in the limit-circle case at b.

To show that it is not in the Dirichlet condition, let

$$f(x) = \int_a^x p^{-1} \qquad (x\in[a,b))\ ;$$

it follows from the present conditions on the coefficients that $f\in\Delta$, but $f\notin D$ because $pf'^2 = p^{-1}\notin L(I)$; hence $\Delta\not\subset D$.

(iii)(b) To show that the Dirichlet condition is satisfied, let $f\in\Delta$ and as usual let f be real; then f satisfies (4.12) and $|q|^{\frac{1}{2}}f\in L^2(I)$; if $p^{\frac{1}{2}}f'\notin L(I)$ then, from (4.12), $(pf')(x)f(x)\to\infty$ as $x\to b-$, and there is a constant $M>0$ and a point $c\in(a,b)$ such that $(pf')(x)f(x)\geq M$ $(x\in[c,b))$; supposing that $(pf')(x)\geq 0$ $(x\in[c,b))$, we obtain (4.14) as before, in particular $(pf')(x)\geq K_1>0$ $(x\in[c,b))$ for some constant K_1; then we have $f'(x)\geq K_1p^{-1}(x)$ (almost all $x\in[c,b)$), whence, on integration,

$$f(x) \geq K_1 \int_c^x p^{-1} + f(c) \qquad (x\in[c,b)) ,$$

so that

$$\int_a^x wf^2 \geq K_1^{\;2} \int_c^x w(t)\{ \int_c^k p^{-1}\}^2 dt + f^2(c) \int_c^x w +$$

$$+ 2f(c)K_1 \int_c^x w(t)\{ \int_c^t p^{-1}\}dt \quad (x\in[c,b)),$$

which contradicts (4.22); thus $p^{\frac{1}{2}}f'\in L(I)$, <u>i.e.</u> $f\in D$; this shows that $\Delta\subset D$ as required.

Next, we show that (4.20) holds, which implies that the differential expression is in the strong limit-point condition. Let $f\in\Delta$, $g\in D$; then (4.15) holds and since $f\in D$ it follows that $\ell = \lim_{b-} pf'g$ exists. If $\ell \neq 0$, then we may suppose that $\ell > 0$ and that there is $c\in(a,b)$ such that $(pf')(x) > 0$, $g(x) > 0$ and $(pf')(x)g(x) \geq \frac{1}{2}\ell$ $(x\in[c,b))$; this leads to (4.14) as in the proof of theorem 2; as above, we obtain a contradiction, and so $\ell = 0$, which is the result (4.20).

<u>Remark</u> In the limit-circle case (iii)(a) of the above theorem, where $\Delta\not\subset D$, there are functions $f\in\Delta$ for which $p^{\frac{1}{2}}f'\notin L^2(I)$ (we have $|q|^{\frac{1}{2}}f\in L^2(I)$, by (4.1)). It can be shown in this situation that if $f\in\Delta$ then $\lim_{b-} pf' = 0$ if and only if $p^{\frac{1}{2}}f'\in L^2(I)$; either of these

conditions is equivalent to the condition $f \in D$, when $f \in \Delta$. (We have shown above in the proof of theorem 2 in the case (a)(β) that $\lim_{b-} pf' = 0$ implies that $p^{\frac{1}{2}}f' \in L^2(I)$ when $f \in \Delta$; to obtain the converse, we may use the argument employed to prove theorem 2 in the case (a)(γ) to show that $\lim_{b-} pf'g = 0$, with $g \in D$, and then take $g(x) = 1$ $(x \in I)$.)

Thus in this case, and also in the case (ii) of theorem 3, we must remove from Δ all functions f for which $\lim_{b-} pf' \neq 0$ when producing a self-adjoint operator that is suitable to the needs of this paper; in the limit-circle case (iii)(a) this is equivalent to the removal from Δ of all functions f for which $p^{\frac{1}{2}}f' \notin L^2(I)$.

5. Special cases and examples In this section we list a number of known results on the inequality (1.1), most being special cases of theorem 1; we then give some illustrative examples. It is to be understood in all cases that the coefficients p, q and w satisfy the basic conditions (2.1) in addition to the conditions specifically listed.

5.1. Consider the regular case, where $I = [a,b]$ is compact and p^{-1}, q, $w \in L(I)$. We have seen above (theorem 2(a)(α)) that the results of theorem 1, including the inequality (1.1), hold with only these minimal conditions on p, q and w.

The inequality was first established in the regular case under these minimal conditions by Amos and Everitt [6], who made use of the theory of compact embedding of one Hilbert function space into another, and of coercive quadratic forms, as developed in [30]. The proof of the inequality given in [6] is lengthy, and it was pointed out by Beesack [9] that a much shorter, elegant proof is possible, based on an integral identity, once one has shown that non-trivial eigenfunctions corresponding to the least eigenvalue of T do not vanish on the interval [a,b]; see also Beesack [10, theorem 2].

Another proof of (1.1), again in the regular case under the minimal conditions, appears in Wray [45, theorem 4], where use is made of the representation theory of sesqui-linear forms in [34, chapter VI] and a spectral identity [45, theorem 2]; the spectral identity could actually be side-stepped, as the methods of section 3 of this paper show. In [45], more general boundary conditions than those in (4.6) and (4.7) are used, but when one restricts them to these Neumann ones the result in [45] is the same as (1.1).

In their paper [11], Bradley and Everitt prove a version of the inequality (1.1) in the regular case under the additional assumptions that $p \in AC[a,b]$, $p' \in L^2[a,b]$ and $q \in L^2[a,b]$; see [11, theorem 1]. These extra assumptions are needed to facilitate a direct construction in the proof of an approximation theorem ([11, section 6]), which is one of the steps in the Sobolev space method.

Hinton [31, theorem 2(i)] proves the spectral identity that appears in [45] in the regular case, under the assumptions that p has a continuous derivative, w is continuous and q is bounded on [a,b]. (Hinton's theory is for differential expressions of any even order; the conditions listed above are for the second-order case and are in our notation.) As in [45], one may readily deduce the inequality (1.1) from the spectral identity, although Hinton is not concerned with this; there are also more general boundary conditions than ours.

The final result we mention in the regular case is that of Sears and Wray [41, theorem 2(ii)], which is the spectral identity once again. The proof of this result in [41] may be said to be a direct proof, but it uses results about the asymptotic behaviour of the eigenvalues and eigenfunctions of T which depend upon the assumption that $p(x) = w(x) = 1 \quad (x \in [a,b])$, and so its scope is rather limited. Again, the boundary conditions are more general than ours.

5.2. The first singular example we consider is the case where

$$\int_a^b (p^{-1} + |q| + w) < \infty$$

and either $a = -\infty$, or $b = \infty$, or both. This is very similar to the regular case.

If $b = \infty$ then the differential expression (1.2) is in the Dirichlet and limit-circle conditions at b, see Amos [3, section 1.4] or (if $a > -\infty$) [4]; similarly if $a = -\infty$. (It follows from the results of [3, section 1.4] that these are the only circumstances under which the Dirichlet and limit-circle conditions are satisfied simultaneously.)

It is shown in [3, section 5.4] and [4] that a change of independent variable is possible here that transforms the differential expression (1.2) into one that is regular, on a compact interval; it is also shown that q satisfies condition (2.10).

It follows from this and the discussion above of the regular case that the conditions of theorem 1 are all satisfied, and so the inequality in this case ([3, theorem 5.4.4]) is a special case of the inequality in the present paper.

The inequality in this case appears also as [46, corollary] when the interval has the form $[a,\infty)$, although in [46] there are more general boundary conditions attached to the self-adjoint operator. (The paper [46] was written without knowledge of the cited theorem of [3].)

5.3. This example, again singular, is due to Bradley and Everitt [11, theorem 2]. Here, I is of the form $[a,\infty)$ and the coefficients p, q and w are required to satisfy $w(x) = 1$ $(x\in I)$, $p\in AC_{loc}(I)$, $p'\in L^2_{loc}(I)$, $q\in L^2_{loc}(I)$ and either $q(x) \geq A$ $(x\in I)$ for some constant A, or both $p(x) = 1$ $(x\in I)$ and $q\in L^r(I)$ for some $r\geq 1$.

This is covered by case (a)(γ) of theorem 2: clearly (4.5a) and (4.5b) are satisfied; when q is bounded below (4.2) holds with $q = q_1$ and we take $q_2 = 0$; in the other case, where $p = 1$ and $q\in L^r(I)$, (4.4) holds with $q_2 = q$, by [22, lemma 1], and we take $q_1 = 0$. ([22, lemma 1] is stated for a smaller set than D_o but readily extends to D_o; it also requires that $x\geq 1$, but the proof of theorem 2 still goes through with this minor extra condition.) Thus, the cited result in [11] falls within the scope of theorem 1.

In [11, section 7], specific examples of the coefficients p and q (w being 1) are given for which the infimum μ of the spectrum of T can be calculated.

5.4. The much earlier results of Putnam [40, theorems (I), (I bis) and (II)] are, essentially, generalised by [11, theorem 2] and hence also by theorem 1 here; see the discussion in [11, section 2].

5.5. In [7, theorem 4] there is another singular version of the inequality (1.1) that is generalised by theorem 1. In that paper, the interval I is $[0,\infty)$, although it is pointed out that the results extend to the more general interval $[a,b)$, b singular. The coefficients p, q and w are subject to the following conditions: $q = q_1 + q_2$, where $q_1\in L_{loc}(I)$, q_1 satisfies condition (2.9) (or (4.2)), $q_2\in L(I)$, $p^{-1}\in L(I)$, $w\notin L(I)$ and

$$\int_0^\infty w(x)\left(\int_x^\infty \{p(t)\}^{-1}dt\right)dx < \infty . \tag{5.1}$$

Again the conditions of case (a)(γ) of theorem 2 are satisfied, and the proof of [7, lemma A] can be modified, to establish that q_2 satisfies (4.4). (As in 5.3 above (4.4) is actually obtained for all $x \geq x_o$ where $x_o \in I$.)

Because (5.1) holds, the spectrum of T is discrete; this seems to be essential for the application of the coercive form argument, which is used in [7], but is unnecessary for the application of theorem 2.

5.6. The result of Penning and Sauer [39] has $I = [0,\infty)$, $p(x) \geq 0$ $(x \in I)$, $p \in L^1_{loc}(I)$, $p^{-1} \in L(I)$, $r \in L(I)$, where $r(x) = \int_x^\infty p^{-1}$ $(x \in I)$, $w(x) = 1$ $(x \in I)$ and $q \in L^\infty(I)$. This result is thus a special case of the result of [7] quoted in 5.5 above, and hence is in turn within the scope of theorem 1.

The paper [39] has a particular importance in that it is the first in which the methods of Hildebrandt [30] are used to tackle the inequality (1.1).

5.7. Yet another singular version of the inequality (1.1) occurs in [24, corollary to theorem 1, with $\alpha = \pi/2$]. The interval is [a,b), condition (2.9) is required to hold, the differential expression (1.2) is taken to be in the Dirichlet condition at the singular end-point b, and it is required also that

$$\lim_{b-} pf'g = 0 \quad (f \in \Delta, g \in D). \tag{5.2}$$

As in the proof of theorem 2, the boundary condition at a in the definition of D(T) is equivalent to the condition $(pf')(a) = 0$, and it follows from this and (5.2) that $D(T) = \{f \in \Delta : (pf')(a) = 0\}$.

We may now deduce from the results in [24] (taking $\alpha = \pi/2$ there) that conditions 2 and 3 of theorem 1 are also satisfied;

thus the cited result in [24], in the special case $\alpha = \pi/2$, follows from theorem 1.

5.8. The final special case of theorem 1 appears in [32, lemma 1(iii)]. That paper deals with differential expressions of any even order, and the interval I may be (a,b) (both end-points singular) or [a,b) (a regular, b singular); we are concerned here only with the second-order case, with the boundary condition $(pf')(a) = 0$ at a when $I = [a,b)$, when the conditions on the coefficients are (in our notation): p' exists and is continuous on I, w is continuous on I, q is bounded on all compact sub-intervals of I and satisfies a condition [32, (H_3), p. 293] which is similar to (2.10) and from which it is easy to deduce that condition 1 of theorem 1 is satisfied.

It is shown in [32] that the differential expression (in the second-order case) is in the strong limit-point and Dirichlet conditions at the singular end-point(s). Hence it follows from the results of [3, section 1.4] that

$$\lim_{b-} pf'g = 0 \quad (f \in \Delta, g \in D),$$

and that the corresponding result holds at a if it is also a singular end-point. If a is a regular end-point, then the boundary condition there in the definition of D(T) is $(pf')(a) = 0$ as before. Thus, D(T) is equal to Δ when $I = (a,b)$ and given by $\{f \in \Delta : (pf')(a) = 0\}$ when $I = [a,b)$. It now follows from the results of [32] that T is in either case self-adjoint and that $D(T) \subset D$. Thus, once more, all the conditions of theorem 1 are met.

5.9. As we mentioned in the introduction, the results in [5, theorem 2] and [12, theorem 1] are beyond the scope of theorem 1; in each case, the only reason for this is that instead of imposing direct conditions on the coefficient q, such as (2.9) or (2.10), to control the degree to which it can become negative, the authors just assume that the operator T is bounded below. However, if one wishes to prove that T is bounded below, one has usually to impose conditions such as those in (2.9) and (2.10) on q, or conditions that imply them; there is no example in which q does not satisfy condition 1 of theorem 1 and yet T is bounded below.

It does not seem to be possible to modify the proof of theorem 1 to deal with the case where condition 1 is replaced by the assumption that T is bounded below.

5.10. A considerable improvement on the results in [5] is given in [3, chapters 2 and 3]; again the results elude theorem 1 above, for the same reason as in 5.9.

5.11. Another approach to the inequality (1.1) is that of Bradley, Hinton and Kauffman [13], which involves the so-called Dirichlet index of the differential expression. In that paper, differential expressions of any even order are considered; in the second-order case, the results of [13] and of this paper are independent of each other; we note in particular that for the example in section 5.2 above (a special case of our result) the Dirichlet index is 2, whilst [13, theorem 3.1] requires that the index have the value 1.

5.12. In [41, theorem 1] and [45, corollary to theorem 1] there appear inequalities in which the Dirichlet integral is present but is not necessarily absolutely convergent. These results require conditions that are quite different from those in theorem 1, and the proofs differ markedly from that given in section 3 above.

5.13. Finally, we cite the paper [26] by Florkiewicz and Rybarski in which inequalities somewhat like that in (1.1) are established. There, the conditions are very different from ours and the results are not closely related to the ones in this paper.

5.14. We now turn to the consideration of examples; we may mention here that all the examples considered in [7], [11], [12] and [39] are included in the results of this paper.

5.15. Consider the situation where the conditions of (a)(β) of theorem 2 are satisfied. We have seen in theorem 3 that this situation includes both limit-point and limit-circle cases; we give specific examples here.

If $I = [1,\infty)$, $q(x) = x^{-2}$ $(x\in I)$, $p(x) = 1$ $(x\in I)$ and $w(x) = x^{-3/2}$ $(x\in I)$, then the conditions of (iii)(b) of theorem 3 are satisfied and so the differential expression (1.2) is in the strong limit-point and Dirichlet conditions at ∞. It may be shown directly that we have the limit-point case at ∞: the differential equation

$$-y''(x) + x^{-2}y(x) = 0 \quad (\text{on } [1,\infty))$$

has solutions $x^{\frac{1}{2}(1\pm\sqrt{5})}$, one of which is not in $L_w^2(I)$.

If $I = [1,\infty)$, $q(x) = x^{-4}$ $(x\in I)$, $p(x) = 1$ $(x\in I)$ and $w(x) = x^{-7/2}$ $(x\in I)$, then the conditions of (iii)(a) of theorem 3 are satisfied and so the differential expression is in the limit-circle condition at ∞ and is not in the Dirichlet condition. To show directly that we have the limit-circle case at ∞, we observe that the differential equation

$$-y''(x) + x^{-4}y(x) = 0 \quad (\text{on } [1,\infty))$$

has solutions $x\cosh(x^{-1})$, $x\sinh(x^{-1})$, both of which are in $L_w^2(I)$.

5.16. Let $I = (-\infty,\infty)$, $p(x) = w(x) = 1$ $(x\in I)$ and $q(x) = x^2$ $(x\in I)$; here, both end-points are singular of course, and the conditions of part (c) of theorem 2 are satisfied. Also, as summarised in [19, section 6], the differential expression (1.2) is in the Dirichlet condition at both $-\infty$ and ∞, and from the proof of theorem 2 we have $D(T) = \Delta$.

The operator has in this case a purely discrete spectrum consisting precisely of the odd positive integers; see [42, section 4.2]. Its lower bound is 1 and the inequality (1.1) here takes the form

$$\int_{-\infty}^{\infty}\{|f'(x)|^2 + x^2|f(x)|^2\}dx \geq \int_{-\infty}^{\infty}|f(x)|^2dx \qquad (f\epsilon D),$$

with equality if and only if $f(x) = Ke^{-\frac{1}{2}x^2}$, for some constant K.

It is worth noting that in this case the inequality admits of an elementary proof.

5.17. Let $I = (0,\infty)$, $p(x) = 1$ $(x\epsilon I)$, $q(x) = 0$ $(x\epsilon I)$ and $w(x) = x^{\alpha}$ $(x\epsilon I)$, where $\alpha\epsilon R$.

Although the inequality (1.1) is not discussed, other aspects of this example are considered in detail in [18]. Since $q = 0$, it follows from theorem 3 that the differential expression (1.2) may here be classified as follows:

(i) $-1<\alpha<\infty$, regular at 0, strong limit-point and Dirichlet at ∞;

(ii) $-3\leq\alpha\leq-1$, strong limit-point and Dirichlet at both end-points;

(iii) $-\infty<\alpha<-3$, strong limit-point and Dirichlet at 0, limit-circle and not Dirichlet at ∞.

This information is also to be found in [18].

If $\alpha>-1$ then the conditions of (a)(β) of theorem 2 are satisfied; otherwise the conditions of (c) hold; the methods employed during the proof of that theorem show that

$$D(T) = \begin{cases} \{f\epsilon\Delta : f'(0) = 0\} \text{ in case (i)}, \\ \quad \Delta \quad \text{in case (ii)}, \\ \{f\epsilon\Delta : \lim_{\infty} f' = 0\} \text{ in case (iii)}. \end{cases}$$

In all cases, T is self-adjoint and the inequality takes the form

$$\int_0^\infty |f'(x)|^2 dx \geq \mu \int_0^\infty x^\alpha |f(x)|^2 dx \qquad (f \in D),$$

where as usual μ is the infimum of $\sigma(T)$.

It is also known that T has no eigenvalues and that its continuous spectrum is $[1/4,\infty)$ when $\alpha = -2$, $[0,\infty)$ otherwise. For these results see [18, sections 9 and 11] in cases (ii) and (iii); in case (i) the argument of [18, section 7] shows that $\sigma(T) \subset [0,\infty)$, and it then follows from remark 5 after [25, theorem 1], and [38, theorem 2, p. 92], that $\sigma(T) = [0,\infty)$ and is purely continuous.

Thus, the result is as follows. Write

$$D = D_\alpha = \{f : (0,\infty) \to C : f \in AC_{loc}(0,\infty) ,$$

$$f' \in L^2(0,\infty) , \quad x^{\alpha/2} f \in L^2(0,\infty)\} .$$

Then

$$\int_0^\infty |f'|^2 \geq 0 \qquad (f \in D_\alpha)$$

for all $\alpha \in R$ except -2, and

$$\int_0^\infty |f'|^2 \geq \frac{1}{4} \int_0^\infty x^{-2} |f(x)|^2 dx \qquad (f \in D_{-2}) . \tag{5.3}$$

In all cases, the lower bound is best possible in the sense that we can find a sequence $\{f_n : n \in N\}$ in D_α with unit norm such that

$$\lim_{n\to\infty} \int_0^\infty |f_n|^2$$

exists and has the given lower bound.

It can be shown that D_{-2} is identical with the set $\{f : [0,\infty) \to C: f \in AC_{loc}[0,\infty), f' \in L^2[0,\infty)$ and $f(0) = 0\}$, so that (5.3) is equivalent to [28, section 7.3, theorem 253], which is a Hardy-Littlewood result of 1932.

5.18. Let $I = (0,1)$, $p(x) = 1$ $(x \in I)$, $q(x) = 0$ $(x \in I)$ and $w(x) =$

$x^{-1}(1-x)^{-1}$ $(x\in I)$. This example comes under case (c) of theorem 2, and the differential expression is in the strong limit-point and Dirichlet conditions at both end-points. The corresponding differential equation

$$-y''(x) = \lambda x^{-1}(1-x)^{-1}y(x) \quad (\text{on } (0,1))$$

is a transformed, particular case of the Gegenbauer equation. We have $D(T) = \Delta$ and it is known that T has discrete spectrum $\{(n+1)(n+2) : n\in N\}$ and that the eigenfunctions corresponding to the least eigenvalue 2 are of the form $Kx(1-x)$, K constant. Thus in this case the inequality takes the form

$$\int_0^1 |f'(x)|^2 dx \geq 2 \int_0^1 x^{-1}(1-x)^{-1}|f(x)|^2 dx \quad (f\in D), \qquad (5.4)$$

with equality if and only if $f(x) = Kx(1-x)$ $(x\in(0,1))$ for some constant, K.

It can be shown that D in this case is equal to the set $\{f : [0,1]\to C : f\in AC[0,1],\ f'\in L^2[0,1] \text{ and } f(0) = f(1) = 0\}$, so that (5.4) is equivalent to another theorem of Hardy and Littlewood, 1932; see [28, p. 193, theorem 262].

5.19. In this final example we illustrate the fact that the results of theorem 1 may fail to be true if there is insufficient control over the degree to which the coefficient q can take negative values.

In [38, theorem 9, p. 235] conditions are given under which every self-adjoint operator derived from the differential expression (1.2) has its continuous spectrum equal to the whole real line R. In such cases the operator T is not bounded below and so the results of the theorem fail.

A specific case, satisfying the conditions of the cited theorem in [38], is given by $I = [0,\infty)$, $p(x) = w(x) = 1$ $(x\in I)$

and $q(x) = - x^{\alpha}$ $(x \in I)$, where $\alpha \in (0,2]$ is a constant. Thus, $q(x) \to -\infty$ as $x \to \infty$.

We note that in this last case, the differential expression (1.2) is in the strong limit-point condition at ∞; see [21, section 4].

6. The approach via the theory of sesqui-linear forms.

6.1. The theorem in section 2 may be proved by making use of the theory of sesqui-linear forms and their associated quadratic forms, as developed in Kato [34, chapter VI, section 1]. This method of proof was used in our earlier paper [24]. We list here the relevant terminology and results and then apply them to obtain an alternative proof of theorem 1. The setting is a complex Hilbert space H, with norm $||\cdot||$ and inner-product $(\cdot,\cdot)$.

Consider a symmetric sesqui-linear form $\tau : D(\tau) \times D(\tau) \to C$, where $D(\tau)$ is a linear manifold of H; we also write $\tau(f) = \tau(f,f)$ $(f \in D(\tau))$, thus defining the quadratic form associated with the sesqui-linear form; the quadratic form is real-valued because the sesqui-linear form is symmetric. (See [34, p. 309] for definitions of the terms "symmetric" and "sesqui-linear".)

A symmetric form τ is said to be bounded below if there is a number $\mu \in R$ such that

$$\tau(f) \geq \mu ||f||^2 \qquad (f \in D(\tau)).$$

In this case, the largest number μ with this property is called the lower bound of τ.

If τ is bounded below it is said to be closed if for any sequence $\{f_n : n \in N\} \subset D(\tau)$ that converges in H to some function f and is such that $\tau(f_m - f_n) \to 0$ as $m,n \to \infty$, we have $f \in D(\tau)$ and $\tau(f - f_n) \to 0$ as $n \to \infty$.

It τ is bounded below and closed, we say that a linear submanifold E of $D(\tau)$ is a <u>core</u> of τ if the restriction of τ to E has closure τ, <u>i.e.</u> $D(\tau)$ is the set of all functions $f \in H$ for which there exists a sequence $\{f_n : n \in N\} \subset E$ such that $||f_n - f|| \to 0$ as $n \to \infty$ and $\tau(f_m - f_n) \to 0$ as $m,n \to \infty$; for such an f one has $\tau(f_n) \to \tau(f)$ as $n \to \infty$.

By combining the results of [34, theorems 2.1 and 2.6, pp. 322-323] we obtain

<u>Proposition</u> <u>Let</u> τ <u>be a symmetric sesqui-linear form in</u> H <u>whose domain</u> $D(\tau)$ <u>is dense in</u> H, <u>and suppose that</u> τ <u>is bounded below and closed. Then there exists a self-adjoint operator</u> $S : D(S) \subset H \to H$ <u>with the following properties</u>:

(i) $D(S) \subset D(\tau)$ <u>and</u>

$$\tau(f,g) = (Sf,g) \qquad (f \in D(S), g \in D(\tau));$$

(ii) $D(S)$ <u>is a core of</u> τ; (6.1)

(iii) <u>if</u> $f \in D(\tau)$, $h \in H$ <u>and</u> $\tau(f,g) = (h,g)$ <u>holds for all</u> g <u>belonging to a core of</u> τ, <u>then</u> $f \in D(S)$ <u>and</u> $Sf = h$;

(iv) S <u>is bounded below and has the same lower bound as has</u> τ.

6.2. Now we apply the above theory to the Dirichlet integral to obtain the results of theorem 1. We suppose that all the conditions of the theorem are satisfied.

Take $H = H_o \equiv L^2_w(I)$ and define the sesqui-linear form τ by $D(\tau) = D$ and

$$\tau(f,g) = \int_a^b \{pf'\bar{g}' + qf\bar{g}\} \qquad (f,g \in D),$$

the bars denoting complex conjugation; its associated quadratic form is given by

$$\tau(f) = \int_a^b \{p|f'|^2 + q|f|^2\} \qquad (f \in D),$$

which is of course the Dirichlet integral. The integrals above are absolutely convergent, from the definition of D in (2.2), and τ is symmetric because p and q are real.

The method of section 3.1 shows that τ is bounded below; in fact

$$\tau(f) \geq -(A+K)||f||_o^2 \qquad (f \in D) ,$$

where A and K are the constants in condition 1 of theorem 1 (but $-(A+K)$ is not necessarily the lower bound of τ). Let τ have lower bound μ, so that

$$\tau(f) \geq \mu||f||_o^2 \qquad (f \in D) . \qquad (6.2)$$

We may now define the Hilbert space H_2 to be the set D with inner-product $(\cdot,\cdot)_2$ and norm $||\cdot||_2$ defined by

$$(f,g)_2 = \tau(f,g) + (1-\mu)(f,g)_o \qquad (f,g \in D) . \qquad (6.3)$$

(That this is indeed an inner-product is an immediate consequence of (6.2) and the properties of τ.) The constant 1 on the right of (6.3) may be replaced by any other positive constant. To obtain the completeness of H_2 in the norm $||\cdot||_2$, we may use the method of [24, section 3]; in [24] the coefficient q is required to satisfy condition (2.9), but since here we have no boundary term at a (see [24, (1,1)]) the method of that paper may still be employed in the present case where q satisfies the more general condition 1 of theorem 1 (in fact the results of [24] hold good with q as in this paper when the constant α, see [24, (1,1)], satisfies $\alpha = 0$ or $\pi/2 \leq \alpha < \pi$).

According to [34, theorem 1.11, p. 314] the completeness of H_2 in the norm $||\cdot||_2$ is equivalent to the closure of τ, and so τ is closed.

A standard argument shows that the domain D is dense in H_o, and so by the proposition above there is a self-adjoint operator S : $D(S) \subset H_o \to H_o$ with the properties in (6.1)(i)-(iv). We identify S with the operator T defined in (2.4) and (2.5) as follows.

By hypothesis, $D(T) \subset D$, and from (3.1) we have

$$\tau(f,g) = (Tf,g)_o \quad (f\epsilon D(T), g\epsilon D) .$$

Hence if $f\epsilon D(T)$ we have $f\epsilon D(S)$ and $Sf = Tf$, by (6.1)(iii), since D is obviously a core of τ. This means that T is a restriction of S, and since both operators are self-adjoint they must be identical. It now follows from (ii) and (iv) of the proposition that T is bounded below with the same lower bound, μ, as has τ, and that D(T) is a core of τ.

Since, from [34, theorem 1.21, p. 317], D(T) is a core of τ if and only if D(T) is dense in H_2 with respect to $||\cdot||_2$, we have thus obtained this density property which is, incidentally, equivalent also to the density of D(T) in H_1 (see section 3.4).

However, now that we are making use of the above theory of quadratic forms, we no longer need to use the fact that D(T) is dense in H_2 (or H_1), and the consequent approximation result in section 3.5, in order to establish the inequality (1.1), since we have already determined that τ is bounded below by the lower bound, μ, of T: this is precisely the inequality (1.1).

To show that μ is the best possible constant in (1.1), and to determine the cases of equality, we proceed as in section 3 again.

References

1. Adams, R. A. : Sobolev spaces (Academic Press, New York, 1975).

2. Akhiezer, N. I. and Glazman, I. M. : Theory of linear operators in Hilbert space (Pitman Publishing, London, 1981).

3. Amos, R. J. : On some problems concerned with integral inequalities associated with symmetric ordinary differential expressions (Ph.D. thesis, University of Dundee, Scotland, 1977).

4. Amos, R. J. : On a Dirichlet and limit-circle criterion for second-order ordinary differential expressions, Quaestiones Mathematicae 3 (1978), 53-65.

5. Amos, R. J. and Everitt, W. N. : On a quadratic integral inequality, Proc. Royal Soc. Edinburgh (A) 78 (1978),241-256.

6. Amos, R. J. and Everitt, W. N. : On integral inequalities associated with ordinary regular differential expressions, Differential Equations and Applications, 237-255 (North Holland, 1978; Mathematical Studies 31; Edited by W. Eckhaus and E. M. de Jager).

7. Amos, R. J. and Everitt, W. N. : On integral inequalities and compact embeddings associated with ordinary differential expressions, Arch. Rat. Mech. Anal. 71 (1979), 15-40.

8. Atkinson, F. V. : Limit-n criteria of integral type, Proc. Royal Soc. Edinburgh (A) 73 (1975), 167-198.

9. Beesack, P. R. : Math. Reviews 80 (1980), 34017.

10. Beesack, P. R. : Minimum properties of eigenvalues - elementary proofs, General Inequalities 2, 109-120 (Birkhäuser Verlag, Basel, 1980; Edited by E. F. Beckenbach).

11. Bradley, J. S. and Everitt, W. N. : Inequalities associated with regular and singular problems in the calculus of variations,

Trans. Amer. Math. Soc. 182 (1973), 303-321.

12. Bradley, J. S. and Everitt, W. N. : A singular integral inequality on a bounded interval, Proc. Amer. Math. Soc. 61 (1976), 29-35.

13. Bradley, J. S., Hinton, D. B. and Kauffman, R. M. : On the minimization of singular quadratic functionals, Proc. Royal Soc. Edinburgh (A) 87 (1981), 193-208.

14. Evans, W. D. : On limit-point and Dirichlet type results for second-order differential expressions, Lecture Notes in Mathematics 564 (1976), 78-92 (Edited by W. N. Everitt and B. D. Sleeman; Springer Verlag, Heidelberg, 1976).

15. Everitt, W. N. : On the strong limit-point condition of second-order differential expressions, Proceedings International Conference on Differential Equations, Los Angeles, 1974; pages 287-307 (Edited by W. A. Harris; Academic Press, New York, 1975).

16. Everitt, W. N. : A note on the Dirichlet condition for second-order differential expressions, Canadian J. Math. 28 (1976), 312-320.

17. Everitt, W. N. : An integral inequality with an application to ordinary differential operators, Proc. Royal Soc. Edinburgh (A) 80 (1979), 35-44.

18. Everitt, W. N. : A note on an integral inequality, Quaestiones Mathematicae 2 (1978), 461-478.

19. Everitt, W. N. : A general integral inequality associated with certain ordinary differential operators, Quaestiones Mathematicae 2 (1978), 479-494.

20. Everitt, W. N. and Giertz, M. : A Dirichlet type result for ordinary differential operators, Math. Ann. 203 (1973), 119-218.

21. Everitt, W. N., Giertz, M. and McLeod, J. B. : On the strong and weak limit-point classification of second-order differential expressions, Proc. London Math. Soc. (3) 29 (1974), 142-158.

22. Everitt, W. N., Giertz, M. and Weidmann, J. : Some remarks on a separation and limit-point criterion of second-order, ordinary differential expressions, Math. Ann. 200 (1973), 335-346.

23. Everitt, W. N., Kwong, M. K. and Zettl, A. : Differential operators and quadratic inequalities with a degenerate weight, (to appear in J. of Math. Analysis and Applications).

24. Everitt, W. N. and Wray, S. D. : A singular spectral identity involving the Dirichlet integral of an ordinary differential expression, (to appear in the Czechoslovak Mathematical Journal).

25. Everitt, W. N. and Zettl, A. : Generalized symmetric ordinary differential expressions I: the general theory, Nieuw Archief voor Wiskunde (3) 27 (1979), 363-397.

26. Florkiewicz, B. and Rybarski, A. : Some integral inequalities of Sturm-Liouville type, Colloq. Math. 36 (1976), 127-141.

27. Gregory, J. : Quadratic form theory and differential equations (Academic Press, New York, 1981).

28. Hardy, G. H., Littlewood, J. E. and Pólya, G. : Inequalities (Cambridge University Press, 1934).

29. Hellwig, G. : Differential operators of mathematical physics (Addison-Wesley, London, 1967).

30. Hildebrandt, S. : Rand-und Eigenwertaufgaben bei stark elliptischen Systemen linearer Differentialgleichungen, Math. Ann. 148 (1962), 411-429.

31. Hinton, D. B. : On the eigenfunction expansions of singular ordinary differential equations, J. Differential Equations 24

(1977), 282-308.

32. Hinton, D. B. : Eigenfunction expansions and spectral matrices of singular differential operators, Proc. Royal Soc. Edinburgh (A) 80 (1978), 289-308.

33. Kalf, H. : Remarks on some Dirichlet type results for semi-bounded Sturm-Liouville operators, Math. Ann. 210 (1974), 197-205.

34. Kato, T. : Perturbation theory for linear operators (first/second edition; Springer Verlag, Heidelberg, 1966/1976).

35. Kwong, M. K. : Note on the strong limit-point condition of second-order differential expressions, Quart. J. Math. (Oxford) (2) 28 (1977), 201-20.

36. Kwong, M. K. : Conditional Dirichlet property of second-order differential expressions, Quart. J. Math. (Oxford) (2) 28 (1977), 329-338.

37. Mitrinović, D. S. : Analytic inequalities (Springer Verlag, Heidelberg, 1970).

38. Naĭmark, M. A. : Linear differential operators Part II (Ungar, New York, 1968).

39. Penning, F. and Sauer, N. : Note on the minimization of $\int_0^\infty \{p(x)|f'(x)|^2 + q(x)|f(x)|^2\}dx$, Research Report UP TW 2, 1976; Department of Applied Mathematics, University of Pretoria, South Africa.

40. Putnam, C. R. : An application of spectral theory to a singular calculus of variations problem, Amer. J. Math. 70 (1948), 780-803.

41. Sears, D. B. and Wray, S. D. : An inequality of C. R. Putnam involving a Dirichlet functional, Proc. Royal Soc. Edinburgh (A) 75 (1976), 199-207.

42. Titchmarsh, E. C. : Eigenfunction expansions Part I (second edition; Oxford University Press, 1962).

43. Walker, P. : A vector-matrix formulation for formally symmetric ordinary differential equations with applications to solutions of integrable-square, J. London Math. Soc. 9 (1974), 151-159.

44. Weidmann, J. : Linear operators in Hilbert spaces (Springer Verlag, Heidelberg, 1980).

45. Wray, S. D. : An inequality involving a conditionally convergent Dirichlet integral of an ordinary differential expression Utilitas Mathematica), 22 (1982), 161-184.

46. Wray, S. D. : On a second-order differential expression and its Dirichlet integral (to appear).

47. Brown, R.G.: A von Neumann factorization of some self-adjoint extensions of positive symmetric differential operators and its application to inequalities, Lecture Notes in Mathematics, vol. 1032 (Edited by W.N. Everitt and R.T. Lewis Springer-Verlag, Heidelberg, 1983).

48. Brown, R.G.: A factorization method for symmetric differential operators and its applications to Dirichlet inequalities and to the Dirichlet index (to appear in the Proceedings of the 1983 International Conference on Differential Equations held at the University of Alabama in Birmingham, U.S.A).

49. Brown, R.G. and Hinton, D.B.: Sufficient conditions for weighted inequalities of sum and product form (in preparation).

50. Krzywicki, A. and Rybarski A.: On some integral inequalities involving Chebyshev weight function, Colloq. Math. 18 (1967), 147-50.

51. Krzywicki, A. Rybarski, A.: On an integral inequality connected with Hardy's inequality, Applicationes Mathematicae (Hugo Steinhaus Memorial Volume) X (1969), 37-41.

Notes

1. We are indebted to R.C. Brown for some very interesting comments on the manuscript:

   (i) would the definition of D(T) in (2.4) be better replaced by

   $D(T) := \{f \in \Delta \cap D$ such that (i) and (ii) hold$\}$?

   (ii) does condition 1 of Theorem 1 imply conditions 2 and 3?

   We hope to consider these points and other matters in a subsequent manuscript.

2. Our attention was drawn to the two papers of A.Kryzwicki and A Rybarski, see [50] and [51], after this manuscript was completed; we hope to comment on these interesting results elsewhere.

WNE and SDW.

Nonoscillation Theorems for Differential Equations with General Deviating Arguments

A. M. Fink and Takaŝi Kusano
(Iowa State University and Hiroshima University)

1. Introduction. The oscillatory and non-oscillatory behavior of the solution of the differential equations

(1) $y^{(n)} = p(t)\,y(t),$

(2) $y^{(n)} = p(t)\,y(g(t)),$

(3) $y^{(n)} = p(t)\,f(y(t))$ and

(4) $y^{(n)} = p(t)\,f(y(g(t)))$

have been discussed by a variety of authors, [1]-[8]. Loosely speaking the existence of oscillating solutions for the equations (2) and (4) require conditions which measure the amount of retardation or advancement that $g(t)$ represents, usually through some integral average, see [9]. On the other hand, non-oscillation results follow from certain moment conditions on the function $p(t)$, see [1], [2], [7] for example.

It is natural to ask if the operator $y^{(n)}$ may be replaced by a more general operator. The equation

(5) $L_n y = p(t)\,y(t)$

which generalizes (1) has been studied in [10] where $L_n y$ is an nth order disconjugate operator. It is our purpose to discuss this generalization of the equation (4). Another way to introduce the in-between derivatives is indicated in [1, Theorem 2].

Specifically, we study the non-oscillation of the equations

$$(A^+) \qquad L_n y(t) + f(t,y(g(t))) = 0$$

$$(A^-) \qquad L_n y(t) - f(t,y(g(t))) = 0$$

where L_n is a disconjugate operator defined by

$$(6) \qquad L_n = \frac{1}{p_n(t)} \frac{d}{dt} \frac{1}{p_{n-1}(t)} \frac{d}{dt} \cdots \frac{d}{dt} \frac{1}{p_1(t)} \frac{d}{dt} \frac{\cdot}{p_0(t)}$$

Surprisingly, any explicity hypothesis on whether $g(t)$ represents a retardation or an advancement is not required for these non-oscillation theorems. The conditions we derive are essentially moment type conditions.

The specific assumptions that we make are

(7)

(a) $p_i : [a,\infty) \to R$, continuous, $p_i(t) > 0$, $0 \le i \le n$, and $\int_a^\infty p_i(t)dt = \infty$, $1 \le i \le n-1$;

(b) $f : [a,\infty) \times R \to R$, continuous, nondecreasing in y, and $y f(t,y) > 0$ for $y \neq 0$;

(c) $g : [a,\infty) \to R$, continuous and $\lim_{t \to \infty} g(t) = \infty$

The operator L_n of the form (6) where the $p_i(t)$ satisfy (7) is said to be in canonical form, see [11]. Trench [11] shows that any operator of the form (6) may be rewritten in an essentially unique way so that (7a) is satisfied.

It is our purpose to find necessary and sufficient conditions for the existence of (non-oscillatory) solutions of $(A^\pm)$ whose growth at ∞ is exactly the same as the principal solutions of $L_n y = 0$. In the case where $L_n y = y^{(n)}$ these principal solutions are the appropriate powers of t.

2. Preliminaries. We will be considering kernels of integral operators which are defined by repeated integrations. To introduce these kernels and the principal solutions we will need a certain amount of notation which will be consistent with that found in [11].

First we define the quasi-derivatives

$$(8)\quad \begin{cases} D^{o}(y\,;\,p_0)(t) = \dfrac{y(t)}{p_0(t)}\,, \\ D^{i}(y\,;\,p_0,\ldots,p_i)(t) = \dfrac{1}{p_i(t)}\,\dfrac{d}{dt}\,D^{i-1}(y\,;\,p_0,\ldots,p_{i-1})(t), \quad 1 \le i \le n. \end{cases}$$

For the functions $q_1, q_2, \ldots$ define

$$(9)\quad \begin{cases} I_0 = 1 \\ I_i(t,s\,;\,q_i,\ldots,q_1) = \int_s^t q_i(r) I_{i-1}(r,s\,;\,q_{i-1},\ldots,q_1)dr, \quad i = 1,2,\ldots \end{cases}$$

For simplicity we put

$$(10)\quad \begin{cases} J_i(t,s) = p_0(t) I_i(t,s\,;\,p_1,\ldots,p_i) & J_i(t) = J_i(t,a), \\ K_i(t,s) = p_n(t) I_i(t,s\,;\,p_{n-1},\ldots,p_{n-i}) & K_i(t) = K_i(t,a). \end{cases}$$

Then $J_0(t),\ J_1(t),\ \ldots,\ J_{n-1}(t)$ form a principal system for $L_n y = 0$, and $K_0(t),\ K_1(t),\ \ldots,\ K_{n-1}(t)$ form a principal system for $M_n y = 0$, where

$$(11)\quad M_n = \frac{1}{p_0(t)}\frac{d}{dt}\frac{1}{p_1(t)}\frac{d}{dt}\cdots\frac{d}{dt}\frac{1}{p_{n-1}(t)}\frac{d}{dt}\frac{\bullet}{p_n(t)}\,.$$

The main property of principal systems are the conditions

$$(12)\quad \lim_{t\to\infty}\frac{J_i(t)}{J_j(t)} = \lim_{t\to\infty}\frac{K_i(t)}{K_j(t)} = 0 \quad \text{for} \quad 0 \le i < j \le n-1.$$

For the reader who is overwhelmed by the notation one can think of the operator $L_n y = y^{(n)}$ in which case one may take $J_i(t) = t^i = K_i(t)$.

It is useful to note that

$$(13)\qquad I_i(t,s\,;q_i,\dots,q_1) = (-1)^i I_i(s,t;\,q_1,\dots,q_i)$$

from which it follows that

$$(14)\qquad I_i(t,s\,;q_i,\dots,q_1) = \int_s^t q_1(r) I_{i-1}(t,r\,;q_i,\dots q_2)dr, \quad \text{and}$$

the generalized Taylor's formula

$$(15)\qquad \begin{aligned} D^i(y\,;p_0,\dots,p_i)(t) &= \sum_{j=i}^{k} D^j(y\,;p_0,\dots,p_j)(s) I_{j-i}(t,s\,;p_{i+1},\dots,p_j) \\ &+ \int_s^t I_{k-i}(t,r\,;p_{i+1},\dots,p_k)p_{k+1}(r)D^{k+1}(y\,;p_0,\dots,p_{k+1})(r)dr. \end{aligned}$$

<u>Lemma</u> (A generalization of Kiguradze's lemma)

If $y(t)$ is a nonoscillatory solution of (A^+) [resp. (A^-)], then there exists an integer $\ell \in \{0,1,\dots,n\}$ such that $\ell \not\equiv n(\mathrm{mod}\,2)$ [resp. $\ell \equiv n(\mathrm{mod}\,2)$] and

$$(16)\qquad \begin{cases} y(t)D^i(y\,;p_0,\dots,p_i)(t) > 0, & 0 \le i \le \ell, \\ (-1)^{i-\ell} y(t)D^i(y\,;p_0,\dots,p_i)(t) > 0, & \ell \le i \le n \end{cases}$$

for all sufficiently large t.

Such a $y(t)$ satisfying (16) is called a (nonoscillatory) solution of degree ℓ. Let N_ℓ be the totality of nonoscillatory solutions of degree ℓ of $(A^\pm)$. If we denote by N the set of all nonoscillatory solutions of $(A^\pm)$, then we have

$$(17) \quad \begin{cases} N = N_1 \cup N_3 \cup \cdots \cup N_{n-1} & \text{for } (A^+) \text{ with } n \text{ even,} \\ N = N_0 \cup N_2 \cup \cdots \cup N_{n-1} & \text{for } (A^+) \text{ with } n \text{ odd,} \\ N = N_0 \cup N_2 \cup \cdots \cup N_n & \text{for } (A^-) \text{ with } n \text{ even,} \\ N = N_1 \cup N_3 \cup \cdots \cup N_n & \text{for } (A^-) \text{ with } n \text{ odd.} \end{cases}$$

Consider N_ℓ for $0 < \ell < n$. For any $y \in N_\ell$ the limits

$$(18) \quad \lim_{t\to\infty} D^\ell(y\,;p_0,\ldots,p_\ell)(t) = \eta_\ell \qquad \text{(finite)}$$

$$(19) \quad \lim_{t\to\infty} D^{\ell-1}(y\,;p_0,\ldots,p_{\ell-1})(t) = \eta_{\ell-1} \qquad \text{(finite or infinite but not zero)}$$

both exist, (see [8]).

A $y \in N_\ell$ is called a <u>maximal solution in N_ℓ</u> if η_ℓ is nonzero, and a <u>minimal solution in N_ℓ</u> if $\eta_{\ell-1}$ is finite. A maximal solution y in N_ℓ grows exactly like $J_\ell(t)$ as $t \to \infty$:

$$(20) \quad \lim_{t\to\infty} \frac{y(t)}{J_\ell(t)} = \eta_\ell \neq 0$$

and a minimal solution y in N_ℓ grows exactly like $J_{\ell-1}(t)$ as $t \to \infty$:

$$(21) \quad \lim_{t\to\infty} \frac{y(t)}{J_{\ell-1}(t)} = \eta_{\ell-1} \neq 0.$$

A <u>maximal solution y in N_0</u> is a solution in N_0 such that

$$(22) \quad \lim_{t\to\infty} D^0(y\,;p_0)(t) = \eta_0 \quad (\neq 0, \text{ finite})$$

and a <u>minimal solution y in N_n</u> is a solution in N_n such that

$$(23) \quad \lim_{t\to\infty} D^{n-1}(y\,;p_0,\ldots,p_{n-1})(t) = \eta_{n-1} \quad (\neq 0, \text{ finite}).$$

Our objective is to give necessary and sufficient conditions for the existence of maximal and minimal solutions in N_ℓ.

3. Main Results. We are now prepared to discuss the existence of maximal and minimal solutions of equations $(A^{\pm})$.

Theorem 1. Let $0 \leq \ell < n$. A necessary and sufficient condition for $(A^{\pm})$ to have a maximal solution in N_ℓ is that

$$(24) \qquad \int^{\infty} K_{n-\ell-1}(t)\,|f(t,cJ_\ell(g(t)))|\,dt < \infty \qquad \text{for some} \quad c \neq 0.$$

(Proof.) (Necessity) Let $y \in N_\ell$. We may suppose without loss of generality that y is eventually positive. From (15) (with $i = \ell$ and $k = n-1$) and with the aid of (13) we have

$$(25) \qquad \begin{aligned} D^\ell(y\,;p_0,\dots,p_\ell)(t) &= \sum_{i=\ell}^{n-1} (-1)^{i-\ell} D^i(y\,;p_0,\dots,p_i)(s) I_{i-\ell}(s,t\,;p_i,\dots,p_{\ell+1}) \\ &+ (-1)^{n-\ell} \int_t^s I_{n-\ell-1}(r,t\,;p_{n-1},\dots,p_{\ell+1}) p_n(r) D^n(y\,;p_0,\dots,p_n)(r)\,dr. \end{aligned}$$

Using (16) and noting that $n-\ell$ is even [resp. odd] for (A^-) [resp. (A^+)], we have from (25)

$$(26) \qquad \begin{aligned} D^\ell(y\,;p_0,\dots,p_\ell)(t) &\geq \int_t^s I_{n-\ell-1}(r,t\,;p_{n-1},\dots,p_{\ell+1}) p_n(r)\, f(r,y(g(r)))\,dr \\ &= \int_t^s K_{n-\ell-1}(r,t)\, f(r,y(g(r)))\,dr \end{aligned}$$

for $s \geq t \geq t_1$ ($t_1 > a$ sufficiently large). Letting $s \to \infty$, we have

$$(27) \qquad D^\ell(y\,;p_0,\dots,p_\ell)(t) \geq \int_t^{\infty} K_{n-\ell-1}(r,t)\, f(r,y(g(r)))\,dr, \quad t \geq t_1,$$

which implies

$$\int_t^\infty K_{n-\ell-1}(r,t)\, f\,(r,y(g(r)))dr < \infty$$

or

$$(28) \qquad \int^\infty K_{n-\ell-1}(r)\, f(r,y(g(r)))dr \ < \infty \ .$$

In view of (20) there are positive constants c_1 and c_2 such that

$$(29) \qquad c_1 J_\ell(t) \leq y(t) \leq c_2 J_\ell(t)$$

provided t is sufficiently large. From (28) and (29) it follows that

$$\int^\infty K_{n-\ell-1}(r)\, f(r,c_1 J_\ell(g(r)))dr < \infty \ .$$

(Sufficiency) Suppose (24) holds for some $c \neq 0$. With no loss of generality we may assume that $c > 0$. Choose $T > a$ so large that $T_0 = \inf\limits_{t \geq T} \min\{g(t),t\} > a$ and

$$(30) \qquad \int_T^\infty K_{n-\ell-1}(t)\, f(t,c\, J_\ell(g(t)))dt < \frac{c}{2} \ .$$

Let $\mathbf{C}$ denote the locally convex space of all continuous functions $y : [T_0,\infty) \to R$ with the compact open topology. Consider a closed convex subset Y_ℓ of $\mathbf{C}$ defined by

$$(31) \qquad Y_\ell = \{y \in \mathbf{C} : \frac{c}{2} J_\ell(t) \leq y(t) \leq c\, J_\ell(t), \ \ t \geq T_0\} \ .$$

Define the operator $\Phi_\ell : Y_\ell \to \mathbf{C}$ by the following formulas:

If $0 < \ell < n$, then

$$(32)\quad \begin{cases} \Phi_\ell y(t) = \frac{c}{2} J_\ell(t) + p_0(t) \int_T^t I_{\ell-1}(t,s;p_1,\ldots,p_{\ell-1}) p_\ell(s) \int_s^\infty \cdot \\ \qquad \cdot\, I_{n-\ell-1}(r,s;p_{n-1},\ldots,p_{\ell+1}) p_n(r) f(r,y(g(r)))\,dr\,ds, \qquad t \geq T, \\ \Phi_\ell y(t) = \frac{c}{2} J\ (t), \qquad T_0 \leq t \leq T. \end{cases}$$

If $\ell = 0$, then

$$(33)\quad \begin{cases} \Phi_0 y(t) = \frac{c}{2} J_0(t) + p_0(t) \int_t^\infty I_{n-1}(s,t;p_{n-1},\ldots,p_1) p_n(s) f(s,y(g(s)))\,ds, \\ \qquad\qquad\qquad\qquad\qquad\qquad\qquad\qquad t \geq T, \\ \Phi_0 y(t) = \frac{c}{2} J_0(t) + p_0(t) \int_T^\infty I_{n-1}(s,T;p_{n-1},\ldots,p_1) p_n(s) f(s,y(g(s)))\,ds, \\ \qquad\qquad\qquad\qquad\qquad\qquad\qquad\qquad T_0 \leq t \leq T. \end{cases}$$

We will show that Φ_ℓ has the properties:

(i) $\Phi_\ell Y_\ell \subset Y_\ell$;

(ii) Φ_ℓ is a continuous operator;

(iii) $\Phi_\ell Y_\ell$ is relatively compact.

(i) $\Phi_\ell Y_\ell \subset Y$. Let $0 < \ell < n$. If $y \in Y_\ell$ and $t \geq T$, then noting that for $T \leq s \leq t$

$$\begin{aligned} &\int_s^\infty I_{n-\ell-1}(r,s;p_{n-1},\ldots,p_{\ell+1}) p_n(r) f(r,y(g(r)))\,dr \\ &\qquad \leq \int_T^\infty K_{n-\ell-1}(r,T) f(r,cJ_\ell(g(r)))\,dr \\ &\qquad \leq \int_T^\infty K_{n-\ell-1}(r) f(r,cJ_\ell(g(r)))\,dr < \frac{c}{2}, \end{aligned}$$

we obtain

$$p_0(t)\int_T^t I_{\ell-1}(t,s;p_1,\ldots,p_{\ell-1})p_\ell(s)\int_s^\infty I_{n-\ell-1}(r,s;p_{n-1},\ldots,p_{\ell+1}) \cdot$$

$$\cdot\ p_n(r)f(r,y(g(r)))drds$$

$$\leq \frac{c}{2}\, p_0(t)\int_T^t I_{\ell-1}(t,s;p_1,\ldots,p_{\ell-1})p_\ell(s)ds$$

$$= \frac{c}{2}\, p_0(t)I_\ell(t,T\,;\,p_1,\ldots,p_\ell)$$

$$\leq \frac{c}{2}\, J_\ell(t) \qquad \text{for} \quad t \geq T.$$

It follows from the above and (32) that

$$(*) \qquad \frac{c}{2}\, J_\ell(t) \leq \Phi_\ell y(t) \leq cJ_\ell(t) \qquad \text{for} \quad t \geq T.$$

It is obvious that (*) holds also for $T_0 \leq t \leq T$, and so Φ_ℓ maps Y_ℓ into Y_ℓ if $0 < \ell < n$. The case where $\ell = 0$ is almost obvious.

(ii) Φ_ℓ is a continuous operator. Let $\{y_k\}$ be a sequence of elements of Y_ℓ which converges to a $y \in Y_\ell$ as $k \to \infty$ in the topology of C. Then for $\ell = 0$

$$|\Phi_0 y_k(t) - \Phi_0 y(t)| \leq p_0(t)\int_T^\infty K_{n-1}(s)\, F_k(s)ds, \qquad t \geq T_0,$$

and for $0 < \ell < n$,

$$|\Phi_\ell y_k(t) - \Phi_\ell y(t)| \leq \delta(t)J_\ell(t)\int_T^\infty K_{n-\ell-1}(s)F_k(s)ds, \qquad t \geq T_0,$$

where $\delta(t) = 0$ for $t \in [T_0,T)$ and $\delta(t) = 1$ for $t \in [T,\infty)$, and

$$F_k(s) = |f(s,y_k(g(s))) - f(s,y(g(s)))|.$$

Since $F_k(s) \leq 2f(s,cJ_\ell(g(s)))$ and $\lim_{k\to\infty} F_k(s) = 0$ for $s \geq T$, from the Lebesgue dominated convergence theorem it follows that $\Phi_\ell y_k \to \Phi_\ell y$ as $k \to \infty$ in the topology of $\mathcal{C}$.

(iii) $\Phi_\ell Y_\ell$ is relatively compact. If $y \in Y_\ell$, then we have for $t \in (T_0,T) \cup (T,\infty)$

$$\left|\frac{d}{dt} D^0(\Phi_0 y\,;\,p_0)(t)\right| \leq \delta(t)p_1(t)\int_T^\infty I_{n-2}(s,a,\,;\,p_{n-1},\dots,p_2)p_n(s)f(s,cJ_0(g(s)))ds,$$

and

$$\left|\frac{d}{dt} D^0(\Phi_\ell y\,;\,p_0)(t)\right| \leq \frac{c}{2}[1+\delta(t)]p_1(t)I_{\ell-1}(t,a\,;\,p_2,\dots,p_\ell), \quad 0 \leq \ell < n.$$

This shows that for each ℓ, $0 \leq \ell < n$, the family of functions $\{\frac{d}{dt}D^0(\Phi_\ell y\,;\,p_0)(t) : y \in Y_\ell\}$ is uniformly bounded on any compact subinterval of $[T_0,\infty)$, and so the family $\{D^0(\Phi_\ell y\,;\,p_0)(t) : y \in Y_\ell\}$ is equicontinuous at every point of (T_0,∞). Now for $t_1,t_2 \in [T_0,\infty)$ we see that

$$\begin{aligned}
|\Phi_\ell y(t_2)-\Phi_\ell y(t_1)| &= |p_0(t_2)D^0(\Phi_\ell y;p_0)(t_2) - p_0(t_1)D^0(\Phi_\ell y\,;\,p_0)(t_1)| \\
&\leq |p_0(t_2) - p_0(t_1)|\,|D^0(\Phi_\ell y\,;\,p_0)(t_2)| \\
&\qquad + p_0(t_1)\,|D^0(\Phi_\ell y\,;\,p_0)(t_2) - D^0(\Phi_\ell y\,;\,p_0)(t_1)| \\
&\leq cJ_\ell(t_2)p_0(t_2)^{-1}|p_0(t_2) - p_0(t_1)| + p_0(t_1)\,|D^0(\Phi_\ell y\,;\,p_0)(t_2) - D^0(\Phi_\ell y\,;\,p_0)(t_1)|,
\end{aligned}$$

and hence that $\Phi_\ell Y_\ell = \{\Phi_\ell y : y \in Y_\ell\}$ is equicontinuous at every point of $[T_0,\infty)$. Since $\Phi_\ell Y_\ell$ is clearly uniformly bounded at every point of $[T_0,\infty)$, it follows that $\Phi_\ell Y_\ell$ is relatively compact.

Therefore, by the Schauder-Tychonoff fixed point theorem, Φ_ℓ has a fixed point

y in Y_ℓ, and a simple computation shows that this fixed point $y = y(t)$ is a solution of degree ℓ of $(A^\pm)$ such that $\lim_{t\to\infty} D^\ell(y\,;p_0,\ldots,p_\ell)(t) = \eta_\ell$ exists and is finite $(\neq 0)$.

<u>Theorem 2</u>. Let $0 < \ell \leq n$. A necessary and sufficient condition for $(A^\pm)$ to have a minimal solution in N_ℓ is that

$$(34)\qquad \int^\infty K_{n-\ell}(t)\,|f(t,c\,J_{\ell-1}(g(t)))|\,dt < \infty \quad \text{for some} \quad c \neq 0.$$

(Proof) (Necessity) Let $y \in N_\ell$ be a minimal solution. We may suppose that y is eventually positive. We have (27). Multiplying (27) by $p_\ell(t)$ and integrating, we get

$$(35)\qquad \begin{aligned} &D^{\ell-1}(y\,;p_0,\ldots,p_{\ell-1})(t) - D^{\ell-1}(y\,;p_0,\ldots,p_{\ell-1})(t_1) \\ &\geq \int_{t_1}^{t} p_\ell(s)\int_s^\infty K_{n-\ell-1}(r,s)f(r,y(g(r)))drds \\ &\geq \int_{t_1}^{t} [\int_{t_1}^{r} p_\ell(s)K_{n-\ell-1}(r,s)ds]f(r,y(g(r)))dr, \quad t \geq t_1. \end{aligned}$$

Now, using (14), we have

$$(36)\qquad \begin{aligned} &\int_{t_1}^{r} p_\ell(s)K_{n-\ell-1}(r,s)ds \\ &= \int_{t_1}^{r} p_\ell(s)p_n(r)I_{n-\ell-1}(r,s\,;p_{n-1},\ldots,p_{\ell+1})ds \\ &= p_n(r)\,I_{n-\ell}(r,t_1;p_{n-1},\ldots,p_{\ell+1},p_\ell) = K_{n-\ell}(r,t_1). \end{aligned}$$

From (35) and (36) we see that

$$(37)\qquad \begin{aligned} &D^{\ell-1}(y\,;p_0,\ldots,p_{\ell-1})(t) - D^{\ell-1}(y\,;p_0,\ldots,p_{\ell-1})(t_1) \\ &\geq \int_{t_1}^{t} K_{n-\ell}(r,t_1)f(r,y(g(r)))dr. \end{aligned}$$

Noting that $\lim_{t\to\infty} D^{\ell-1}(y\,;p_0,\ldots,p_{\ell-1})(t) = \eta_{\ell=1} \neq 0$ (finite) and that there are constants c_1, c_2 such that

$$(38)\qquad c_1 J_{\ell-1}(t) \leq y(t) \leq c_2 J_{\ell-1}(t)$$

for all sufficiently large t, we conclude from (37) that

$$\int_{t_1}^{\infty} K_{n-\ell}(r,t_1) f(r, c_1 J_{\ell-1}(g(r)))dr < \infty .$$

(Sufficiency) Suppose (34) holds for some $c \neq 0$. We may assume that $c > 0$. Take $T > a$ so large that $T_0 = \inf_{t\geq T} \min\{g(t),t\} > a$ and

$$(39)\qquad \int_T^{\infty} K_{n-\ell}(t) f(t, c J_{\ell-1}(g(t)))dt < \frac{c}{2} .$$

Consider a closed convex subset Z_ℓ of $\mathbf{C}$ defined by

$$(40)\qquad Z_\ell = \{y \in \mathbf{C} : \frac{c}{2} J_{\ell-1}(t) \leq y(t) \leq cJ_{\ell-1}(t), \quad t \geq T_0\} .$$

Define the operator $\Psi_\ell : Z_\ell \to \mathbf{C}$ by the following formulas:

If $\ell = 1$, then

$$(41)\begin{cases} \Psi_1 y(t) = c\,p_0(t) - p_0(t)\int_t^{\infty} I_{n-1}(s,t\,;p_{n-1},\ldots,p_1)p_n(s)f(s,y(g(s)))ds, \\ \qquad\qquad\qquad\qquad\qquad\qquad\qquad\qquad\qquad\qquad\qquad \text{for } t \geq T, \\ \Psi_1 y(t) = c\,p_0(t) - p_0(t)\int_T^{\infty} I_{n-1}(s,T\,;p_{n-1},\ldots,p_1)p_n(s)f(s,y(g(s)))ds, \\ \qquad\qquad\qquad\qquad\qquad\qquad\qquad\qquad\qquad\qquad\qquad \text{for } T_0 \leq t \leq T. \end{cases}$$

If $2 \leq \ell < n$, then

$$(42)\begin{cases} \Psi_\ell y(t) = cJ_{\ell-1}(t) - p_0(t)\int_T^{t} I_{\ell-2}(t,s\,;p_1,\ldots,p_{\ell-2})p_{\ell-1}(s)\int_s^{\infty} \cdot \\ \qquad\qquad \cdot\, I_{n-\ell}(r,s\,;p_{n-1},\ldots,p_\ell)p_n(r)f(r,y(g(r)))drds, \quad t \geq T, \\ \Psi_\ell y(t) = cJ_{\ell-1}(t), \quad T_0 \leq t \leq T. \end{cases}$$

If $\ell = n$ (which occurs only for (A^-)), then

$$(43)\begin{cases} \Psi_n y(t) = \frac{c}{2} J_{n-1}(t) + p_0(t)\int_T^t I_{n-1}(t,s\,;p_1,\dots,p_{n-1})p_n(s)f(s,y(g(s)))ds, \\ \qquad\qquad\qquad\qquad\qquad\qquad\qquad\qquad\qquad\qquad t \geq T, \\ \Psi_n y(t) = \frac{c}{2} J_{n-1}(t), \quad T_0 \leq t \leq T. \end{cases}$$

Again we can show that (i) $\Psi_\ell Z_\ell \subset Z_\ell$, (ii) Ψ_ℓ is a continuous operator, and (iii) $\Psi_\ell Z_\ell$ is relatively compact. It follows from the Schauder-Tychonoff fixed-point theorem Ψ_ℓ has a fixed point y in Z_ℓ. It is easy to verify that this fixed point $y = y(t)$ is a solution of degree ℓ of $(A^\pm)$ such that

$\lim_{t\to\infty} D^{\ell-1}(y\,;p_0,\dots,p_{\ell-1})(t) = \eta_{\ell-1}$ exists and is finite $(\neq 0)$.

4. Further comments and results.

The above results easily extend to equations of the forms:

$$L_n y(t) + f(t,y(g_1(t)),\dots,y(g_m(t))) = 0,$$

$$L_n y(t) - f(t,y(g_1(t)),\dots,y(g_m(t))) = 0.$$

Consider the particular cases of $(A^\pm)$:

$$(B^\pm) \qquad y^{(n)}(t) \pm f(t,y(g(t))) = 0\,.$$

Since $p_0(t) = p_1(t) = \dots = p_n(t) = 1$, we have $J_i(t\,,\,s) = \frac{(t-s)^i}{i!}$, and we can take

$$J_i(t) = K_i(t) = t^i/i!, \quad 0 \leq i \leq n-1.$$

Corollary 1. Let $0 \leq \ell < n$. A necessary and sufficient condition for $(B^\pm)$ to have a maximal solution in N_ℓ is that

$$\int^\infty t^{n-\ell-1}\,|f(t,c\,g^\ell(t))|dt < \infty \qquad \text{for some } c \neq 0\,.$$

Corollary 2. Let $0 < \ell \leq n$. A necessary and sufficient condition for $(B^\pm)$ to have a minimal solution in N_ℓ is that

$$\int^{\infty} t^{n-\ell} \, |f(t, c\, g^{\ell-1}(t))| dt < \infty \qquad \text{for some} \quad c \neq 0.$$

The following corollary is a consequence of Theorems 1 and 2.

Corollary . Equation $(A^{\pm})$ has a nonoscillatory solution $y(t)$ such that $\lim_{t\to\infty} [y(t)/J_i(t)] = \text{constant} \neq 0$ for some i, $0 \leq i \leq n-1$, if and only if for that i,

$$(44) \qquad \int^{\infty} K_{n-i-1}(t) |f(t, cJ_i(g(t)))| dt < \infty \qquad \text{for some} \quad c \neq 0.$$

Remark. Corollary for (A^{+}) was proved by Kitamura & Kusano [5].

Consider the differential equation

$$(C) \qquad L_n y(t) + F(t, y(g(t))) = 0,$$

where L_n and g are as before and $F : [a,\infty) \times R \to R$ is a continuous function. No sign condition is imposed on F. For such an equation (C) the structure of the nonoscillatory solutions is not as simple as for $(A^{\pm})$. However, under certain assumptions on F, we are able to give conditions which guarantee the existence of a solution which is asymptotic to a $J_i(t)$, $0 \leq i \leq n-1$, as $t \to \infty$.

Theorem 3. Suppose that

$$|F(t,y)| \leq \phi(t, |y|) \qquad \text{for} \quad (t,y) \in [a,\infty) \times R,$$

where $\phi : [a,\infty) \times R_+ \to R_+$, $R_+ = [0,\infty)$, is a continuous function which is nondecreasing in the second variable.

If $0 \leq i \leq n-1$ and

$$\int^{\infty} K_{n-i-1}(t) \phi(t, |c| J_i(g(t))) dt < \infty \qquad \text{for some} \quad c \neq 0,$$

then equation (C) has a nonoscillatory solution $y(t)$ such that $\lim_{t\to\infty} [y(t)/J_i(t)] = \text{const.} \neq 0.$

Proof. Choose $T > a$ so large that $T_0 = \inf_{t \geq T} \min\{g(t),t\} > a$ and

$$\int_T^\infty K_{n-i-1}(t)\phi(t,|c|J_\ell(g(t)))dt < \frac{c}{4}.$$

Let Y_i be a closed convex subset of $\mathbf{C}$ defined by

$$Y_i = \{y \in \mathbf{C} : |y(t)| \leq |c|J_i(t), \quad t \geq T_0\},$$

and define the operator $\Phi_i : Y_i \to \mathbf{C}$ by (32) and (33) with ℓ replaced by i. Then, proceeding as in the proof of Theorems 1 and 2, it can be shown that (i) $\Phi_i Y_i \in Y_i$, (ii) Φ_i is continuous and (iii) $\Phi_i Y_i$ is relatively compact. Therefore Φ_i has a fixed point y in Y_i, which becomes a desired non-oscillatory solution of (C) with the required asymptotic property.

Remark. This considerably generalizes a recent result of Tong [12] for the second order equation $y'' + f(t,y) = 0$.

Example. Consider the ODE

$$y^{(n)} + q(t)\, y^{2N} \cos y = 0, \tag{45}$$

where $N \geq 1$ is an integer and $q : [0,\infty) \to R$ is a continuous function. Here $F(t,y) = q(t)\, y^{2N} \cos y$, and we can take $\phi(t,\eta) = |q(t)|\eta^{2N}$. By Theorem 3, if

$$\int_0^\infty |q(t)|\, t^{n-i-1+2Ni}\, dt < \infty, \quad 0 \leq i \leq n-1,$$

then (45) has a solution $y(t)$ such that $\lim_{t\to\infty} [y(t)/t^i] = \text{const.} \neq 0$. In partifular if

$$\int_0^\infty |q(t)|t^{2N(n-1)}\, dt < \infty,$$

then for each i, $0 \leq i \leq n-1$, (45) has a solution $y_i(t)$ such that $\lim_{t\to\infty} [y_i(t)/t^i] = \text{const.} \neq 0$.

REFERENCES

[1] A. M. Fink, Monotone solutions to certain differential equations, C. R. Math. Rep. Acad. Sci. Canada 1(1979), 241-244.

[2] K. E. Foster and R. C. Grimmer, Nonoscillatory solutions of higher order differential equations, J. Math. Anal. Appl. 71(1979), 1-17.

[3] -----, Nonoscillatory solutions of higher order delay operations, J. Math. Anal. Appl. 77(1980), 150-164.

[4] Y. Kitamura, On nonoscillatory solutions of functional differential equations with a general deviating argument, Hiroshima Math. J. 8(1978), 49-62.

[5] Y. Kitamura and T. Kusano, Nonlinear oscillation of higher-order functional differential equations with deviating arguments, J. Math. Anal. Appl. 77(1980), 100-119.

[6] K. Kreith and T. Kusano, Extremal solutions of general nonlinear differential equations, Hiroshima Math. J. 10(1980), 141-152.

[7] D. L. Lovelady, On the asymptotic classes of a superlinear differential equation, Czechoslovak Math. J. 27(1977), 242-245.

[8] D. L. Lovelady, On the nonoscillatory solutions of a sublinear even order equation, J. Math. Anal. Appl. 57(1977), 36-40.

[9] T. Kusano, On even-order functional differential equations with advanced and retarded arguments, J. Diff. Equations 44(1982),

[10] T. Kusano, M. Naito, and K. Tanaka, Oscillatory and asymptotic behavior of solutions of a class of linear ordinary differential equations, Proc. Royal Soc. of Edin. 90A(1981), 25-40.

[11] W. F. Trench, Canonical forms and principal systems for general disconjugate equations, Trans. Amer. Math. Soc. 189(1974), 319-327.

[12] Jingcheng Tong, The asymptotic behavior of a class of nonlinear differential equations of second order, Proc. Amer. Math. Soc. 84(1982), 235-236.

A. M. Fink
Department of Mathematics
Iowa State University
Ames, Iowa 50011, U.S.A.

Takaŝi Kusano
Department of Mathematics
Faculty of Science
Hiroshima University
Hiroshima 730, Japan

Self-Adjoint 4-th Order Boundary Value Problem in the Limit - 4 Case

by

Charles T. Fulton[1]

and

Allan M. Krall[2]

SYNOPSIS. The 4-th order symmetric differential expression L has 4 linearly independent boundary values in the Lim - 4 case. Here we show that these boundary values can be characterized as limits of certain Wronskian combinations with solutions of $Lu = 0$. Boundary conditions which generate self-adjoint eigenvalue problems are classified. To demonstrate the use of this formulation of boundary conditions, the $4^{\underline{th}}$ order Legendre equation is considered in the case when the eigenfunctions are $P_n(x)$.

1. INTRODUCTION. During the past twenty-five years continuing work has vastly extended the work begun by Herman Weyl [14] on singular Sturm-Liouville systems. The large group of students of Titchmarsh has been particularly influential in this progress. Led by Atkinson [1] and Everitt [4-9], systems have been explored; higher order problems have been extensively studied. A substantial theory now is in existence, showing in particular that if the system is n-dimensional, or if the order of the symmetric expression involved is n, then there are at least $[\frac{n}{2}]$ and perhaps as many as n solutions in $L^2[a,b]$, where $[a,b]$ is the interval of definition of the problem.

In order to develop a self-adjoint operator in $L^2[a,b]$, however, the domain of operator must, among other things, be restricted by employing boundary conditions, which, even in Weyl's original second order situation assumed a rather cumbersome form. Recently, Fulton [11] has given an effective and practical form of second order limit-circle boundary conditions.

For higher order scalar equations a number of possibilities arise. If the order of differential expression is n, then there may be from $[\frac{n}{2}]$ to n solutions of the associated differential equation in $L^2[a,b]$. For each m in excess of $[\frac{n}{2}]$ a boundary condition is required. Exactly how they are found and used does not appear to be at all clear in any of the literature to date.

In this paper, therefore, we begin their description and use by considering the 4th order Lim - 4 case, which seems to be the simplest nontrivial situation. Virtually everything we do can be extended to higher order problems without essential difficulty. The case $n = 4$, however, shows the complexities without overwhelming the reader with details.

2. 4th Order Limit - 4 Boundary Value Problems.

W.N. Everitt [4] has given a discussion of the Sturm-Liouville theory for 4th order differential equations in the case of two regular endpoints. Here we employ a formulation of Lim - 4 boundary conditions which enables us to extend his analysis to the case when there are two singular endpoints, both belonging to the Lim - 4 Case. This represents a straight forward generalization of the limit circle boundary conditions used by Fulton [11]. By means of an initial value problem similar to that used by Fulton [11, p. 55, Theorem 1 (ii)] in the second order case, it becomes possible to formulate a "regularized" version of the singular eigenfunction expansion theory for 4th order equations having two Lim - 4 endpoints. For the case of separated boundary conditions this means that an entire function $W(\lambda)$ (analogous to Everitt's function in [4 equa. (83)]) having its zeros on the eigenvalues of the associated self-adjoint boundary value problem can be introduced in a direct fashion, thus obviating the need for so-called '$m(\lambda)$' - functions which are usually nonuniquely defined in Lim - 3 and Lim - 4 cases.

Following Everitt [4] we consider the general 4th order equation in the Liouville Normal Form,

$$Lf = f^{(4)} - (q_1 f')' + (q_0 f) = \lambda f.$$

We note that Green's formula may be written as

$$\int_{x_1}^{x_2} [\bar{g}Lf - fL\bar{g}]dx = [f,\bar{g}](x)\Big|_{x_1}^{x_2},$$

where the bracket symbol is defined by

$$[f,g](x) = (f'''g - fg''') - (f''g' - f'g'')$$

$$- q_1(f'g - fg')$$

$$= [g,g',g'',g'''] \begin{pmatrix} 0 & -q_1 & 0 & 1 \\ q_1 & 0 & -1 & 0 \\ 0 & 1 & 0 & 0 \\ -1 & 0 & 0 & 0 \end{pmatrix} \begin{pmatrix} f \\ f' \\ f'' \\ f''' \end{pmatrix}$$

$$= G^T MF,$$

G,M,F being defined as the matrices given above. Let $\mathcal{D}_1$ be the space of elements $f \in L^2(a,b)$ satisfying

a. f',f'',f''' are absolutely continuous on compact subsets of (a,b).

b. Lf exists a.e. and is in $L^2(a,b)$.

On the space $\mathcal{D}_1$ we may define a vector valued transformation S by choosing a real fundamental system $\{u_1,u_2,u_3,u_4\}$ of $Lu = 0$ satisfying

$$\det U = \det \begin{pmatrix} u_1 & u_2 & u_3 & u_4 \\ u_1' & u_2' & u_3' & u_4' \\ u_1'' & u_2'' & u_3'' & u_4'' \\ u_1''' & u_2''' & u_3''' & u_4''' \end{pmatrix} = W_x(u_1,u_2,u_3,u_4) = 1$$

and putting for $f \in \mathcal{D}_1$,

$$(Sf)(x) = \begin{pmatrix} (Sf)_1(x) \\ (Sf)_2(x) \\ (Sf)_3(x) \\ (Sf)_4(x) \end{pmatrix} = U^{-1}(x) \begin{pmatrix} f(x) \\ f'(x) \\ f''(x) \\ f'''(x) \end{pmatrix}$$

<u>Theorem 1</u>.

$$(Sf)(x) = \begin{pmatrix} (Sf)_1(x) \\ (Sf)_2(x) \\ (Sf)_3(x) \\ (Sf)_4(x) \end{pmatrix} = \begin{pmatrix} W_x(f,u_2,u_3,u_4) \\ W_x(u_1,f,u_3,u_4) \\ W_x(u_1,u_2,f,u_4) \\ W_x(u_1,u_2,u_3,f) \end{pmatrix}$$

<u>Proof</u>: By definition, $U \cdot (Sf) = F$, so the result follows by use of Cramer's rule.

We note a well known determinental identity associated with the bracket quantities arising in Green's formula (cf. Everitt [5, p. 374]). Namely, if $\{f_1,f_2,f_3,f_4\}$ and $\{g_1,g_2,g_3,g_4\}$ are any system of C^3-functions we have

$$[f_i,g_j](x) = G_j^T M F_i, \quad G_j^T = [g_j,g'_j,g''_j,g'''_j]; \quad F_i^T = [f_i,f'_i,f''_i,f'''_i],$$

and therefore the matrix with the above as the $ij^{\underline{th}}$ entry satisfies

$$\det([f_i,g_j](x)) = (\det G^T)(\det M)(\det F) = W_x(g_1,g_2,g_3,g_4) \cdot W_x(f_1,f_2,f_3,f_4)$$

where F and G are the Wronskian matrices for $\{f_1,f_2,f_3,f_4\}$ and $\{g_1,g_2,g_3,g_4\}$ respectively.

Theorem 2. Let $-\infty \leqq a < b \leqq \infty$ be singular Lim - 4 endpoints for L. Then for all $f \in \mathcal{D}_1$ the limits

$$(Sf)(a) = \lim_{x \to a} (Sf)(x), \quad (Sf)(b) = \lim_{x \to b} (Sf)(x)$$

exist.

Proof: Since all solutions of $Lu = 0$ are square integrable near a and b and since $Lf \in L^2(a,b)$ for $f \in \mathcal{D}_1$ it follows from Green's formula that the limits

$$[f,u_j](a) = \lim_{x \to a} [f,u_j](x)$$

and

$$[f,u_j](b) = \lim_{x \to b} [f,u_j](x)$$

exist. To show that $(Sf)_1(a)$ and $(Sf)_1(b)$ exist we can take $g_j = u_j$, $j = 1,2,3,4$, and $f_1 = f$, $f_i = u_i$, $i = 2,3,4$ in the above determinent identity to get

$$(Sf)_1(x) = 1 \cdot W_x(f,u_2,u_3,u_4)$$

$$= \det \begin{pmatrix} [f,u_1] & [f,u_2] & [f,u_3] & [f,u_4] \\ [u_2,u_1] & [u_2,u_2] & [u_2,u_3] & [u_2,u_4] \\ [u_3,u_1] & [u_3,u_2] & [u_3,u_3] & [u_3,u_4] \\ [u_4,u_1] & [u_4,u_2] & [u_4,u_3] & [u_4,u_4] \end{pmatrix} .$$

Since the limits of the quantities in the matrix on the right hand side exist as $x \to a,b$, it follows that the limits of the left hand side exist also.

Theorem 3. For any two elements $f,g \in \mathcal{D}_1$, Green's formula on the singular interval (a,b) is

$$\int_a^b [\bar{g}(Lf) - f(L\bar{g})]dx = (Sg)^*(b)A(Sf)(b) - (Sg)^*(a)A(Sf)(a),$$

where

$$A = U^T MU = ([u_i,u_j]).$$

Proof: In the right side of Green's formula on $[x_1,x_2]$ with $a < x_1 < x_2 < b$ we may write the bracket quantity in terms of S by using $F = U \cdot Sf$ and $G^* = (U \cdot Sg)^* = (Sg)^* U^T$ to obtain

$$[f,\bar{g}](x) = (Sg)^*(x)U^T(x)M(x)U(x)(Sf)(x).$$

Since the limits of $(Sf)(x)$ and $(Sg)(x)$ exist as $x \to a,b$ and since A is a constant matrix, we may pass to the limits $x_1 \to a$ and $x_2 \to b$ to obtain Green's formula for the singular interval.

3. The Initial Value Problem at a Lim - 4 Endpoint

The transformation S when restricted to solutions $f = f(x,\lambda)$ of the equation $Lf = \lambda f$ carries them over to solutions of a first order system similar to that used by Fulton in the 2nd. order case [11, p. 54]. As before, an initial value problem for the system can be posed at the singular endpoints a and b by means of which solutions of $Lf = \lambda f$ may be defined by their limiting behaviour as $x \to a,b$. This represents a generalization of Theorem 1 of [11] to the 4th. order equation.

To illustrate the generality of the conversion to system form we first give a second proof of Theorem 2 above. For $f \in \mathcal{D}_1$ put $g = Lf \in L^2(a,b)$. Then the 4th. order equation $Lf = g$ can be written in system form as

$$F' = \begin{pmatrix} f \\ f' \\ f'' \\ f''' \end{pmatrix}' = \begin{pmatrix} 0 & 1 & 0 & 0 \\ 0 & 0 & 1 & 0 \\ 0 & 0 & 0 & 1 \\ -q_0 & q_1' & q_1 & 0 \end{pmatrix} \begin{pmatrix} f \\ f' \\ f'' \\ f''' \end{pmatrix} + \begin{pmatrix} 0 \\ 0 \\ 0 \\ g \end{pmatrix} = AF + G.$$

The Wronskian matrix U satisfies $U' = AU$, while its inverse $V = U^{-1}$ satisfies the adjoint equation $V' = -VA$. Equating matrix elements in the adjoint equation a calculation shows that V has the general form

$$V = \begin{pmatrix} -v_1''' + q_1 v_1' & v_1'' - q_1 v_1 & -v_1' & v_1 \\ -v_2''' + q_1 v_2' & v_2'' - q_1 v_2 & -v_2' & v_2 \\ -v_3''' + q_1 v_3' & v_3'' - q_1 v_3 & -v_3' & v_3 \\ -v_4''' + q_1 v_4' & v_4'' - q_1 v_4 & -v_4' & v_4 \end{pmatrix}$$

where v_i, $i = 1,2,3,4$, are solutions of $Lu = 0$, and are therefore in $L^2(a,b)$. For given u_i, $i = 1,2,3,4$, it is possible to represent V in terms of u_i by solving $V \cdot U = I$ for v_i. However, for our purposes it will be sufficient to know that $v_i \in L^2(a,b)$.

In the above equation we make the change of variable

$$Y(x) = (Sf)(x) = V(x) \cdot F(x),$$

which gives

$$Y' = V'F + VF' = -VAF + V[AF + G] = VG = H,$$

where $H = (v_1g, v_2g, v_3g, v_4g)^T$. Since v_i and g are in $L^2(a,b)$ the products are integrable near a and b, so it follows that $Y(x) = (Sf)(x)$ has limits as $x \to a,b$.

Definition 1: For $f \in \mathcal{D}_1$ we define linear functionals on $\mathcal{D}_1$ by

$$B_+(f) = (B_1^+(f), B_2^+(f), B_3^+(f), B_4^+(f))^T = (Sf)(b)$$
$$B_-(f) = (B_1^-(f), B_2^-(f), B_3^-(f), B_4^-(f))^T = (Sf)(a)$$

Note: It follows from Dunford and Schwartz [3, p.1302, Thm. 27] that $B_i^+(f)$, $i = 1,2,3,4$, are linearly independent boundary values for L at $x = b$ and $B_i^-(f)$, $i = 1,2,3,4$, are linearly independent boundary values for L at $x = a$.

With this notation the Green's formula on (a,b) becomes

$$\int_a^b [\bar{g}(Lf) - f(L\bar{g})dx$$
$$= (B_-^*(g), B_+^*(g)) \begin{pmatrix} -A & 0 \\ 0 & A \end{pmatrix} \begin{pmatrix} B_-(f) \\ B_+(f) \end{pmatrix}.$$

We now restrict S to operate on solutions of the homogeneous equation $Lf = \lambda f$. In this case the above first order system can be written as

$$F' = AF + \lambda QF$$

where

$$Q = \begin{pmatrix} 0 & 0 & 0 & 0 \\ 0 & 0 & 0 & 0 \\ 0 & 0 & 0 & 0 \\ 1 & 0 & 0 & 0 \end{pmatrix}$$

Under the change of variable

$$Y(x) = (Sf)(x) = V(x) \cdot F(x) = U^{-1}(x)F(x)$$

we now obtain

$$Y' = \lambda VQF = \lambda VQUY = \lambda BY$$

where

$$B = VQU = (v_i u_j).$$

Since B consists of a matrix whose components are products of square integrable solutions of $Lu = 0$ it follows that $\|B(x)\| \in L_1(a,b)$. Hence the assumptions of Theorems 1.1 and 1.2 of Hartman [12; p. 273] are satisfied (in the linear case), so we have:

Theorem 4. (i) Let $z_i \in \mathbb{C}$, $i = 1,2,3,4$. Then for each $\lambda \in \mathbb{C}$ there exists a unique solutions $Y = Y(x,\lambda)$ of the initial value problem

$$Y' = \lambda BY, \quad Y(a,\lambda) = (z_1,z_2,z_3,z_4)^T.$$

Equivalently, we may say that the solution $f(x,\lambda)$ of $Lf = \lambda f$, corresponding to $Y(x,\lambda)$ via $(f(x,\lambda), f'(x,\lambda), f''(x,\lambda), f'''(x,\lambda))^T = U(x) \cdot Y(x,\lambda)$, is the unique solution of the equivalent "terminal value" problem

$$Lf = \lambda f$$

$$B_-(f) = \lim_{x \to a} \begin{pmatrix} W_x(f,u_2,u_3,u_4) \\ W_x(u_1,f,u_3,u_4) \\ W_x(u_1,u_2,f,u_4) \\ W_x(u_1,u_2,u_3,f) \end{pmatrix} = \begin{pmatrix} z_1 \\ z_2 \\ z_3 \\ z_4 \end{pmatrix}$$

(ii) The solutions $Y(x,\lambda)$ and $f(x,\lambda)$ defined in part (i) are entire functions of λ for $x \in (a,b)$.

The basic transformation which brings the equation $Lf = \lambda f$ into the above modified form for $Y = Sf$ extends without difficulty to higher order equations and systems of equations. For first order systems the transformed system corresponding to the above has, for example, been given by F.V. Atkinson [1; p. 297, Equa. (9.11.6)] who employs it to give a proof of the fact that the limit-circle classification is independent of λ.

In the next section we formulate general self-adjoint boundary conditions under which $\mathcal{D}_1$ is restricted to be the domain of a self-adjoint operator associated with L. In the case of separated boundary conditions, we are required to have two boundary conditions at $x = a$ and two at $x = b$, and these may be written as linear combinations of $B_i^-(f)$, and $B_i^+(f)$, $i = 1,2,3,4$, respectively. The significance of Theorem 4 lies in the fact that by choosing z_i, $i = 1,2,3,4$, appropriately it is possible to define a pair of solutions $\varphi_1(x,\lambda)$ and $\varphi_2(x,\lambda)$ by "initial" or "terminal" conditions at $x = a$ so that $\{\varphi_1,\varphi_2\}$ satisfy the two boundary conditions at $x = a$, together with the relation

$$[\varphi_1(x,\lambda), \varphi_2(x,\lambda)] = 0.$$

Similarly, a pair of solutions $\{X_1(x,\lambda),X_2(x,\lambda)\}$, also entire in λ, may be defined by "initial" conditions at $x = b$ via Theorem 4 (at $x = b$) so as to satisfy the two boundary conditions at $x = b$, together with the relation

$$[X_1(x,\lambda),X_2(x,\lambda)] = 0.$$

It follows that the associated eigenfunction expansion theory can be accomplished in the Lim - 4 case in straightforward analogy to Everitt's discussion in [4] of the expansion theory for the regular 4th. order Sturm-Liouville problem. This approach has the advantage over more classical methods since the nonuniqueness of the Lim - 4 'm(λ)-functions' is circumvented and replaced by the entire function,

$$W(\lambda) = W_x(\varphi_1,\varphi_2,X_1,X_2),$$

which is automatically parametrized by the parameters associated with the Lim - 4 boundary conditions. A complete discussion of Legendre's equation in the second order case has been given by Fulton [10]. As an example to illustrate the general theory in the 4th. order case we consider the Legendre-squared operator, and identify those Lim - 4 boundary conditions which give rise to $P_n(x)$ as the associated eigenfunctions. A complete characterization of all eigenfunction expansions which could arise by different choices of boundary conditions could possibly be done along the lines of Fulton [10], but would require knowledge of "connection" formulae expressing Frobenius solutions near one endpoint in terms of the Frobenius solutions near the other end-point, which do not appear to be available in the literature.

4. Self-adjoint Boundary Value Problems

The previous section established that

$$\int_a^b \bar{g}(Lf) - f(L\bar{g})dx = (B_-^*(g), B_+^*(g)) \begin{pmatrix} -A & 0 \\ 0 & A \end{pmatrix} \begin{pmatrix} B_-(f) \\ B_+(f) \end{pmatrix} .$$

It is our purpose now to show how to separate the expressions above into constraints for boundary value problems and to classify which are self-adjoint.

Let A and B be $m \times 4$ matrices; $0 \leqq m \leqq 8$, such that the rank of $(A:B)$ is m. Let C and D be complimentary $(8 - m) \times 4$ matrices such that $\begin{pmatrix} A & B \\ C & D \end{pmatrix}$ has rank 8. Then $\begin{pmatrix} A & B \\ C & B \end{pmatrix}^{-1}$ exists and can be represented

by $\begin{pmatrix} -\tilde{A}^* & -\tilde{C}^* \\ \tilde{B}^* & \tilde{D}^* \end{pmatrix}$, where $\tilde{A}$ and $\tilde{B}$ are $m \times 4$ matrices with rank of $(\tilde{A}:\tilde{B})$ equal to m and $\tilde{C}$ and $\tilde{D}$ are $(8 - m) \times 4$ matrices.

Since

$$\begin{pmatrix} -\tilde{A}^* & -\tilde{C}^* \\ \tilde{B}^* & \tilde{D}^* \end{pmatrix} \begin{pmatrix} A & B \\ C & D \end{pmatrix} = \begin{pmatrix} I & 0 \\ 0 & I \end{pmatrix} ,$$

then

$$\begin{pmatrix} \mathcal{A}\tilde{A}^* & \mathcal{A}\tilde{C}^* \\ \mathcal{A}\tilde{B}^* & \mathcal{A}\tilde{D}^* \end{pmatrix} \begin{pmatrix} A & B \\ C & D \end{pmatrix} = \begin{pmatrix} -\mathcal{A} & 0 \\ 0 & \mathcal{A} \end{pmatrix} .$$

Consequently, inserting this in the previous Green's formula, we have:

<u>Theorem 5</u>. <u>Green's formula can be written as</u>

$$\int_a^b [\bar{g}(Lf) - f(\bar{L}g)]dx =$$

$$[\tilde{A}A^* B_-(g) + \tilde{B}A^* B_+(g)]^* \cdot [AB_-(f) + BB_+(f)]$$
$$+ [\tilde{C}A^* B_-(g) + \tilde{D}A^* B_+(g)]^* \cdot [CB_-(f) + DB_+(f)].$$

DEFINITION. <u>We denote by</u> $\mathcal{D}$ <u>those elements</u> f <u>in</u> $L^2(a,b)$ <u>with the following properties</u>:

a. f', f'', f''' <u>exist and</u> f''' <u>is absolutely continuous on compact subsets of</u> (a,b).

b. Lf <u>exists a.e. and is in</u> $L^2(a,b)$.

c. $AB_-(f) + BB_+(f) = 0$

DEFINITION. <u>We define the differential operator</u> $\mathcal{L}$ <u>by setting</u> $\mathcal{L}f = Lf$ <u>for all</u> f <u>in</u> $\mathcal{D}$.

THEOREM 6. <u>In</u> $L^2(a,b)$ <u>the adjoint to</u> $\mathcal{L}$, $\mathcal{L}^*$, <u>is determined by the following</u>: $\mathcal{D}^*$ <u>consists of those elements</u> g <u>in</u> $L^2(a,b)$ <u>with the following properties</u>:

a. g', g'', g''' <u>exist and</u> g''' <u>is absolutely continuous on compact subsets of</u> (a,b)

b. Lg <u>exists a.e. and is in</u> $L^2(a,b)$.

c. $\tilde{C}A^* B_-(g) + \tilde{D}A^* B_+(g) = 0$.

<u>For all</u> g <u>in</u> $\mathcal{D}^*$, <u>then</u> $\mathcal{L}^* g = Lg$.

The proof of these statements follows the usual well known arguments. Green's formula is employed.

Self-adjointness can now be specified.

Theorem 7. A necessary and sufficient condition for L to be self-adjoint is

$$AA^{-1}A^{*} = BA^{-1}B^{*}.$$

Again the proof is well known. We cite Coddington and Levinson's classic book [2].

6. An Example. The Legendre Squared Operator.

The familiar Legendre operator $\ell y = ((1-t^2)y')'$, defined on $(-1,1)$ is in the limit circle case at both ends and has been discussed by Fulton [10]. Its square gives a fourth order operator, which is also of interest, since the Legendre polynomials may still serve as eigenfunctions.

It is easy to see that

$$Ly = \ell^2 y = ((1-t^2)y'')'' - 2((1-t^2)y')',$$

and that the equation $Ly = 0$ has Frobenius indicial roots of $0,0,1$ and 1. Thus L is in the limit-4 case at ± 1. Four solutions are

$$\begin{aligned} y_1 &= 1, \\ y_2 &= \ell n(1-t), \\ y_3 &= \ell n(1+t), \\ y_4 &= \ell n(1+t)\ell n(1-t) - \sum_{k=1}^{\infty} \frac{2k\left(\frac{1+t}{2}\right)^{k+1}}{(k+1)^2}. \end{aligned}$$

All are in $L^2(-1,1)$.

In order to discuss the boundary value problems associated with L, however, a Liouville transformation (Everitt [4]) is helpful. If we let $t = \sin x$, $f = (\cos^{1/2}x)y$, then $-\frac{\pi}{2} \le x \le \frac{\pi}{2}$ and Ly is transformed into

$$Lf = f^{(iv)} + (\tfrac{1}{2}[\sec^2 x + 1]f')' + (\tfrac{25}{16}\sec^4 x - \tfrac{7}{8}\sec^2 x + \tfrac{1}{16})f .$$

Further y is in $L^2(-1,1)$ if and only if f is in $L^2(-\frac{\pi}{2}, \frac{\pi}{2})$.

Solutions of $Lf = 0$ are now given by

$$u_1 = \cos^{1/2}x$$

$$u_2 = \cos^{1/2}x \, \ell n(1 - \sin x)$$

$$u_3 = \cos^{1/2}x \, \ell n(1 + \sin x)$$

$$u_4 = -\frac{1}{2}\cos^{1/2}x \int_0^x \frac{\cos s}{1 - \sin s}\left[\int_0^s \frac{\cos t}{(1 + \sin t)^2}\, \ell n(1 - \sin t)\,dt\right]ds .$$

Here u_4 has been found by reduction of order and not by Frobenius' method. We have:

$$W[u_1, u_2, u_3, u_4] = 1.$$

It is easy to show that $f = \sin^n x \; \cos^{1/2}x$ (the equivalent of the n-th power of t) satisfies $B_4^-(f) = 0$ and $B_3^-(f) = 0$. Likewise, at the other end, $B_4^+(f) = 0$ and $B_2^+(f) = 0$. Thus the boundary value problem which has as eigenfunctions the transformed Legendre polynomials

$$f_n = \cos^{1/2}x \;\; P_n(\sin x)$$

is given by

$$Lf = f^{(iv)} + (\tfrac{1}{2}[\sec^2 x + 1]f')' + (\tfrac{25}{16}\sec^4 x - \tfrac{7}{8}\sec^2 x + \tfrac{1}{16})f = \lambda f,$$

$$B_4^-(f) = 0, \qquad B_3^-(f) = 0$$

$$B_4^+(f) = 0, \qquad B_2^+(f) = 0.$$

A simple computation shows that the matrices $A, B, \mathcal{A}$ are given by

$$A = \begin{pmatrix} 0 & 0 & 1 & 0 \\ 0 & 0 & 0 & 1 \\ 0 & 0 & 0 & 0 \\ 0 & 0 & 0 & 0 \end{pmatrix} \qquad B = \begin{pmatrix} 0 & 0 & 0 & 0 \\ 0 & 0 & 0 & 0 \\ 0 & 1 & 0 & 0 \\ 0 & 0 & 0 & 1 \end{pmatrix}$$

$$\mathcal{A} = \begin{pmatrix} 0 & 0 & 0 & -\frac{1}{2} \\ 0 & 0 & 2 & 0 \\ 0 & -2 & 0 & 0 \\ \frac{1}{2} & 0 & 0 & 0 \end{pmatrix} \qquad \mathcal{A}^{-1} = \begin{pmatrix} 0 & 0 & 0 & 2 \\ 0 & 0 & -\frac{1}{2} & 0 \\ 0 & \frac{1}{2} & 0 & 0 \\ -2 & 0 & 0 & 0 \end{pmatrix}$$

and that $A\mathcal{A}^{-1}A^* = B\mathcal{A}^{-1}B^* = 0$. Thus the Legendre squared operator L is self-adjoint when its domain is constrained by the boundary conditions listed above.

REFERENCES

1. F.V. Atkinson, DISCRETE AND CONTINUOUS BOUNDARY VALUE PROBLEMS, Academic Press, New York, 1964.

2. E.A. Coddington and N. Levinson, THEORY OF ORDINARY DIFFERENTIAL EQUATIONS, McGraw-Hill, New York, 1955.

3. N. Dunford and J.T. Schwartz, Linear operators, Part II, Interscience, New York, 1963.

4. W.N. Everitt, "The Sturm-Liouville problem for fourth order differential equations," Quart. J. Math. Oxford (2), 8(1957), 146-160.

5. ____________, "Self-adjoint boundary value problems on finite intervals," J. London Math. Soc. 37(1962), 372-384.

6. ____________, "Fourth order singular differential equations," Math. Annalen, 149(1963), 320-340.

7. ____________, "Singular differential equations I: The even order case," Math. Annalen, 1956(1964), 9-24.

8. W.N. Everitt and V.K. Kumar, "On the Titchmarsh-Weyl theory of ordinary symmetric differential expressions I: The general theory," N. Arch. Wisk. (3), 24(1976), 1-48.

9. ____________________________, "On the Titchmarsh-Weyl theory of ordinary symmetric differential expressions II: The odd order case," N. Arch. Wisk. (3), 24(1976), 109-145.

10. C.T. Fulton, "Expansions in Legendre functions," Quart. J. Math. Oxford, to appear.

11. ___________, "Paramet rizations of Titchmarsh's $m(\lambda)$-functions in the limit circle case," Trans. Amer. Math. Soc., 229(1977), 51-63.

12. P. Hartman, ORDINARY DIFFERENTIAL EQUATIONS, John Wiley and Sons, New York, 1964.

13. E.C. Titchmarsh, EIGENFUNCTION EXPANSIONS ASSOCIATED WITH SECOND ORDER DIFFERENTIAL EQUATIONS, PART I, Oxford University Press, 1962.

14. H. Weyl, "Ueber gewöhnliche Differentialgleichungen mit singularitäten und die zugehörigen Entwicklungen," Math. Annalen, 69(1910), 220-269.

Factorization and the Friedrichs extension for ordinary differential operators

Robert M. Kauffman
Department of Mathematics
Western Washington University, Bellingham, Washington 98225, USA

0. Introduction

If M is a formally symmetric ordinary differential expression such that, for some positive real number ε, $(Mf,f) \geq \varepsilon(f,f)$ for all f in $C_0^\infty(a,\infty)$, the Friedrichs extension H of the restriction of M to $C_0^\infty(a,\infty)$ is a natural self-adjoint operator in $L_2(a,\infty)$ to associate with M. This operator is constructed by functional analytic means (see Dunford and Schwartz [4], p.1240). For any f in the domain of H or $H^{1/2}$, very nice convergence results are available for the spectral decomposition. (See Hinton [5] and [6], and Kauffman [7].) In addition, the square root of H is associated in a natural way with the problem of minimization of quadratic functionals. (See Bradley-Hinton-Kauffman [2].) In all of these applications, in order to deal with concrete problems one must know something about the domain of H or of $H^{1/2}$. This paper studies the question for arbitrary M, by proving that M may be factored in the form $M = L^+L$, with L limit-point in the general non-symmetric sense defined below.

The general study of the domain of the Friedrichs extension and its square root was apparently begun by Bennewitz [1], in the more general context of pairs of differential operators. He developed a good deal of functional-analytic machinery for the study of the domain of H and of $H^{1/2}$ in this very general context. In particular, the idea of studying the Dirichlet index and relating it to the domain of H and of $H^{1/2}$ appears first in this paper.

Hinton [5], and [6], working independently of Bennewitz, gave a characterization of the domains of H and $H^{1/2}$ in the weighted Hilbert space $L_2^W(a,b)$, when $M = \frac{1}{W}\sum_0^N (-1)^i D^i p_i D^i$, and $p_N > 0$, and when M is limit-point. Although these papers deal with much more general differential operators and boundary conditions, in the case of the above Hinton's results give the domain of $H^{1/2}$ as $\{f \mid p_i^{1/2} f^{(i)} \in L_2(a,\infty)$ for $i = 0, \ldots, N$ and $f^{(i)}(a) = 0$ for $i = 0, \ldots, N-1\}$. Kauffman [7] extended Hinton's results, in the positive coefficient case with $W=1$, to the case where the <u>Dirichlet index</u>, or dimension of the set of all f such that $Mf = 0$ and $p_i^{1/2} f^{(i)} \in L_2(a,\infty)$ for all $i \leq N$, is equal to N. (This fact also follows from the results of Bennewitz referred to earlier.) Kauffman then showed that for a large number of well-behaved p_i, including all polynomials, the Dirichlet index is N, even though M need not be limit-point. The question of whether the Dirichlet index need <u>always</u> be N in the positive-coefficient case discussed above is unsolved, but Robinette [9] extended Kauffman's results to a considerably larger class of coefficients. Bradley-Hinton-Kauffman [2] extended those results to the case where the coefficients are allowed to be negative, but not very negative compared to the positive part of M.

It is clear that for the general case of any M such that $(Mf,f) \geq \varepsilon(f,f) > 0$ for all f in $C_0^\infty(a,\infty)$, the methods of Hinton, Kauffman and Bradley-Hinton-Kauffman no longer apply. The coefficients may be too negative in this case. This paper treats this case by using the well-known factorization $M = L^+L$. It is obvious that the domain of $H^{1/2}$ is the domain of the minimal operator $T_0(L)$, but this is not an explicit characterization, because even when L is known, the domain of $T_0(L)$ is not usually obvious. If L is <u>limit-point</u>, in a sense to be defined below, the domain of $T_0(L)$ is the set of all f such that $f^{(i)}(a) = 0$ $(0 \leq i \leq N-1)$ and $Lf \in L_2(a,\infty)$. This is explicit, if L is known, and is analogous to the conditions obtained earlier in the positive coefficient case. We are then led to ask "When is $M = L^+L$, with L limit-point?"

The somewhat surprising answer is that if $M = \Sigma_0^N (-1)^i D^i p_i D^i$, with p_i real and C^∞ and $p_N(x) > 0$ for all x in $[a,\infty)$, and $(Mf,f) \geq \varepsilon(f,f)$ for all f in $C_0^\infty(a,\infty)$, where ε is a positive real number, then M has the factorization $M = L^+L$, where L is limit-point. In this paper we prove this theorem, and apply it to characterize the domains of H and of $H^{1/2}$ for a fairly large class of self-adjoint operators H associated with M.

1. The domain of H and of $H^{1/2}$

We consider the problems discussed in the introduction, allowing more general boundary conditions at the end-point a.

Definition 1.1. Let domain R be the set of restrictions to $[a,\infty)$ of elements f of $C_0^\infty(-\infty,\infty)$ such that $\langle f(a), f^{(1)}(a), \ldots, f^{(2N-1)}(a)\rangle \varepsilon S$, where S is an N-dimensional subspace of complex 2N-space K^{2N}. Let $Rf = Mf$ for all f in domain R, where $M = \Sigma_0^N (-1)^i D^i p_i D^i$, with each p_i in $C^\infty[a,\infty)$, $p_N(x) > 0$ for all x in $[a,\infty)$, p_i real-valued, and $D = d/dx$. ($C^\infty[a,\infty)$ denotes the set of functions g such that g is infinitely differentiable at all x in (a,∞) and $g^{(i)}$ has a continuous extension to $[a,\infty)$ for each i.) Assume that $(Mf,f) \geq \varepsilon(f,f)$ for all f in $C_0^\infty(a,\infty)$, for some positive constant ε. Assume that $(Rf,g) = (f,Rg)$ for all f and g in domain R. Let domain R_0 be $C_0^\infty(a,\infty)$, and let $R_0 f = Mf$ for all f in domain R_0.

Lemma 1.2. R is bounded below.

Proof: $\bar{R}_0$ is the minimal operator $T_0(M)$ in $L_2(a,\infty)$ associated with the differential expression M. Since $(R_0 f,f) \geq \varepsilon(f,f)$ for all f in domain R_0, it follows that $\bar{R}_0$ has closed range. Therefore any finite dimensional extension of $\bar{R}_0$ also has closed range, since the direct sum of a closed subspace and a finite-dimensional subspace is another closed subspace. If $T_1(M)$ is the maximal operator in $L_2(a,\infty)$ associated with M, then $T_1(M)$ is a finite dimensional extension of $T_0(M)$. Hence $\bar{R}$ is a finite dimensional extension of $T_0(M)$, and therefore range $\bar{R}$ is closed. It follows that the deficiency indices of $\bar{R}$ are equal, and therefore $\bar{R}$ has a self-adjoint extension Q which is contained in $T_1(M)$. Since Q is a self-adjoint extension of $T_0(M)$, it follows that Q is bounded below. Hence $\bar{R}$ is bounded below.

Notation 1.3. Let H_R denote the Friedrichs extension of R. Let H_0 denote the Friedrichs extension of $T_0(M)$. Let H_{00} be the restriction of H_0 to $\{f \in \text{domain } H_0 | f^{(i)}(a) = 0 \text{ for } i \leq 2N-1\}$. (Note that H_{00} will not be $T_0(M)$ unless M is limit-point.) If R is as above, then R + bI is defined by domain (R + bI) = domain R, and (R + bI)f = Rf + bf, for any real number b.

Lemma 1.4. Suppose, for some $\varepsilon > 0$, $((R + bI)f,f) \geq \varepsilon(f,f)$ for all f in domain R. Then domain H_{R+bI} is the set of all f in $L_2(a,\infty)$ such that

a) f is in the domain of R^* (the Hilbert space adjoint of R);

b) there exists a sequence f_n in domain R with f_n converging to f in $L_2(a,\infty)$ and with f_n Cauchy in the norm given by $||g||^2_{R+bI} = ((R + bI)g,g)$.

Proof: This follows from the construction of the Friedrichs extension (see Dunford and Schwartz [4], p.1240).

Lemma 1.5. $H_{00} \subset H_R$. Also, if $M = L^+L$, with $L = \Sigma_0^N b_i D^i$, with b_i in $C^\infty[a,\infty)$, then domain H_0 is contained in the domain of the minimal operator $T_0(L)$ corresponding to L in $L_2(a,\infty)$.

Proof: Integration by parts shows that $H_{00} \subset R^*$. But, if f is in domain H_0, there is a sequence f_n in $C_0^\infty(a,\infty)$ with f_n converging to f in $L_2(a,\infty)$ and with f_n Cauchy in the norm $||g||^2_{R_0} = (Mg,g)$. Thus, by Lemma 1.4 domain H_{00} is contained in domain H_{R+bI}, since f_n is clearly also Cauchy in the norm $|| \ ||_{R+bI}$. However, it is easy to see that $H_{R+bI} = H_R + bI$. The first assertion of the lemma is proved.

But, if f and f_n are as above, it is clear from the factorization $M = L^+L$ that Lf_n is also a cauchy sequence. Hence f is in the domain of $T_0(L)$. The lemma is proved.

Definition 1.6. Let L be given by $L = \Sigma_0^N b_i D^i$, with $b_i \in C^\infty[a,\infty)$ and with $b_N(x) \neq 0$ for any x in $[a,\infty)$. Then L is said to be <u>limit-point</u> if $T_1(L)$ is an N-dimensional extension of $T_0(L)$.

Remark: This definition, which agrees with the usual definition for symmetric L, is discussed in Kauffman-Read-Zettl [8]. The next lemma is proved there.

Lemma 1.7. Let L be as in Definition 1.6. Then, if range $T_0(L)$ is closed, L is limit-point if and only if nullity $T_1(L)$ + nullity $T_1(L^+) = N$.

Lemma 1.8. Suppose that $M = L^+L$, with $L = \Sigma_0^N b_i D^i$, with each b_i in $C^\infty[a,\infty)$ and with $b_N(x) \neq 0$ for any x in $[a,\infty)$. Then domain H_0 = domain $T_1(M) \cap$ domain $T_0(L)$. If, in addition, L is limit-point, then domain H_0 is the set of all f such that

a) $f^{(i)}(a) = 0$ for $(i \leq N - 1)$;

b) $f \in$ domain $T_1(M)$;

c) $Lf \in L_2(a,\infty)$.

Proof: If f is in domain H_0 then, by Lemma 3, f is in domain $T_0(L)$. Clearly f is in domain $T_1(M)$.

If f is in both domain $T_1(M)$ and domain $T_0(L)$, then there is a sequence f_n in $C_0^\infty(a,\infty)$ such that Lf_n is Cauchy and f_n converges to f in $L_2(a,\infty)$. Since $(L^+Lf_n, f_n) = ||Lf_n||^2$, it follows that f satisfies condition b) of Lemma 1.4. Since $R_0^* = T_0(M)^* = T_1(M)$, it follows that f satisfies condition a) of Lemma 1.4. Hence f is in the domain of H_0.

If L is limit-point, and f satisfies a) and c), then by corollary 4.7, p. 18 [8], it follows that $f \in$ domain H_0 if f also satisfies b).

Lemma 1.9. Domain H_R = domain H_{00} + linear span $\{f_i\}$, where f_i is the restriction to $[a,\infty)$ of an element of $C_0^\infty(-\infty,\infty)$ such that $\langle f_i(a), f_i^{(1)}(a), \ldots, f_i^{(2N-1)}(a)\rangle = V_i$, where the vectors V_i form a basis for S.

Proof: The <u>Fredholm</u> <u>index</u> of an ordinary differential operator T with closed range in $L_2[a,\infty)$ is the nullity of T minus the deficiency of range T in $L_2(a,\infty)$. Clearly, if T is self-adjoint, index T = 0.

We first observe that H_0 is an N-dimensional extension of H_{00}. To see this, note that for any g such that g is the restriction to $[a,\infty)$ of an element of $C_0^\infty(-\infty,\infty)$, and such that $g^{(i)}(a) = 0$ for $i \le N - 1$, there is well-known to exist a sequence g_n in $C_0^\infty(a,\infty)$ such that $g_n^{(i)}$ converges to $g^{(i)}$ in $L_2(a,\infty)$ for all $i \le N$. (One way to see this is to note that such a g must necessarily be in domain $T_0(D^N)$.) By Lemma 1.4, it follows that g is in domain H_0. By adding N functions ϕ_i such that each ϕ_i is the restriction to $[a,\infty)$ of an element of $C^\infty(-\infty,\infty)$, and such that $\phi_i^{(j)}(a) = 0$ for $j \le N - 1$, and $\phi_i^{(N+j)}(a) = \delta_{ij}$ for $0 \le j \le N - 1$, we build up domain H_0 from domain H_{00}.

It is well-known, and clear from linear algebra, that the index of an N-dimensional extension of T is N + index T. Since index $H_0 = 0$, it follows that index $H_{00} = -N$.

However, H_R contains H_{00}, and index H_R is also 0. If H_R' is the restriction of H_R to H_{00} + linear span $\{f_i\}_{i=1}^N$, then $H_R' \subseteq H_R$ and index $H_R' = 0$, since H_R' is an N-dimensional extension of H_{00}. Thus $H_R' = H_R$.

Lemma 1.10. Suppose $M = L^+L$, with L limit-point. Then for any R, domain H_R is the set of all f such that

a) $\langle f(a), f^{(1)}(a), \ldots, f^{(2N-1)}(a)\rangle \in S$;

b) $f \in$ domain $T_1(M)$;

c) $Lf \in L_2(a,\infty)$.

Proof: It follows from Lemma 1.9 and Lemma 1.8 that f satisfies a), b) and c) if f is in the domain of H_R.

We now show that if f satisfies a), b) and c), then f is in domain H_R. Let $\hat{g}(a) = \langle g(a), \ldots, g^{(2N-1)}(a)\rangle$. Construct complex numbers c_i such that, if $f - \Sigma_1^N c_i f_i = g$, then $\hat{g}(a) = 0$, where $\{f_i\}_{i=1}^N$ is a set of N functions which are restrictions to $[a,\infty)$ of elements of $C_0^\infty(-\infty,\infty)$, such that $\{\hat{f}_i(a)\}$ is a basis for S. g is in the domain of $T_1(M)$ and $T_1(L)$. By Lemma 1.8, g is in domain H_0 and therefore domain H_{00}. By Lemma 1.9, it follows that f is in domain H_R.

Notation 1.11. Let domain H_1 be the set of all f of the form $f = f_0 + g$, where g is the restriction to $[a,\infty)$ of an element of $C_0^\infty(-\infty,\infty)$, and where f is in domain H_{00}. Let H_1 be the restriction of M to domain H_1.

Theorem 1.12. Let $M = L^+L$. Select b such that $((H_R + bI)f,f) \geq \varepsilon(f,f)$ for all f in domain H_R, where ε is a positive real number. Then domain $(H_R + bI)^{1/2}$ is the set of all f of type $f = f_0 + f_1$, where f_0 is in the domain of $T_0(L)$ and f_1 is the restriction to $[a,\infty)$ of an element of $C_0^\infty(-\infty,\infty)$, such that $\langle f_1(a), f_1^{(1)}(a), \ldots, f_1^{(2N-1)}(a)\rangle \varepsilon$ S.

Proof: Let W be the set of all f such that $f = f_0 + f_1$, with f_0 and f_1 as above. Note that domain $(H_R + bI)^{1/2}$ is the set of all g such that there exists a sequence g_n in domain $H_R + bI$ with g_n Cauchy in the norm $||\phi||_1 = ((H_R + bI)\phi,\phi)^{1/2}$. By Lemma 1.4, we see that the sequence g_n may be selected in domain R.

If f_0 is in domain $T_0(L)$, there is a sequence g_n in $C_0^\infty(a,\infty)$ such that $||L(g_n - f_0)||$ converges to zero. Hence Lg_n is a Cauchy sequence. It follows that g_n is Cauchy in the norm $||\ ||_1$. Hence f_0 is in domain $(H_R+bI)^{1/2}$. Since domain H_R is contained in domain $(H_R+bI)^{1/2}$, it follows that W is contained in domain $(H_R+bI)^{1/2}$.

If g is in the domain of R, select f_1 as above such that $(g-f_1)^{(i)}(a) = 0$ for all $i \leq 2N - 1$. Then $g - f_1$ is in domain $T_0(L)$. It follows that g is in W. Hence, for any f in domain $(H_R+bI)^{1/2}$, there is a sequence g_n in W such that $(H_R+bI)^{1/2}g_n$ converges to $(H_R+bI)^{1/2}f$. If the restriction of $(H_R+bI)^{1/2}$ to W is closed, it follows directly that W = domain $(H_R+bI)^{1/2}$.

Since W is a finite-dimensional extension of domain $T_0(L)$, and since a finite-dimensional extension of a closed operator is also closed, we need only show that the restriction of $(H_R+bI)^{1/2}$ to domain $T_0(L)$ is closed. As we saw above, it suffices to show that the domain of the closure of the restriction of $(H_R+bI)^{1/2}$ to $C_0^\infty(a,\infty)$ is contained in domain $T_0(L)$ This follows easily, since a sequence

ϕ_n of elements of $C_0^\infty(a,\infty)$ is Cauchy in $\| \|_1$ if and only if $L\phi_n$ is Cauchy in $L_2(R)$. The theorem is proved.

Theorem 1.13. Suppose $M = L^+L$, with L limit-point. Suppose that b is a real number such that $((H_R + bI)f,f) \geq \varepsilon(f,f)$ for all f in domain H_R, where ε is a positive real number. Then domain $(H_R+bI)^{1/2}$ is the set of all f in domain $T_1(L)$ such that $\langle f(a), \ldots, f^{(N-1)}(a)\rangle$ is in $\pi^N S$, where π^N is the projection on the first N coordinates.

Proof: It is clear that any element in domain $(H_R+bI)^{1/2}$ is in the desired form, by Theorem 1.12. We must show that any f in domain $T_1(L)$ such that $\langle f(a), \ldots, f^{(N-1)}(a)\rangle$ is in $\pi^N S$ is of the form $f = f_0 + f_1$, with f_0 in domain $T_0(L)$ and with f_1 the restriction to $[a,\infty)$ of an element of $C_0^\infty(-\infty,\infty)$, such that $\langle f_1(a), f_1'(a), \ldots, f^{2N-1}(a)\rangle$ is in S.

Let $\langle f(a), f^{(1)}(a), \ldots, f^{(N-1)}(a)\rangle = \pi^N V$, for some V in S. Select f_2 to be the restriction to $[a,\infty)$ of an element of $C_0^\infty(-\infty,\infty)$ such that $\langle f_2(a), \ldots, f_2^{(2N-1)}(a)\rangle = V$. Since L is limit-point, it follows from Corollary 4.7, p. 18, Kauffman-Read-Zettl [8], that $f-f_2$ is in domain $T_0(L)$. The theorem is proved.

Theorem 1.14. Suppose $M = \sum_0^N (-1)^i D^i p_i D^i$, where $p_i \in C^\infty[a,\infty)$ for all i, $p_N(x) > 0$ for all x in $[a,\infty)$, and each p_i is a real-valued member of $C^\infty[a,\infty)$. Suppose that $(Mf,f) \geq \varepsilon(f,f)$ for all f in $C_0^\infty(a,\infty)$, where ε is a positive real number. Then $M = L^+L$, where $L = \sum_0^N a_i D^i$, with each a_i in $C^\infty[a,\infty)$, with $a_N = (p_N)^{1/2}$, and where L is limit-point.

Remark: There are many factorizations of M of the form Q^+Q, and in some of them, Q may not be limit-point. For example, if $M = Q^2$, where Q is a symmetric limit-circle expression, M satisfies the hypotheses of the theorem. What the theorem says is that, even though M may be limit-circle, there still exists at least one factorization $M = L^+L$ with L limit-point.

Proof of Theorem 1.14:

We have seen that the nullity of H_1 is N. Let $y_1, \ldots, y_N$ be a basis for the null space of H_1. We show that the Wronskian

$W(y_1, \ldots, y_N)$ is never zero. Suppose that $W(y_1, \ldots, y_N)(b) = 0$ for some b. Then, for some complex numbers c_i, $\Sigma_1^N (c_i y_i)^j (b) = 0$ for all $j = 0, \ldots, N - 1$. Let $f = \Sigma_1^N c_i y_i$. $f = f_0 + g$, with f_0 in domain H_{00}, and with g the restriction to $[a,\infty)$ of an element of $C_0^\infty(-\infty,\infty)$. Suppose $M = Q^+Q$. Then, by Lemma 1.8, f_0 is in domain $T_0(Q)$. If Q_b is the restriction of the differential expression Q to $[b,\infty)$, it follows directly that, if $\theta(x) = 0$ for $x \leq b + 1$, $\theta(x) = 1$ for $x \geq b + 2$, and $\theta \in C^\infty(-\infty,\infty)$, then θf_0 is in the domain of the minimal operator $T_0(Q_b)$ corresponding to Q_b in $L_2(b,\infty)$. However, it is clear that the restriction of $f - \theta f$ to $[b,\infty)$ is also in the domain of $T_0(Q_b)$. Thus f_b is in the domain of $T_0(Q_b)$, where f_b is the restriction of f to $[b,\infty)$. It is clear that f_b is in the domain of $T_1(M_b)$. It follows from Lemma 1.8 that f is in the domain of the Friedrichs extension $(H_b)_0$ of the restriction R_b of M to $C_0^\infty(b,\infty)$. But $Mf = 0$. This is a contradiction, which arose from the existence of b. It follows that $W(y_1, \ldots, y_N)$ never vanishes.

Define L by $Ly = (p_N^{1/2}/W(y_1, \ldots, y_N))\, W(y_1, \ldots, y_N, y)$. By Theorem 19, p. 80, Coppel [3], $M = L^+L$, provided it can be shown that $B_M(y_i,y_j)(0) = 0$ for all i and j, where B_M is the Lagrange bilinear form for the differential expression M. However, this is easy to see, because $B_M(y_i,y_j)$ is constant, and y_i and y_j agree outside a compact set with elements of the domain of the symmetric operator H_0.

We have seen that $M = L^+L$. We need only show that L is limit-point. To prove this, note that $\{y_i\}_{i=1}^N$ is a basis for $\{f | Lf = 0\}$, and each y_i agrees outside a compact set with an element of domain H_0. By Lemma 1.8, each y_i therefore agrees outside a compact set with an element of domain $T_0(L)$.

Since $(Mf,f) \geq \varepsilon(f,f)$ for all f in $C_0^\infty(a,\infty)$, it follows that for all f in domain $T_0(L)$, $||Lf|| \geq \varepsilon^{1/2}||f||$. Hence range $T_0(L)$ is closed. Therefore, to prove that L is limit-point, we need only show that there is no non-trivial square-integrable solution f to the equation $L^+f = 0$.

Suppose that f is such a solution. Then f is in the domain of $T_1(L^+)$. Since $T_1(L^+) = T_0(L)^*$, and since y_i agrees outside a compact set with an element of domain $T_0(L)$, it follows that $B_L(y_i,f)$ vanishes at infinity for all i. Since $B_L(y_i,f)$ is constant, it follows that $B_L(y_i,f)(a) = 0$ for all i. However, if $z_i = \langle y_i(a), \ldots, y_i^{(N-1)}(a)\rangle$, the z_i form a basis for complex N-space, since $W(y_1, \ldots, y_N)(a) \neq 0$. Hence $f^{(i)}(a) = 0$ for all $i \leq N - 1$. It follows that $f(x) = 0$ for all x, as we desired to show. Therefore L is limit-point. The theorem is proved.

References

1. C. Bennewitz, "Spectral theory for pairs of differential operators," Ark. Math 15(1977) 33-61.

2. J. S. Bradley, D. B. Hinton, and R. M. Kauffman, "On the minimization of singular quadratic functionals," Proc. Roy. Soc. Edinburgh 87A(1981) 193-208.

3. W. A. Coppel, "Disconjugacy," Lecture notes in Math. 220, Springer-Verlag, New York, 1971.

4. N. Dunford and J. T. Schwartz, "Linear Operators," II, Interscience, New York, 1963.

5. D. B. Hinton, "On the eigenfunction expansion of singular ordinary differential equations," J. Differential Equations 24(1977) 282-308.

6. D. B. Hinton, "Eigenfunctions and spectral matrices of singular differential operators," Proc. Roy. Soc. Edinburgh 80A(1978) 289-309.

7. R. M. Kauffman, "The number of Dirichlet solutions to a class of ordinary differential equations," J. Differential Equations 31 (1979) 117-129.

8. R. M. Kauffman, T. T. Read, A. Zettl, "The deficiency index problem for powers of ordinary differential expressions," Lecture Notes in Mathematics 621, Springer-Verlag, New York (1977).

9. J. Robinette, "On the Dirichlet index of singular differential operators," M.S. thesis, Univ. of Tennessee, 1980.

EIGENVALUE PROBLEMS AND THE RIEMANN ZETA FUNCTION II

Ian Knowles

1. The main objective of the paper is to present a detailed account of the proof of the following theorem on the zero-free region for the classical Riemann zeta function $\zeta(s)$:

<u>Theorem</u> ([2]) There exists a kernel $k(r,t,s)$ such that

(i) the integral equation

$$u(r,s) = 1 + \int_r^\infty k(r,t,s)\, u(t,s)dt \tag{1.1}$$

has a unique bounded solution, $u(r,s)$, for all s with $\operatorname{Re} s > 1$;

(ii) if (1.1) has a solution $u(r,s)$ for all s in a region G contained in the strip $\{s : \frac{1}{2} < \operatorname{Re} s \le 1\}$ and contiguous to the half-plane $\operatorname{Re} s > 1$, then $\zeta(s) \ne 0$ for $s \in G$.

Here, the kernel $k(r,t,s)$ is given explicitly by equation (4.19) of §4. The proof, which is somewhat lengthy, is given in detail in §4. The key idea behind the proof involves the construction (§3) of an eigenvalue problem E_n (see (3.4-5)) with eigenvalue equation of the form $\xi_n(s) = 0$, where ξ_n is the nth partial product of a certain Euler product expression involving $\zeta(s)$ (see equation (3.1)).

2. Before proceeding, it is convenient to gather together a number of preliminary results and notations.

Let $\{p_i\}$ denote the sequence of positive prime numbers listed in increasing order beginning with $p_1 = 2$. Let $\theta > 1$ be a fixed real number and let $\{\delta_n : n = 1,2,\dots\}$ be any sequence of positive numbers. We define the sequence $\{\varepsilon(n) : n = 1,2,\dots\}$ by

$$\varepsilon(1) = 1 \ , \quad \varepsilon(2) = \frac{1}{p_1^{\theta}} \ , \tag{2.1}$$

and for $n \geq 1$

$$\varepsilon(2^n+k) = \frac{1}{p_{n+1}^{\theta}} \, \varepsilon(k) \ , \quad 1 \leq k \leq 2^{n-1} \tag{2.2a}$$

$$= \frac{1+\delta_n}{p_{n+1}^{\theta}} \, \varepsilon(k) \ , \quad 2^{n-1}+1 \leq k \leq 2^n \ . \tag{2.2b}$$

<u>Lemma 2.1</u> For integers $n \geq 1$ and $q \geq 0$,

$$\text{(i)} \quad \varepsilon(1+2^{n-1}+q.2^{n+1}) = \frac{1}{p_n^{\theta}} \, \varepsilon(1+q.2^{n+1}) \ , \tag{2.3}$$

$$\text{(ii)} \quad \varepsilon(1+3.2^{n-1}+q.2^{n+1}) = \frac{1+\delta_n}{p_n^{\theta}} \, \varepsilon(1+2.2^{n-1}+q.2^{n+1}) \ . \tag{2.4}$$

<u>Proof</u> We use induction on q, with n fixed. Observe that when $q = 0$, (i) follows directly from (2.2). Assume that (i) is true for all q^* with $0 \leq q^* < 2^r$, for some $r \geq 0$, and let $2^r \leq q < 2^{r+1}$. Then $q = 2^r + q^*$ where $0 \leq q^* < 2^r$. Also

$$\frac{\varepsilon(1+2^{n-1}+q.2^{n+1})}{\varepsilon(1+q.2^{n+1})} = \frac{\varepsilon(2^{r+n+1}+k_1)}{\varepsilon(2^{r+n+1}+k_2)} \tag{2.5}$$

where $k_1 = 1 + 2^{n-1} + q^*.2^{n+1}$ and $k_2 = 1 + q^*.2^{n+1}$. As $q^* \leq 2^r - 1$ we have $q^*.2^{n+1} + 2^{n-1} < 2^{n+r+1}$ and hence $k_2 < k_1 < 2^{n+r+1}$. We also have that either $k_1 \leq 2^{n+r}$ or $k_2 \geq 2^{n+r} + 1$. This is obvious if $q^*.2^{n+1} \geq 2^{n+r}$ or $2^{n-1} + q^* . 2^{n+1}$. Otherwise, if $q^*.2^{n+1} < 2^{n+r}$, then $q^* \leq 2^{r-1} - 1$; consequently, $q^*.2^{n+1} + 2^{n-1} \leq 2^{n+r} - 2^{n+1} + 2^{n-1} < 2^{n+r}$, and so $k_1 \leq 2^{n+r}$. If $2^{n-1} + q^*.2^{n+1} \geq 2^{n+r}$, we have $4q^* \geq 2^{r+1} - 1$, and thus $4q^* \geq 2^{r+1}$; consequently $2^{n+1}q^* \geq 2^{r+n}$ and so $k_2 \geq 2^{n+r} + 1$. Using (2.2) in (2.5) now gives

$$\frac{\varepsilon(1+2^{n-1}+q.2^{n+1})}{\varepsilon(1+q.2^{n+1})} = \frac{\varepsilon(1+2^{n-1}+q^*.2^{n+1})}{\varepsilon(1+q^*.2^{n+1})}$$

and the truth of (i) for any q with $2^r \leq q < 2^{r+1}$ then follows by the induction assumption.

The proof of (ii) proceeds in a similar fashion. When $q = 0$ we have, from (2.2b)

$$\frac{\varepsilon(1+2^{n-1}+2^n)}{\varepsilon(1+2^n)} = \frac{\varepsilon(1+2^{n-1})}{\varepsilon(1+2^n)} \cdot \frac{(1+\delta_n)}{p_{n+1}^{\theta}}$$

$$= \frac{1+\delta_n}{p_n^{\theta}} \quad , \quad \text{by (2.2a).}$$

Assume that (ii) is true for all q^* with $0 \le q^* < 2^r$, for some $r \ge 0$, and let $2^r \le q < 2^{r+1}$. Then, setting $q = 2^r + q^*$,

$$\frac{\varepsilon(1+3.2^{n-1}+q.2^{n+1})}{\varepsilon(1+2^n+q.2^{n+1})} = \frac{\varepsilon(2^{n+r+1}+k_3)}{\varepsilon(2^{n+r+1}+k_4)} \tag{2.6}$$

where $k_3 = q^*.2^{n+1} + 2^n + 2^{n-1} + 1$ and $k_4 = q^*.2^{n+1} + 2^n + 1$. Observe that $q^*.2^{n+1} + 2^n + 2^{n-1} \le 2^{n+r+1} - 2^{n-1} < 2^{n+r+1}$ and hence $k_4 < k_3 < 2^{n+r+1}$. As before we either have $k_3 \le 2^{n+r}$ or $k_4 \ge 2^{n+r} + 1$. This is obvious when either $q^*.2^{n+1} + 2^n \ge 2^{n+r}$ or $q^*.2^{n+1} + 2^n + 2^{n-1} < 2^{n+r}$. If $q^*.2^{n+1} + 2^n < 2^{n+r}$, we have $2q^* < 2^r - 1$ and so $2q^* \le 2^r - 2$, or $q^* \le 2^{r-1}$. In this case $q^*.2^{n+1} + 2^n + 2^{n-1} \le 2^{n+r} - 2^{n-1} < 2^{n+r}$ and so $k_3 \le 2^{n+r}$. If $q^*.2^{n+1} + 2^n + 2^{n-1} \ge 2^{n+r}$, we have $2q^* \ge 2^r - 1$ and so $q^*.2^{n+1} + 2^n \ge 2^{n+r}$; it then follows that $k_4 \ge 2^{n+r} + 1$. From (2.2) and (2.6) and the induction assumption we now have

$$\frac{\varepsilon(1+3.2^{n-1}+q.2^{n+1})}{\varepsilon(1+2^n+q.2^{n+1})} = \frac{\varepsilon(k_3)}{\varepsilon(k_4)} = \frac{1+\delta_n}{p_{n+1}^{\theta}}$$

for any q with $2^r \le q < 2^{r+1}$, as required.

Define the sequence $\{w_n^{(r)}\}$, $n \ge 0$, $0 \le r \le n$, by

$$w_n^{(r)} = \varepsilon(1) + \varepsilon(1+2^r) + \ldots + \varepsilon(1+k.2^r) + \ldots + \varepsilon(1+2^n-2^r). \tag{2.7a}$$

We set $w_n^{(0)} = w_n$. Notice that $w_n^{(n-1)} = \varepsilon(1) + \varepsilon(1+2^{n-1})$, and $w_n^{(n)} = \varepsilon(1)$.

<u>Lemma 2.2</u> For $3 \le n < \infty$ and $r = 0,1$,

(i) $p_n^{\theta}\left(w_n^{(r)} - w_{n-1}^{(r)}\right) = w_{n-1}^{(r)} + \delta_{n-1}\left(w_{n-1}^{(r)} - w_{n-2}^{(r)}\right)$;

(ii) $$w_n^{(r)} = \left(1 + \frac{1}{p_n^{\theta}}\right)w_{n-1}^{(r)} + \frac{\delta_{n-1}}{p_n^{\theta}}\left(w_{n-1}^{(r)} - w_{n-2}^{(r)}\right) . \tag{2.8}$$

Also, for $n \geq 1$,

$$\text{(iii)} \quad w_n \leq \prod_{r=1}^{n} \left(1 + \frac{1 + \delta_{r-1}}{p_r^\theta}\right) . \tag{2.9}$$

Proof Observe that (i) and (ii) are equivalent. To prove (i) for $r = 0$, we have from (2.2) that

$$w_n - w_{n-1}$$

$$= \varepsilon(2^{n-1}+1) + \ldots + \varepsilon(2^{n-1}+2^{n-2}) + \varepsilon(2^{n-1}+2^{n-2}+1) + \ldots + \varepsilon(2^{n-1}+2^{n-1})$$

$$= \frac{1}{p_n^\theta} [\{\varepsilon(1) + \ldots + \varepsilon(2^{n-2})\} + (1+\delta_{n-1})\{\varepsilon(2^{n-2}+1) + \ldots + \varepsilon(2^{n-1})\}]$$

$$= \frac{1}{p_n^\theta} [w_{n-2} + (1+\delta_{n-1})(w_{n-1}-w_{n-2})] .$$

The argument for $r = 1$ is similar. We prove (iii) by induction. Noticing that (2.9) is true for $n = 1,2$, we assume that (2.9) holds for some value $n = k \geq 2$. By part (ii) above

$$w_{k+1} = \left(1 + \frac{1}{p_{k+1}^\theta}\right)w_k + \frac{\delta_k}{p_{k+1}^\theta}(w_k - w_{k-1})$$

$$\leq \left(1 + \frac{1}{p_{k+1}^\theta}\right) \prod_{r=1}^{k} \left(1 + \frac{1 + \delta_{r-1}}{p_r^\theta}\right) + \frac{\delta_k}{p_{k+1}^\theta} \prod_{r=1}^{k} \left(1 + \frac{1 + \delta_{r-1}}{p_r^\theta}\right)$$

$$= \prod_{r=1}^{k+1} \left(1 + \frac{1 + \delta_{r-1}}{p_r^\theta}\right)$$

and (2.9) is therefore true for all $n \geq 1$. □

As the partial product on the right of (2.9) is convergent for $\theta > 1$, it follows that the series $\sum_{r=1}^{\infty} \varepsilon(r)$ is convergent. We define

$$w_\infty^{(r)} = \sum_{k=0}^{\infty} \varepsilon(1+k.2^r) . \tag{2.7b}$$

Lemma 2.3 (i) For $1 \leq n < \infty$ and $1 \leq r \leq n-1$ set $v_n^{(r)} = w_n^{(r-1)} - w_n^{(r)}$. Then

$$v_n^{(r)} = \frac{1}{p_r^\theta} [w_n^{(r)} + \delta_r(w_n^{(r)} - w_n^{(r+1)})] ;$$

(ii) For $r \geq 1$,

$$p_r^\theta[w_\infty^{(r-1)} - w_\infty^{(r)}] = w_\infty^{(r)} + \delta_r\left(w_\infty^{(r)} - w_\infty^{(r+1)}\right) .$$

<u>Proof</u> We only consider the case $n < \infty$; the case $n = \infty$ is similar.

$$v_n^{(r)} = w_n^{(r-1)} - w_n^{(r)}$$

$$= \varepsilon(1+2^{r-1}) + \varepsilon(1+3.2^{r-1}) + \ldots + \varepsilon(1+2^n-2^{r-1})$$

$$= \{\varepsilon(1+2^{r-1}) + \varepsilon(1+5.2^{r-1}) + \ldots + \varepsilon(1+2^n-3.2^{r-1})\}$$

$$+ \{\varepsilon(1+3.2^{r-1}) + \varepsilon(1+7.2^{r-1}) + \ldots + \varepsilon(1+2^n-2^{r-1})\}$$

$$= \frac{1}{p_r^\theta}\{\varepsilon(1) + \varepsilon(1+2^{r+1}) + \ldots + \varepsilon(1+k.2^{r+1}) + \ldots + \varepsilon(1+2^n-2^{r+1})\}$$

$$+ \frac{1+\delta_r}{p_r^\theta}\{\varepsilon(1+2^r) + \varepsilon(1+2^r+2^{r+1}) + \ldots + \varepsilon\left(1+2^r+(2^{n-r-1}-1)2^{r+1}\right)\}$$

(from lemma 2.1 with $n = r$)

$$= \frac{1}{p_r^\theta}\{w_n^{(r)} + \delta_r\left(w_n^{(r)} - w_n^{(r+1)}\right)\} .$$

To prove (ii) we again make use of Lemma 2.1 .

$$w_\infty^{(r-1)} - w_\infty^{(r)} = \sum_{q=0}^{\infty} \varepsilon\left(1 + (2q+1)2^{r-1}\right)$$

$$= \sum_{q \text{ even}} \varepsilon\left(1 + (2q+1)2^{r-1}\right) + \sum_{q \text{ odd}} \varepsilon\left(1 + (2q+1)2^{r-1}\right)$$

$$= \sum_{m=0}^{\infty} \varepsilon(1 + 2^{r-1} + m.2^{r+1}) + \sum_{m=0}^{\infty} \varepsilon(1 + 3.2^{r-1} + m.2^{r+1})$$

$$= \frac{1}{p_r^\theta}\sum_{m=0}^{\infty} \varepsilon(1 + m.2^{r+1}) + \frac{1+\delta_r}{p_r^\theta}\sum_{m=0}^{\infty} \varepsilon(1 + 2^r + m.2^{r+1})$$

$$= \frac{1}{p_r^\theta} w_\infty^{(r)} + \frac{\delta_r}{p_r^\theta}\left(w_\infty^{(r)} - w_\infty^{(r+1)}\right)$$

as required.

<u>Lemma 2.4</u> (i) For $r \geq 3$,

$$w_r w_{r-1}^{(1)} - w_{r-1} w_r^{(1)} = \frac{\delta_{r-1}}{p_r^\theta}\left(w_{r-1} w_{r-2}^{(1)} - w_{r-2} w_{r-1}^{(1)}\right) .$$

(ii) For fixed $r \geq 2$ and $1 \leq k \leq r-1$,

$$w_r^{(k-1)} w_\infty^{(k)} - w_r^{(k)} w_\infty^{(k-1)} = \frac{\delta_k}{p_k^\theta} \left(w_r^{(k)} w_\infty^{(k+1)} - w_r^{(k+1)} w_\infty^{k} \right) .$$

<u>Proof</u> (i) $w_r w_{r-1}^{(1)} - w_{r-1} w_r^{(1)}$

$$= \{(1 + \frac{1}{p_r^\theta}) w_{r-1} + \frac{\delta_{r-1}}{p_r^\theta} (w_{r-1} - w_{r-2})\} w_{r-1}^{(1)}$$

$$- w_{r-1}\{(1 + \frac{1}{p_r^\theta}) w_{r-1}^{(1)} + \frac{\delta_{r-1}}{p_r^\theta} \left(w_{r-1}^{(1)} - w_{r-2}^{(1)}\right)\}$$

$$= \frac{\delta_{r-1}}{p_r^\theta} \left(w_{r-1} w_{r-2}^{(1)} - w_{r-2} w_{r-1}^{(1)} \right) .$$

(ii) $w_r^{(k-1)} w_\infty^{(k)} - w_r^{(k)} w_\infty^{(k-1)}$

$$= \left(w_r^{(k)} + v_r^{(k)}\right) w_\infty^{(k)} - w_r^{(k)} \left(w_\infty^{(k)} + v_\infty^{(k)}\right) \quad \text{by (2.7c)}$$

$$= v_r^{(k)} w_\infty^{(k)} - w_r^{(k)} v_\infty^{(k)}$$

$$= \frac{1}{p_k^\theta} [w_r^{(k)} + \delta_k \left(w_r^{(k)} - w_r^{(k+1)}\right)] w_\infty^{(k)} - \frac{1}{p_k^\theta} [w_\infty^{(k)} + \delta_k \left(w_\infty^{(k)} - w_\infty^{(k+1)}\right)] w_r^{(k)}$$

$$= \frac{\delta_k}{p_k^\theta} [w_r^{(k)} w_\infty^{(k+1)} - w_r^{(k+1)} w_\infty^{(k)}] . \qquad \square$$

For $n \geq 1$ and $1 \leq r \leq n+1$ define the sequences $\{\alpha_r^{(n)}\}$ and $\{\beta_r^{(n)}\}$ as follows:

$$\alpha_1^{(n)} = 0 \tag{2.10}$$

$$\beta_1^{(n)} = - w_n \tag{2.11}$$

$$\alpha_r^{(n)} = - \frac{w_r^{(1)} - w_{r-1}^{(1)}}{w_r - w_{r-1}} w_n , \quad 2 \leq r \leq n+1 \tag{2.12}$$

$$\beta_r^{(n)} = - \frac{w_{r-1}^{(1)}}{w_{r-1}} w_n , \quad 2 \leq r \leq n+1 . \tag{2.13}$$

Lemma 2.5 For $n \geq 1$ and $1 \leq r \leq n$

$$\text{(i)} \quad \frac{\beta_{r+1}^{(n)} - \alpha_r^{(n)}}{\beta_r^{(n)} - \alpha_r^{(n)}} = \frac{w_{r-1}}{w_r} \; ; \qquad \text{(ii)} \quad \frac{\beta_{r+1}^{(n)} - \beta_r^{(n)}}{\beta_r^{(n)} - \alpha_r^{(n)}} = - \frac{w_r - w_{r-1}}{w_r} \; ;$$

$$\text{(iii)} \quad \frac{\alpha_r^{(n)} - \alpha_{r+1}^{(n)}}{\beta_r^{(n)} - \alpha_r^{(n)}} = - \frac{1}{p_{r+1}^{\theta}} \frac{w_{r-1}}{w_{r+1} - w_r} \; ;$$

$$\text{(iv)} \quad \frac{\beta_r^{(n)} - \alpha_{r+1}^{(n)}}{\beta_r^{(n)} - \alpha_r^{(n)}} = \frac{1 + \delta_r}{p_{r+1}^{\theta}} \frac{w_r - w_{r-1}}{w_{r+1} - w_r} \; .$$

Proof From the definitions (2.10-12) we have for $2 \leq r \leq n$ that

$$\frac{\beta_{r+1}^{(n)} - \alpha_r^{(n)}}{\beta_r^{(n)} - \alpha_r^{(n)}} = \frac{- \dfrac{w_r^{(1)}}{w_r} + \dfrac{w_r^{(1)} - w_{r-1}^{(1)}}{w_r - w_{r-1}}}{- \dfrac{w_{r-1}^{(1)}}{w_{r-1}} + \dfrac{w_r^{(1)} - w_{r-1}^{(1)}}{w_r - w_{r-1}}}$$

$$= \frac{w_{r-1}}{w_r} \; .$$

Also, when $r = 1$,

$$\frac{\beta_2^{(n)} - \alpha_1^{(n)}}{\beta_1^{(n)} - \alpha_1^{(n)}} = \frac{w_1^{(1)}}{w_1} = \frac{w_0}{w_1} \; .$$

Part (ii) follows directly from (i), as we have

$$\frac{\beta_{r+1}^{(n)} - \alpha_r^{(n)}}{\beta_r^{(n)} - \alpha_r^{(n)}} - \frac{\beta_{r+1}^{(n)} - \beta_r^{(n)}}{\beta_r^{(n)} - \alpha_r^{(n)}} = 1 \; .$$

To prove (iv), we have from (2.10-12) for $2 \leq r \leq n$,

$$\frac{\beta_r^{(n)} - \alpha_{r+1}^{(n)}}{\beta_r^{(n)} - \alpha_r^{(n)}} = \frac{w_r - w_{r-1}}{w_{r+1} - w_r} \cdot \frac{w_{r-1}\left(w_{r+1}^{(1)} - w_r^{(1)}\right) - w_{r-1}^{(1)}\,(w_{r+1} - w_r)}{-w_{r-1}^{(1)}(w_r - w_{r-1}) + w_{r-1}\left(w_r^{(1)} - w_{r-1}^{(1)}\right)}$$

$$= \frac{w_r - w_{r-1}}{w_{r+1} - w_r} \cdot \frac{1}{p_{r+1}^{\theta}} \cdot \frac{w_{r-1}\{w_r^{(1)} + \delta_r\left(w_r^{(1)} - w_{r-1}^{(1)}\right)\} - w_{r-1}^{(1)}\{w_r + \delta_r(w_r - w_{r-1})\}}{w_{r-1}w_r^{(1)} - w_r w_{r-1}^{(1)}}$$

(by Lemma 2.2 (i))
$$= \frac{1 + \delta_r}{p_{r+1}^{\theta}} \frac{w_r - w_{r-1}}{w_{r+1} - w_r} \; .$$

When $r = 1$ we have

$$\frac{\beta_1^{(n)} - \alpha_2^{(n)}}{\beta_1^{(n)} - \alpha_1^{(n)}} = 1 - \frac{w_2^{(1)} - w_1^{(1)}}{w_2 - w_1}$$

$$= \frac{\varepsilon(4)}{w_2 - w_1}$$

$$= \frac{1 + \delta_1}{p_2^\theta} \cdot \frac{w_1 - w_0}{w_2 - w_1}$$

by (2.2b).

Finally,

$$\frac{\beta_r^{(n)} - \alpha_{r+1}^{(n)}}{\beta_r^{(n)} - \alpha_r^{(n)}} - \frac{\alpha_r^{(n)} - \alpha_{r+1}^{(n)}}{\beta_r^{(n)} - \alpha_r^{(n)}} = 1$$

and hence

$$\frac{\alpha_r^{(n)} - \alpha_{r+1}^{(n)}}{\beta_r^{(n)} - \alpha_r^{(n)}} = \frac{1 + \delta_r}{p_{r+1}^\theta} \; \frac{w_r - w_{r-1}}{w_{r+1} - w_r} - 1$$

$$= \frac{(w_r - w_{r-1})(1+\delta_r) - p_{r+1}^\theta(w_{r+1} - w_r)}{w_{r+1} w_r \, p_{r+1}^\theta}$$

$$= - \frac{w_{r-1}}{w_{r+1} - w_r} \cdot \frac{1}{p_{r+1}^\theta}$$

by Lemma 2.2 (i) again. □

<u>Lemma 2.6</u> For $2 \le r \le n+1$,

(i) $$\alpha_r^{(n)} - \beta_r^{(n)} = \frac{\delta_1 \, \delta_2 \cdots \delta_{r-1}}{(p_1 \, p_2 \cdots p_r)^\theta} \cdot \frac{w_n}{w_{r-1}(w_r - w_{r-1})} \; ;$$

(ii) $$\beta_{r+1}^{(n)} - \beta_r^{(n)} = \frac{\delta_1 \, \delta_2 \cdots \delta_{r-1}}{(p_1 \, p_2 \cdots p_r)^\theta} \cdot \frac{w_n}{w_r w_{r-1}} \; .$$

<u>Proof</u> (i) By (2.12-13)

$$\frac{1}{w_n}\left(\alpha_r^{(n)} - \beta_r^{(n)}\right) = \frac{w_{r-1}^{(1)}}{w_{r-1}} - \frac{w_r^{(1)} - w_{r-1}^{(1)}}{w_r - w_{r-1}}$$

$$= \frac{w_r w_{r-1}^{(1)} - w_{r-1} w_r^{(1)}}{w_{r-1}(w_r - w_{r-1})}$$

$$= \frac{\delta_{r-1}}{p_r^{\theta}} \cdot \frac{\delta_{r-2}}{p_{r-1}^{\theta}} \cdot \ldots \cdot \frac{\delta_2}{p_3^{\theta}} \; \frac{w_2 w_1^{(1)} - w_1 w_2^{(1)}}{w_{r-1}(w_r - w_{r-1})}$$

by Lemma 2.4 (i)

$$= \frac{\delta_{r-1}\,\delta_{r-2}\ldots\delta_1}{(p_r\,p_{r-1}\ldots p_1)^{\theta}} \cdot \frac{1}{w_{r-1}(w_r - w_{r-1})} \; .$$

(ii) By (i) above and Lemma 2.5 (ii)

$$\beta_{r+1}^{(n)} - \beta_r^{(n)} = \left(\alpha_r^{(n)} - \beta_r^{(n)}\right) \cdot \frac{(w_r - w_{r-1})w_n}{w_r}$$

$$= \frac{\delta_1\,\delta_2\ldots\delta_{r-1}}{(p_1\,p_2\ldots p_r)^{\theta}} \cdot \frac{w_n}{w_r w_{r-1}} \; .$$

3. In this section we assume n to be a fixed positive integer, and construct a boundary value problem, E_n , with eigenvalue equation essentially of the form $\xi_n(s) = 0$, where ξ_n is the n-th partial product of the function

$$\frac{\zeta(s+\theta)}{\zeta(2s+2\theta)} = \prod_{j=1}^{\infty} \left(1 + \frac{1}{p_j^{s+\theta}}\right) , \quad \mathrm{Re}(s+\theta) > 1 \; . \tag{3.1}$$

Noting that $\alpha_r^{(n)} - \beta_r^{(n)} > 0$, $n \geq 1$, $1 \leq r \leq n+1$, by Lemma 2.6 (i), we define $a_r^{(n)}$, $n \geq 1$, $0 \leq r \leq n+1$, by $a_0^{(n)} = 0$, $a_{n+1}^{(n)} = \infty$, **and for** $1 \leqslant r \leqslant n$

$$a_r^{(n)} = a_{r-1}^{(n)} + \frac{\ell n\,(p_r)}{\alpha_r^{(n)} - \beta_r^{(n)}} \; . \tag{3.2}$$

For fixed $n \geq 1$ the set $\{a_r^{(n)} : 0 \leq r \leq n\}$ clearly forms a partition of the interval $[0,\infty)$. Define functions $b_n(x)$ and $c_n(x)$ on $[0,\infty)$ by

$$b_n(x) = b_r^{(n)} \tag{3.3a}$$

$$c_n(x) = c_r^{(n)} \tag{3.3b}$$

for $a_{r-1}^{(n)} \leq x < a_r^{(n)}$, $n \geq 1$, $1 \leq r \leq n+1$, where $b_r^{(n)} = \alpha_r^{(n)} + \beta_r^{(n)}$ and $c_r^{(n)} = \alpha_r^{(n)}\,\beta_r^{(n)}$. The eigenvalue problem E_n consists of the equation

$$y'' - s\, b_n(x) y' + s^2 c_n(x) y = 0 \tag{3.4}$$

together with the boundary conditions

$$y(0) = 0 \,, \tag{3.5a}$$

$$y(x) \sim \exp\left(\beta_{n+1}^{(n)}\, sx\right) \quad \text{as} \quad x \to \infty \,. \tag{3.5b}$$

On $[a_{r-1}^{(n)}, a_r^{(n)}]$ any solution of (3.4) may be written in the form

$$y(x,s) = A_r^{(n)}(s) \exp\left(\alpha_r^{(n)}\, sx\right) + B_r^{(n)}(s) \exp\left(\beta_r^{(n)}\, sx\right) \tag{3.6}$$

where $1 \le r \le n+1$. Condition (3.5b) is satisfied if we choose $A_{n+1}^{(n)} = 0$ and $B_{n+1}^{(n)} = 1$; this choice also fixes the solution $y(x,s)$, which we henceforth denote by $\phi_n(x,s)$. In general, for this solution we then have

$$A_{n+2-k}^{(n)}(s) = \sum_{i=1}^{2^{k-1}} C_{k,i}^{(n)} \exp\left(\eta_{k,i}^{(n)}\, s\right) \tag{3.7a}$$

$$B_{n+2-k}^{(n)}(s) = \sum_{i=1}^{2^{k-1}} D_{k,i}^{(n)} \exp\left(\nu_{k,i}^{(n)}\, s\right) \tag{3.7b}$$

for certain constants $C_{k,i}^{(n)}$, $D_{k,i}^{(n)}$, $\eta_{k,i}^{(n)}$, and $\nu_{k,i}^{(n)}$, and for $1 \le k \le n+1$. Set $C_{1,1}^{(n)} = 0$, $D_{1,1}^{(n)} = 1$, and $\eta_{1,1}^{(n)} = \nu_{1,1}^{(n)} = 0$.

The condition that s be an eigenvalue of E_n is now $\phi_n(0,s) = 0$, or

$$A_1^{(n)}(s) + B_1^{(n)}(s) = 0 \,. \tag{3.8}$$

Our next task is to rewrite the coefficients in (3.7a-b) in terms of the sequence $\varepsilon(n)$ defined in §2.

Consider the solution $\phi_n(x,s)$ and its derivative at a boundary point $a_r^{(n)}$, $1 \le r \le n$. We have

$$A_{r+1}^{(n)} \exp\left(\alpha_{r+1}^{(n)}\, s\, a_r^{(n)}\right) + B_{r+1}^{(n)} \exp\left(\beta_{r+1}^{(n)}\, s\, a_r^{(n)}\right)$$
$$= A_r^{(n)} \exp\left(\alpha_r^{(n)}\, s\, a_r^{(n)}\right) + B_r^{(n)} \exp\left(\beta_r^{(n)}\, s\, a_r^{(n)}\right) \tag{3.9a}$$

and

$$\alpha_{r+1}^{(n)} A_{r+1}^{(n)} \exp\left(\alpha_{r+1}^{(n)}\, s\, a_r^{(n)}\right) + \beta_{r+1}^{(n)} B_{r+1}^{(n)} \exp\left(\beta_{r+1}^{(n)}\, s\, a_r^{(n)}\right)$$
$$= \alpha_r^{(n)} A_r^{(n)} \exp\left(\alpha_r^{(n)}\, s\, a_r^{(n)}\right) + \beta_r^{(n)} B_r^{(n)} \exp\left(\beta_r^{(n)}\, s\, a_r^{(n)}\right). \tag{3.9b}$$

From these equations it follows that

$$A_r^{(n)} = \frac{\alpha_{r+1}^{(n)} - \beta_r^{(n)}}{\alpha_r^{(n)} - \beta_r^{(n)}} \exp[(\alpha_{r+1}^{(n)} - \alpha_r^{(n)})s\ a_r^{(n)}]A_{r+1}^{(n)}$$

$$+ \frac{\beta_{r+1}^{(n)} - \beta_r^{(n)}}{\alpha_r^{(n)} - \beta_r^{(n)}} \exp[(\beta_{r+1}^{(n)} - \alpha_r^{(n)})s\ a_r^{(n)}]B_{r+1}^{(n)} \tag{3.10a}$$

$$B_r^{(n)} = \frac{\alpha_r^{(n)} - \alpha_{r+1}^{(n)}}{\alpha_r^{(n)} - \beta_r^{(n)}} \exp[(\alpha_{r+1}^{(n)} - \beta_r^{(n)})s\ a_r^{(n)}]A_{r+1}^{(n)}$$

$$+ \frac{\alpha_r^{(n)} - \beta_{r+1}^{(n)}}{\alpha_r^{(n)} - \beta_r^{(n)}} \exp[(\beta_{r+1}^{(n)} - \beta_r^{(n)})s\ a_r^{(n)}]B_{r+1}^{(n)} . \tag{3.10b}$$

Thus if

$$A_{n+1-k}^{(n)}(s) = \sum_{i=1}^{2^k} C_{k+1,i}^{(n)} \exp(\eta_{k+1,i}^{(n)}\ s) \tag{3.11a}$$

$$B_{n+1-k}^{(n)}(s) = \sum_{i=1}^{2^k} D_{k+1,i}^{(n)} \exp(\nu_{k+1,i}^{(n)}\ s) \tag{3.11b}$$

we have, for $1 \le k \le n$

$$\begin{aligned} C_{k+1,i}^{(n)} &= \frac{\alpha_{n+2-k}^{(n)} - \beta_{n+1-k}^{(n)}}{\alpha_{n+1-k}^{(n)} - \beta_{n+1-k}^{(n)}} C_{k,i}^{(n)} , \quad 1 \le i \le 2^{k-1} , \\ &= \frac{\beta_{n+2-k}^{(n)} - \beta_{n+1-k}^{(n)}}{\alpha_{n+1-k}^{(n)} - \beta_{n+1-k}^{(n)}} D_{k,i-2^{k-1}}^{(n)} , \quad 2^{k-1} + 1 \le i \le 2^k \end{aligned} \tag{3.12}$$

$$\begin{aligned} D_{k+1,i}^{(n)} &= \frac{\alpha_{n+1-k}^{(n)} - \alpha_{n+2-k}^{(n)}}{\alpha_{n+1-k}^{(n)} - \beta_{n+1-k}^{(n)}} C_{k,i}^{(n)} , \quad 1 \le i \le 2^{k-1} \\ &= \frac{\alpha_{n+1-k}^{(n)} - \beta_{n+2-k}^{(n)}}{\alpha_{n+1-k}^{(n)} - \beta_{n+1-k}^{(n)}} D_{k,i-2^{k-1}}^{(n)} , \quad 2^{k-1} + 1 \le i \le 2^k \end{aligned} \tag{3.13}$$

$$\begin{aligned} \eta_{k+1,i}^{(n)} &= \eta_{k,i}^{(n)} + (\alpha_{n+2-k}^{(n)} - \alpha_{n+1-k}^{(n)})a_{n+1-k}^{(n)} , \quad 1 \le i \le 2^{k-1} \\ &= \nu_{k,i-2^{k-1}}^{(n)} + (\beta_{n+2-k}^{(n)} - \alpha_{n+1-k}^{(n)})a_{n+1-k}^{(n)} , \quad 2^{k-1} + 1 \le i \le 2^k \end{aligned} \tag{3.14}$$

$$\nu^{(n)}_{k+1,i} = \eta^{(n)}_{k,i} + \left(\alpha^{(n)}_{n+2-k} - \beta^{(n)}_{n+1-k}\right)a^{(n)}_{n+1-k}\,, \quad 1 \le i \le 2^{k-1}$$

$$= \nu^{(n)}_{k,i-2^{k-1}} + \left(\beta^{(n)}_{n+2-k} - \beta^{(n)}_{n+1-k}\right)a^{(n)}_{n+1-k}\,, \quad 2^{k-1}+1 \le i \le 2^{k}. \tag{3.15}$$

Lemma 3.1

(i) For $1 \le k \le n$, $1 \le i \le 2^k$

$$\nu^{(n)}_{k+1,i} = \eta^{(n)}_{k+1,i} + \left(\alpha^{(n)}_{n+1-k} - \beta^{(n)}_{n+1-k}\right)a^{(n)}_{n+1-k}\,. \tag{3.16}$$

(ii) For $k \ge 2$, $1 \le i \le 2^{k-2}$

$$\eta^{(n)}_{k+1,4i-2} = \eta^{(n)}_{k+1,4i} - \ln(p_n) \tag{3.17}$$

$$\nu^{(n)}_{k+1,4i-2} = \nu^{(n)}_{k+1,4i} - \ln(p_n)\,. \tag{3.18}$$

(iii) For $k \ge 2$, $1 \le i \le 2^{k-2}$,

$$\eta^{(n)}_{k+1,4i} - \eta^{(n)}_{k+1,2^k} = \eta^{(n-1)}_{k,2i} - \eta^{(n-1)}_{k,2^{k-1}} \tag{3.19}$$

$$\nu^{(n)}_{k+1,4i} - \nu^{(n)}_{k+1,2^k} = \nu^{(n-1)}_{k,2i} - \nu^{(n-1)}_{k,2^{k-1}}\,. \tag{3.20}$$

Proof (i) This follows directly from (3.14-15).

(ii) We use induction on k. For $k = 2$

$$\eta^{(n)}_{3,2} - \eta^{(n)}_{3,4} = \left(\beta^{(n)}_{n} - \alpha^{(n)}_{n}\right)a^{(n)}_{n} + \left(\alpha^{(n)}_{n} - \beta^{(n)}_{n}\right)a^{(n)}_{n-1}$$

$$= -\ln(p_n)$$

$$= \nu^{(n)}_{3,2} - \nu^{(n)}_{3,4}\,.$$

Assume that, for $1 \le i \le 2^{k-3}$

$$\eta^{(n)}_{k,4i-2} = \eta^{(n)}_{k,4i} - \ln(p_n) \tag{3.21}$$

$$\nu^{(n)}_{k,4i-2} = \nu^{(n)}_{k,4i} - \ln(p_n)\,. \tag{3.22}$$

Then from (3.14), for $1 \le i \le 2^{k-3}$

$$\eta^{(n)}_{k+1,4i-2} - \eta^{(n)}_{k+1,4i} = \eta^{(n)}_{k,4i-2} - \eta^{(n)}_{k,4i} = -\ln(p_n)$$

by (3.21), and for $2^{k-3}+1 \le i \le 2^{k-2}$,

$$\eta^{(n)}_{k+1,4i-2} - \eta^{(n)}_{k+1,4i} = \nu^{(n)}_{k,4i-2-2^{k-1}} - \nu^{(n)}_{k,4i-2^{k-1}} = -\ln(p_n)$$

by (3.22), with similar results for $\nu^{(n)}_{k+1,4i-2} - \nu^{(n)}_{k+1,4i}$, using (3.15).

(iii) Again we use inducation on k. For $k = 2$ the result is trivially true. Assume now that for $1 \le i \le 2^{k-3}$,

$$\eta^{(n)}_{k,4i} - \eta^{(n)}_{k,2^{k-1}} = \eta^{(n-1)}_{k-1,2i} - \eta^{(n-1)}_{k-1,2^{k-2}} \tag{3.23}$$

$$\nu^{(n)}_{k,4i} - \nu^{(n)}_{k,2^{k-1}} = \nu^{(n-1)}_{k-1,2i} - \nu^{(n-1)}_{k-1,2^{k-2}} . \tag{3.24}$$

Assume first that $1 \le i \le 2^{k-3}$. From (3.14)

$$\begin{aligned}
&\eta^{(n)}_{k+1,4i} - \eta^{(n)}_{k+1,2^k} \\
&= \eta^{(n)}_{k,4i} - \nu^{(n)}_{k,2^{k-1}} + \left(\alpha^{(n)}_{n+2-k} - \beta^{(n)}_{n+2-k}\right) a^{(n)}_{n+1-k} \\
&= \left(\eta^{(n)}_{k,4i} - \eta^{(n)}_{k,2^{k-1}}\right) + \left(\eta^{(n)}_{k,2^{k-1}} - \nu^{(n)}_{k,2^{k-1}}\right) + \left(\alpha^{(n)}_{n+2-k} - \beta^{(n)}_{n+2-k}\right) a^{(n)}_{n+1-k} \\
&= \eta^{(n)}_{k,4i} - \eta^{(n)}_{k,2^{k-1}} - \ln(p_{n+2-k})
\end{aligned} \tag{3.25}$$

by (3.16) and (3.2).

Also

$$\begin{aligned}
\eta^{(n-1)}_{k,2i} - \eta^{(n-1)}_{k,2^{k-1}} &= \eta^{(n-1)}_{k-1,2i} - \nu^{(n-1)}_{k-1,2^{k-2}} + \left(\alpha^{(n-1)}_{n+2-k} - \beta^{(n-1)}_{n+2-k}\right) a^{(n-1)}_{n+1-k} \\
&= \left(\eta^{(n-1)}_{k-1,2i} - \eta^{(n-1)}_{k-1,2^{k-2}}\right) + \left(\eta^{(n-1)}_{k-1,2^{k-2}} - \nu^{(n-1)}_{k-1,2^{k-2}}\right) \\
&\qquad + \left(\alpha^{(n-1)}_{n+2-k} - \beta^{(n-1)}_{n+2-k}\right) a^{(n-1)}_{n+1-k} \\
&= \eta^{(n-1)}_{k-1,2i} - \eta^{(n-1)}_{k-1,2^{k-2}} - \ln(p_{n+2-k}) \\
&= \eta^{(n)}_{k+1,4i} - \eta^{(n)}_{k+1,2^k}
\end{aligned}$$

by (3.23) and (3.25). When $2^{k-3} + 1 \le i \le 2^{k-2}$ we have by (3.14),

$$\begin{aligned}
\eta^{(n)}_{k+1,4i} - \eta^{(n)}_{k+1,2^k} &= \nu^{(n)}_{k,4i-2^{k-1}} - \nu^{(n)}_{k,2^{k-1}} \\
&= \nu^{(n-1)}_{k-1,2i-2^{k-2}} - \nu^{(n-1)}_{k-1,2^{k-2}} \\
\text{by (3.24)} \qquad &= \eta^{(n-1)}_{k,2i} - \eta^{(n-1)}_{k,2^{k-1}}
\end{aligned}$$

by (3.14) again. The truth of (3.20) may be established in a similar way. □

By Lemma 2.5, the equations (3.12-13) become

$$C_{k+1,i}^{(n)} = \frac{1+\delta_{n+1-k}}{p_{n+2-k}^{\theta}} \cdot \frac{w_{n+1-k} - w_{n-k}}{w_{n+2-k} - w_{n+1-k}} C_{k,i}^{(n)} , \quad 1 \le i \le 2^{k-1}$$

$$= \frac{w_{n+1-k} - w_{n-k}}{w_{n+1-k}} D_{k,i-2^{k-1}}^{(n)} , \quad 2^{k-1} + 1 \le i \le 2^k \tag{3.26}$$

$$D_{k+1,i}^{(n)} = \frac{1}{p_{n+2-k}^{\theta}} \frac{w_{n-k}}{w_{n+2-k} - w_{n+1-k}} C_{k,i}^{(n)} , \quad 1 \le i \le 2^{k-1}$$

$$= \frac{w_{n-k}}{w_{n-k+1}} D_{k,i-2^{k-1}}^{(n)} , \quad 2^{k-1} + 1 \le i \le 2^k . \tag{3.27}$$

Define $E_{k,i}^{(n)}$ and $F_{k,i}^{(n)}$ by,

$$C_{k,i}^{(n)} = \frac{w_{n-k+2} - w_{n-k+1}}{w_n} E_{k,i}^{(n)} \tag{3.28}$$

$$D_{k,i}^{(n)} = \frac{w_{n-k+1}}{w_n} F_{k,i}^{(n)} . \tag{3.29}$$

It follows directly from (3.26-27) that

$$E_{k+1,i}^{(n)} = \frac{1+\delta_{n+1-k}}{p_{n+2-k}^{\theta}} E_{k,i}^{(n)} , \quad 1 \le i \le 2^{k-1}$$

$$= F_{k,i}^{(n)} , \quad 2^{k-1} + 1 \le i \le 2^k \tag{3.30}$$

$$F_{k+1,i}^{(n)} = \frac{1}{p_{n+2-k}^{\theta}} E_{k,i}^{(n)} , \quad 1 \le i \le 2^{k-1}$$

$$= F_{k,i}^{(n)} , \quad 2^{k-1} + 1 \le i \le 2^k . \tag{3.31}$$

<u>Lemma 3.2</u>

(i) $E_{k,i}^{(n)} = F_{k,i}^{(n)} = 0$ if i is odd.

(ii) For $k \ge 2$ and $1 \le i \le 2^{k-2}$,

$$E_{k+1,4i}^{(n)} = E_{k,2i}^{(n-1)} \tag{3.32}$$

$$F_{k+1,4i}^{(n)} = F_{k,2i}^{(n-1)} . \tag{3.33}$$

(iii) For $k \geq 4$, $1 \leq i \leq 2^{k-4}$,

$$E^{(n)}_{k,8i-2} = \frac{1}{p_n^{\theta}} E^{(n)}_{k,8i} \tag{3.34}$$

$$E^{(n)}_{k,8i-6} = \frac{1+\delta_{n-1}}{p_n^{\theta}} E^{(n)}_{k,8i-4} \tag{3.35}$$

$$F^{(n)}_{k,8i-2} = \frac{1}{p_n^{\theta}} F^{(n)}_{k,8i} \tag{3.36}$$

$$F^{(n)}_{k,8i-6} = \frac{1+\delta_{n-1}}{p_n^{\theta}} F^{(n)}_{k,8i-4} . \tag{3.37}$$

<u>Proof</u> (i) This is easily established by induction on k (with n fixed), using the fact that $C^{(n)}_{1,1} = E^{(n)}_{1,1} = 0$.

(ii) When $k = 2$ we have

$$E^{(n)}_{3,4} = E^{(n-1)}_{2,2} = 1 = F^{(n-1)}_{2,2} = F^{(n)}_{3,4} .$$

Assume that for some k we have

$$E^{(n)}_{k,4i} = E^{(n-1)}_{k-1,2i} \tag{3.38}$$

$$F^{(n)}_{k,4i} = F^{(n-1)}_{k-1,2i} . \tag{3.39}$$

For $1 \leq i \leq 2^{k-3}$ we have from (3.30),

$$E^{(n)}_{k+1,4i} = \frac{1+\delta_{n+1-k}}{p_{n+2-k}^{\theta}} E^{(n)}_{k,4i} \tag{3.40}$$

$$E^{(n-1)}_{k,2i} = \frac{1+\delta_{n+1-k}}{p_{n+2-k}^{\theta}} E^{(n-1)}_{k-1,2i}$$

$$= E^{(n)}_{k+1,4i}$$

by (3.38) and (3.40). Similarly, if $1 + 2^{k-3} \leq i \leq 2^{k-2}$,

$$E^{(n)}_{k+1,4i} = F^{(n)}_{k,4i}$$

$$= F^{(n-1)}_{k-1,2i} \qquad \text{by (3.39)}$$

$$= E^{(n-1)}_{k,4i} \qquad \text{by (3.30) and (3.39)} .$$

One can also deduce (3.33) in like manner. The result then follows by an induction on k.

(iii) For $k = 4$ we have

$$E^{(n)}_{4,6} = \frac{1}{p_n^\theta} E^{(n)}_{4,8} = \frac{1}{p_n^\theta}$$

$$E^{(n)}_{4,2} = \frac{1+\delta_{n-1}}{p_n^\theta} E^{(n)}_{4,4} = \frac{1+\delta_{n-2}}{p_{n-1}^\theta} \cdot \frac{1+\delta_{n-1}}{p_n^\theta}$$

$$F^{(n)}_{4,6} = \frac{1}{p_n^\theta} F^{(n)}_{4,8} = \frac{1}{p_n^\theta}$$

$$F^{(n)}_{4,2} = \frac{1+\delta_{n-1}}{p_n^\theta} F^{(n)}_{4,4} = \frac{1}{p_{n-1}^\theta} \cdot \frac{1+\delta_{n-1}}{p_n^\theta} .$$

Assume that (3.34-37) hold with $k-1$ replacing k, for $1 \le i \le 2^{k-5}$. Then from (3.30) for $1 \le i \le 2^{k-5}$

$$\begin{aligned} E^{(n)}_{k,8i-2} &= \frac{1+\delta_{n+2-k}}{p_{n+3-k}^\theta} E^{(n)}_{k-1,8i-2} \\ &= \frac{1}{p_n^\theta} \cdot \frac{1+\delta_{n+2-k}}{p_{n+3-k}^\theta} \cdot E^{(n)}_{k-1,8i} \\ &= \frac{1}{p_n^\theta} E^{(n)}_{k,8i} \quad ; \end{aligned}$$

for $2^{k-5}+1 \le i \le 2^{k-4}$, we have

$$\begin{aligned} E^{(n)}_{k,8i-2} &= F^{(n)}_{k-1,8i-2} && \text{by (3.30),} \\ &= \frac{1}{p_n^\theta} F^{(n)}_{k-1,8i} && \text{from the induction assumption,} \\ &= \frac{1}{p_n^\theta} E^{(n)}_{k,8i} && \text{by (3.30) again.} \end{aligned}$$

Thus (3.34) holds. One can establish (3.25-7) similarly. □

We now define $L^{(n)}_r(s)$ and $M^{(n)}_r(s)$ by

$$A^{(n)}_r(s) = \exp[s\, \eta^{(n)}_{n+2-r,2^{n+1-r}}] \frac{w_r - w_{r-1}}{w_n} L^{(n)}_r(s) \tag{3.41}$$

$$B^{(n)}_r(s) = \exp[s\, \nu^{(n)}_{n+2-r,2^{n+1-r}}] \frac{w_{r-1}}{w_n} M^{(n)}_r(s) . \tag{3.42}$$

It follows from (3.7a,b), (3.28-9), and (3.41-2) that

$$L_r^{(n)}(s) = \sum_{i=1}^{2^{n+1-r}} E_{n+2-r,i}^{(n)} \exp[s\{\eta_{n+2-r,i}^{(n)} - \eta_{n+2-r,2^{n+1-r}}^{(n)}\}] \tag{3.43}$$

and

$$M_r^{(n)}(s) = \sum_{i=1}^{2^{k+1-r}} F_{n+2-r,i}^{(n)} \exp[s\{\nu_{n+2-r,i}^{(n)} - \nu_{n+2-r,2^{n+1-r}}^{(n)}\}] . \tag{3.44}$$

Lemma 3.3

(i) $$L_r^{(r)}(s) = 1 , \tag{3.45a}$$

$$L_r^{(r+1)}(s) = 1 + \frac{1+\delta_r}{p_{r+1}^{\theta+s}} , \tag{3.45b}$$

and for $n \geq r+2$,

$$L_r^{(n)}(s) = \left(1 + \frac{1}{p_n^{\theta+s}}\right)L_r^{(n-1)}(s) + \frac{\delta_{n-1}}{p_n^{\theta+s}}[L_r^{(n-1)}(s) - L_r^{(n-2)}(s)] . \tag{3.45c}$$

(ii) $$M_r^{(r)}(s) = 1 , \tag{3.46a}$$

$$M_r^{(r+1)}(s) = 1 + \frac{1}{p_{r+1}^{\theta+s}} , \tag{3.46b}$$

and for $n \geq r+2$

$$M_r^{(n)}(s) = \left(1 + \frac{1}{p_n^{\theta+s}}\right)M_r^{(n-1)}(s) + \frac{\delta_{n-1}}{p_n^{\theta+s}}[M_r^{(n-1)}(s) - M_r^{(n-2)}(s)] . \tag{3.46c}$$

Proof (i)

$$L_r^{(r)}(s) = E_{2,2}^{(r)} = 1$$

$$L_r^{(r+1)}(s) = E_{3,2}^{(r+1)} \exp[s\{\eta_{3,2}^{(r+1)} - \eta_{3,4}^{(r+1)}\}] + E_{3,4}^{(r+1)}$$

$$= \frac{1+\delta_r}{p_{r+1}^{\theta+s}} + 1 .$$

Observe that, as $E_{k,i}^{(n)} = 0$ for i odd,

$$L_r^{(n)}(s) = \sum_{i=1}^{2^{n-r}} E_{n+2-r,2i}^{(n)} \exp[s\{\eta_{n+2-r,2i}^{(n)} - \eta_{n+2-r,2^{n+1-r}}^{(n)}\}]$$

$$= \sum_{i=1}^{2^{n-r-1}} E^{(n)}_{n+2-r,4i} \exp[s\{\eta^{(n)}_{n+2-r,4i} - \eta^{(n)}_{n+2-r,2^{n+1-r}}\}]$$

$$+ \sum_{\substack{i=3,\\ i\equiv 3 \pmod 4}}^{2^{n-r}} E^{(n)}_{n+2-r,2i} \exp[s\{\eta^{(n)}_{n+2-r,2i} - \eta^{(n)}_{n+2-r,2^{n+1-r}}\}]$$

$$+ \sum_{\substack{i=1,\\ i\equiv 1 \pmod 4}}^{2^{n-r}} E^{(n)}_{n+2-r,2i} \exp[s\{\eta^{(n)}_{n+2-r,2i} - \eta^{(n)}_{n+2-r,2^{n+1-r}}\}]$$

$$= \sum_{i=1}^{2^{n-r-1}} E^{(n-1)}_{n+1-r,2i} \exp[s\{\eta^{(n-1)}_{n+1-r,2i} - \eta^{(n-1)}_{n+1-r,2^{n-r}}\}]$$

$$+ \frac{1}{p_n^{\theta+s}} \sum_{i=1}^{2^{n-r-2}} E^{(n)}_{n+2-r,8i} \exp[s\{\eta^{(n)}_{n+2-r,8i} - \eta^{(n)}_{n+2-r,2^{n+1-r}}\}]$$

$$+ \frac{1+\delta_{n-1}}{p_n^{\theta+s}} \sum_{\substack{i=2,\\ i\equiv 2 \pmod 4}}^{2^{n-r}} E^{(n)}_{n+2-r,2i} \exp[s\{\eta^{(n)}_{n+2-r,2i} - \eta^{(n)}_{n+2-r,2^{n+1-r}}\}]$$

by (3.17), (3.19), (3.32), (3.34-5),

$$= L_r^{(n-1)}(s) + \frac{1}{p_n^{\theta+s}} L_r^{(n-2)}(s) + \frac{1+\delta_{n-1}}{p_n^{\theta+s}} [L_r^{(n-1)}(s) - L_r^{(n-2)}(s)]$$

as required.

(ii) The proof for (3.46a-c) follows along similar lines, using (3.18), (3.20), (3.33) and (3.36-7).

<u>Lemma 3.4</u>

(i) $$L_r^{(r+2)}(s) = \left(1 + \frac{1+\delta_r}{p_{r+1}^{\theta+s}}\right)\left(1 + \frac{1}{p_{r+2}^{\theta+s}}\right) + \frac{\delta_{r+1}(1+\delta_r)}{(p_{r+1}p_{r+2})^{\theta+s}} \tag{3.47}$$

and for $n \geq r+3$

$$L_r^{(n)}(s) = \left(1 + \frac{1+\delta_r}{p_{r+1}^{\theta+s}}\right)\left(1 + \frac{1}{p_{r+2}^{\theta+s}}\right) \cdots \left(1 + \frac{1}{p_n^{\theta+s}}\right)$$

$$+ \frac{\delta_{r+1}(1+\delta_r)}{(p_{r+1}p_{r+2})^{\theta+s}} \left(1 + \frac{1}{p_{r+3}^{\theta+s}}\right) \cdots \left(1 + \frac{1}{p_n^{\theta+s}}\right)$$

$$+ \frac{\delta_{r+2}}{(p_{r+2}p_{r+3})^{\theta+s}} [p_{r+2}^{\theta+s}(L_r^{(r+2)}(s) - L_r^{(r+1)}(s))](1+\frac{1}{p_{r+4}^{\theta+s}}) \dots (1+\frac{1}{p_n^{\theta+s}})$$

$$\vdots$$

$$+ \frac{\delta_{n-1}}{(p_{n-1}p_n)^{\theta+s}} [p_{n-1}^{\theta+s}(L_r^{(n-1)}(s) - L_r^{(n-2)}(s)] . \tag{3.48}$$

(ii) $$M_r^{(r+2)}(s) = (1 + \frac{1}{p_{r+1}^{\theta+s}})(1 + \frac{1}{p_{r+2}^{\theta+s}}) + \frac{\delta_{r+1}}{(p_{r+1}p_{r+2})^{\theta+s}} \tag{3.49}$$

and for $n \geq r + 3$,

$$M_r^{(n)}(s) = (1 + \frac{1}{p_{r+1}^{\theta+s}})(1 + \frac{1}{p_{r+2}^{\theta+s}}) \dots (1 + \frac{1}{p_n^{\theta+s}})$$

$$+ \frac{\delta_{r+1}}{(p_{r+1}p_{r+2})^{\theta+s}} (1 + \frac{1}{p_{r+3}^{\theta+s}}) \dots (1 + \frac{1}{p_n^{\theta+s}})$$

$$+ \frac{\delta_{r+2}}{(p_{r+2}p_{r+3})^{\theta+s}} [p_{r+2}^{\theta+s} (M_r^{(r+2)}(s) - M_r^{(r+1)}(s))](1 + \frac{1}{p_{r+4}^{\theta+s}}) \dots (1 + \frac{1}{p_n^{\theta+s}})$$

$$\vdots$$

$$+ \frac{\delta_{n-1}}{(p_{n-1}p_n)^{\theta+s}} [p_{n-1}^{\theta+s} (L_r^{(n-1)}(s) - L_r^{(n-2)}(s))] . \tag{3.50}$$

These identities follow easily by induction on n (for fixed r), using the results of Lemma 3.3.

We now define functions $H_r^{(n)}(s)$, $n \geq r + 1$, by $H_r^{(r+1)}(s) = 1$ and

$$H_r^{(n)}(s) = \frac{p_n^{\theta+s} [L_r^{(n)}(s) - L_r^{(n-1)}(s)]}{\left(1 + \frac{1 + \delta_r}{p_{r+1}^{\theta+s}}\right)\left(1 + \frac{1}{p_{r+2}^{\theta+s}}\right) \dots \left(1 + \frac{1}{p_{n-1}^{\theta+s}}\right)}, \quad n \geq r + 2 . \tag{3.51}$$

Define also

$$U_r^{(n)}(s) = \left(1 + \frac{1 + \delta_r}{p_{r+1}^{\theta+s}}\right) \left(1 + \frac{1}{p_{r+2}^{\theta+s}}\right) \dots \left(1 + \frac{1}{p_n^{\theta+s}}\right) . \tag{3.52}$$

Then (3.48) becomes, for $n \geq r + 3$

$$L_r^{(n)}(s) = U_r^{(n)}(s)[1 + \frac{\delta_{r+1}}{\left(1 + \frac{p_{r+1}^{\theta+s}}{1+\delta_r}\right)\left(1 + p_{r+2}^{\theta+s}\right)} + \sum_{j=r+2}^{n-1} \frac{\delta_j H_r^{(j)}(s)}{(1 + p_j^{\theta+s})(1 + p_{j+1}^{\theta+s})}] . \tag{3.53}$$

Moreover, setting $\Psi_r^{(n)}(s) = L_r^{(n)}(s)[U_r^{(n)}(s)]^{-1}$ we have from (3.45c) that for $n \geq r + 3$,

$$H_r^{(n)}(s) = \Psi_r^{(n-1)}(s) + \frac{\delta_{n-1} H_r^{(n-1)}(s)}{1 + p_{n-1}^{\theta+s}}$$

$$= [1 + \frac{\delta_{r+1}}{\left(1 + \frac{p_{r+1}^{\theta+s}}{1+\delta_r}\right)\left(1 + p_{r+2}^{\theta+s}\right)} + \sum_{j=r+2}^{n-2} \frac{\delta_j H_r^{(j)}(s)}{(1 + p_j^{\theta+s})(1 + p_{j+1}^{\theta+s})}] + \frac{\delta_{n-1} H_r^{(n-1)}(s)}{(1 + p_{n-1}^{\theta+s})} \tag{3.54}$$

using (3.53).

In a similar fashion define $K_r^{(n)}(s)$, $n \geq r + 1$, by $K_r^{(r+1)}(s) = 1$ and

$$K_r^{(n)}(s) = \frac{p_n^{\theta+s}[M_r^{(n)}(s) - M_r^{(n-1)}(s)]}{(1 + \frac{1}{p_{r+1}^{\theta+s}})(1 + \frac{1}{p_{r+2}^{\theta+s}}) \ldots (1 + \frac{1}{p_{n-1}^{\theta+s}})}, \quad n \geq r + 2. \tag{3.55}$$

Define

$$V_r^{(n)}(s) = \prod_{j=r+1}^{n} (1 + \frac{1}{p_j^{\theta+s}}) . \tag{3.56}$$

Then (3.50) becomes, for $n \geq r + 3$

$$M_r^{(n)}(s) = V_r^{(n)}(s)[1 + \sum_{j=r+1}^{n-1} \frac{\delta_j K_r^{(j)}(s)}{(1 + p_j^{\theta+s})(1 + p_{j+1}^{\theta+s})}] . \tag{3.57}$$

Finally, setting $\Phi_r^{(n)}(s) = M_r^{(n)}(s)[V_r^{(n)}(s)]^{-1}$, we have from (3.46c) and (3.52) that for $n \geq r + 3$

$$K_r^{(n)}(s) = \Phi_r^{(n-1)}(s) + \frac{\delta_{n-1} K_r^{(n-1)}(s)}{(1 + p_{n-1}^{\theta+s})}$$

$$= [1 + \sum_{j=r+1}^{n-2} \frac{\delta_j K_r^{(j)}(s)}{(1 + p_j^{\theta+s})(1 + p_{j+1}^{\theta+s})}] + \frac{\delta_{n-1} K_r^{(n-1)}(s)}{(1 + p_{n-1}^{\theta+s})} . \tag{3.58}$$

For later use we note

<u>Lemma 3.5</u> Assume that the numbers δ_r, $r \geq 1$, are all equal to some fixed positive constant δ. Then for each real number σ_0 with $\sigma_0 + \theta > \frac{1}{2}$ there is a value $\delta = \delta(\sigma_0)$ and a constant C such that $|K_r^{(n)}(s)| \leq C$ and $|H_r^{(n)}(s)| \leq C$ for all n and r, $n \geq r + 1$, and all complex numbers s with $\text{Re}(s+\theta) \geq \sigma_0 + \theta$.

<u>Proof</u> We omit the proof for $H_r^{(n)}(s)$, as it is similar to that for $K_r^{(n)}(s)$ given below. We consider only values of $C \geq 2$, and values of $\delta \leq 1$. One can easily verify that $K_r^{(r+2)}(s)$ is bounded for $\text{Re}(s+\theta) > \frac{1}{2}$. Choose δ so that

$$\delta[4 + \sum_{j=2}^{\infty} \frac{1}{(p_j^{\theta+\sigma_0} - 1)(p_{j+1}^{\theta+\sigma_0} - 1)}] \leq \tfrac{1}{2} . \tag{3.59}$$

Proceeding by induction, let us assume that $|K_r^{(m)}(s)| \leq C$ for $r + 1 \leq m \leq n-1$. Then by (3.53), for $\text{Re}(s+\theta) > \sigma_0 + \theta$

$$|K_r^{(n)}(s)| \leq 1 + C\,\delta \sum_{j=r+1}^{n-2} \frac{1}{(p_j^{\theta+\sigma_0} - 1)(p_{j+1}^{\theta+\sigma_0} - 1)} + \frac{C\,\delta}{(p_{n-1}^{\theta+\sigma_0} - 1)}$$

$$\leq 1 + C\,\delta[\sum_{j=2}^{\infty} \frac{1}{(p_j^{\theta+\sigma_0} - 1)(p_{j+1}^{\theta+\sigma_0} - 1)} + 4] ,$$

on noting that $p_{n-1}^{\theta+\sigma_0} - 1 > p_2^{\frac{1}{2}} - 1 > \frac{1}{4}$,

$$\leq \frac{C}{2} + \frac{C}{2} = C ,$$

as required. □

We observe that the solution $\phi_n(x,s)$ defined after equation (3.6) is now defined explicitly by means of equations (3.41), (3.42), (3.52), (3.53), (3.56) and (3.57). The eigenvalues of (3.4-5b) are now the values s such that $\phi_n(0,s) = A_1^{(n)}(s) + B_1^{(n)}(s) = 0$; we shall not however need to make use of this information in the sequel.

4. We next consider the effect of formally letting $n \to \infty$ in the differential equation (3.4). Henceforth, for simplicity, we assume that the constants δ_n introduced at the beginning of §2 are all equal to a fixed constant $\delta > 0$. Observe that for fixed $r \geq 0$ $w_n^{(r)} \to w_\infty^{(r)}$ as $n \to \infty$ (see (2.7b)). Set $w_\infty^{(0)} = w_\infty$. It is clear from the formulae (2.10-13) that for fixed $r \geq 1$ the sequences $\{\alpha_r^{(n)}\}$, $\{\beta_r^{(n)}\}$, $\{b_r^{(n)}\}$, $\{c_r^{(n)}\}$ and $\{a_r^{(n)}\}$ all converge to finite limits, which we denote by α_r, β_r, b_r, c_r and a_r, respectively. Observe also that $b_r = \alpha_r + \beta_r$, $c_r = \alpha_r\beta_r$, and that $a_0 = 0$ and

$$a_r = a_{r-1} + \frac{\ln (p_r)}{\alpha_r - \beta_r}, \quad r \geq 1, \tag{4.1}$$

from (3.2). Furthermore, it follows from Lemma 2.6(i) that for $r \geq 2$,

$$\alpha_r - \beta_r = \frac{\delta^{r-1}}{(p_1p_2 \cdots p_{r-1})^\theta} \cdot \frac{w_\infty}{w_{r-1}[w_{r-1} + \delta(w_{r-1}-w_{r-2})]} \tag{4.2}$$

and

$$\beta_{r+1} - \beta_r = \frac{\delta^{r-1}}{(p_1p_2 \cdots p_r)^\theta} \cdot \frac{w_\infty}{w_{r-1}\, w_r}. \tag{4.3}$$

We also have

$$\begin{aligned}
\beta_r + w_\infty^{(1)} &= -\frac{w_{r-1}^{(1)}}{w_{r-1}} w_\infty + w_\infty^{(1)} \\
&= \frac{-w_{r-1}^{(1)} w_\infty + w_\infty^{(1)} w_{r-1}}{w_{r-1}} \\
&= \frac{\delta^{r-2}}{w_{r-1}} \cdot \frac{\left(w_{r-1}^{(r-2)} w_\infty^{(r-1)} - w_{r-1}^{(r-1)} w_\infty^{(r-2)}\right)}{(p_1p_2 \cdots p_{r-2})^\theta},
\end{aligned}$$

by Lemma 2.4(ii),

$$= -\frac{\delta^{r-1}}{(p_1p_2 \cdots p_r)^\theta} \cdot \frac{w_\infty^{(r)} + \delta\left(w_\infty^{(r)} - w_\infty^{(r+1)}\right)}{w_{r-1}}, \tag{4.4}$$

by Lemma 2.3(ii). Finally we have

$$a_r = \sum_{j=1}^{r} (a_j - a_{j-1})$$

$$= \sum_{j=1}^{r} \frac{\ln (p_j)}{\alpha_j - \beta_j} \text{ , by (4.1) ,}$$

$$= O\left(\sum_{j=2}^{r} \ln(p_j)\, \delta^{1-j} (p_1 p_2 \cdots p_{j-1})^{\theta}\right) \text{ , by (4.2),}$$

$$= O\left(\frac{(p_1 p_2 \cdots p_{r-1})^{\theta} \ln(p_r)}{\delta^{r-1}} \sum_{j=2}^{r-1} \frac{1}{(p_j p_{j+1} \cdots p_{r-1})^{\theta}}\right)$$

$$= O\left(\frac{(p_1 p_2 \cdots p_{r-1})^{\theta} \ln(p_r)}{\delta^{r-1}}\right) \tag{4.5}$$

where the order constant does not depend on r.

Setting $\gamma = - w_{\infty}^{(1)}$, it follows from (4.2) and (4.4) that $\alpha_r \to \gamma$ and $\beta_r \to \gamma$ as $r \to \infty$.

Define functions $b(x)$ and $c(x)$ on $(0,\infty)$ by $b(x) = b_r$ and $c(x) = c_r$ on $(a_{r-1}, a_r]$, $r \geq 1$. Formally letting $n \to \infty$ in the differential equation (3.4) we obtain the equation

$$y'' - sb(x)y' + s^2 c(x)y = 0 \text{ ,} \tag{4.6}$$

while the boundary condition (2.12b) becomes

$$y(x,s) \sim \exp(\gamma s x) \quad \text{as} \quad x \to \infty \text{ .} \tag{4.7}$$

It is convenient at this stage to transform (4.6-7) by means of the change of dependent variable $y(x,s) = e^{\gamma s x} z(x,s)$:

$$z'' - s\tilde{b}(x)z' + s^2 \tilde{c}(x)z = 0 \tag{4.8}$$

$$z(x,s) \sim 1 \text{ ,} \quad \text{as} \quad x \to \infty \tag{4.9}$$

where $\tilde{b}(x) = b(x) - 2\gamma$ and $\tilde{c}(x) = c(x) - \gamma b(x) + \gamma^2$. Equation **(4.8)** may be rewritten as

$$\frac{d}{dx}\left[\exp\left(-s \int_0^x \tilde{b}\right)\frac{dz}{dx} \right] + s^2 \tilde{c}(x) \exp\left(-s \int_0^x \tilde{b}\right) \cdot z = 0 \text{ .} \tag{4.10}$$

Set

$$r(x) = \int_0^x \tilde{b}(u)du \tag{4.11}$$

and

$$u(r,s) = z\big(x(r), s\big) \text{ .} \tag{4.12}$$

Then, on changing the independent variable from x to r in (4.10) we have

$$\frac{d}{dr}\left[e^{-sr}\,\tilde{b}\big(x(r)\big)\frac{du}{dr}\right] + s^2\,\frac{\tilde{c}\big(x(r)\big)}{\tilde{b}\big(x(r)\big)}\,e^{-sr}\,u(r,s) = 0\,. \tag{4.13}$$

Lemma 4.1 [1] If $u(r,s)$ is a solution of (4.13) satisfying $u(r,s) \to 1$ and $e^{-sr}\,\tilde{b}\big(x(r)\big)\frac{du}{dr} \to 0$ as $r \to \infty$ then $u(r,s)$ satisfies the integral equation

$$u(r,s) = 1 - s^2 \int_r^\infty \int_t^\infty e^{-s(w-t)}\,\frac{\tilde{c}\big(x(w)\big)}{\tilde{b}\big(x(t)\big)\tilde{b}\big(x(w)\big)}\,u(w,s)\,dw\,dt\,. \tag{4.14}$$

Furthermore (4.14) has a unique bounded solution for values $s = \sigma + i\tau$ such that

$$\int_0^\infty \int_t^\infty e^{-\sigma(w-t)} \left| \frac{\tilde{c}\big(x(w)\big)}{\tilde{b}\big(x(w)\big)\tilde{b}\big(x(t)\big)} \right| dw\,dt < \infty\,. \tag{4.15}$$

Proof The latter assertion may be proved by means of the relevant Neumann series expansion in the space of bounded continuous functions on $[r_0,\infty)$, r_0 suitably large, equipped with the usual supremum norm. The former assertion follows by an integration of equation (4.13). □

Lemma 4.2 The integral equation (4.14) has a unique bounded solution $u(r,s)$ whenever $\mathrm{Re}(s + \theta) > 1$.

Proof By the last lemma it is enough to show that (4.15) holds when $\sigma + \theta > 1$.

Observe first that on $(a_{n-1}, a_n]$, $\tilde{c}(x) = \tilde{c}_n$, where

$$\begin{aligned}
\tilde{c}_n &= \alpha_n\beta_n - \gamma(\alpha_n + \beta_n) + \gamma^2 \\
&= (\beta_n - \gamma)^2 + (\beta_n - \gamma)(\alpha_n - \beta_n) \\
&= (\beta_n - \gamma)(\alpha_n - \beta_n)\left[1 + \frac{\beta_n - \gamma}{\alpha_n - \beta_n}\right] \\
&\sim (\beta_n - \gamma)(\alpha_n - \beta_n) \quad \text{as} \quad n \to \infty
\end{aligned} \tag{4.15}$$

by (4.2) and (4.4). Also, on the same interval $\tilde{b}(x) = \tilde{b}_n$, where

$$\begin{aligned}
\tilde{b}_n &= 2(\beta_n - \gamma) + (\alpha_n - \beta_n) \\
&= (\alpha_n - \beta_n)\left[1 + \frac{2(\beta_n - \gamma)}{\alpha_n - \beta_n}\right] \\
&\sim \alpha_n - \beta_n
\end{aligned} \tag{4.16}$$

by (4.2) and (4.4) again. Now, setting $r(a_n) = r_n$, we have

$$\int_0^\infty \int_t^\infty e^{-\sigma(w-t)} \frac{|\tilde{c}(x(w))|}{\tilde{b}(x(w))\tilde{b}(x(t))} \, dwdt$$

$$= \sum_{n=1}^{\infty} \int_{r_{n-1}}^{r_n} e^{\sigma t}[\tilde{b}(x(t))]^{-1} \int_t^\infty e^{-\sigma w} \frac{|\tilde{c}(x(w))|}{\tilde{b}(x(w))} \, dwdt$$

$$= \sum_{n=1}^{\infty} \int_{r_{n-1}}^{r_n} e^{\sigma t}[\tilde{b}(x(t))]^{-1} \int_t^{r_n} e^{-\sigma w} \frac{|\tilde{c}(x(w))|}{\tilde{b}(x(w))} \, dwdt$$

$$+ \sum_{n=1}^{\infty} \int_{r_{n-1}}^{r_n} e^{\sigma t}[\tilde{b}(x(t))]^{-1} \int_{r_n}^\infty e^{-\sigma w} \frac{|\tilde{c}(x(w))|}{\tilde{b}(x(w))} \, dwdt$$

$$= S_1 + S_2 ,$$

where

$$S_1 = - \sum_{n=1}^{\infty} \frac{\tilde{c}_n}{\sigma \tilde{b}_n^2} \int_{r_{n-1}}^{r_n} e^{\sigma t} \{e^{-\sigma r_n} - e^{-\sigma t}\} dt$$

$$= \sum_{n=1}^{\infty} \frac{\tilde{c}_n}{\sigma^2 \tilde{b}_n^2} [1 - e^{-\sigma(r_n - r_{n-1})}] + \sum_{n=1}^{\infty} \frac{\tilde{c}_n (r_n - r_{n-1})}{\sigma \tilde{b}_n^2} .$$

Here

$$r_n - r_{n-1} = \int_{a_{n-1}}^{a_n} \tilde{b}(v) dv$$

$$= \tilde{b}_n (a_n - a_{n-1})$$

$$\sim \ln (p_n) , \tag{4.17}$$

as $n \to \infty$, by (4.1) and (4.16). Thus $e^{-\sigma(r_n - r_{n-1})} \sim p_n^{-\sigma}$ as $n \to \infty$. As $\tilde{c}_n \tilde{b}_n^{-2} = O(p_n^{-\theta})$ as $n \to \infty$, by **(4.2)** and (4.4), it is clear that $S_1 < \infty$ for $\sigma + \theta > 1$. Also, assuming $\sigma < 0$ (the case $\sigma > 0$ is similar),

$$S_2 = \sum_{n=1}^{\infty} \int_{r_{n-1}}^{r_n} e^{\sigma t} \tilde{b}_n^{-1} [\sum_{k=n+1}^{\infty} \int_{r_{k-1}}^{r_k} e^{-\sigma w} \frac{\tilde{c}_k}{\tilde{b}_k} dw] dt$$

$$= \sum_{n=1}^{\infty} \int_{r_{n-1}}^{r_n} e^{\sigma t} \tilde{b}_n^{-1} [\sum_{k=n+1}^{\infty} \frac{\tilde{c}_k}{(-\sigma)\tilde{b}_k} \{e^{-\sigma r_k} - e^{-\sigma r_{k-1}}\}] dt$$

$$= \sum_{n=1}^{\infty} \frac{1}{(-\sigma)\tilde{b}_n} [e^{\sigma r_n} - e^{\sigma r_{n-1}}] \cdot \sum_{k=n+1}^{\infty} \frac{\tilde{c}_k e^{-\sigma r_k}}{(-\sigma)\tilde{b}_k} \{1 - e^{\sigma(r_k - r_{k-1})}\}$$

$$= O\Big(\sum_{n=1}^{\infty} \frac{e^{\sigma r_{n-1}}}{\tilde{b}_n} \sum_{k=n+1}^{\infty} \frac{\tilde{c}_k e^{-\sigma r_k}}{\tilde{b}_k} \Big) ,$$

where

$$r_m = \sum_{j=1}^{m} (r_j - r_{j-1})$$

$$\sim \ln(p_1 p_2 \ldots p_m)$$

as $m \to \infty$, by (4.1) and (4.16). Thus

$$S_2 = O\Big(\sum_{n=1}^{\infty} (p_1 p_2 \ldots p_{n-1})^{\theta+\sigma} \sum_{k=n+1}^{\infty} (p_1 p_2 \ldots p_k)^{-(\theta+\sigma)} \Big)$$

$$= O\Big(\sum_{n=1}^{\infty} (p_n p_{n+1})^{-(\theta+\sigma)} \Big)$$

$$< \infty ,$$

when $\sigma + \theta > \frac{1}{2}$. □

Observe that, as the double integral on the right of (4.14) is now absolutely convergent for $\mathrm{Re}(s+\theta) > 1$, it follows by Fubini's theorem that for these values of s (4.14) may be reformulated as

$$u(r,s) = 1 + \int_r^{\infty} k_1(r,t,s)\, u(t,s)\, dt \qquad (4.18)$$

where

$$k_1(r,t,s) = -s^2 e^{-st} \frac{\tilde{c}(x(t))}{\tilde{b}(x(t))} \int_r^t \frac{e^{sv}}{\tilde{b}(x(v))} dv , \quad r \le t$$

$$= 0 , \qquad r > t .$$

The kernel $k(r,t,s)$ occurring in the statement of the main theorem (§1) is now given by

$$k(r,t,s) = k_1(r,t,s-\theta) . \qquad (4.19)$$

We consider now the solutions $\phi_n(x,s)$, studied in the previous section, when n tends to ∞. From (3.53) and (3.57) we have

$$L_r^{(n)}(s) = U_r^{(n)}(s)\ \Psi_r^{(n)}(s) \tag{4.20a}$$

$$M_r^{(n)}(s) = V_r^{(n)}(s)\ \Phi_r^{(n)}(s)\ . \tag{4.20b}$$

For fixed r, we have, for $\mathrm{Re}(s+\theta) > 1$, $\lim_{n\to\infty} U_r^{(n)}(s) = U_r(s)$ and $\lim_{n\to\infty} V_r^{(n)}(s) = V_r(s)$ where, by (3.1)

$$U_r(s) = \frac{\zeta(s+\theta)}{\zeta(2s+2\theta)} \cdot \frac{\left(1 + \frac{1+\delta}{p_{r+1}^{\theta+s}}\right)}{\left(1 + \frac{1}{p_1^{\theta+s}}\right) \cdots \left(1 + \frac{1}{p_{r+1}^{\theta+s}}\right)} \tag{4.21a}$$

and

$$V_r(s) = \frac{\zeta(s+\theta)}{\zeta(2s+2\theta)} \cdot \frac{1}{\left(1 + \frac{1}{p_1^{\theta+s}}\right) \cdots \left(1 + \frac{1}{p_r^{\theta+s}}\right)}\ . \tag{4.21b}$$

Also by Lemma 3.5, it is clear that the function sequences $\{\Psi_r^{(n)}(s)\}$ and $\{\Phi_r^{(n)}(s)\}$ are absolutely convergent as $n \to \infty$ in any half plane $\mathrm{Re}(s+\theta) > \sigma_0 + \theta > \frac{1}{2}$, provided the fixed constant $\delta = \delta(\sigma_0)$ is chosen small enough. We set

$$\Psi_r(s) = \lim_{n\to\infty} \Psi_r^{(n)}(s) \tag{4.22a}$$

$$\Phi_r(s) = \lim_{n\to\infty} \Phi_r^{(n)}(s)\ . \tag{4.22b}$$

From (3.15) we have

$$\begin{aligned} &\nu^{(n)}_{n+2-r,2^{n+1-r}} \\ &= \sum_{k=1}^{n+1-r} \left(\nu^{(n)}_{k+1,2^k} - \nu^{(n)}_{k,2^k-1}\right) \\ &= \sum_{k=1}^{n+1-r} \left(\beta^{(n)}_{n+2-k} - \beta^{(n)}_{n+1-k}\right) a^{(n)}_{n+1-k} \\ &= \sum_{k=r}^{n} \left(\beta^{(n)}_{k+1} - \beta^{(n)}_{k}\right) a^{(n)}_{k}\ . \end{aligned} \tag{4.23}$$

Also, from (3.14)

$$\eta^{(n)}_{n+2-r,2^{n+1}-r} = \nu^{(n)}_{n+1-r,2^{n}-r} + \left(\beta^{(n)}_{r+1} - \alpha^{(n)}_{r}\right)a^{(n)}_{r}$$

$$= \sum_{k=r+1}^{n} \left(\beta^{(n)}_{k+1} - \beta^{(n)}_{k}\right)a^{(n)}_{k} + \left(\beta^{(n)}_{r+1} - \alpha^{(n)}_{r}\right)a^{(n)}_{r}$$

by (4.23)

$$= \sum_{k=r}^{n} \left(\beta^{(n)}_{k+1} - \beta^{(n)}_{k}\right)a^{(n)}_{k} - \ell n(p_r) \tag{4.24}$$

by (3.2).

Thus from Lemma 2.6(ii) and the analogue of (4.5) for $a^{(n)}_r$ we have

$$\left(\beta^{(n)}_{k+1} - \beta^{(n)}_{k}\right)a^{(n)}_{k} = O\left(\frac{\ell n\ k}{k^{\theta}}\right) \tag{4.25}$$

and therefore the series $\sum_k \left(\beta^{(n)}_{k+1} - \beta^{(n)}_{k}\right)a^{(n)}_{k}$ **are uniformly bounded in n,** as $\theta > 1$. Notice also that from (4.3) and (4.5) we deduce that

$$\sum_{k=1}^{\infty} \left(\beta_{k+1} - \beta_{k}\right)a_{k} < \infty \ . \tag{4.26}$$

We are now in a position to let $n \to \infty$ in the formula (3.6) for the solution $\phi_n(x,s)$ on $[a^{(n)}_{r-1}, a^{(n)}_{r})$. First note that for fixed r, the sequence $\{a^{(n)}_r\}$ is decreasing in n. Consequently, if $x \in (a_{r-1}, a_r]$ is fixed, then there is an integer N such that $x \in [a^{(n)}_{r-1}, a^{(n)}_{r})$ for all $n \geq N$. From (3.6), (3.41), (3.42) it is clear that for $\mathrm{Re}(s+\theta) > 1$, $\lim_{n\to\infty} \phi_n(x,s) = \phi(x,s)$ where

$$\phi(x,s) = A_r(s)\exp(\alpha_r\ sx) + B_r(s)\exp(\beta_r\ \mathbf{sx})\ ; \tag{4.27}$$

here,

$$A_r(s) = p_r^{-s}\exp\left[s\sum_{k=r}^{\infty}(\beta_{k+1} - \beta_k)a_k\right] \cdot \frac{w_r - w_{r-1}}{w_\infty}\ U_r(s)\ \Psi_r(s)$$

$$= \frac{1}{p_r^{\theta+s}}\exp\left[s\sum_{k=r}^{\infty}(\beta_{k+1} - \beta_k)a_k\right]\frac{w_{r-1} + \delta_{r-1}(w_{r-1} - w_{r-2})}{w_\infty}\ U_r(s)\ \Psi_r(s) \tag{4.28}$$

for $r \geq 2$, and

$$B_r(s) = \exp\left[s\sum_{k=r}^{\infty}(\beta_{k+1} - \beta_k)a_k\right] \cdot \frac{w_{r-1}}{w_\infty} \cdot V_r(s)\ \Phi_r(s) \tag{4.29}$$

and the functions $\Psi_r(s)$ and $\Phi_r(s)$ (and therefore also $\phi(x,s)$, $s \neq 1$) are analytic in an[y] region $\mathrm{Re}(s+\theta) > \sigma_0 + \theta > \frac{1}{2}$, provided the constant $\delta = \delta(\sigma_0)$ is chosen small enough.

We define the function $\psi(x,s)$ by

$$\psi(x,s) = e^{-\gamma sx}\,\phi(x,s)\ . \tag{4.30}$$

Lemma 4.4 The function $\psi(x(r),s)$ is the unique bounded solution of the integral equation (4.14) when $\mathrm{Re}(s+\theta) > 1$.

Proof By Lemma 4.2, Lemma 4.1 and equation (4.12), it is enough to show that $\psi(x,s)$ satisfies the differential equation (4.8) and, for $\mathrm{Re}(s+\theta) > 1$, the conditions $\psi(x,s) \to 1$ and $\exp\left(-s\int_0^x \tilde{b}(t)dt\right)\psi'(x,s) \to 0$ as $x \to \infty$. The first of these assertions follows easily if we can show that $\phi(x,s)$ satisfies the differential equation (4.6); but this is true because on each interval $(a_{r-1},a_r]$ ϕ is a linear combination of solutions of (4.6), and the compatibility at the end-points is established by letting n tend to infinity in the identities (3.9a-b).

From (4.30) we have for $x \in (a_{r-1},a_r]$,

$$\psi(x,s) = A_r(s)\exp[(\alpha_r-\gamma)sx] + B_r(s)[\exp\,(\beta_r-\gamma)sx] \tag{4.31}$$

and hence,

$$\psi'(x,s) = s(\alpha_r-\gamma)A_r(s)\exp[(\alpha_r-\gamma)sx] + s(\beta_r-\gamma)B_r(s)\exp[(\beta_r-\gamma)sx]. \tag{4.32}$$

Thus, for $x \in (a_{r-1},a_r]$ and $s = \sigma + i\tau$ fixed

$$\begin{aligned} &|\psi(x,s) - 1| \\ &\le |A_r(s)|\exp[(\alpha_r-\gamma)\sigma x] + |B_r(s)||\exp(\beta_r-\gamma)sx - 1| + |B_r(s) - 1| \\ &\le O(1)|A_r(s)|\exp[(\alpha_r-\beta_r)\sigma a_{r-1}] + |B_r(s)||\exp[(\beta_r-\gamma)sx] - 1| + |B_r(s) - 1| \end{aligned} \tag{4.33}$$

as $\sigma < 0$ and $\alpha_r - \gamma \sim \alpha_r - \beta_r$ as $r \to \infty$. From the series representations for $U_r(s)$, $V_r(s)$, $\Psi_r(s)$ and $\Phi_r(s)$ it is clear that $U_r\Psi_r$ and $V_r\Phi_r$ both tend to 1 as $r \to \infty$. Consequently, from (4.28-9) $A_r \to 0$ and $B_r \to 1$ as $r \to \infty$. Also, $(\alpha_r-\beta_r)a_{r-1} = O(1)$ as $r \to \infty$, by (4.25-6). Finally

$$
\begin{aligned}
&|\exp[(\beta_r-\gamma)sx] - 1| \\
&\le |\exp[(\beta_r-\gamma)\sigma x] \cos[(\beta_r-\gamma)\tau x] - 1| + |\exp[(\beta_r-\gamma)\sigma x] \sin[(\beta_r-\gamma)\tau x]| \\
&\le \{\exp[(\beta_r-\gamma)\sigma x] - 1\} + |\cos[(\beta_r-\gamma)\tau x] - 1| \\
&\qquad + \exp[(\beta_r-\gamma)\sigma x] \, |\sin[(\beta_r-\gamma)\tau x]| \\
&\le \{\exp[(\beta_r-\gamma)\sigma a_r] - 1\} + 2\sin^2[(\beta_r-\gamma)\tau x/2] \\
&\qquad + \exp[(\beta_r-\gamma)\sigma a_r] \, |\sin[(\beta_r-\gamma)\tau x]| \\
&\le \{\exp[(\beta_r-\gamma)\sigma a_r] - 1\} + 2^{-1}|(\beta_r-\gamma)\tau a_r|^2 \\
&\qquad + |(\beta_r-\gamma)\tau a_r| \exp[(\beta_r-\gamma)\sigma a_r] \quad . \qquad (4.34)
\end{aligned}
$$

Here

$$
\begin{aligned}
|(\beta_r-\gamma)a_r| &= |(\beta_r-\gamma)(a_r-a_{r-1}) + (\beta_r-\gamma)a_{r-1}| \\
&= \left|(\beta_r-\gamma)\frac{\ln(p_r)}{(\alpha_r-\beta_r)} + (\beta_r-\gamma)a_{r-1}\right| \\
&= O\left(\frac{\ln(p_r)+1}{p_r}\right) \qquad (4.35)
\end{aligned}
$$

by (4.2), (4.4) and (4.5). It is now not difficult to see that $\psi(x,s) \to 1$ as $x \to \infty$ from (4.33-5). A similar computation shows that $\exp\left(-s\int_0^x \tilde{b}(t)dt\right) \psi'(x,s) \to 0$ as $x \to \infty$. We have

$$
\begin{aligned}
&|\exp[-s\int_0^x \tilde{b}(t)dt]| \\
&= \exp[-\sigma \int_0^{a_{r-1}} \tilde{b}(t)dt] \,.\, \exp[-\sigma \int_{a_{r-1}}^x \tilde{b}_r \, dt] \\
&= O\left(\frac{1}{(p_1p_2 \cdots p_{r-1})^\sigma} \,.\, \exp[-\sigma \tilde{b}_r(x-a_{r-1})]\right)
\end{aligned}
$$

as $\tilde{b}_k \sim \alpha_k - \beta_k$ for $k \to \infty$,

$$
= O\left(\frac{1}{(p_1p_2 \cdots p_{r-1})^\sigma} \,.\, \exp[-\sigma(\alpha_r-\beta_r)x]\right) \,. \qquad (4.36)
$$

Hence

$$
|\psi'(x,s) \,.\, \exp[-s\int_0^x \tilde{b}(t)dt]|
$$

$$= O\left(|s|\ |\alpha_r - \gamma|\ |A_r(s)| \cdot \frac{1}{(p_1 p_2 \cdots p_{r-1})^{\sigma}} + \right.$$

$$\left. + |s|\ |\beta_r - \gamma|\ |B_r(s)| \cdot \frac{1}{(p_1 p_2 \cdots p_{r-1})^{\sigma}} \cdot \exp[-\sigma a_r(\alpha_r - \beta_r)] \right)$$

$$= O\left(\frac{1}{(p_1 p_2 \cdots p_r)^{\theta+\sigma}} \right) = o(1) \quad \text{as} \quad r \to \infty . \qquad \square$$

Thus if $u(r,s)$ denotes the solution of the integral equation (4.18) we have

$$\psi(x(r),s) = u(r,s) \tag{4.37}$$

for $\operatorname{Re}(s+\theta) > 1$. Recall from (4.27-30) that (provided the constant $\delta = \delta(\sigma_0)$ is chosen small enough) the function $\psi(x(r),s)$ is analytic in any half plane $\operatorname{Re}(s+\theta) > \sigma_0 + \theta > \frac{1}{2}$. The proof of part (ii) of the theorem is now almost immediate. If $u(r,s)$ continues analytically into a sub-region G of the strip $\frac{1}{2} < \operatorname{Re}(s+\theta) \leq 1$, it follows that (on choosing δ suitably small) $\psi(x(r),s)$ satisfies (4.18) for $s \in G$. In particular, for these values of s, $\psi(x(r),s)$ cannot be identically zero. It then follows from (4.27-29) and (4.21a-b) that $\zeta(s+\theta) \neq 0$ for $s \in G$, as required.

Acknowledgements

The author is particularly grateful to F V Atkinson, P Deift, P Sarnak, and R A Smith for helpful and stimulating discussions on topics related to and bearing on this paper.

References

[1] F V Atkinson, private communication.

[2] I W Knowles, Eigenvalue problems and the Riemann zeta function I, to appear, Proc. Conf. Ordinary and Partial Differential Equations, Dundee 1982, Springer-Verlag. LMN 964, pp.308-405

[3] H Rademacher, Topics in Analytic Number Theory, Springer-Verlag, 1973.

Boundedness Criteria for Hyperbolic
Characteristic Initial Value Problems

K. Kreith[1] and C.A. Swanson[2]

1. Introduction. In the case of ordinary differential equations there is a well developed theory of necessary and sufficient conditions for solutions of

$$(1.1)\qquad \frac{d^n u}{dx^n} \pm f(x,u) = 0 \quad ; \qquad 0 \leq x < \infty$$

to have an asymptotically constant solution. For $n = 2$ such results go back to Atkinson [1] and Nehari [6], while far reaching ODE generalizations have been established by several authors [2],[3],[5], with the most general formulation (including certain functional differential equations) being due to Kitamura and Kusano [2]. Related questions for elliptic equations have been studied by Swanson [8],[9], and Kreith and Swanson [4].

The purpose of this note is to consider a generalization in another direction, namely that of hyperbolic partial differential equations of the form

$$(1.2)\qquad D_1^m D_2^n u \pm f(x,y,u) = 0$$

where f is continuous and satisfies $uf(x,y,u) \geq 0$ for all (x,y,u) in $\mathbb{R}^+ \times \mathbb{R}^+ \times \mathbb{R}$. In (1.2) we use the notation $D_1^m = \frac{\partial^m}{\partial x^m}$, $D_2^n = \frac{\partial^n}{\partial y^n}$.

By way of motivation it is useful to sketch a now standard technique used to establish the existence of bounded solutions of ODE's. Noting

---

1) Research supported by the National Science Foundation, MCS-8002130.

2) Research supported by the Natural Sciences and Engineering Research Council of Canada, Grant A3105.

that (1.1) can be written as an integral equation

$$(1.3) \qquad u(x) = \sum_{j=0}^{n-1} b_j \frac{x^j}{j!} \pm \int_x^\infty \frac{(x-\xi)^{n-1}}{(n-1)!} f(\xi,u(\xi))d\xi,$$

we seek conditions on $f(x,u)$ which assure that (1.3) has a solution $u(x) \sim b_0$ as $x \to \infty$.

Such conditions are readily established in terms of the functional

$$(1.4) \qquad \Phi[u] \equiv b_0 \pm \int_x^\infty \frac{(x-\xi)^{n-1}}{(n-1)!} f(\xi,u(\xi))d\xi$$

whose domain is contained in the Banach space $B(X)$ of functions $w(x)$ which are continuous in $[X,\infty)$ and for which $\|w\| = \sup_{X \le x < \infty} |w(x)|$ is finite. It is readily shown [3] that if $f(x,u)$ is monotone in u and satisfies

$$(1.5) \qquad \int^\infty x^{n-1} f(x,b_0)dx < \infty$$

for some constant b_0, then one can apply the Schauder fixed point Theorem to Φ in the closed, bounded, convex subset

$$S(X) = \{w \in B(X) \mid c_1 \le w(x) \le c_2 \text{ for } x \ge X\},$$

where c_1 and c_2 are constants satisfying $c_1 < b_0 < c_2$. When such a fixed point solution is extended to $[0,X)$ it yields a bounded solution of (1.1) on $[0,\infty)$.

In §2 we show that these techniques can be extended to certain characteristic initial value problems for (1.2), but that in this setting the question of <u>global</u> boundedness introduces considerations not encountered in the ODE case. Necessary conditions for the existence of bounded solutions of (1.2) are established in §3, and we again find certain analogies to known results for (1.1). Finally, §4 deals with generalizations to more than two variables, to more general equations, and to different types of asymptotic behavior.

2. Boundedness of Solutions. For the sake of definiteness, in this section we restrict our discussion of sufficiency conditions to the equation

$$D_1^m D_2^n u + f(x,y,u) = 0, \tag{2.1}$$

the equation with opposite sign allowing an analogous treatment. A well posed problem associated with (2.1) is determined by initial data of the form

$$D_1^i u(X,y) = q_i(y) \quad ; i = 0,\dots,m-1 \tag{2.2}$$

$$D_2^j u(x,Y) = p_j(x) \quad ; j = 0,\dots,n-1 \tag{2.3}$$

for fixed $(X,Y) \in \mathbb{R}^+ \times \mathbb{R}^+$. The compatibility conditions

$$p_j^{(i)}(X) = q_i^{(j)}(Y) \tag{2.4}$$

for $0 \le i \le m-1$, $0 \le j \le n-1$ assure that the $D_1^i D_2^j u(x,y)$ are continuous at (X,Y).

In connection with the boundedness criteria to be established below, it will suffice to deal with the case

$p_j(x) = 0$ on $[X,\infty)$ for $j = 1,\dots,n-1$;

(2.5) $\qquad q_i(y) = 0$ on $[Y,\infty)$ for $i=1,\dots,m-1$;

$p_0(x)$ and $q_0(y)$ continuous on $[X,\infty)$ and $[Y,\infty)$, respectively, with $p_0(X) = q_0(Y)$.

A solution of (2.1)- (2.3) then corresponds to a fixed point of the functional

$$\Phi[u] \equiv p_0(x) + q_0(y) - p_0(X) + \int_X^x\int_Y^y \frac{(x-\xi)^{m-1}}{(m-1)!}\frac{(y-\eta)^{n-1}}{(n-1)!} f(\xi,\eta,u(\xi,\eta))\,d\xi\,d\eta. \tag{2.6}$$

To assure that p_0 and q_0 remain compatible for arbitrarily large values of X and Y we also assume the existence of a continuous strictly increasing function $h(x)$ satisfying $h(0) = 0$, $\lim_{x\to\infty} h(x) = \infty$, and for which $p_0(x) = q_0(h(x))$ for $0 \le x < \infty$.

Denoting the quadrant $\{(x,y)\,|\,x \ge X,\ y \ge Y\}$ by $R(X,Y)$, we consider the Banach space $B(X,Y)$ of functions $w(x,y)$ which are continuous in $R(X,Y)$ and for which $\|w\| = \sup_{R(X,Y)} |w(x,y)|$ is finite. The existence of a solution of (2.1) which is bounded in $R(X,Y)$ can now be established by showing that (2.6) has a fixed point in a bounded subset of $B(X,Y)$.

2.1 Theorem. *Let $p_0(x)$ and $q_0(y)$ be positive and bounded away from 0 for $x \ge X_0$, $y \ge Y_0$. Suppose $f(x,y,u)$ is positive and monotone in u in $R(X_0,Y_0) \times \mathbb{R}^+$, where $Y_0 = h(X_0)$. If*

$$\int^\infty\int^\infty x^{m-1}y^{n-1}f(x,y,p_0(x)+q_0(y)-p_0(X_0))\,dx\,dy < \infty$$

then for every $\varepsilon_0 > 0$ there exist $X \ge X_0$ and $Y = h(X) \ge Y_0$ and a solution

$u(x,y)$ *of* (2.1)-(2.3) *such that*

$$|u(x,y) - (p_0(x) + q_0(y) - p_0(X))| < \varepsilon_0.$$

for all $(x,y) \in R(X,Y)$.

Proof. Supposing $f(x,y,u)$ to be monotonic increasing in u, we choose $\varepsilon > 0$ sufficiently small so that $\varepsilon < \varepsilon_0$ and $p_0(x)+q_0(y)-p_0(X) > \varepsilon$ in $R(X_0,Y_0)$. Letting $U(x\ v) = p_0(x)+q_0(y)-p_0(X)$ and noting that

$$(2.7) \qquad u(x,y) = U(x,y) - \frac{\varepsilon}{2} - \int_y^\infty\int_x^\infty \frac{(x-\xi)^{m-1}}{(m-1)!}\,\frac{(y-\eta)^{n-1}}{(n-1)!}\, f(\xi,\eta,u(\xi,\eta))\,d\xi\, d\eta$$

is also a solution of (2.1), we choose X and $Y = h(X)$ sufficiently large so that $X \geq X_0$, $Y \geq Y_0$, and

$$(2.8) \qquad \int_Y^\infty\int_X^\infty \frac{\xi^{m-1}}{(m-1)!}\,\frac{\eta^{n-1}}{(n-1)!}\, f(\xi\eta,U(\xi,\eta))\,d\xi\, d\eta < \frac{\varepsilon}{4}.$$

Defining the closed, convex, and bounded subset $S(X,Y,\varepsilon)$ of $B(X,Y)$ by

$$S(X,Y,\varepsilon) = \{w \in B(X,Y): \|w - (U - \tfrac{\varepsilon}{2})\| \leq \tfrac{\varepsilon}{2}\},$$

we have $U(x,y) \geq w(x,y) > 0$ for all $w \in S(X,Y,\varepsilon)$. Since $f(x,y,u)$ is positive and monotonic increasing in u for $u > 0$, it follows that (2.8) remains valid when U is replaced by any $w \in S(X,Y,\varepsilon)$. Consequently, if $\Phi[u]$ denotes the right side of (2.7), then Φ is a mapping of $S(X,Y,\varepsilon)$ into itself. The techniques of [5] now generalize readily to establish the continuity and compactness of Φ. It therefore follows from the Schauder fixed point Theorem that (2.7) has a solution which belongs to $S(X,Y,\varepsilon)$.

In case t is monotone decreasing in u, an analogous argument applies in the subset of $B(X,Y)$ characterized by $\|w-(U+\frac{\varepsilon}{2})\| \leq \frac{\varepsilon}{2}$.

A natural choice for $U(x,y)$ is a positive constant b_0. This leads to the following special case of Theorem 2.1.

2.2 Corollary. *Suppose* $f(x,y,u)$ *is positive for* $u > 0$ *and monotone in* u. *If there exists a constant* $b_0 > 0$ *such that*

$$\int^{\infty}\int^{\infty} x^{m-1}y^{n-1}f(x,y,b_0)dxdy < \infty, \tag{2.9}$$

then there exist $X \geq 0$ *and* $Y \geq 0$ *such that* (2.1) *has a solution which is bounded in* $R(X,Y)$.

One essential difference between the ordinary and hyperbolic cases concerns the question of global boundedness. Specifically, in the ODE case the existence of a solution bounded in $[X,\infty)$ implies the existence of solution bounded in $[0,\infty)$; however, in the hyperbolic case a solution bounded in $R(X,Y)$ may, when extended to $R(0,0)$ become unbounded in $R(0,Y) - R(X,Y)$ or in $R(X,0) - R(X,Y)$. However, if

$$\int^{\infty}\int_0^{X} x^{m-1}y^{n-1}f(x,y,b_0)dxdy < \infty \qquad \text{for all } X > 0 \tag{2.10}$$

and

$$\int^{\infty}\int_0^{Y} x^{m-1}y^{n-1}f(x,y,b_0)dydx < \infty \qquad \text{for all } Y > 0 \tag{2.11}$$

Then the fixed point argument of Theorem 2.1 can be applied in $R(0,Y) - R(X,Y)$ and $R(X,0) - R(X,Y)$ as well as in $R(X,Y)$. These observations are summarized as follows.

2.3 Theorem. *Suppose* $f(x,y,u)$ *is positive for* $u > 0$ *and monotone in* u. *If there exists a constant* $b_0 > 0$ *such that* (2.9), (2.10), *and* (2.11) *are satisfied, then* (2.1) *has a solution which is bounded in* $\mathbb{R}^+ \times \mathbb{R}^+$.

By way of example, we note that $f(x,y,u) = x^\alpha y^\beta\, p(u)$ satisfies the hypotheses of Theorem 2.3 if $\alpha < -m$, $\beta < -n$ and $p(u)$ is locally bounded.

3. Necessary Conditions. For the ordinary differential equation (1.1) (with $u\, f(x,u) > 0$ for $u \neq 0$) [3] provides a rather complete theory of necessary, as well as sufficient, conditions for the existence of bounded positive solutions. However, such "necessary conditions" depend heavily on whether n is even or odd as well as the sign in (1.1). Specifically, if n is even

$$(3.1) \qquad \frac{d^n u}{dx^n} - f(x,u) = 0$$

always has a "monotone solution" satisfying $(-1)^k u^{(k)}(x) > 0$ for $0 \leq k \leq n$, and such solutions are necessarily bounded. However, if n is odd then a bounded positive solution of (3.1) must satisfy $(-1)^k u^{(k)}(x) < 0$ for $1 \leq k \leq n$, and the existence of such a bounded solution requires that (1.5) be satisfied for some constant $b_0 > 0$. When n is odd similar results hold when (3.1) is replaced by

$$(3.2) \qquad \frac{d^n u}{dx^n} + f(x,u) = 0.$$

In the case of hyperbolic equations

$$(3.3) \qquad D_1^m D_2^n\, u \pm f(x,y,u) = 0,$$

the corresponding considerations of signs and parity seem to preclude a simple set of necessary conditions for the existence of bounded solutions. However, by restricting attention to bounded solutions which are uniformly positive (i.e. bounded away from zero for large values of x), we are able to establish necessary conditions for problems of the form (2.1)-(2.3) reminiscent of those which exist for (1.1).

One such result deals with

$$(3.4)\qquad \begin{aligned} &D_1^{2}D_2^{n}\, u + f(x,y,u) = 0 \\ &u(x,Y) = C_2 \;;\; D_2^{j}u(x,Y) = 0 \text{ for } j=1,\ldots,n-1 \text{ and } x \geq X. \end{aligned}$$

3.1 Theorem. *Let $f(x,y,u)$ be positive for $(x,y,u) \in R(0,0) \times \mathbb{R}^{+}$ and monotone in u. If (3.4) has a solution which is bounded and uniformly positive in $R(x_0,Y)$ for some $x_0 \geq X$, then there exists a positive constant c such that*

$$(3.5)\qquad \int^{\infty}\!\!\int^{\infty} xy^{n-1} f(x,y,c)\,dx\,dy < \infty.$$

Proof. From the characteristic data of (3.4) it is clear that $(D_1^{2}D_2^{j}u)(x,Y) = 0$ for $j=0,\ldots,n-1$. Integrating n times with respect to y we get

$$(3.6)\qquad D_1^{2}u(x,y) = -\int_Y^y \frac{(y-\eta)^{n-1}}{(n-1)!}\, f(x,\eta,u(x,\eta))\,d\eta \leq 0,\; x \geq x_0 \;\; y \geq Y.$$

Integration with respect to x over (x_0,x) gives

$$(3.7)\qquad D_1u(x,y) - D_1u(x_0,y) = -\int_{x_0}^{x}\int_Y^y \frac{(y-\eta)^{n-1}}{(n-1)!}\, f(\xi,\eta,u(\xi,\eta))\,d\eta\, d\xi,$$

$$(3.8) \qquad u(x,y)-u(x_0,y)-(x-x_0)D_1u(x_0,y)=-\int_{x_0}^{x}\int_{Y}^{y}(x-\xi)\frac{(y-\eta)^{n-1}}{(n-1)!}f(\xi,\eta,u(\xi,\eta))d\eta d\xi .$$

Writing $\xi-x = (\xi-x_0)+(x_0-x)$ and using (3.7), we can rewrite (3.8) as

$$(3.9) \qquad u(x,y)-u(x_0,y)=(x-x_0)D_1u(x,y)+\int_{x_0}^{x}\int_{Y}^{y}(\xi-x_0)\frac{(y-\eta)^{n-1}}{(n-1)!}f(\xi,\eta,u(\xi,\eta,u(\xi,\eta))d\eta d\xi .$$

Since $D_1^{\;2}u(x,y) \leq 0$ for $x \geq x_0$, $y \geq Y$ by (3.6), $D_1u(x,y) \geq 0$ for all $x > x_0$, $y \geq Y$; in fact, if $D_1u(x_1,y) < 0$ for some $x_1 > x_0$, then for $x > x_1$, since $D_1u(x,y)$ is nonincreasing in x,

$$u(x,y)-u(x_1,y) = \int_{x_1}^{x}(D_1u)(t,y)dt \leq (D_1u)(x_1,y)\ (x-x_1).$$

Since we are assuming that $u(x_1,y)$ is bounded above, $u(x,y)$ would be negative for x sufficiently large. Therefore (3.9) shows, since $0 < u(x,y) \leq k$ for all $x \geq x_0$, $y \geq Y$, that the integral on the right side of (3.9) is a positive bounded function of x,y for all $x \geq x_0$, $y \geq 2Y$. Let $x_2 = 2x_0$. Then $\xi \geq x_2$ implies that $\xi-x_0 \geq \frac{1}{2}\xi$. Also, $Y \leq \eta \leq \frac{1}{2}y$ implies that $Y \leq \eta \leq y-\eta$.
Then

$$0 \leq \frac{1}{2}\int_{x_2}^{x}\int_{Y}^{y/2}\xi\eta^{n-1}f(\xi,\eta,u(\xi,\eta))d\eta d\xi$$

$$(3.10) \qquad \leq \int_{x_2}^{x}\int_{Y}^{y/2}(\xi-x_0)(y-\eta)^{n-1}f(\xi,\eta,u(\xi,\eta))d\eta d\xi$$

$$\leq \int_{x_0}^{x}\int_{Y}^{y}(\xi-x_0)(y-\eta)^{n-1}f(\xi,\eta,u(\xi,\eta,u(\xi,\eta))d\eta d\xi \leq K$$

for all $x \geq x_2$ $y \geq 2Y$.

Since $u(\xi,\eta)$ is _uniformly_ positive and bounded for $\xi \geq x_0$, $\eta \geq Y$ by hypothesis, and $x_2 = 2x_0$, there exist positive constants c_1 and c_2 such that

$$c_1 \leq u(\xi,\eta) \leq c_2 \quad \text{for all} \quad \xi \geq x_2,\ \eta \geq Y.$$

Since $f(x,y,u)$ is monotone in u for each fixed x,y, (3.10) implies that

$$\theta \leq \int_{x_2}^{x}\int_{Y}^{y/2} \xi\eta^{n-1} f(\xi,\eta,c)\,d\xi\,d\eta \leq \int_{x_2}^{x}\int_{Y}^{y/2} \xi\eta^{n-1} f(\xi,\eta,u(\xi,\eta))\,d\eta\,d\xi$$
$$\leq 2K$$

for all $x \geq x_2$, $y/2 \geq Y$, and consequently

$$\int_{x_2}^{\infty}\int_{Y}^{\infty} xy^{n-1} f(x,y,c)\,dx\,dy < \infty,$$

where $c = c_1$ or $c = c_2$ according as $f(x,y,u)$ is nondecreasing or nonincreasing, respectively. This proves the theorem.

The assumption that $f(x,y,u) > 0$ for $u > 0$ can in some instances be deleted in connection with necessary conditions for uniformly positive bounded solutions. For the case $m = 1$, $n \geq 1$:

$$\text{(3.11)} \quad \begin{aligned} &D_1 D_2^{\,n}\, u + f(x,y,u) = 0 \\ &u(x,Y) = C_2;\quad D_2^{\,j} u(x,Y) = 0 \text{ for } \quad j = 1,\ldots,n-1 \text{ and } x \geq X \end{aligned}$$

we establish the following.

3.2 Theorem. _Under the assumption of Theorem 3.1, except that f may now assume negative values for $u > 0$, suppose (3.11) has a solution which is bounded and uniformly positive for some $x_0 \geq X$. Then there exists a positive_ constant c such that

$$(3.12)\qquad -\int^{\infty}\int^{\infty} y^{n-1} f(x,y,c)\,dxdy < +\infty.$$

Proof. Integrating (3.11) n times with respect to y gives

$$D_1u(x,y) = -\int_Y^y \frac{(y-\eta)^{n-1}}{(n-1)!} f(x,\eta,u(x,\eta))\,d\eta,\quad x \geq x_0 \quad y \geq Y.$$

Then integration over (x_o,x) gives

$$u(x,y)-u(x_0,y) = -\int_{x_0}^{x}\int_Y^y \frac{(y-\eta)^{n-1}}{(n-1)!} f(\xi,\eta,u(\xi,\eta))\,d\eta d\xi,$$

and (3.12) follows easily as in the proof of Theorem 3.1.

In the general case $m \geq 1$, $n \geq 1$ we get the following weaker theorem for

$$D_1^mD_2^nu + f(x,y,u) = 0$$

$$(3.13)\qquad u(x,Y) = C_2\ ;\ D_2^ju(x,Y) = 0 \text{ for } j=1,\dots,n-1, \text{ and } x \geq X.$$

$$u(X,y) = C_1\ ;\ D_1^iu(X,y) = 0 \text{ for } i=1,\dots,m-1, \text{ and } y \geq Y.$$

3.3 Theorem. Let f be as in Theorem 3.1, but now allowed to assume negative values, and let $u(x,y)$ be a uniformly positive, bounded solution of (3.13) in $R(X,Y)$. Then there exists a positive constant c so that

$$(3.14)\qquad -\int_X^{\infty}\int_Y^{\infty} x^{m-1}y^{n-1} f(x,y;c)\,dxdy < +\infty.$$

Proof. With D_1^2 replaced by D_1^m, (3.6) is unchanged, and integration of this m times with respect to x over (X,x) gives, in view of the initial conditions at X,

$$u(x,y) - C_1 = - \int_X^x \int_Y^y \frac{(x-\xi)^{m-1}}{(m-1)!} \frac{(y-\eta)^{n-1}}{(n-1)!} f(\xi,\eta,u(\xi,\eta))d\eta d\xi .$$

Again (3.14) follows as in the proof of Theorem 3.1.

4. Generalizations. The results of §2 and §3 seem to allow a direct generalization to higher dimensions by considering

$$D_1^{n_1} \ldots D_k^{n_k} u + f(x_1,\ldots,x_k,u) = 0$$

in place of (2.1). Another generalization of great physical importance is to regard $D_1D_2\, u + f(x,y,u) = 0$ as a canonical form of $u_{tt} - u_{ss} + F(s,t,u) = 0$ and to seek corresponding boundedness criteria for

$$(4.1) \qquad u_{tt} - \Delta u + F(\underline{s},t,u) = 0$$

in three space variables. One difficulty here is that one cannot assign arbitrary initial data on a characteristic cone in $\mathbb{R}^3$, and for that reason it is more appropriate to consider the Cauchy problem associated with (4.1). In connection with the nonlinear Klein-Gordon equation

$$(4.2) \qquad u_{tt} - \Delta u + m^2 u + F(u) = 0$$

such boundedness criteria are established in [7].

Another generalization would involve consideration of more general equations of the form

$$(4.3) \qquad D_1^m D_2^n g(x,y)u + f(x,y,u) = 0$$

or more complicated versions thereof. Since we now have

$D_1^m D_2^n \dfrac{p_0(x)+q_0(y)-p_0(X)}{g(x,y)} = 0$, it is natural to seek solutions asymptotic to

$$U(x,y) = \frac{p_0(x)+q_0(y)-p_0(X)}{g(x,y)} ,$$

and the condition assuring their existence in some $R(X,Y)$ is again

$$\int^{\infty}\int^{\infty} x^{m-1}y^{n-1}f(x,y,U(x,y))dxdy < \infty.$$

Finally, we note that the ODE theory for (1.1) deals not just with boundedness but more generally with asymptotic behavior corresponding to an asymptotically ordered set of solutions of $\frac{d^n u}{dx^n} = 0$. An appropriate generalization of this theory is a topic we hope to address in a later paper.

REFERENCES

1. F.V. Atkinson, On second-order non-linear oscillations, Pacific J. Math. 5(1955),643-647.
2. Y. Kitamura and T. Kusano, Nonlinear oscillations of higher order functional differential equations with deviating arguments, J. Math. Anal. Appl. 77(1980),100-119.
3. K. Kreith, Extremal solutions for a class of non linear differential equations, Proc. Amer. Math. Soc. 79(1980),415-421.
4. K. Kreith and C.A. Swanson, Asymptotic solutions of semilinear Schrödinger equations, to appear.
5. T. Kusano and M. Naito, Nonlinear oscillation of fourth order differential equations, Canad. J. Math. 28(1976),840-852.
6. Z. Nehari, On a class of nonlinear second order differential equations, Trans. Amer. Math. Soc. 95(1980),101-123.
7. W. Strauss, Nonlinear invariant wave equations, Springer Lecture Notes in Physics #73, Springer-Verlag, Heidelberg, 1978.
8. C. A. Swanson, Bounded positive solutions of semilinear Schrödinger equations, SIAM J. Math. Anal. 13(1982), in press.
9. ______________, Extremal positive solutions of semilinear Schrödinger equations, to appear.

MATRIX RICCATI INEQUALITY AND OSCILLATION OF SECOND ORDER DIFFERENTIAL SYSTEMS

By Man Kam Kwong

I. Introduction

We are interested in second order differential systems of the following form

$$(P(t)\bar{x}'(t))' + Q(t)\,\bar{x}(t) = 0 \qquad \text{on} \quad (a,b) \tag{1.1}$$

where $-\infty \le a < b \le \infty$, $' \equiv \frac{d}{dt}$, $\bar{x}(t)$ is an n-vector function of t and $P(t) = (p_{ij}(t))_{1\le i,j\le n}$ and $Q(t) = (q_{ij}(t))_{1\le i,j\le n}$ are $n \times n$ matrix functions of t. We shall impose further conditions on P and Q.

Let us denonte by $\underline{D}$ the set of all positive $n \times n$ diagonal matrices

$$\underline{D} = \{D = \begin{pmatrix} d_1(t) & & 0 \\ & \ddots & \\ 0 & & d_n(t) \end{pmatrix} : d_i(t) > 0,\ i = 1,2,\dots,n\}.$$

by $\underline{P}(\underline{P}^+)$ the set of all nonnegative (positive) matrices

$$\underline{P}(\underline{P}^+) = \{A = (a_{ij}):\ a_{ij} \ge 0\ (>0),\ 1 \le i,\ j \le n\},$$

and by $\underline{EP}$ the set of all essentially nonnegative matrices, i.e. all off-diagonal elements are nonnegative,

$$\underline{EP} = \{A = (a_{ij}):\ a_{ij} \ge 0,\ 1 \le i \ne j \le n\}.$$

It is obvious that if $D \in \underline{D}$ and $A \in \underline{P}$, $\underline{P}^+$, $\underline{EP}$ then DA and $AD \in \underline{P}$, $\underline{P}^+$, $\underline{EP}$ respectively.

By an abuse of language we say that a matrix function $A(t)$ of t belongs to one of the above sets on an interval $[c,d]$ if for each $t \in [c,d]$, $A(t)$ belongs to the set.

Let us now assume that

$$\begin{aligned} &P \in \underline{D} \ \text{ on } (a,b) \text{ and is continuously differentiable} \\ &\text{and } Q \in \underline{EP} \ \text{ on } (a,b) \text{ and is continuous.} \end{aligned} \tag{1.2}$$

Notice that we have imposed smoothness conditions on P and Q. This is done just for the sake of convenience. The theory of differential equations only requires that P^{-1} and Q be locally integrable, but then the equation has to be interpreted using the concept of quasi-derivatives.

The equation (1.1) is said to be conjugate in some subinterval $I \subset (a,b)$ if there is a solution $\bar{x}(t)$ of (1.1) such that

$$\bar{x}(c) = \bar{x}(d) = 0$$

for two distinct points c,d in I. The interval I can be open or closed at either endpoint. In the contrary case (1.1) is said to be disconjugate in I. Given a point c, the point $\hat{c} > c$ is conjugate to c if

$$\hat{c} = \inf \{x \in (c,b): \ (1.1) \text{ is conjugate in } [c,x]\}.$$

The conjugate point $\hat{c}$ is not defined if (1.1) is disconjugate in $[c,b)$. The scalar case has been very thoroughly studied. Under different assumptions on the coefficients (P is real positive definite and Q is real symmetric), the vector case has also been well studied, especially by Reid, [8,9], Hinton and Lewis [6], Etgen [5], Lewis and Wright [7]; see also references cited in the above references. The nonsymmetric case has only attracted attention recently. In [4], Cheng developed the theory for $P =$ the identity matrix and $Q \in \underline{P}$ using the Perron theory of nonnegative matrices. In a series of papers, [1-3], Ahmad and Lazer considered the case $P =$ identity and $Q \in \underline{EP}$ and developed a similar thoery.

In the scalar case, the most important tools used in the study of oscillation are calculus of variation and Riccati equation. Recently A. Zettl and the author observed that the theory of differential and integral inequalities coupled with the Riccati equation approach can lead to rather strong results, see the article by the author and A. Zettl in this book.

The most general results in differential and integral inequalities do not have exact analogs in higher dimensions. It is the purpose of the present paper to show that, however, in the case of inequalities of the Riccati type, certain differential and integral inequality type results can be extended and these are fortunately the ones needed to study the vector oscillation theory.

Associated with the vector equation (1.1) is the matrix equation

$$(PX')' + QX = 0 \qquad \text{on} \quad (a,b), \tag{1.3}$$

where X is an $n \times n$ matrix function of t.

Suppose X is a solution of (1.3) and it is nonsingular in a subinterval (c,d), then the matrix $R = -PX'X^{-1}$ satisfies, on (c,d), the Riccati matrix equation

$$R'(t) = R(t)\, P^{-1}(t)\, R(t) + Q(t). \tag{1.4}$$

Conversely any solution R of (1.4) defined and continuous on (c,d) determines a family of matrix solutions of (1.3), given by solutions of the linear equation of first order

$$X'(t) = -P^{-1}(t)\, R(t)\, X(t). \tag{1.5}$$

Any one in this family is uniquely determined by its value at some point in (c,d).

In section 2 we consider differential inequalities of the Riccati type. In section 3 we relate the results in section 2 to the oscillation theory of (1.1). In section 4 we state a few oscillation criteria.

2. Riccati inequalities

Inequalities between matrices are interpreted componentwise. At an endpoint of a closed interval, differential equations or inequalities are interpreted using one-sided derivatives. All matrix functions are assumed continuous.

Lemma 1 Suppose on an interval $[t_0, t_1]$, we have

$$Y'(t) \geq \alpha(t)\, Y(t) \tag{2.1}$$

$$Y(t_0) \geq 0 \tag{2.2}$$

where α is an integrable scalar function and $Y(t)$ an $n \times n$ matrix. Then

$$Y(t) \geq 0 \quad \text{for all} \quad t \in [t_0,t_1].$$

If furthermore either $Y(t_0) > 0$ or strict inequality holds in (2.1) for at least a point in $[t_0,t]$, then $Y(t) > 0$.

Proof. Although stated in a vector form the result is actually scalar in nature when we look at each component separately. The scalar result is well known and easy to prove.

Lemma 2 Let $A(t)$, $B(t) \in \underline{EP}$ and $Y(t)$ satisfy the inequalities

$$Y'(t) \geq A(t)\, Y(t) + Y(t)\, B(t) \qquad \text{on} \quad [t_0,t_1] \tag{2.3}$$
$$Y(t_0) > 0 \ (\geq 0)$$

Then $Y(t) \in \underline{P}^+(\underline{P})$ on $[t_0,t_1]$.

Proof. We can find a constant β so large that $A(t) + \beta I$ and $B(t) + \beta I \in \underline{P}$, where I is the $n \times n$ identity matrix. The inequality (2.3) becomes
$Y'(t) \geq (A(t) + \beta I)\, Y(t) + Y(t)\, (B(t) + \beta I) - 2\beta\, Y(t) \geq Y(t)$ on $[t_0,\tau]$.
where τ is the smallest number at which $Y(\tau) \in \underline{P}$. The result then follows from Lemma 1 and the technique used in the proof of the next lemma.

Lemma 3 Let R satisfy the Riccati equation (1.4) on an interval $[c,d]$ with $P \in \underline{D}$ and $Q \in \underline{EP}$. If $R(c) \in \underline{EP}$ then $R(t) \in \underline{EP}$ on $[c,d]$.

Proof. We can find two positive constants β and γ so large that $R(c) + \gamma \geq 0$ and $Q(t) \geq -\beta I$, $t \in [c,d]$. Let $Y(t) = R(t) + (\gamma+\beta(t-c))\, I$. Then Y satisfies the equation.

$$Y'(t) = Y(t)\, P^{-1}(t)\, Y(t) - (\gamma+\beta(t-c))\, P^{-1}(t)\, Y(t) - Y(t)\, P^{-1}(t)\, (\gamma+\beta(t-c))$$
$$+ Q(t) + \beta I \qquad \text{on} \quad [c,d]. \tag{2.4}$$

At $t = c$, $Y(c) \geq 0$. Suppose that we have $Y(c) > 0$ instead, we claim that $Y(t) \in \underline{P}^+$ on $[c,d]$. Suppose that this is false and τ is the smallest number at which $Y(\tau) \in \underline{P}$ but not $\underline{P}^+$. Then on $[c,\tau]$, $Y(t) \geq 0$ and the following inequality follows from (2.4)

$$Y'(t) \geq A(t)\, Y(t) + Y(t)\, A(t) \quad \text{on} \quad [c,\tau], \text{ where } A(t) = -(\gamma+\beta(t-c))P^{-1}(t) \in \underline{EP}.$$

By Lemma 2, $Y(t) \in \underline{P}^+$ on $[c,\tau]$ contradicting the assumption that $Y(\tau) \notin \underline{P}^+$. When we have only $Y(c) \geq 0$, it is obvious that a continuity argument will lead to the desired conclusion.

<u>Theorem 1</u>. Let $P(t)$, $Q(t)$ satisfy (1.2) on $[c,d]$ and $\bar{Q}(t)$ be continuous on $[c,d]$ such that

$$\bar{Q}(t) \geq Q(t) \qquad t \in [c,d]. \tag{2.5}$$

Let $R(t)$ satisfy the Riccati equation (1.4) on $[c,d]$ and $\bar{R}(t)$ satisfy a similar Riccati equation on $[c,\bar{d}]$

$$\bar{R}'(t) = \bar{R}(t)\, P^{-1}(t)\, \bar{R}(t) + \bar{Q}(t). \tag{2.6}$$

If furthermore at the initial point c

$$\bar{R}(c) \geq R(c) \in \underline{EP} \tag{2.7}$$

then

$$\bar{R}(t) \geq R(t) \qquad \text{for all} \quad t \in [c,d] \cap [c,\bar{d}]. \tag{2.8}$$

<u>Proof</u>. By Lemma 3, $R(t)$, $\bar{R}(t) \in \underline{EP}$. Let $U(t) = \bar{R}(t) - R(t)$. Then U satisfies the inequality

$$\begin{aligned} U'(t) &= U(t)\, P^{-1}(t)\, R(t) + R(t)\, P^{-1}(t)\, U(t) + U(t)\, P^{-1}(t)\, U(t) + \bar{Q}(t) - Q(t) \\ &\geq A(t)\, U(t) + U(t)\, B(t) \end{aligned} \tag{2.9}$$

on any $[c,\tau]$ on which $U(t) \geq 0$, where $A(t) = R(t)\, P^{-1}(t)$ and $B(t) = P^{-1}(t)\, R(t)$ are both in $\underline{EP}$. At the initial point c, $U(c) \geq 0$. As in the proof of the last lemma, the stronger initial condition $U(c) > 0$ together with Lemma 2 implies that $U(t) > 0$ in $[c,d] \cap [c,\bar{d}]$. A continuity argument then completes the proof.

Similar techniques give the following result which we state without proof.

Lemma 4 Let $X(t) \in \underline{P}$ satisfy (1.5) on $[c,d]$ where $P \in \underline{D}$ and $R(t) \in \underline{EP}$. Let $R^*(t)$ be defined on $[c,d]$ such that

$$R^*(t) \geq R(t) \quad \text{on} \quad [c,d] \tag{2.10}$$

and $X^*(t)$ satisfies

$$X^{*\prime}(t) = -P^{-1}(t)\, R^*(t)\, X^*(t) \quad \text{on} \quad [c,d] \tag{2.11}$$

and

$$0 \leq X^*(d) \leq X(d). \tag{2.12}$$

Then

$$X^*(t) \leq X(t) \quad t \in [c,d]. \tag{2.13}$$

Furthermore with the notation $R(t) = (r_{ij}(t))$ and similar ones for R^*, X and X^*, if $r_{ii}{}^*(t) > r_{ii}(t)$ for one of the diagonal elements, then

$$x^*_{ii}(c) < x_{ii}(c). \tag{2.14}$$

3. Conjugacy of (1.1) on a closed subinterval

Let us first elucidate the relation among solutions of the second order system (1.1), those of (1.3) and those of the Riccati equation (1.4). We assume in this section that (1.2) is satisfied.

Given a matrix solution $X(t)$ of (1.3), each of its columns, or more generally any linear combination of its columns, is a solution of (1.1). If $X(t_0)$ is singular at a point t_0, so that there exists a nonzero constant vector $\bar{c}$ such that $X(t_0)\bar{c} = 0$, then $X(t)\bar{c}$ is a solution of (1.1) that vanishes at t_0. Unfortunately if X becomes singular at two distinct points t_0 and t_1, there is no guarantee that the same vector $\bar{c}$ will make both $X(t_0)\bar{c}$ and $X(t_1)\bar{c}$ vanish.

As pointed out in section 1, given any solution $X(t)$ of (1.3) that is nonsingular in an interval, $R(t) = -P(t)\,X'(t)\,X^{-1}(t)$ satisfies the Riccati equation (1.4). However, in general $R(t)$ may not be in $\underline{EP}$. If it is, then Theorem 1 or other results in section 2 can apply.

On the other hand given any solution $\bar{x}(t)$ of (1.1) such that no component of $\bar{x}(t_0)$ is zero at a point t_0, then we can consider the solution of (1.4) satisfying the initial condition $R(t_0) = (r_{ij})$ such that $r_{ij} = 0$ when $i \neq j$, and $r_{ii} = -p_i(t_0)\, x_i'(t_0)/x_i(t_0)$ where $x_i(t_0)$ is the ith component of $\bar{x}(t_0)$ and $x_i'(t_0)$ that of $\bar{x}'(t_0)$. Then $R(t0) \in \underline{EP}$ and by Lemma 3, $R(t) \in \underline{EP}$ in $[t_0,t_1]$ for any t_1 as long as $R(t)$ is defined and continuous. The original $\bar{x}(t)$ can be recovered, at least in a neighborhood of t_0, from $R(t)$ by solving (1.5) for $X(t)$, or more directly by solving the vector equation

$$\bar{x}'(t) = -P^{-1}(t)\, R(t)\, \bar{x}(t) \tag{3.1}$$

with the initial condition $\bar{x}(t_0)$.

The following simple observation is the key to relating the oscillation theory and results in section 2.

<u>Lemma 5</u> If each element of $R(t) \in \underline{EP}$ satisfies

$$r_{ij}(t)/p_i(t) \le M \tag{3.2}$$

on an interval $[t_0,t_1] \subset (a,b)$, where M is some constant, then any nontrivial solution of (3.1) satisfies the inequality

$$\max_{1\le i\le n} |x_i(t_1)| \ge \exp\{-Mn(t_1-t_0)\} \max_{1\le i\le n} |x_i(t_0)|. \tag{3.3}$$

<u>Proof</u>. We can first show that the result is true when $\max_{1\le i\le n} |x_i(t)|$ is attained by one component over the entire interval $[t_0.t_1]$. Without loss of generality we may take $x_1(t) = \max_{1\le i\le n} |x_i(t)|$, $t \in [t_0,t_1]$. Then the first component of equation (3.1) leads to the differential inequality

$$x_1'(t) \ge -M\, x_1(t) - (r_{1j}(t)/p_i(t))|\, x_i(t_0)| \ge -Mnx_1(t)$$

from which (3.3) follows easily.

In case $\max_{1\le i\le n} |x_i(t)|$ is attained by different components at different t in $[t_0,t_1]$, we simply decompose $[t_0,t_1]$ into a countable union of open intervals plus their endpoints and obtain an inequality of the form (3.3) for each open component and then combine them together.

Lemma 6 If equation (1.1) has a nontrivial solution such that $\bar{x}(c) = \bar{x}(d) = 0$, $a < c < d < b$, then there exists no EP solution of the Riccati equation (1.4) defined on $[c,d]$.

Proof. Suppose there were an EP solution $R^*(t)$ of (1.4) on $[c,d]$. In particular $R^*(t)$ is defined and bounded above and below. Consider the solution $\bar{x}_\varepsilon(t)$ of (1.1) that satisfies the initial conditions

$$\bar{x}_\varepsilon(c) = 0$$
$$\bar{x}'_\varepsilon(c) = \bar{x}'(c) + \bar{\varepsilon}$$

where $\bar{\varepsilon} = \begin{pmatrix} \varepsilon \\ \varepsilon \\ \vdots \\ \varepsilon \end{pmatrix}$ and $\varepsilon > 0$ is a small number. For ε small enough, say $\varepsilon < \varepsilon_0$, each component of $\bar{x}'_\varepsilon(c)$ is different from zero. Thus $-p_i(t)\, x'_{\varepsilon i}(t)/x_{\varepsilon i}(t) \to -\infty$ as $t \to c+$ and $\varepsilon < \varepsilon_0$. Thus for any $\varepsilon < \varepsilon_0$, we can find a $t_\varepsilon > c$, yet so close to c that $\rho_{\varepsilon i} = -p_i(t_\varepsilon)\, x'_i(t_\varepsilon)/x_i(t_\varepsilon)$, for each $i = 1,\ldots,n$, is less than the lower bound of all the elements of $R^*(t)$ on $[c,d]$. Let $R_\varepsilon(t)$ be the solution of (1.4) with the initial condition

$$R_\varepsilon(t_\varepsilon) = \begin{pmatrix} \rho_{\varepsilon 1} & & 0 \\ & \ddots & \\ 0 & & \rho_{\varepsilon n} \end{pmatrix}.$$

As explained before Lemma 5, $R_\varepsilon(t)$ and $x_\varepsilon(t)$ are connected by the equation.

$$\bar{x}'_\varepsilon(t) = -P^{-1}(t)\, R_\varepsilon(t)\, \bar{x}_\varepsilon(t). \tag{3.4}$$

By the choice of t_ε, we have $R_\varepsilon(t_\varepsilon) \le R^*(t_\varepsilon)$. Since $R(t)$ and $R^*(t)$ both satisfy the same equation (1.4), Theorem 1 shows that $R_\varepsilon(t) \le R^*(t)$ for all

$t \in (t_\varepsilon, d]$. Thus $R_\varepsilon(t)$ is bounded above, independent of ε. We claim that this leads to a contradiction.

By continuity, $\lim_{\varepsilon\to 0} \bar{x}_\varepsilon(t) = \bar{x}(t)$ for every t. In particular $\lim_{\varepsilon\to 0} \bar{x}_\varepsilon(d) = 0$ and $\lim_{\varepsilon\to 0} \bar{x}_\varepsilon(t_0) = \lim_{\varepsilon\to 0} \bar{x}(t_0)$ for some fixed $t_0 \in (c,d)$. Thus

$$\max_{1\le i\le n} |x_{\varepsilon i}(d)| \Big/ \max_{1\le i\le n} |x_{\varepsilon i}(t_0)| \to 0 .$$

This, according to Lemma 5, contradicts the fact that $R_\varepsilon(t)$ is bounded above, and the proof is complete.

Eventually we wish to show that the converse of Lemma 6 is also true. To this end, we define a sequence of solutions of (1.4) and study the limit function. For each positive integer n, let $R_n(t) \in \underline{EP}$ be the solution of (1.4) satisfying the initial condition.

$$R_n(c) = -nI . \tag{3.5}$$

Since $R_n(c) \le R_m(c)$ for $n \ge m$, by Theorem 1 $\{R_n(t)\}_{n\ge n_0}$ for fixed t is a nonincreasing sequence, as long as $R_n(t)$ is defined for all $n \ge n_0$. Let $r_{n,ij}(t)$ denote the ijth element of $R_n(t)$. For $i \ne j$, $r_{n,ij}(t) \ge 0$. Thus $\lim_{n\to\infty} r_{n,ij}(t)$ exists and is non-negative.

The ith diagonal element satisfies the following differential inequality which follows obviously from (1.4):

$$r'_{n,ii}(t) \ge r_{n,ii}^2(t)/p_i(t) + q_{ii}(t). \tag{3.6}$$

Consider the differential equation

$$r'(t) = r^2(t)/p_i(t) + q_{ii}(t) \tag{3.7}$$

which is related to the second order differential equation

$$(p_i(t)\, x'(t))' + q_{ii}(t)\, x(t) = 0 \tag{3.8}$$

in the usual way. Take any nontrivial solution of (3.8) such that $x(c) = 0$. Then the function $r(t) = -p_i(t)\, x'(t)/x(t)$ defines a solution of (3.7) in some interval (c,d) such that $\lim_{t\to c+} r(t) = -\infty$. The standard comparison result in the theory of differential inequalities yields

$$r_{n,ii}(t) \geq r(t).$$

Hence the sequence $\{r_{n,ii}(t)\}_{n\geq n_0}$ is bounded below and there exists a finite limit.

Thus we conclude that $\lim_{n\to\infty} R_n(t)$ exists for every t in an interval $(c,\hat{c})$. The right endpoint $\hat{c}$ can be chosen as the supremum of the number t such that $R_n(t)$ is defined for some n. By the result of continuity with respect to initial condition in the theory of differential equation, we know that the limit function $R(t)$ is a solution of (1.4) in (c,d).

Lemma 7 Let $R(t)$ be defined on $(c,\hat{c})$ as above. Let $X(t)$ be any solution of (1.5) on (c,d). Then $\lim_{t\to c+} X(t) = 0$.

Proof. Let $R_n(t)$ be defined as above. Pick a point t_0 in (c,d). Let $X_n(t)$ be the solution of (1.5) with $R_n(t)$ in place of $R(t)$ and $X_n(t)$ satisfies the initial condition $X_n(t_0) = X(t_0)$. Since $R_n(t_0) \to R(t_0)$, $X_n'(t_0) \to X'(t_0)$ as $n \to \infty$. Since $X(t)$ and $X_n(t)$ are all solutions of (1.3), continuity results show that $X_n(c) \to X(c)$ and $X_n'(c) \to X'(c)$. By the definition of X_n, $X_n(c) = -R_n^{-1}(c)\, P(c)\, X_n'(c) = \frac{1}{n} P(c)\, X_n'(c) \to 0$. Hence $X(c) = 0$.

Lemma 8 If there does not exist an EP solution of (1.4) defined on $[c,d]$, then (1.1) has a nontrivial solution such that $\bar{x}(c) = \bar{x}(\hat{c}) = 0$ for some $\hat{c} \leq d$.

Proof. Let us define $R(t)$ as we did before Lemma 7 and let $(c,\hat{c})$ be the maximal interval of definition of $R(t)$. We claim that $\hat{c} \leq d$. In fact if $\hat{c} > d$, then $R(d)$ is finite. By the definition of $R(t)$ we know that $R_n(d)$ must be defined for large n. This contradicts the hypothesis. At $\hat{c}$, R is not defined. This can only occur when $X(\hat{c})$ becomes singular where $X(t)$ is any nontrivial solution of

(1.5). We can choose one such solution using some initial condition $X(t_0)$ which is nonsingular. Since $X(\hat{c})$ is singular, we can find a nonzero constant vector $\bar{c}$ such that $X(\hat{c})\bar{c} = 0$. Now $X(t)\bar{c}$ is a solution of (1.1) which vanishes at $t = c$ (by Lemma 7) and at $t = \hat{c}$.

Combining Lemmas 6 and 8 we have.

Theorem 2. Suppose that (1.2) holds and $[c,d] \subset (a,b)$. The following are equivalent:

(i) Equation (1.1) is conjugate on $[c,d]$

(ii) There exists no EP solution of the Riccati equation (1.4) on $[c,d]$

(iii) Let $R(t)$ be any solution of the Riccati equation (1.4) such that $R(c) \in$ EP. Then $R(t)$ blows up to ∞ (i.e. some element of $R(t)$ diverges to ∞) as t approaches some number $\hat{c}$ $(\leq d)$.

(iv) Let $X(t)$ be any solution of (1.3) such that $-P(c)\, X'(c)\, X^{-1}(c) \in$ EP then $X(t)$ must be singular at some point in $[c,d]$.

Also obvious from the proof of Lemma 8 is the following characterization of conjugate points, which together with Theorem 1 imply the monotonity of the conjugation operation.

Theorem 3. Suppose (1.2) holds. Let $c \in (a,b)$. Then its conjugate point $\hat{c}$, if it exists, is characterised as the left endpoint of the maximal interval of definitions of the limit function $R(t)$ as defined before Lemma 7. Alternatively, let $X(t)$ be the solution of (1.3) such that $X(c) = 0$, $X'(c) = I$. Then $\hat{c}$ is the smallest number to the right of c at which $X(\hat{c})$ is singular. A third characterization uses $R_n(t)$ as defined before Lemma 7. Let $[c,d_n)$ be the maximal interval of definition of $R_n(t)$. Then $\{d_n\}$ forms a nondecreasing sequence and $\hat{c} = \lim_{n\to\infty} d_n$. The conjugate point $\hat{c}$ is a nondecreasing function of c, i.e. $c_1 \leq c_2$ implies that $\hat{c}_1 \leq \hat{c}_2$.

The following comparison theorem, first proved by Ahmad and Lazer for the case $P(t) = I$ is a simple consequence of Theorems 1 and 2.

Consider another equation

$$(P(t)\ \bar{y}'(t))' + Q^*(t)\ \bar{y}(t) = 0 \quad \text{on} \quad (a,b). \tag{3.9}$$

Theorem 4. Suppose that $Q^*(t) \in$ EP on (a,b) and is continuous. If (1.1) is conjugate on a subinterval $[c,d]$ and $Q^*(t) \geq Q(t)$ for all $t \in [c,d]$, then (3.9) is conjugate on $[c,d]$.

Proof. By hypothesis, $\hat{c}$, the conjugate point of c with respect to equation (1.1), is less than or equal to d and is defined as the limit $\lim_{n\to\infty} d_n$ where d_n is defined as in Theorem 3. Let d_n^* be defined in exactly the same way as d_n is defined but using equation (3.9) instead. By Theorem 1, $d_n^* \leq d_n$. Since the conjugate point of c with respect to (3.9) is defined as $\hat{c}^* = \lim d_n^*$ we have $\hat{c}^* \leq \hat{c}$.

4. Oscillation of (1.1)

Equation (1.1) is said to be oscillatory near b if every point $c \in (a,b)$ has a conjugate point $\hat{c}$. Alternatively, (1.1) is oscillatory near b if it is conjugate in (c,b) for every c. In contrary to the scalar case, this does not imply that there exists a nontrivial solution $\bar{x}(t)$ such that $\bar{x}(t_1) = \bar{x}(t_2) = \dots = 0$ for a sequence of points $t_1 < t_2 < \dots \to b$.

The following result, although resembles Theorem 2, requires (just for the only if part) extra effort to establish because the right end point of the interval is singular. We omit the proof since it is not of much interest.

Theorem 5. Suppose that (1.2) holds. Equation (1.1) is oscillatory near b if and only if there exists no EP solution of the Riccati equation (1.4) on (c,b) for any number $c \in (a,b)$.

The following is a simple consequence of the comparison theorem.

Theorem 6. Let $\{i_1,i_2,\dots,i_s\}$ be a subset of $\{1,2,\dots,n\}$. Let $\tilde{P}$ and $\tilde{Q}$ be obtainedrespectively from P and Q by deleting the i_kth row and the i_kth column $(k=1,\dots,s)$. If the reduced system of second order differential equations

$$(\tilde{P}(t)\ \bar{y}'(t))' + \tilde{Q}(t)\ \bar{y}(t) = 0 \quad \text{on} \quad (a,b) \tag{4.1}$$

(where $\bar{y}(t)$ is an (n-s) vector function of t) is oscillatory near b, so is (1.1).

In particular if for any i $(1 \le i \le n)$, the scalar equation

$$(p_i\ (t)\ y'(t))' + q_{ii}(t)\ y(t) = 0 \quad \text{on} \quad (a,b) \tag{4.2}$$

is oscillatory, so is (1.1).

Proof. Let $\hat{Q}$ be obtained from Q by replacing all the off-diagonal elements of the i_kth row and i_kth column $(k=1,\ldots,s)$ by zeros. The oscillation of (4.1) implies that of

$$(P(t)\ \bar{z}'(t))' + \hat{Q}(t)\ \bar{z}(t) = 0 \quad \text{on} \quad (a,b) \tag{4.3}$$

since the system of equations (4.3) are decomposable into $s + 1$ independent systems, those involving the i_kth equation and that involving the rest. By the Comparison Theorem, the oscillation of (4.3) implies that of (1.1). This completes the proof.

We have at our disposal all known oscillation criteria for scalar second order equations. If any diagonal element of Q satisfies any of such criteria, then (1.1) is oscillatory. On the other hand, even if none of the diagonal elements give rise to oscillatory scalar equations, the off diagonal elements of Q can still play a role and cause oscillation. The following simple example, however, shows that elements on one side of the diagonal alone is not sufficient. Consider

$$\bar{x}''(t) + \begin{pmatrix} t^{-3} & f(t) \\ 0 & t^{-3} \end{pmatrix} \bar{x}(t) = 0 \qquad t \in (0,\infty),$$

where $f(t)$ is any function of t, probably going to ∞ as $t \to \infty$. The associated Riccati equation is

$$R'(t) = R^2(t) + \begin{pmatrix} t^{-3} & f(t) \\ 0 & t^{-3} \end{pmatrix}.$$

Since Q is upper triangular, the Riccati equation has an upper triangular solution. Componentwise the Riccati system becomes

$$r_{ii}' = r_{ii}^2 + t^{-3} \qquad i = 1,2 \tag{4.4}$$

and

$$r_{12}' = (r_{11}+r_{22})\, r_{12} + f(t). \tag{4.5}$$

It is well known that (4.4) has a solution on (c,∞) for some c. We may take $r_{11} = r_{22}$. Equation (4.5) is linear in r_{12} and so its solutions exist on (c,∞). Thus the original equation is not oscillatory no matter how large $f(t)$ is.

The following oscillation criterion holds. For simplicity we let $P(t) = I$.

Theorem 7. Let $P(t) = I$ and $(a,b) = (0,\infty)$. Suppose $\lim_{t\to\infty} \int_0^t q_{kk}(s)\, ds$ exists for two distinct integers $k = i$ and $k = j$ $(1\le i<j\le n)$ and the function

$$f_\tau(t) = e^{-2t}\left(\int_\tau^t e^s q_{ij}(s)ds\right)\left(\int_\tau^t e^s q_{ji}(s)ds\right) + q_{ii}(t)$$

is such that the scalar equation

$$y''(t) + f_\tau(t)\, y(t) = 0 \qquad t \in (\tau,\infty) \tag{4.6}$$

is oscillatory near ∞, for all $\tau > 0$. Then (1.1) is oscillatory.

Proof. We may assume that the system (1.1) is only two dimensional (after deleting all rows and columns but the kth and jth ones) and we let $i = 1$, $j = 2$. The Riccati equation (1.4), written out component by component, becomes

$$r_{11}'(t) = r_{11}^2(t) + r_{12}(t)\, r_{21}(t) + q_{11}(t) \tag{4.7}$$

$$r_{12}'(t) = (r_{11}(t)+r_{22}(t))\, r_{12}(t) + q_{12}(t) \tag{4.8}$$

$$r_{21}'(t) = (r_{11}(t)+r_{22}(t))\, r_{21}(t) + q_{21}(t) \tag{4.9}$$

$$r_{22}'(t) = r_{22}^2(t) + r_{12}(t)\, r_{21}(t) + q_{22}(t). \tag{4.10}$$

If $\lim_{t\to\infty} \int_0^t r_{12}(s)\, r_{21}(s)\, ds = \infty$, then (4.7) has no solution on $[c,\infty)$ for any c, and the equation will be oscillatory. If $\lim_{t\to\infty} \int_0^t r_{12}(s)\, r_{21}(s)\, ds$ exists, then it is well known that either (4.7) has no solution on $[c,\infty)$ for any c, in which case the equation is oscillatory, or $\lim_{t\to\infty} r_{11}(t) = 0$. Similarly, $\lim_{t\to\infty} r_{22}(t) = 0$ unless the equation is already oscillatory. Thus for $\tau \in (0,\infty)$, $r_{11}(t) + r_{22}(t) \le 1$ for $t \ge \tau$. Equations (4.8) and (4.9) yield the inequalities

$$r_{12}'(t) \ge -r_{12}(t) + q_{12}(t) \qquad t \ge \tau$$

$$r_{21}'(t) \ge -r_{21}(t) + q_{21}(t) \qquad t \ge \tau,$$

since $r_{12}(t)$ and $r_{21}(t)$ are nonnegative. Solving these differential inequalities we have $r_{12}(t)\, r_{21}(t) + q_{11}(t) \ge f_\tau(t)$. By hypothesis, equation (4.7) cannot have a solution on $[c,\infty)$ for any c. Thus equation (1.1) is oscillatory.

An example to which Theorem 7 applies is

$$\bar{x}''(t) + \begin{pmatrix} -t^{-3} & t^{-1} \\ t^{-1+\varepsilon} & -t^{-3} \end{pmatrix} \bar{x}(t) = 0 \qquad t \in (0,\infty), \tag{4.11}$$

where ε is any positive number.

If $q_{ii}(t)$ and $q_{jj}(t)$ in the hypothesis of Theorem 7 are nonnegative, then a stronger result holds.

<u>Theorem 8</u> If for two distinct integers $k = i$ and $k = j$ $(1 \le i < j \le n)$ we have $q_{kk}(t) \ge 0$ and the function

$$f_\tau(t) = \lambda t^{-4} \left(\int_\tau^t s^2 q_{ij}(s) ds \right) \left(\int_\tau^t s^2 q_{ji}(s) ds \right) + q_{ii}(t)$$

where λ is some positive constant less than one, is such that (4.6) is oscillatory near ∞ for all $\tau > 0$, then (1.1) is oscillatory.

<u>Proof</u>. The proof is similar to that of Theorem 7. Equation (4.7) now gives the inequality

$$r_{11}'(t) \ge r_{11}^2(t)$$

from which we see that $r_{11}(t) \geq -(t-t_0)^{-1}$ for some t_0 and for t large enough. A similar inequality holds for $r_{22}(t)$. Putting these estimates in (4.8) and (4.9) we obtain the estimates

$$r_{12}(t) \geq (t-t_0)^{-2} \int_{\tau}^{t} (s-t_0)^2 q_{ij}(s)ds \qquad \geq t^{-2} \int_{\tau}^{t} (s-t_0)^2 q_{ij}(s)ds$$

$$r_{21}(t) \geq (t-t_0)^{-2} \int_{\tau}^{t} (s-t_0)^2 q_{ij}(s)ds$$

$$\geq t^{-2} \int_{\tau}^{t} (s-t_0)^2 q_{ij}(s)ds\,.$$

For s large enough $(s-t_0)^2 \geq \lambda s^2$. We can then complete the proof as before

An example to which Theorem 8 applies is

$$\bar{x}''(t) + \begin{pmatrix} t^{-3} & t^{-2} \\ t^{-2+\varepsilon} & t^{-3} \end{pmatrix} \bar{x}(t) = 0. \qquad (4.12)$$

Notice the improvement on the power of t for $q_{12}(t)$ and $q_{21}(t)$ over the previous example (4.11).

It is of course interesting to know if Theorem 7 can be improved to narrow the gap between the two examples. A detailed analysis of the asymptotic decay of the solutions of the equations (4.7) and (4.10) may be needed to answer this question.

We raise a few questions to end the paper. As noted in Theorem 2, the conjugation operation is nondecreasing. Is it strictly increasing? Does a comparison theorem involving varying the coefficient $P(t)$ exist? Can the theory be further extended to equations in which the coefficient $P(t)$ is not diagonal?

REFERENCES

1. Shair Ahmad and Alan C. Lazer, "An N-dimensional extension of the Sturm separation and comparison theory to a class of nonselfadjoint systems", SIAM J. Math. Anal. 9(1978), 1137-1150.

2. ___________, "Positive operators and Sturmian theory of nonselfadjoint second order systems", Nonlinear Equations in Abstract Spaces, Academic Press, (1978) 25-42.

3. ___________, "Oscillation criteria for second-order differential systems", Proc. Amer. Math. Soc. 71(1978), 247-252.

4. Sui-Sen Cheng, "On the n-dimensional harmonic oscillation", Ann. Mat. Pura Appl. (4)119(1979), 247-258.

5. E.T. Etgen and J.F. Pawlowski, "Oscillation criteria for second order self-adjoint differential systems", Pac. J. Math. 66(1976), 99-110.

6. D.B. Hinton and R.T. Lewis, "Oscillation theory for generalized second-order differential equations", Rocky Moun. J. of Math. 10(1980), 751-766.

7. R.T. Lewis and L.C. Wright, "Comparison and oscillation criteria for self-adjoint vector-matrix differential equations", Pac. J. Math. (to appear)

8. W.T. Reid, Sturmain Theory, Springer-Verlag (in press)

9. ___________, "Monotoneity properties of solutions of hermitian Riccati differential equations", SIAM J. Math. Anal. 1(1970), 195-213.

Department of Mathematical Sciences
Northern Illinois University
DeKalb, IL 60115

A New Approach to Second Order Linear Oscillation Theory

Man Kam Kwong and A. Zettl

Dedicated to F. V. Atkinson, in admiration

1. Introduction

In this paper we survey some new results obtained recently by the authors in the oscillation theory of the equation

$$(py')' + qy = 0 \tag{1.1}$$

on the interval $[0,a)$, $0 < a \le \infty$. Throughout, the real-valued measureable coefficients are assumed to satisfy the minimal conditions

$$p \ge 0,\ p^{-1} = 1/p \quad \text{and} \quad q \in L^1(0,b) \quad \text{for all} \quad b \in (0,a). \tag{1.2}$$

With (1.2), the equation (1.1) is regular at each point except possibly the right endpoint a. We assume that a is a singular point, i.e. either

(i) $a = \infty$

or (ii) $a < \infty$ and at least one of p^{-1}, q is not integrable in $(0,a)$.

Otherwise it is well known that the equation is nonoscillatory. The left endpoint is chosen to be 0 just for convenience. It can be any other point $< a$. The study of zeros of solutions of (1.1) dates back at least to the fundamental comparison theorem of Sturm [21]. Since this result is usually stated, even in the recent literature, under stronger conditions than (1.2), we mention the most general form of it here for the sake of completeness. Consider a second equation

$$(p_1z')' + q_1z = 0 \quad \text{on} \quad [0,a). \tag{1.3}$$

THEOREM 1. (Sturm Comparison Theorem) Suppose $0 \le p_1(t) \le p(t)$, $q_1(t) \ge q(t)$ a.e. in $[c,d] \subset [0,a)$ and both pairs of coefficients (p,q), (p_1,q_1) satisfy (1.2). If y is a nontrivial solution of (1.3) such that $y(c) = 0 = y(d)$, then every solution z of (1.3) has a zero in $[c,d]$. Furthermore, if in addition, $p_1(t) < p(t)$ or $q_1(t) > q(t)$ on a set of positive Lebesgue measure, then z must have a zero in the open interval (c,d).

Proof. A proof using differential inequalities is given in Everitt-Kwong-Zettl [4], in which a stronger form of the comparison theorem can also be found. See also Atkinson [1].

Remark 1. The theory of quasi-differential expressions allows us to study equation (1.1) even without the restriction $p \ge 0$ in (1.2). In particular, solutions of initial value problems for (1.1) exist and are unique even when p changes sign on the interval — as long as $p^{-1} \in L^1_{loc}$ and (1.1) is treated as a quasi-differential equation. However, it is not clear what form the comparison theroem takes when p and p_1 are allowed to change sign.

Remark 2. Relaxing the usual assumption that the coefficients are continuous in the Comparison Theorem considerably broadens the class of equations whose solutions can be explicitly found and thus used for comparison purpose. In particular p or q or both can be step functions in which case solutions can be computed explicitly. Or q many be allowed to blow up to ∞ at certain points, though not too rapidly.

A nontrivial solution of (1.1) is said to be oscillatory if it has an infinite number of zeros on $[0,a)$. Since the zeros are isolated this can occur only when there are zeros arbitrarily close to a. It follows from the Comparison Theorem that all nontrivial solutions are oscillatory if one is.

In this case, we say that equation (1.1) is oscillatory, or $\underline{0}$ for short. In the contrary case, all nontrivial solutions are nonoscillitory and we say that (1.1) is nonoscillatory. The primary problem in the oscillation theory is to obtain sufficient/necessary conditions on the coefficients that will place the equation in one of these two classifications. Obviously the criteria involves only the behavior of p and q "near" the singular point a.

Since Sturm's work appeared, an enormous number of papers have been written on the subject and hundreds of theorems established. Several approaches proved useful. Among these the more wellknown ones are the Riccati equation approach and the calculus of variation approach. We shall discuss the latter in the next section.

It is straight forward to see that if y is a nontrivial solution of (1.1), then the function $r(t) = -(py')(t)/y(t)$ is defined on any interval which does not include any zeros of y and the Riccati differential equation holds

$$r'(t) = q(t) + r^2(t)/p(t). \tag{1.4}$$

Equivalently we have the Riccati integral equation

$$r(t) = r(t_0) + \int_{t_0}^{t} q(s)ds + \int_{t_0}^{t} \frac{r^2(s)}{p(s)} ds. \tag{1.5}$$

THEOREM 2 The following are equivalent:

(i) Equation (1.1) is not $\underline{0}$.

(ii) Equation (1.4), or (1.5) has an absolutely continuous solution defined on $[c,a)$ for some $c \in [0,a)$.

(iii) There exists an absolutely continuous function $\bar{r}(t)$ defined on $[c,a)$ for some $c \in [0,a)$ such that it satisfies the differential inequality

$$\bar{r}'(t) \geq q(t) + \bar{r}^2(t)/p(t). \tag{1.6}$$

(iv) The coefficient q admits the decomposition $q = q_1 + q_2$ with

$$(\alpha + \int_c^t q_2(s)ds)^2 \leq -p(t)q_1(t)$$

for $c \leq t < a$ and for some constant α.

That (i) and (ii) are equivalent is almost obvious from the way (1.4) is derived. Condition (iii) was first state by Wintner and is a consequence of (ii) and the Sturm Comparison Theorem. Condition (iv) is first stated by Read (see Kauffman, Read and Zettl [8]). It is a reformulation (take $\bar{r}(t) = \alpha + \int_c^t q_2(s)ds$) of (iii) in a way more convenient to use in some situations. Condition (iv) has been very successfully applied in [20] by Read to obtain discreteness conditions for the spectrum of (1.1)

The concept of a differential inequality is present in the statement of condition (iii) but the classical proof of the theorm does not involve the modern theory of differential inequalities as expounded in, for instance, W. Walter [22] or Lakshmikantham and Leela [18]. This theory has been used by a few authors but never fully exploited. The authors recently discovered that it is in fact a very powerful tool when coupled with the following observation.

The oscillation of equation (1.1) is equivalent to the nonexistence of a function $r(t)$ satisfying condition (ii) of Theorem 2. On the other hand, solutions of the Riccati equation (1.4) always exist locally. Thus given any initial condition at $t = c$, there is always a solution to (1.4) in a neighborhood of c. A close examination reveals that the only reason why this solution cannot be extended to the whole interval $[c,a)$ is that it blows up to ∞ at some point less than a.

Thus in order to prove that (1.1) is $\underline{0}$, we need to show that given any initial point c, the solution to (1.4) grows to ∞ before reaching a. The theory of differential and integral inequalities comes in handy in the estimation of the growth of such solutions. As an illustration, let us give a differential inequality proof of the well known

THEOREM 3 (Leighton-Wintner-Fite) If $\int_0^a \frac{ds}{p(s)} = \lim_{t \to a} \int_0^t q(s)ds = \infty$, then (1.1) is $\underline{0}$.

Proof. Let $r(t)$ be a solution of (1.4) with inital condition $r(c) = r_0$. From (1.5) we see that for t_1 large enough and for all $t \geq t_1$,

$$r(t) = \left[r(c) + \int_c^t q(s)ds + \int_c^{t_1} \frac{r^2(s)}{p(s)} ds\right] + \int_{t_1}^t \frac{r^2(s)}{p(s)} ds \geq 1 + \int_{t_1}^t \frac{r^2(s)}{p(s)} ds.$$

In particular $r(t) \geq 0$ for all $t \geq t_1$. Thus the integrand in the last integral is a nondecreasing function of r, as long as $t > t_1$. This allows us to use standard results in the theory of integral inequalities, see Walter [22], to conclude that $r(t) \geq \bar{r}(t)$ for all $t(>t_1)$ at which both r and $\bar{r}$ are defined, where $\bar{r}(t)$ is the solution of the integral equation

$$\bar{r}(t) = 1 + \int_{t_1}^t \frac{\bar{r}^2(s)}{p(s)} ds.$$

By differentiating this equation and solving the resulting differential equation we obtain

$$\bar{r}(t) = \left[1 - \int_{t_1}^t \frac{ds}{p(s)}\right]^{-1}$$

It follows that $\bar{r}(t)$, and hence also $r(t)$, blows up to ∞ before reaching a. Thus equation (1.1) is $\underline{0}$.

Using such techniques the authors have been able to obtain new results as well as to extend many old ones. In section 2 we describe a useful new principle established using this approach and give some of its applications. In section 3 we present a sample of our other recent results.

One more result we want to mention is an algorithm for the construction of all disconjugate (and hence nonoscillatory in the second order case) expressions of any order. For details see Zettl [23].

2. The Telescoping Principle

Oscillation preserving transformations involving changing the dependent and independent variables have been widely used in the theory of oscillation. In [16] we introduced a new construction that transforms an equation into one which oscillates less. In other words, if the transformed equation is $\underline{0}$, then the original equation is $\underline{0}$.

Let $S = \bigcup_{i=1}^{\infty} (a_i, b_i)$ be a union of disjoint open subintervals of $[0,a)$. Let

$$\tau = \tau(t) = \mu([0,t]-S) \tag{2.1}$$

$$A_i = \tau(a_i) \quad i=1,2,... \tag{2.2}$$

$$A = \tau(a) \tag{2.3}$$

where μ denotes the Lebesgue measure.

The effect of τ is to shrink each interval (a_i, b_i) to its left end-point. Thus the interval $[0,a)$ has been "shrunk" to another interval $[0,A)$. Let D_a denote the set of locally integrable functions on $[0,a)$. We construct a transformation T_S from D_a to D_A as follows:
For $f \in D_a$, $F = T_S(f)$ is defined by

$$\begin{cases} F(\tau) = f(t) & \text{if} \quad \tau=\tau(t),\ \tau \neq A_i,\ i = 1,2,3,\ldots \\ F(A_i) = f(a_i). \end{cases} \tag{2.4}$$

The function F is obtained from f by a telescoping action — collapsing each interval (a_i, b_i) to its left endpoint.

In [16] the following theorem was proved as a corollary of a more general theorem under the additional condition that the coefficients p and q are piecewise continuous. The same assertions under weaker conditions on p and q can be proved in exactly the same way using more general results on differential inequalities in the Caratheodory sense. In the following we give an alternative proof of the theorem using Calculus of variation.

<u>THEOREM 4</u> (Telescoping Principle [16]) Let the real-valued coefficients p and q satisfy (1.2). Let $S = \bigcup_{i=1}^{\infty} (a_i, b_i)$ be given such that

$$\int_{a_i}^{b_i} q(t)dt \geq 0,\ i=1,2,3,\ldots \tag{2.5}$$

Let $p_1 = T_S(p)$ and $q_1 = T_S(q)$. If the equation

$$(p_1 z')' + q_1 z = 0 \quad \text{on} \quad [0,A) \tag{2.6}$$

is $\underline{0}$, then equation (1) is also $\underline{0}$.

Proof. We actually can prove more. Suppose a nontrivial solution z has two consecutive zeros, say at τ_1 and τ_2. Let t_1 be such that $\tau(t_1) = \tau_1$ if $\tau_1 \neq A_i$ and $t_1 = a_i$ if $\tau_1 = A_i$. Let t_2 be such that $\tau(t_2) = \tau_2$ if $\tau_2 \neq A_i$ and $t_2 = b_1$ if $\tau_2 = A_i$. We claim that any solution y of (1.1) has at least one zero in $[t_1, t_2]$. The calculus of variation apprach to oscillation theory tells us that it suffices to exhibit an absolutely continuous function u defined on $[t_1,t_2]$ such that $u(t_1) = u(t_2) = 0$ and

$$\int_{t_1}^{t_2} [p(s)\, u'^2(s) - q(s)\, u^2(s)]ds \le 0 \tag{2.7}$$

From the linear operator theoretic point of view, this is equivalent to showing that the differential operator generated by the differential expression $-(py')' - qy$ with boundary conditions $y(t_1) = y(t_2) = 0$ has a negative eigenvalue. By assumption,

$$0 = -\int_{\tau_1}^{\tau_2} [(p_1 z')' + q_1 z]z d\tau$$

$$= \int_{\tau_1}^{\tau_2} [p_1 z'^2 + q_1 z^2]d\tau.$$

Now define u on $[t_1,t_2]$ to be

$$u(t) = \begin{cases} z(\tau(t)) & t \notin S \\ \text{constant} & t \in (a_i,b_i). \end{cases}$$

The constant in each (a_i,b_i) is chosen so that u is continuous. It is obvious that (2.5) implies that (2.7) holds with this definition of u. This completes the proof.

Theorem 4 can be used to construct new examples of $\underline{0}$ equations. Start with any known $\underline{0}$ equation (2.1). Choose a sequence of points $A_i \to \infty$. Cut the plane at each vertical line $\tau = A_i$ and pull the two half planes apart forming a gap of arbitrary finite length. Now fill each gap with an arbitrary positive piecewise continuous function p and an arbitrary piecewise continuous function q whose integral over the length of the gap is nonnegative. Denote the new cofficients constructed this way by p and q. Then equation (1.1) is $\underline{0}$.

In particular, this construction can be applied to known "borderline" cases. The first example illustrates this point.

Example 1. Let $p_1(\tau) = 1$, $q_1(\tau) = c\,\tau^{-2}$, $c > 4^{-1}$, $1 \le \tau < \infty$.

(Here we have replaced $[0,\infty)$ by $[1,\infty)$. Since we are interested in oscillation near ∞ we can do this.) Let $\Lambda_i = i$, $i=2,3,4,\ldots$ and let each of the gaps be of length one. Let $p = 1$, $q = 0$ in each gap. The graph of q is sketched below.

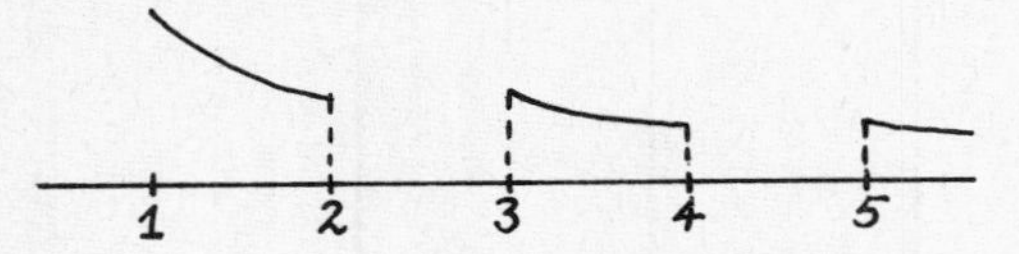

Since (2.1) is $\underline{0}$ by the well known result of Kneser, it follows from Theorem 4 that (1.1) is $\underline{0}$.

Of course this is just one of a wide variety of examples of $\underline{0}$ equations (1.1) which can be constructed this way: Each gap can be made of arbitrary length; p can be an arbitrary positive piecewise continuous function in each gap; q can be an arbitrary piecewise continuous function in each gap whose integral over the gap is nonnegative. Other borderline cases between oscillation and nonoscillation can be "streched" in the same way.

Example 2. Let $p_1(\tau) = 1$, $q_1(\tau) = \sin\tau$. Then it is known, see section 3 below, that (2.1) is oscillatory. Let the graph of q be as shown

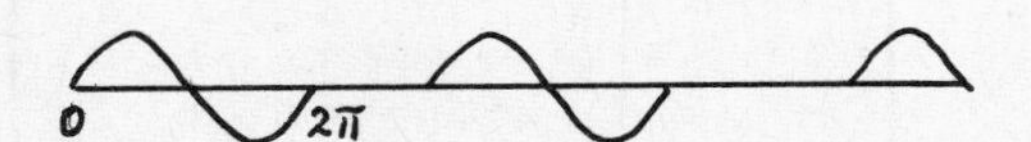

The coefficients p_1 and q_1 are extended to the gaps as before. The resulting equation is $\underline{0}$.

The most interesting application of the telescoping principle, however, is for extending known sufficient conditions: instead of verifying the conditions on a whole half-line, they need only be checked on the shrunk interval $[0,A)$ as long as (2.5) is satisfied.

We mention a few such improvements. However, it is our intention here only to illustrate the technique - not to give the most general result possible. The reader is referred to the papers by Kwong and Zettl [16] for more details. We use arabic numerals to list some best known sufficient criteria and use roman numerals for their extentions.

1. (Opial [19]) Assume $p = 1$ and $a = \infty$. Let q be integrable, perhaps conditionally, i.e. $\lim_{\tau\to\infty} \int_0^\tau q(s)ds$ exists and is finite. Let $Q(t) = \lim_{\tau\to\infty} \int_t^\tau q(s)ds$. Then

$$\int_t^\infty Q^2 ds \geq \frac{1+\varepsilon}{4} Q(t) \geq 0 \tag{2.8}$$

for some positive constant ε implies $\underline{O}$.

I. Opial's criterion bears the restriction that $Q(t) \geq 0$. In case Q takes negative values, the following extension [16] holds:
Let $Q_+(t)$ denote $\max(Q(t),0)$. If Q_+ is not identically zero in $[T,\infty)$ for any $T > 0$ and

$$\int_t^\infty Q_+^2(s)ds \geq \frac{1+\varepsilon}{4} Q(t), \tag{2.9}$$

then (1.1) is $\underline{O}$. Note that (2.9) holds automatically at each point where $Q(t)$ is negative.

2. (Hartman [5]) With the notations from 1, a result of Hartman states that

$$\int_0^{\infty} \exp\left[-4\int_0^t Q_+(s)ds\right]dt < \infty \Rightarrow \underline{O}. \tag{2.10}$$

II. Let $S = \{t \in [0,\infty): Q(t) \geq 0\}$ and T be any set of infinite Lebesgue measure containing S. Then, see [16],

$$\int_T \exp\left[-4\int_0^t Q_+(s)ds\right]dt < \infty \Rightarrow \underline{O}. \tag{2.11}$$

3. (Hille [7]) With the notations from 1, if

$$q(t) \geq 0 \text{ and } \limsup_{t\to\infty} tQ(t) > 1 \Rightarrow \underline{O}. \tag{2.12}$$

III. If q changes sign, Hille's result need modification. The telescoping principle leads to the following extension which reduces to Hille's criterion in case $q \geq 0$:

$$\limsup_{\lambda\to 0} \lambda\mu(\{t \in [0,\infty): Q(t) \geq \lambda\}) > 1 \Rightarrow \underline{O} \tag{2.13}$$

where μ denotes the Lebesgue measure.

4. (Hawking - Penrose [6]) Let $p(t)\equiv 1$. We assume that equation (1.1) is defined on $(-\infty,\infty)$ instead of just $[0,\infty)$. If $q(t) \geq 0$ but $q \not\equiv 0$ then there exists a nontrivial solution of (1.1) having two zeros in $(-\infty,\infty)$.

IV. The nonnegativity of $q(t)$ can be relaxed to the following condition: Let $\bar{Q}(t) = \int_0^t q(s)ds$. For every t, the sets $S = \{s \in [t,\infty): \bar{Q}(s) \geq \bar{Q}(t)\}$ and $S_- = \{s \in (-\infty,t]: \bar{Q}(s) \leq \bar{Q}(t)\}$ have infinite Lebesgue measure, and $q \not\equiv 0$.

The proof depends on a stronger form of the Telescoping Principle proved in [16]. First suppose there is a point t_1 at which $y'(t_1) < 0$. Consider the equation on $[t_1,\infty)$. By shrinking the open set $\{s \in [t_1,\infty): \bar{Q}(s) < \bar{Q}(t_1)\}$

we obtain a new equation for which $\int_{t_1}^{t} q_1(s)ds \geq 0$. The same argument as in the proof of Theorem 3 shows that the solution of the transformed equation having the same initial values of y at t_1 must have a zero to the right of t_1. The stronger telescoping principle then yields the conclusion that y must have a zero to the right of t_1. The same argument applied to $(-\infty,t_1]$ shows that if $y'(t_1) > 0$ at some t_1, then y has a zero in $(-\infty,t_1]$.

Since $q \not\equiv 0$, there is a point t_0 such that $q \not\equiv 0$ on both $(-\infty,t_0)$ and $[t_0,\infty)$. Since the set $\{s \in [t_0,\infty): \bar{Q}(s) \geq \bar{Q}(t_0)\}$ has infinite measure, it contains a point t_1 such that $q \not\equiv 0$ on $[t_0,t_1]$. Let y be the soltuion of (1.1) with the initial conditions $y(t_0) = 1$, $y'(t_0) = 0$. The Riccati equation (1.5) shows that $y'(t_1) < 0$. Hence y has a zero to the right of t_1 as proved above. Similarily, y has another zero in $(-\infty,t_0)$.

It is easy to see that the above result can be extended to equation (1.1) with general p as long as $\int_0^c \frac{ds}{p(s)} = \int_c^a \frac{ds}{p(s)} = \infty$, for some $c \in [0,a)$ and the measure of a set S is interpreted as $\int_S \frac{ds}{p(s)}$.

3. Further Oscillation Criteria

Our new approach yields other far-reaching results, some of which are described in this section. Again the reader is referred to the original paper for the most general result.

Assume that $\int_0^a \frac{ds}{p(s)} = \infty$. A function f is said to be asymptotically constant (in the L^2 sense with respect to the weight $1/p$) if the following two numbers are finite and equal.

$$\bar{\alpha} = \sup \{\lambda \in (-\infty,\infty): \int_0^a \frac{[f(s)-\lambda]_+^2 \, ds}{p(s)} = \infty\} \tag{3.1}$$

$$\underline{\alpha} = \inf\{\lambda \in (-\infty,\infty) = \int_0^a \frac{[f(s)-\lambda]_-^2\, ds}{p(s)} = \infty\} \tag{3.2}$$

where $[x]_+ = \max\{x,0\}$ and $[x] = \max\{-x,0\}$ for any real number x. If $\underline{\alpha} = \bar{\alpha} = \infty$, we say that f is asymptotically large. If $\underline{\alpha} \neq \bar{\alpha}$, we say that f is asymptotically oscillatory. The following two results were established in [17].

<u>THEOREM 5</u> Let $\bar{Q}(t) = \int_0^t q(s)ds$, and define $\bar{\alpha}$, $\underline{\alpha}$ for $\bar{Q}$ instead of f using (3.1), (3.2). If $\underline{\alpha} \neq -\infty$ and $\bar{Q}$ is asymptotically oscillatory, then (1.1) is <u>O</u>. If $\underline{\alpha} = -\infty$, the following additional condition is needed:

For some λ, the set $S = \{t \in [0,\infty): \bar{Q}(t) \geq \lambda\}$ is such that

$$\int_S \frac{ds}{p(s)} = \infty.$$

(A much weaker condition was used in [17].)

<u>THEOREM 6</u> If $\bar{Q}$ is asymptotically constant, then the results of Opial, Hartman and Hille (1,2,3 of section 2 as well as their extentions I, II, and III) and other similar results continue to hold if $\bar{Q}(t)$ is interpreted as $\alpha - \bar{Q}(t)$ where $\alpha = \bar{\alpha} = \underline{\alpha}$.

It can be proved that if $\lim_{t\to\infty} \bar{Q}(t)$ exists, then $\bar{Q}$ is asymptotically constant and $\bar{\alpha} = \underline{\alpha} = \lim_{t\to\infty} \int_0^t q(s)ds$ so that $Q(t) = \int_t^\infty q(s)ds$.

Theorems 5 and 6 are the basis of a theory of classification of (1.1) according to oscillation which parallels the well known theory (using the averaging method) proposed by Coles and Willet [2]. For details see [17].

An improvement of Theorem 3 was established in [16] by combining the proof of Theorem 3 with the telescoping principle.

THEOREM 7 For λ a real number, let

$$S(\lambda) = \{t \in [0,a): \int_0^t q(s)ds \geq \lambda\}.$$

If there exists an increasing sequence of numbers $\lambda_1 < \lambda_2 < \lambda_3 < \ldots < \lambda_n < \ldots \to \infty$ such that

$$\lambda_n \int_{S(\lambda_n)} \frac{ds}{p(s)} \geq 1 \quad n = 1,2,\ldots \tag{3.3}$$

then (1.1) is 0.

Notice that we do not require $\int_0^a \frac{ds}{p(s)}$ to be divergent. In particular, the criterion works for equations on a compact interval and $p = 1$. The well known criteria of Hille [7] have the following extension which is analogous to Theorem 4.

THEOREM 8 ([19]) Suppose $\bar{Q}(t) = \int_0^t q(s)ds$ is asymptotically constant and define $Q(t) = \alpha - \bar{Q}(t)$ as in Theorem 6. Let

$$S(\lambda) = \{t \in [0,\infty): Q(t) \geq \lambda\}.$$

If we have either

$$\limsup_{\lambda \to 0+} \int_{S(\lambda)} \frac{ds}{p(s)} > 1 \tag{3.4}$$

or

$$\liminf_{\lambda \to 0+} \lambda \int_{S(\lambda)} \frac{ds}{p(s)} > \frac{1}{4} \tag{3.5}$$

then (1.1) is 0.

The first part can be proved in the same way as Theorem 7 is proved. The second part follows from a rearrangement principle proved in [9], in which the results are stated only for $p \equiv 1$ and $a = \infty$. By a standard transformation of the independent variable we can see that the results can be generalized to the form given below.

Let f be a continuous function defined on $[0,a)$. We define its nonincreasing rearrangement f^* with respect to the weight $\frac{1}{p}$, to be a nonincreasing function such that for any real number λ, $\int_{T^*} \frac{ds}{p(s)} = \int_T \frac{ds}{p(s)}$

where $T^* = \{t \in [0,a): f^*(t) < \lambda\}$ and $T = \{t \in [0,a): f(t) < \lambda\}$.

<u>THEOREM 9</u> (Rearrangement Principle [9]) Let $q_1(t) \geq 0$ be such that the equation (1.3) is <u>0</u> and $\int_0^a q_1(t) < \infty$. Define $Q_1(t) = \int_t^\infty q_1(s)ds$. Now suppose that $\bar{Q}(t) = \int_0^t q(s)ds$ is asymptotically constant and define $Q(t) = \alpha - \bar{Q}(t)$ (as in Theorem 6). If the nonincreasing rearrangement $Q^*(t) \geq Q_1(t)$, then (1.1) is <u>0</u>.

Intuitively speaking, if we rearranged the function Q into a monotone nonincreasing function and the resulting second order equation is oscillatory then the original equation is also oscillatory.

An interesting corollary of Theorems 7 and 8 is:

<u>THEOREM 10</u> Let $a = \infty$, $p(t) \equiv 1$. Suppose for some λ, $\int_S \frac{ds}{p(s)} < \infty$ with $S = \{t \in [0,\infty): \bar{Q}(t) \geq \lambda\}$. (In the contrary case, (1.1) is oscillatory by Theorem 7). If for some real number $\gamma \in (0,1)$, $\int_S [\bar{Q}(t) - \lambda]^\gamma dt = \infty$, then (1.1) is <u>0</u>. Suppose $\bar{Q}$ is asymptotically constant and $Q(t) = \alpha - \bar{Q}(t)$. Define $\tilde{Q}(t) = \max\{0, \min\{Q(t), 1\}\}$. If for some $\gamma > 1$, $\int_0^\infty \tilde{Q}^\gamma(t)dt = \infty$, then (1.1) is oscillatory.

An interesting example to which the second part of the above theorem applies is $q(t) = t^{-\gamma} \sin t$, $0 < \gamma < 1$.

A well known Theorem of Zlamal [24] has also been extended with the following being a special case.

<u>THEOREM 11</u> ([16]) Let $p(t) \equiv 1$ and $a = \infty$. Equation (1.1) is <u>0</u> if there exists a continuous nondecreasing function $f: [0,\infty) \to [1,\infty)$ such that

$$\int_0^\infty \frac{f'^2(t)}{f(t)}\, dt < \infty \tag{3.6}$$

and

$$\lim_{t\to\infty} \int_0^t f(s)\, q(s) ds = \infty . \tag{3.7}$$

Hartman's result (2 of section 2) has the following extension.

<u>THEOREM 12</u> ([17]) If

$$\int_0^x \exp\left[-4\int_0^t \frac{Q_+(s)}{p(s)}\, ds\right] \frac{dt}{p(t)} = o\left[\ln\left(\int_0^x \frac{ds}{p(s)}\right)\right],$$

then (1.1) is <u>0</u> .

The following is an interesting comparison theorem.

<u>THEOREM 13</u> ([11]) Let $f: [0,\infty) \to [1,\infty)$ be a C^1 function such that for large t

$$2\, p(t)\, f'(t) - 3\int_0^t p(s)\, \frac{f'^2(s)}{f(s)}\, ds \tag{3.8}$$

is nondecreasing. If (1.1) is <u>0</u>, then so is

$$(p(t)\, z'(t))' + f(t)\, q(t)\, z(t) = 0 \quad t \in [0,\infty).$$

Examples of f satisfying the hypotheses of the theorem are $\beta t^{\frac{1}{2}} + t^{-3/2} \sin t$ for any $\beta \in (1/2, 1)$ and C^2 functions such that $f(t) \geq 1$ and $3\, p(t)\, f'^2(t) \geq 2(p(t)\, f'(t))'\, f(t)$ for large t, in particular those such that $f(t) \geq 1$ and pf' is nonincreasing (this particular case is due to Erbe [3]).

The technique of differential and integral inequalities has also been applied to the derivation of Lyapunov type inequalities for oscillation [10], and oscillation criteria for super and sublinear equation of Emden-Fowler type, [12-15].

REFERENCES

1. F. V. Atkinson, Discrete and Continuous Boundary Value Problems, Academic Press, N.Y., 1964.

2. W. J. Coles and D. Willett, "Summability criteria for oscillation of second order linear differential equations, "Annali di mat, pura et appl. 79(1968), 391-398.

3. L. Erbe, "Oscillation Theorems for second order linear differential equations," Pacific J. Math. 35(1970), 337-343.

4. W. N. Everitt, Man Kam Kwong and A. Zettl, "Oscillation of eigenfunctions of weighted regular Sturm-Liouville problems," J. London Math. Soc. (to appear).

5. P. Hartman, Ordinary Differential Equations, 1970.

6. S. W. Hawking and R. Penrose, "The singularities of gravity collapse and cosmology," Proc. Roy. Soc. Lond. A 314(1970), 529-548.

7. E. Hille, "Nonoscillation theorems," Trans. Amer. Math. Sec. 64(1948), 234-252.

8. R. M. Kauffman, T. T. Read and A. Zettl, The Deficiency Index Problem for Power of Ordinary Differential Expressions, Lecture Notes in Maths. 621, Springer-Verlag, 1977.

9. Man Kam Kwong, "On certain Riccati integral equations and second-order linear oscillation, "J. Math. Anal. Appl. 85(1982), 315-330.

10. _____________, "Lyapunov inequality of disfocality," J. Math. Anal. Appl. 83(1981), 486-494.

11. ______________, "On certain comparison theorems for second order linear oscillation, "Proc. Amer. Math. Soc. 84(1982), 539-542.

12. Man Kam Kwong and James S. W. Wong, "On an oscillation theorem of Belohorec," SIAM J. of Math. Ana. (to appear).

13. ______________, "On the oscillation and nonoscillation of second order sublinear equations, "Proc. Amer. Math. Soc. (to appear).

14. ______________, "An application of integral inequality to second order nonlinear oscillation," J. Diff. Eq. (to appear).

15. ______________, "Linearization of second order nonlinear oscillation theorems," preprint.

16. Man Kam Kwong and A. Zettl, "Differential inequality and second order linear oscillation," J. Diff. Eq. (to appear).

17. ______________, "Asymptotically constant functions and second order linear oscillation." J. Math. Anal. Appl. (to appear).

18. V. Lakshmikantham and S. Leela, Differential and Integral Inequalities, Theory and Applications, vol. I, Academic Press, N.Y. 1969

19. Z. Opial, "Sur les integrales oscillantes de l'equation differentielle $u'' + f(t)\ u = o$, "Ann. Polonici Math. 4(1958), 308-313.

20. T. T. Read, "Factorization and discrete spectra for second-order differential expressions," J. Diff. Eq. 35(1980), 388-406.

21. J. C. F. Sturm, "Memoire sur les equations differentielles lineaires du second ordre," J. de Math. Pure et appl, (1836), 106-186.

22. W. Walter, Differential and Integral Inequalities, Springer-Verlag, N.Y., 1970.

23. A. Zettl, "An algorithm for the construction of all disconjugate operators," Proc. Roy. Soc. Edinburgh, 75(1976), 33-40.

24. M. Zlamal, "Oscillation criterions," Cas. Pest Mat. a Pis. 75(1950), 213-217.

Department of Mathematical Sciences
Northern Illinois University
DeKalb, IL 60115, U.S.A.

Small amplitude limit cycles of polynomial differential equations

N.G. Lloyd

Department of Pure Mathematics, University College of Wales, Aberystwyth.

1. A short survey

We consider two-dimensional autonomous differential systems

$$\dot{x} = P(x,y), \qquad \dot{y} = Q(x,y), \tag{1}$$

where P and Q are polynomials. Let S be the collection of such systems and, for $n > 1$, let S_n be the subset of S with P and Q of degree at most n. We are interested in the number of limit cycles; in particular, we seek estimates in terms of n for the number of limit cycles that equations in S_n can have. This is a famous and long standing question: it is part of the sixteenth of Hilbert's list of forty problems. More precisely, let $\pi(S)$ be the number of limit cycles of $S \in S$; for $n > 1$ define

$$H_n = \sup\{\pi(S);\ S \in S_n\}.$$

Hilbert's problem is to estimate H_n in terms of n.

Very little is known about the numbers H_n - it has not even been proved that they are finite for any n. There has been a recent upsurge of interest in the whole question, but progress remains slow. It is a case of accumultating new knowledge little by little, mainly by considering special cases; this paper is no exception to this strategy.

The first significant contribution in this area was that of Bautin in 1952 who proved in [1]* that $H_2 \geq 3$. Though Petrovskii and Landis published two papers in the mid-fifties ([11],[12]) in one of which they suggested that $H_2 = 3$ and in the other of which they gave explicit upper bounds for $H_n (n > 2)$, they eventually withdrew their proofs [13]. Thereafter no clear advances were made until 1979. Then Shi and other Chinese mathematicians ([14],[15],[16]) gave examples of quadratic systems with at

---

*[Bautin's paper was translated in 1954 by Professor Atkinson under the auspices of the American Mathematical Society.]

least four limit cycles, so showing that $H_2 \geq 4$. Several classes of such systems are described in simple terms in [2].

One effective way to approach the whole question is to first work in a neighbourhood of an appropriate type of critical point; suitable perturbations of the equations are then sought which give rise to the bifurcation of limit cycles out of the critical point. Such limit cycles are termed 'local' or 'small-amplitude'. In this paper we adopt this local approach and do not consider more general global questions.

For the technique to work, we have to start with a so-called fine focus.

Definition The point p is a fine focus of (1) if it is a critical point of (1) and is a centre for the linearised system at p:

$$\dot{\xi} = \left[\frac{\partial(P,Q)}{\partial(x,y)}(p)\right]\xi \qquad (\xi \in \mathbb{R}^2).$$

We shall therefore suppose to start with that the origin is a critical point of (1) and that it is of focus type. By a suitable linear change of co-ordinates we can write our system in the form

$$\left.\begin{aligned}\dot{x} &= \lambda x + y + p(x,y)\\ \dot{y} &= -x + \lambda y + q(x,y)\end{aligned}\right\}, \tag{2}$$

where p and q are polynomials all of whose terms are of degree at least 2. The origin is a fine focus precisely when $\lambda = 0$.

It is now natural to write (2) in polar form:

$$\left.\begin{aligned}\dot{r} &= \lambda r + F(r,\theta)\\ \dot{\theta} &= -1 + G(r,\theta)\end{aligned}\right\}, \tag{3}$$

where $\quad F(r,\theta) = \cos\theta.p(r\cos\theta, r\sin\theta) + \sin\theta.q(r\cos\theta, r\sin\theta)$

and $\quad G(r,\theta) = r^{-1}[\cos\theta.q(r\cos\theta, r\sin\theta) - \sin\theta.p(r\cos\theta, r\sin\theta)]$.

At least locally we have what may be described as 'swirling flow'; in the region in which $\dot{\theta} \neq 0$, we can write (3) as

$$\frac{dr}{d\theta} = \frac{\lambda r + F(r,\theta)}{-1 + G(r,\theta)}. \tag{4}$$

In fact, for quadratic systems $\dot{\theta} \neq 0$ inside the largest limit cycle encircling the critical point 0 (see Coppel [4]); thus in this case, equation (4) holds in a region containing all limit cycles encircling 0.

For system (3) the so-called return map is defined in a neighbourhood of 0; this is the mapping H defined by

$$H : x \longmapsto r(2\pi;0,x),$$

where $r(\theta;\theta_o,\rho)$ is the solution of (3) satisfying $r(\theta_o;\theta_o,\rho) = \rho$. Limit cycles of (3) correspond to zeros of h, where

$$h(x) = H(x) - x.$$

Clearly $h(0) = 0$ and, if $x_1 = r(\pi;0,x_o)$, then $h(x_o) = 0$ implies that $h(x_1) = 0$. Hence 0 has odd multiplicity as a zero of the real analytic function h. Its multiplicity is greater than one if and only if 0 is a fine focus. If the multiplicity is 2k+1, then perturbation of the functions p and q can produce at most k limit cycles.

Definition The origin is a *fine focus of order k* of (3) if 0 is a zero of h of multiplicity (exactly) 2k+1.

Suppose now that p and q are of degree n. We first have to find the largest integer k - k_n, say - for which (2) can have a fine focus of order k, and then to produce perturbations of the system *remaining in the class* S_n which produce as many small-amplitude limit cycles as possible. If this maximum number is μ_n, one might expect that $\mu_n = k_n$. Bautin's contribution [1] was to show that $\mu_2 = k_2 = 3$ (see also [2],[3] for related work). All this is, of course, in a neighbourhood of one critical point; the next task would be to work with several critical points simultaneously, and to consider the co-existence of fine foci. This is done in [2] for n = 2, but is not considered in this paper.

Using a slight variant of a classical technique (described in [5] and [9], for example) an algorithm for calculating k_n and μ_n is presented in [3]. It involves some very complicated calculations with polynomials in the coefficients of p and q which have to be implemented on a computer. Without further refinement the calculations rapidly outgrow the storage capacity of most computers. So far it has been possible to complete the calculations for equations in the subset S_3' of S_3 consisting of those systems (2) with p and q being homogeneous cubics. In this case at most five small-amplitude limit cycles can be produced around any one critical point.

Very briefly, the technique referred to is to seek a Liapunov function V such that

$$\frac{dV}{dt} = \eta_2 r^2 + \eta_4 r^4 + \ldots, \tag{5}$$

where the η_j are polynomials in the coefficients of p and q. That this is possible is a classical result. Let I be the ideal generated by the η_j (over the ring of polynomials in the coefficients of p and q); I has a finite basis $B = \{L(0),L(1),\ldots,L(m)\}$, say. The L(j) are called Liapunov quantities, though the basis B is not of course unique. It can be shown that 0 is a fine focus of (2) of order k if and only if $L(0) = L(1) = \ldots = L(k-1) = 0$, but $L(k) \neq 0$. To find the integer k_n it is therefore necessary to find B and then to seek the largest k with the property that $L(0) = L(1) = \ldots = L(k) = 0$ implies $L(m) = 0$ for $m > k$. When all

$L(j)$ are zero, the origin is a centre. Having found B and k_n, one starts with a system for which $L(0) = \ldots = L(k_n-1) = 0$ and then in a sequence of perturbations of the coefficients arranges successively that $L(k_n-1) \neq 0$, $L(k_n-2) \neq 0$, ..., $L(0) \neq 0$ - at least in the best cases. The perturbations are so chosen that

$$L(j)\, L(j-1) < 0 \qquad (j=1,\ldots,k_n-1) \tag{6}$$

and

$$|L(0)| \ll |L(1)| \ldots \ll |L(k_n-1)|. \tag{7}$$

This leads to k_n level curves of V each of which is never tangent to the vector field. By (6) the stability of the origin is reversed at each stage, so that a limit cycle is thrown out; (7) ensures that those limit cycles already produced are not destroyed. We said that this is what happens in the 'best cases': it cannot always be arranged that all the L_j with $0 < j < k_n$ are non-zero. In that eventuality fewer than k_n limit cycles are produced.

In this paper we introduce a different way of approaching the question of small-amplitude limit cycles, with the consequence that the computations are less onerous and the prospects of further progress enhanced.

2. A transformation

We now concentrate on systems such as (2) in which p and q are homogeneous polynomials: we write $\mathcal{S}_n'$ for the corresponding subset of $\mathcal{S}$. The algorithm described in [3] is more easily implemented in such cases - simply because of the number of coefficients occurring.

If p and q are of degree n, the form (4) becomes

$$\frac{dr}{d\theta} = r\,\frac{(\lambda+r^{n-1}f(\theta))}{-1+r^{n-1}g(\theta)}, \tag{8}$$

where f and g are homogeneous polynomials in $\cos\theta$ and $\sin\theta$ of degree $n+1$. We now make the transformation

$$\rho = \frac{r^{n-1}}{1 - r^{n-1}g(\theta)} \tag{9}$$

(see Neto [10] for the case $n = 2$). After some manipulation we have

$$\frac{d\rho}{d\theta} = \alpha(\theta)\rho^3 - \beta(\theta)\rho^2 - \lambda(n-1)\rho, \tag{10}$$

where

$$\left.\begin{aligned} \alpha(\theta) &= -(n-1)g(\theta)[f(\theta)+\lambda g(\theta)] \\ \beta(\theta) &= (n-1)f(\theta) - g'(\theta) + 2\lambda(n-1)g(\theta). \end{aligned}\right\} \tag{11}$$

and

Note that (10) is a cubic in ρ for all values of n; α and β are homogeneous polynomials in $\cos\theta$ and $\sin\theta$, α of degree $2(n+1)$ and β of degree $n+1$.

The transformation (9) maps $r = 0$ to $\rho = 0$, $r > 0$ to $\rho > 0$ and a neighbourhood of $r = 0$ to a neighbourhood of $\rho = 0$. If $\rho > 0$ and small enough (specifically, $1 + \rho g(0) > 0$ for all θ), we can invert (9): r is the positive $(n-1)$th root of

$$\frac{\rho}{1 + \rho g(\theta)}.$$

With regard to the differential equations we are considering, the critical point at the origin of (2) corresponds to the constant solution $\rho = 0$ of (10) and small-amplitude limit cycles of (2) correspond to 2π-periodic solutions of (10) with ρ small and positive.

Our problem has therefore been reduced to a consideration of the positive 2π-periodic solutions of (10). This equation has a striking resemblance to the subject of earlier papers ([6],[7],[8]) written in an apparently unrelated context. It is natural, therefore, to consider the complexified form of (10) and regard θ as time (which we shall denote by τ in this paper):

$$\dot{z} = \alpha(\tau)z^3 - \beta(\tau)z^2 - \lambda(n-1)z. \tag{10)'$$

It is convenient to denote the set equations (10)' when α and β are of the form (11) by C_n.

In [6] we were concerned with the number of periodic solutions of equations

$$\dot{z} = z^N + p_1(\tau)z^{N-1} + \ldots + p_N(\tau),$$

where $p_1, \ldots, p_N$ are continuous periodic functions (all of the same period) and z is complex-valued. It was shown that every equation of the form

$$\dot{z} = z^3 + p_1(\tau)z^2 + p_2(\tau)z + p_3(\tau)$$

has exactly three periodic solutions when multiplicity is taken into account. If $p(\tau) \neq 0$, the same is true of the equation

$$\dot{z} = p(\tau)z^3 + p_1(\tau)z^2 + p_2(\tau)z + p_3(\tau),$$

simply by introducing the new independent variable

$$\tau_1 = \int_0^\tau p(u)du.$$

The same conclusion does not hold for (10)', for $\alpha(\tau)$ has zeros.

As in [6] we introduce the holomorphic function

$$q(c) = z(2\pi;0,c) - c,$$

where $z(t;t_o,c)$ is the solution of (10)' satisfying $z(t_o;t_o,c) = c$; q is defined on an open subset of $\mathbb{C}$ containing the origin. Solutions of (10)' of period 2π correspond to the zeros of q.

Having introduced a complex-valued dependent variable, we can exploit the properties of holomorphic functions. Thus (10)' has either finitely many or an open set of 2π-periodic solutions, the latter possibility corresponding to a centre of (2). By applying Rouché's theorem to q we find that, provided account is taken of multiplicity, the number of solutions of a given period, if finite, is preserved under perturbation of the equations (see [6] for details).

<u>Definition</u> The *multiplicity* of a periodic solution ϕ of (10)' is the multiplicity of $\phi(0)$ as a zero of q.

Let ϕ be a 2π-periodic solution of (10)' of multiplicity k. Perturbing the equation splits ϕ into a number of 2π-periodic solutions the sum of whose multiplicities is k. Some perturbations will split ϕ into exactly k simple periodic solutions, but it is a separate question whether this can be achieved while remaining in the class C_n of equations.

When discussing perturbations we are implicitly supposing that an appropriate topology has been defined on the class of all equations: it is usual to take the topology of uniform convergence on compact sets. However, we note that systems in C_n are determined by 2n+2 real coefficients (those of p and q in (2)). Now it is easily checked that the form (2) is preserved under a rotation of axes; by choosing a suitable rotation we can impose one linear condition on the coefficients of our equations, so that C_n can be identified with $\mathbb{R}^{2n+2}$ (including λ).

The real axis is invariant for (10)', so a solution is either real or has a non-vanishing imaginary part. If $\phi(t)$ is a periodic solution which is not real, then its complex conjugate $\overline{\phi}(t)$ is another such solution, and is of the same multiplicity as ϕ. Since $z \equiv 0$ is a solution, a real solution cannot take both positive and negative values. We shall see that $z = 0$ is a simple periodic solution (that is, of multiplicity 1) if and only if $\lambda \neq 0$; thus $z = 0$ is a multiple solution of (10)' when the origin is a fine focus of the original system (2).

In our case $\alpha(\tau)$ is a polynomial in $\cos\tau$ and $\sin\tau$ of even degree (namely 2n+2), while $\beta(\tau)$ is of degree n+1. When n is even the real periodic solutions of (10)' occur in pairs, for $-\phi(t+\pi)$ is such a solution whenever $\phi(t)$ is, and, being of opposite signs, these are distinct solutions. When n is odd, the right hand side of (10)' has least period π; we are therefore seeking 2-harmonics (periodic solutions whose least period is twice that of the equation). If ϕ is a solution of least period 2π, then $\phi(t)$ and $\phi(t+\pi)$ are distinct periodic solutions. The possibility of solutions of least period π remains.

3. The multiplicity of the origin

In this section we use the technique described in Section 3 of [6] to compute the multiplicity of the solution $z = 0$ of (10)'. We shall recover the Liapunov quantities L(j) described earlier.

We observe in passing that the multiplicity of a non-trivial periodic solution ϕ of (10)' is that of $\zeta = 0$ as a solution of

$$\dot{\zeta} = \alpha(\tau)\zeta^3 + (3\alpha(\tau)\phi(\tau) - \beta(\tau))\zeta^2 + (3\alpha(\tau)(\phi(\tau))^2 - 2\beta(\tau)\phi(\tau) - \lambda(n-1))\zeta.$$

To see this, we simply write $\zeta = z - \phi(\tau)$ in (10)'.

Returning to the solution $z = 0$ of (10)' we write $z(\tau;0,c)$ as a power series

$$a_1(\tau)c + \dots + a_n(\tau)c^n + \dots, \tag{12}$$

and substitute into the equation. We obtain a set of differential equations for the a_i; the initial conditions are

$$a_1(0) = 1, \quad a_j(0) = 0 \quad (j > 1).$$

The multiplicity is 1 if $a_1(2\pi) \neq 1$, and $k > 1$ if $a_1(2\pi) = 1$, $a_j(2\pi) = 0$ $(j=2,\dots,k-1)$ and $a_k(2\pi) \neq 0$.

The equation for a_1 is

$$\dot{a}_1 = -\lambda(n-1)a_1$$

whence
$$a_1(\tau) = \exp(-\lambda(n-1)\tau).$$

Hence $z = 0$ is simple if and only if $\lambda \neq 0$ (that is, the origin is not a fine focus of (2)). We shall now suppose that $\lambda = 0$, whence $a_1(\tau) \equiv 1$. Then

$$\left.\begin{aligned} \dot{a}_2 &= -\beta \\ \dot{a}_3 &= \alpha - 2a_2\beta, \\ \dot{a}_4 &= 3a_2\alpha - (a_2^2 + 2a_3)\beta, \\ &\dots\dots, \end{aligned}\right\} \tag{13}$$

where we omit the argument τ throughout. Hence, if $z = 0$ is not simple, we have, for example,

$$a_2(\tau) = -\int_0^\tau \beta(u)\,du$$

$$a_3(\tau) = \int_0^\tau \alpha(u)\,du + \left(\int_0^\tau \beta(u)\,du\right)^2$$

$$a_4(\tau) = -\int_0^\tau \{3\alpha(u)\int_0^u \beta(v)\,dv + 2\beta(u)\int_0^u \alpha(v)\,dv\}du - \left(\int_0^\tau \beta(u)\,du\right)^3.$$

We have already noted that $\beta(\tau)$ is a polynomial in $\cos\tau$ and $\sin\tau$ all of whose terms are of degree $n+1$. If n is even then all these terms necessarily have zero mean value; we conclude immediately that $a_2(2\pi) = 0$, whence the multiplicity of $z = 0$ cannot be 2. Indeed we can proceed to prove inductively that when n is even the multiplicity of $z = 0$ is odd.

We exploit the fact that for all ℓ and m, the indefinite integral of $\cos^\ell\theta \cdot \sin^m\theta$ is a sum of terms of the same form, except that when ℓ and m are both even, there is an additional term $\nu\theta$ (ν a constant), a so-called secular term. Suppose, then, that n is even and that $a_j(2\pi) = 0$ for $j=2,\dots,k-1$ with k even; we show that $a_k(2\pi) = 0$ necessarily. For $j=2,\dots,k-1$, the a_j have zero mean value, so do not contain secular terms. We see that a_j is a sum of terms $\cos^\ell\theta\ \sin^m\theta$ with $\ell+m$ of opposite parity to j. We prove this also by induction: suppose that the statement holds for suffices strictly less than j. Now a_j is of the form $\alpha A_j - \beta B_j$, where

$$A_j = \sum_{\substack{p+q+r=j \\ p,q,r\geq 1}} a_p a_q a_r \tag{14}$$

and
$$B_j = \sum_{\substack{p+q=j \\ p,q \geq 1}} a_p a_q. \tag{15}$$

If j is even, then in summation (14) either p,q,r are all even or two of them are odd. By the induction hypothesis, the terms of A_j are all of odd degree. In the summation (15), p and q are either both even or both odd, so that the terms of B_j are all of even degree. Since α and β are of even and odd degree, respectively, $\dot{a}_j$ is a sum of terms of the form $\cos^\ell\theta \sin^m\theta$ with $\ell+m$ odd; the same is true of a_j. By similar reasoning, when j is odd, $\dot{a}_j$ is the sum of terms of this form with $\ell+m$ even; since it has no secular term, the same is true of a_j. We can now deduce that a_k is a sum of terms $\cos^\ell\theta \sin^m\theta$ with $\ell+m$ odd. Hence $a_k(2\pi) = 0$, as required.

We observed in the last paragraph of Section 2 that when n is even the non-trivial periodic solutions of (10)' occur in pairs ; either as complex conjugates or as real solutions of opposite signs. We can summarise the situation for n even as follows:

Theorem. Suppose that n is even. The number of non-trivial 2π-periodic solutions of (10)' is even; the multiplicity of $z = 0$ is odd. If the multiplicity of $z = 0$ is $2k+1$, then a perturbation of the original system (2) produces at most k small-amplitude limit cycles.

Remark When n is odd all we can say is that if the multiplicity of $z = 0$ as a solution of (10)' is k, then a perturbation of (2) produces at most k-1 small-amplitude limit cycles (the solution $z = 0$ is, of course, preserved).

A knowledge of the multiplicity of the origin for (10)' gives us an upper bound for the number of small-amplitude limit cycles of (2). We must also address the question of whether this number can in fact be achieved by perturbations which remain in the class S_n' (or C_n if we refer to (10)'). We proceed as follows:

First of all we seek k such that for no equation in C_n can $z = 0$ be of multiplicity greater than k; suppose that E_o is an equation of the form (10)' for which $z=0$ has multiplicity exactly k - so that $a_j(2\pi) = 0$ for $j=2,\ldots,k-1$ and $a_k(2\pi) \neq 0$. We note that the origin attracts or repels neighbouring positive real solutions according as $a_k(2\pi) < 0$ or > 0. Next we perturb the coefficients of E_o so as to preserve the sign of $a_k(2\pi)$ and arrange that for the largest possible ℓ strictly less than k

$$a_\ell(2\pi) \neq 0, \quad a_j(2\pi) = 0 \quad (j=2,\ldots,\ell-1).$$

If n is even, k and ℓ are necessarily even; if n is odd $\ell = k-1$ is possible. For the perturbed equation (E_1, say), the origin then has multiplicity ℓ; thus exactly $k-\ell$ positive 2π-periodic solutions have been produced (counting multiplicity). We have to make sure that at least one of these is real and positive. One way of so doing is to arrange that a (2π) is much less than $a_\ell(2\pi)$ and of the opposite sign, so that the stability of origin with respect to the real axis is reversed. This process is repeated as many times as possible, so producing at most k small-amplitude limit cycles.

4. Particular cases

Before looking at the specific cases n = 2 and n = 3, we indicate how to calculate the quantities $a_i(2\pi)$ (essentially the 'Liapunov quantities'). It is convenient to write

$$\tilde{A}(\tau) = \int_0^\tau \alpha(u)\,du$$
$$\tilde{B}(\tau) = \int_0^\tau \beta(u)\,du.$$

From equations (13), we see that if $\lambda = 0$, $a_2(\tau) = -\tilde{B}(\tau)$.
Thus $\dot{a}_3 = \alpha + 2\tilde{B}\beta$, whence $a_3 = \tilde{A} + \tilde{B}^2$. Continuing,

$$\dot{a}_4 = -3\alpha\tilde{B} - 2\tilde{A}\beta - 3\tilde{B}^2\beta,$$

whence, by judicious integration by parts,

$$a_4 = -3\tilde{A}\tilde{B} - \tilde{B}^3 + \int\tilde{A}\beta.$$

The next step gives

$$\dot{a}_5 = 3\alpha\tilde{A} + 6\alpha\tilde{B}^2 + 7\tilde{A}\tilde{B}\beta + 3\tilde{B}^3\beta - 2\beta\int\tilde{A}\beta,$$

so that

$$a_5 = \frac{3}{2}\tilde{A}^2 + 6\tilde{A}\tilde{B}^2 + \frac{4}{3}\tilde{B}^4 - 5\int\tilde{A}\tilde{B}\beta - 2\int\beta\int\tilde{A}\beta.$$

It is now clear how to proceed.

If $\lambda = a_2(2\pi) = 0$, we have $a_3(2\pi) = \tilde{A}$ and

$$a_4(2\pi) = \int_0^{2\pi}\tilde{A}\beta.$$

If $\lambda = a_2(2\pi) = a_3(2\pi) = 0$, then

$$a_5(2\pi) = -3\int_0^{2\pi}\tilde{A}\tilde{B}\beta.$$

Thus we pick up the successive Liapunov quantities.

When n = 2, it is convenient, as explained in [2], to use the canonical form

$$\left.\begin{aligned} \dot{x} &= \lambda x + y + Ax^2 + (B+2D)xy + Cy^2 \\ y &= -x + \lambda y + Dx^2 + (E-2A)xy - Dy^2. \end{aligned}\right\} \qquad (16)$$

This form is obtained by a rotation of co-ordinates. The somewhat surprising forms of the coefficients are chosen so that the divergence of the right hand side takes a simple form (in this case $2\lambda+B+E$): a system with identically zero divergence has a centre.

From (14), we have

$$f(\theta) = A\cos^3\theta + (B+3D)\cos^2\theta\sin\theta + (C+E-2A)\cos\theta\sin^2\theta - D\sin^3\theta$$

and

$$g(\theta) = D\cos^3\theta + (E-3A)\cos^2\theta\sin\theta - (B+3D)\cos\theta\sin^2\theta - C\sin^3\theta .$$

After some tedious, but elementary, calculations, we deduce that $a_1(2\pi) = \exp(-2\pi\lambda)$ and if $\lambda = 0$, then

$$a_3(2\pi) = \frac{\pi}{4} B(A+C).$$

If $\lambda = a_3(2\pi) = 0$, we have that $a_5(2\pi)$ is a constant multiple of

$$-DE(A+C)(E-5(A+C))$$

and if $\lambda = a_3(2\pi) = a_5(2\pi) = 0$, $a_7(2\pi)$ is a constant multiple of

$$DE(A+C)^2((A+2C)C+D^2).$$

It is shown in [2] how to construct systems with four limit cycles from a knowledge of the quantities $a_3(2\pi), a_5(2\pi)$ and $a_7(2\pi)$. We remark that the signs here are opposite to those in [2]; this is because we are using θ as the independent variable and $\dot\theta < 0$.

Now for the case $n = 3$. Following [3] we use the canonical form

$$\left.\begin{aligned} \dot x &= \lambda x + y + Ax^3 + Bx^2y + (C-3H)xy^2 + Dy^3 \\ \dot y &= -x + \lambda y + Ex^3 + (F_1-C-3A)x^2y + (G-B)xy^2 + Hy^3 . \end{aligned}\right\} \qquad (17)$$

Then

$$f(\theta) = A\cos^4\theta + (B+E)\cos^3\theta\sin\theta + (F_1-3A-3H)\cos^2\theta\sin^2\theta + (D+G-B)\cos\theta\sin^3\theta + H\sin^4\theta$$

and

$$g(\theta) = E\cos^4\theta + (F_1-C-4A)\cos^3\theta\sin\theta + (G-2B)\cos^2\theta\sin^2\theta + (-C+4H)\cos\theta\sin^3\theta - D\sin^4\theta .$$

Again suppressing the unpleasant details, we have $a_1(2\pi) = \exp(-4\pi\lambda)$. If $\lambda = 0$, then

$$a_2(2\pi) = -\frac{3}{2}\pi F_1.$$

If $F_1 = 0$, we can rotate co-ordinates so that (17) retains its form but with $C = 0$. We then find that the successive Liapunov quantities are

$$G(A-H),\ AG(G+5(D+E)),\ AG(D+E)(B+4D+E),\ -AG(D+5)(A^2-DE).$$

It is shown in [3] that there are at most five small amplitude limit cycles about the origin, and classes of equations are displayed which have the full complement of five limit cycles.

We note that the results we have outlined for systems in S_n' $(n > 2)$ apply only to small-amplitude limit cycles encircling the origin. At other critical points, the system will not be of the required form.

Remark The technique described in the last section is very similar to the more well-known method described in Section 1. The point of approaching the problem in this way is that what happens becomes conceptually clearer, and, more significantly perhaps, the computations become substantially less onerous.

References

[1] N.N. BAUTIN, 'On the number of limit cycles which appear with the variation of coefficients from an equilibrium position of focus or centre type', Mat. Sb.N.S. 30 (72) (1952) 181-196 (in Russian); Amer. Math. Soc. Transl. No.100 (1954), and in Amer. Math. Soc. Transl. (1) 5 (1962) 396-413.

[2] T. BLOWS and N.G. LLOYD, 'On the number of limit cycles of quadratic systems', preprint.

[3] T. BLOWS and N.G. LLOYD, 'The number of limit cycles of certain cubic differential systems', preprint.

[4] W.A. COPPEL, 'A survey of quadratic systems', J. Differential Equations 2 (1966) 293-304.

[5] F. GRÜBBER and K.-D. WILLAMOWSKI, 'Liapunov approach to multiple Hopf bifurcation', J. Math. Anal. Appl. 71 (1979) 333-350.

[6] N.G. LLOYD, 'The number of periodic solutions of the equation $\dot{z} = z^N + p_1(t)z^{N-1} + \ldots + p_N(t)$', Proc. London Math. Soc. (3) 27 (1973) 667-700.

[7] N.G. LLOYD, 'On analytic differential equations', Proc. London Math. Soc. (3) 30 (1975) 430-444.

[8] N.G. LLOYD, 'On a class of differential equations of Riccati type', J. London Math. Soc. (2) 10 (1975) 1-10.

[9] V.V. NEMYTSKII and V.V. STEPANOV, Qualitative theory of differential equations (Princeton University Press, 1960).

[10] A.L. NETO, 'On the number of solutions of the equation $dx/dt = \sum_{j=0}^{n} a_j(t)x^j$, $0 \leq t \leq 1$, for which $x(0) = x(1)$', Invent. Math. 59 (1980) 67-76.

[11] I.G. PETROVSKII and E.M. LANDIS, 'On the number of limit cycles of the equation $dy/dx = P(x,y)/Q(x,y)$, where P and Q are polynomials of the second degree', Mat. Sb.N.S. 37 (79) (1955), 209-250 (in Russian); Amer. Math. Soc. Transl. (2) 10 (1958) 177-221.

[12] I.G. PETROVSKII and E.M. LANDIS, 'On the number of limit cycles of the equation $dy/dx = P(x,y)/Q(x,y)$, where P and Q are polynomials, Mat. Sb.N.S. 43 (85) (1957) 149-168 (in Russian); Amer. Math. Soc. Transl. (2) 14 (1960) 181-200.

[13] I.G. PETROVSKII and E.M. LANDIS, 'Corrections to the articles "On the number of limit cycles of the equations $dy/dx = P(x,y)/Q(x,y)$ where P and Q are polynomials of the second degree" and "On the number of limit cycles of the equation $dy/dx = P(x,y)/Q(x,y)$ where P and Q are polynomials"', Mat. Sb.N.S. 48 (90) (1959) 253-255 (in Russian).

[14] QIN YUANXUN, SHI SONGLING and CAI SUILIN, 'On limit cycles of planar quadratic systems', Sci. Sinica 25 (1982) 41-50.

[15] SHI SONGLING, 'A concrete example of the existence of four limit cycles for plane quadratic systems', Sci. Sinica 23 (1980) 153-158.

[16] SHI SONGLING, 'A method of constructing cycles without contact around a weak focus', J. Differential Equations 41 (1981) 301-312.

Acknowledgement

This paper was formulated in connection with my visit to the Symposium on Differential Equations held in Dundee during 1981-82. I am grateful to Professor Everitt for the invitation to participate and to the Science and Engineering Research Council for the financial support (GR/B/78083) which enabled me to attend. The work described in the paper has been supported by an SERC research grant (GR/B/57156).

Regular Growth and Zero-Tending Solutions

Jack W. Macki

This research was supported in part by the National Research Council of Canada under Operating Grant NRC-A-3053. This paper was written while the author was a guest of the Istituto di Matematica Applicata "G. Sansone" of the University of Florence, Italy. The author expresses his sincere appreciation to Professors P. Zecca and P. Zezza for their kind hospitality.

Our purpose is to present a survey, to some extent historical, of results on the existence of zero-tending solutions ($\lim_{x\to\infty} y(x) = 0$) of

$$(1) \qquad y'' + q(x)y = 0,$$

under the hypothesis

$$(\mathbf{A}) \qquad 0 < q(x) \text{ is continuous, nondecreasing, and } \lim_{x\to\infty} q(x) = +\infty$$

We are especially interested in tracing the evolution of the concept of _regular growth_.

Theorem 1 (Milloux)

_If_ $q(x)$ _is differentiable and satisfies_ (**A**) _then there exists a zero-tending solution of_ (1).

The earliest proof of this theorem is due to Milloux [12]. The assumption that $q(x)$ is differentiable is easy to remove (cf. Hartman [4]). Milloux also provided an example to show that one cannot conclude that <u>all</u> solutions go to zero. He in essence uses the step function

$$q(x) = (n+1)^2 \quad \text{for} \quad x \in [\pi \sum_1^n \frac{1}{k}, \pi \sum_1^{n+1} \frac{1}{k}] ,$$

with solution

$$y(x) = \cos[\pi q^{1/2}(x)(x-m_n)]$$

on $[m_n, m_{n+1}]$, where $m_n = \sum_1^n \frac{1}{k}$. Short proofs of this theorem can be found in the paper of Prodi [14], the book of Hartman [4] (p. 511), and the paper of Macki and Muldowney [8].

The question "when do <u>all</u> solutions tend to zero?", which is our primary interest in this paper, goes back at least to a paper by Wiman [19] in 1917 and an independent paper of Biernacki [3] in 1933. Notice that (1) is oscillatory under (**A**), so for a given solution $y(x)$ we may label the zeros $\{x_n\}$, $x_n \to \infty$, and the amplitudes $y_n = \max_{[x_n, x_{n+1}]} |y(x)| = y(z_n)$.

<u>Theorem 2</u> (Wiman)

<u>Assume that</u> $q(x)$ <u>is differentiable, that</u> (**A**) <u>holds, and that</u>

$$\lim_{\xi \to \infty} \frac{q(\xi + hq^{-1/2}(\xi))}{q(\xi)} = 1 , \tag{2}$$

$$\lim_{\xi \to \infty} \frac{q'(\xi + hq^{-1/2}(\xi))}{q'(\xi)} = 1, \quad \text{for} \quad |h| < k, \tag{3}$$

<u>$0 < k$ a fixed constant. Then all solutions tend to zero, in fact</u> $y_n = c[q(z_n)]^{-1/4+o(1)}$.

We call a function $q(x)$ satisfying (2) "Wiman regular". Wiman's assumption, then, is that $q(x)$ and $q'(x)$ are Wiman regular. Notice that the value of k is not important, since for example

$$\{q(\xi+2kq^{-1/2}(\xi))/q(\xi)] = \frac{q(\xi+2kq^{-1/2}(\xi))}{q(\xi+kq^{-1/2}(\xi))} \cdot \frac{q(\xi+kq^{-1/2}(\xi))}{q(\xi)} .$$

<u>Theorem 3</u> (Biernacki)

<u>If</u> $q(x)$ <u>is differentiable and satisfies</u> (**A**), <u>then</u>: $0 \leqslant q'(x)$ <u>and decreasing implies</u> $\lim_{x\to\infty} y(x) = 0$, <u>and</u> $\limsup_{x\to\infty} |y(x)|q^{1/2}(x) > 0$ <u>for all solutions</u> $y(x)$ <u>of</u> (1). <u>The same conclusion holds if</u> $0 \leqslant q'(x)$ <u>is increasing with</u>

$$\lim_{x\to\infty} \frac{q(x+q^{-1/2}(x))}{q(x)} = 1.$$

In both of the theorems above the term $\xi + kq^{-1/2}(\xi)$ is related to the location of the next zero of a solution $y_\xi(x)$ which vanishes at ξ, since by the Sturm Theorems this zero must occur before $\xi + \pi q^{-1/2}(\xi)$ when $q(x)$ is increasing. Biernacki's proof rests on his clever use of a theorem of Borel (which has an elegant proof due to Nevanlinna [13]). This theorem asserts that every increasing $f(x)$ is Wiman regular off a relatively small exceptional set.

Theorem (Borel-Nevanlinna-Biernacki)

If $0 < f(x)$ is increasing with $\lim_{x\to\infty} f(x) = +\infty$, then $f(x) \leq f(x+kf^{-1/2}(x)) \leq [1+o(1)]f(x)$ as $x\to+\infty$ for $0 \leq k \leq 4$ on the complement of an exceptional set E, where the measure of E satisfies $|E| \leq 4 + \int_0^\infty \exp(-u^2/2)dx$. ($E$ of course depends on $f(\cdot)$).

In both the Biernacki and Wiman results we see the first germ of a concept of "regularity" for $q(x)$. In 1935, G. Armellini [1] published the first paper to use the phrase "regular growth". It is interesting that Armellini was apparently unaware of the work of Wiman, Biernacki and Milloux, in fact he states that he discovered his result in the course of his researches in celestial mechanics. Armellini's definition of regular growth uses the notion of the density of a collection of intervals:

Let $E = \bigcup_1^\infty [a_k, b_k]$ be a family of intervals with $a_k < b_k < a_{k+1} \to +\infty$. Then the density of E is

$$\delta(E) = \limsup_{n\to\infty} \frac{1}{b_n} \sum_{k=1}^{n} (b_k - a_k) .$$

Definition 1 (Armellini)

$q(x)$ satisfying (**A**) is of irregular growth if for each $\varepsilon > 0$ there is a family $E(\varepsilon) = \bigcup_1^\infty [a_k, b_k]$ of intervals with $\delta(E) < \varepsilon$ and

$$\sum_{k=1}^{\infty} [\log q(a_{k+1}) - \log q(b_k)] \equiv \int_{E^c} d(\log q(x)) < \infty$$

Otherwise we say that $q(x)$ is of regular growth.

Intuitively, $\log q(x)$ does most of its (infinite) growth on arbitrarily small sets. The simplest example of irregular growth is a step function, e.g.

$$\log q(x) = F_n \text{ on } [n,n+1] \quad \text{with} \quad F_n \leq F_{n+1} \to +\infty .$$

Here we can take $E(\varepsilon) = \bigcup_{n=1}^{\infty} (n - \frac{1}{n^2}, n + \frac{1}{n^2})$, then $\delta(E) = 0$ and $\int_{E^c} d(\log q(x)) = 0$.

<u>Theorem 4</u> (Armellini-Tonelli)

<u>If</u> $q(x)$ <u>is differentiable, satisfies</u> (**A**) <u>and is of A-regular growth, then all solutions of</u> (1) <u>satisfy</u> $\lim_{x\to\infty} y(x) = 0$.

Armellini's proof depends on a careful analysis of the relation between the decreasing "amplitude" function

$$V(x) = \{\frac{(y')^2}{q(x)} + y^2\}$$

and the growth of $q(x)$ (notice that at a maximum $|y_n| = |y(z_n)|$, we have $V^{1/2}(z_n) = |y(z_n)|$, thus $y(x) \to 0$ if and only if $V(x) \to 0$). We will not give the proof, since we shall shortly present a more general result. Armellini's proof had a small gap which was quickly filled by Tonelli [17]. At the same time as Tonelli's paper (in fact in the same publication), Sansone [15] presented a completely independent proof of the Armellini-Tonelli Theorem, deducing it as a corollary to a stronger theorem. For some reason, Sansone chose not to mention his strongest result in his well-known text [16], choosing instead to state and prove the Armellini-Tonelli result stated above. He also mentions one corollary of his strong theorem, without at all describing the theorem. The result has been that this important theorem has been unknown for almost 50 years. As we shall show below, it to some extent provides a unifying element for many more recent studies of zero-tending solutions. In the following definition, it helps the intuition to think of $\{x_n\}$ as the zeros of an oscillatory solution of (1).

Definition 2.

$q(x)$ satisfying (**A**) is of S-irregular growth if there exists an unbounded sequence $0 < x_1 < x_2 < \ldots$ such that

(i) $\pi q^{-1/2}(x_{n+1}) \leq x_{n+1} - x_n \leq \pi q^{-1/2}(x_n)$,

(ii) $$\limsup_{n\to\infty} \frac{q(x_n)}{q(x_{n+1})} = 1$$

(iii) for all $0 < \rho < 1/2$, with $\Delta_k \equiv \pi q^{-1/2}(x_k)$,

$$\sum_{k=1}^{\infty} \left[\frac{q(x_k+\rho\Delta_k)}{q(x_k-\rho\Delta_k)} - 1\right] < +\infty$$

Remarks. 4(i) clearly implies that $x_{n+1} - x_n \leq x_n - x_{n-1}$, and $x_{n+1} - x_n \to 0$ as $n \to \infty$. 4(i)(ii) imply that $\limsup (x_{n+1}-x_n)/(x_n-x_{n+1}) = 1$. These three consequences of 4(i)(ii) were in fact used by Sansone [15] with 4(iii) to define irregular growth. Our conditions 4(i)(ii) are more restrictive, hence our class of functions of S-regular growth (for which all solutions tend to zero) is larger. Of course the above definition is rather impractical to use, but as we show below, it leads directly to some practical criteria.

Theorem 5

If $q(x)$ *satisfying* (**A**) *is of S-regular growth then all solutions of* (1) *tend to zero.*

Proof: See the Appendix.

Corollary 5.1 (Sansone [15], [16])

If $q(x)$ *is differentiable, satisfies* (**A**), *and for each sequence* $0 \leqslant x_1 < x_2 < \dots < x_n \to +\infty$ *satisfying* 4(i)(ii) *we have*

$$\sum_{k=1}^{\infty} (x_{n+1}-x_n) \frac{q'(\xi_n)}{q(\xi_n)} = +\infty \text{ , where}$$

$$q'(\xi_n)q^{-1}(\xi_n) \equiv \min_{[x_n, x_{n+1}]} q'(x)q^{-1}(x)$$

then $q(x)$ *is of S-regular growth (hence all solutions of* (1) *tend to zero*).

Corollary 5.2 (Sansone [15])

If $q(x)$ *is differentiable, satisfies* (**A**), *and* $q'(x) \geqslant a^2 > 0$, $\int^{\infty} q^{-1}(x)dx = +\infty$, *then all solutions of* (1) *tend to zero*.

We now show that Sansone's result to some extent unifies the various existing concepts of regular growth. In addition to the Armellini definition (which Sansone demonstrated yields a less general theorem), there are definitions due to Hartman [6] in 1959 and McShane [9] in 1966. The definition given by Hartman in his book [4] has an apparent typographical error - it appears it was intended to give the Armellini definition, which differs from the definition in [6].

Definition 3 (Hartman)

$q(x)$ satisfying (**A**) is said to be of H-irregular growth if for each $\varepsilon > 0$ there is an unbounded sequence $t_0 < t_1 < \dots$ so that if $E = \bigcup_0^\infty (t_{2k}, t_{2k+1})$, then

(i) $\int_E d(\log q(x)) < \infty$,

(ii) $\lim_{n\to\infty} (t_{2n+1} - t_{2n})\, q^{1/2}(t_{2n}) = \gamma \in (\pi - \varepsilon, \pi)$,

(iii) $\limsup_{n\to\infty} \dfrac{t_{2n} - t_{2n-1}}{t_{2n-1} - t_{2n-2}} < \varepsilon$.

Theorem 6 (Hartman [6])

If $q(x)$ satisfies (**A**) and is of H-regular growth, then all solutions of (1) tend to zero.

Definition 4 (M^c^Shane)

$q(x)$ satisfying (**A**) is of M-irregular growth if for each $\varepsilon > 0$ there are unbounded sequences $a_1 < b_1 < c_1 < a_2 < b_2 < c_2 < \dots$, such that

(i) $\int_{b_i}^{c_j} q^{1/2}(x)dx > \pi - \varepsilon$,

(ii) $\int_{c_j}^{a_{j+1}} q^{1/2}(x)dx < \varepsilon$,

(iii) $\int_{a_j}^{b_j} (b_j - x) q(x)dx < \varepsilon^2$,

(iv) $\sum [\log q(c_j) - \log q(b_j)] \equiv \int_{E^c} d(\log q(x)) < \infty$,

$(E = \bigcup_j (c_j, b_{j+1})$.

Theorem 7 (M^c^Shane [9])

If $q(x)$ satisfying (**A**) is of M-regular growth, then all solutions of (1) tend to zero.

Corollary 7.1

If $q(x)$ satisfies (**A**) and if there exists a constant K such that $q'(t)q^{-2}(t) \leq Kq'(s)q^{-2}(s)$ for all $t > s$ sufficiently large, then all solutions of (1) tend to zero.

Both McShane and Hartman show that their representative theorems include the Armellini-Tonelli result, and McShane gives an example to show that there is an M-regular function which is not A-regular. In fact the following theorem shows that Sansone's result includes Hartman's. We do not have examples to show that Sansone's result is more general.

Theorem 8

If $q(x)$ satisfies (**A**) and is either A-regular or H-regular, then $q(x)$ is S-regular. If $q(x)$ satisfies (**A**), is Wiman-regular and is M-regular, then $q(x)$ is S-regular.

Proof: Since Hartman and McShane have respectively shown that A-regularity implies both H-regularity and M-regularity, we need only show (a) H-regular implies S-regular, and (b) M-regular plus Wiman regular implies S-regular.

(a) Let $q(x)$ satisfy (**A**) and be S-irregular. Then there exists a sequence $\{x_n\}$ satisfying 4(i)(ii)(iii). We must show that for each $\varepsilon > 0$ there is a sequence $\{t_n\}_1^\infty$ such that

5(i)(ii)(iii) hold. 4(iii) implies that the infinite product

$$p = \prod_{k=1}^{\infty} \frac{q(x_n+\rho\Delta_n)}{q(x_n-\rho\Delta_n)}$$ converges with $p > 1$, for any

$0 < \rho < 1/2$. Then $\sum_{k=0}^{\infty} [\log q(x_n+\rho\Delta_n) - \log q(x_n-\rho\Delta_n)]$

converges to $\log p$. Thus 5(i) holds, if we define

$t_{2k} = x_k - \rho\Delta_k$, $t_{2k+1} = x_k + \rho\Delta_k$, $k = 0,1,2,\ldots$. Then the desired 5(ii) becomes $\lim_{n\to\infty} (t_{2n+1}-t_{2n})q^{1/2}(t_{2n}) =$

$\lim_{n\to\infty} 2\pi\rho\ q^{-1/2}(t_{2n})q^{1/2}(t_{2n}) = 2\pi\rho$. Given $\varepsilon > 0$ we can

clearly choose ρ $(0,1/2)$ so that $2\pi\rho$ $(\pi-\varepsilon,\pi)$.

Condition 5(iii) becomes

$$0 \leq \limsup_{n\to\infty} \frac{(x_n-\rho\Delta_n) - (x_{n-1}+\rho\Delta_{n-1})}{x_{n-1} + \rho\Delta_{n-1} - (x_{n-1}-\rho\Delta_{n-1})} =$$

$$= \limsup_{n\to\infty} \frac{(x_n-x_{n-1}) - \pi\rho[q^{-1/2}(x_{n-1}) + q^{-1/2}(x_n)]}{2\pi\rho q^{-1/2}(x_{n-1})} =$$

$$\leq \limsup_{n\to\infty} \frac{\pi - \pi\rho - q^{1/2}(x_{n-1})q^{-1/2}(x_n)\pi\rho}{2\pi\rho},$$

using 4(i),

$$\leq \{1 - \rho - \rho \limsup_{n\to\infty} q^{1/2}(x_{n-1})q^{-1/2}(x_n)\}/2\rho$$

$$= (1-2\rho)/\rho \quad \text{by } 4(ii).$$

Therefore we can choose ρ $(0,1/2)$ so that 5(iii) holds.

(b) Suppose that $q(x)$ is S-irregular. We choose

$$a_j = x_j - \Delta_j/2, \; b_j = x_j - \rho\Delta_j,$$
$$c_j = a_{j+1} = x_{j+1} - \Delta_{j+1}/2.$$

where $0 < \rho < 1/2$ is to be determined. Then for 6(i) we have

$$\int_{b_j}^{c_j} q^{1/2}(x)\,dx \geq$$

$$\geq q^{1/2}(x_j-\rho\Delta_j)[x_{j+1} - x_j + \rho\pi q^{-1/2}(x_j) - \pi q^{-1/2}(x_{j+1})/2]$$

$$\geq \pi q^{1/2}(x_j-\rho\Delta_j)[\rho q^{-1/2}(x_j) + q^{-1/2}(x_{j+1})/2] ,$$

where we have used 4(i). Therefore we can choose $0 < \rho < 1/2$ so as to obtain 6(i) <u>if</u> we can show $\lim_{j\to\infty} q(x_j)q^{-1}(x_j-\rho\Delta_j) = \lim_{j\to\infty} q(x_j)q^{-1}(x_{j+1}) = 1$. The first follows from the convergence of the series of nonnegative terms $\sum[q(x_j)q^{-1}(x_j-\rho\Delta_j) -1]$ which is dominated by the series in 4(iii). The second is just Wiman regularity (2), which has been assumed. Concerning 6(ii) we have $\int_{c_j}^{a_{j+1}} q^{1/2}(x)dx = 0$. Finally, 6(iii) becomes

$$\int_{a_j}^{b_j} (x_j-\rho\Delta_j-x)q(x)dx \leq q(x_j-\rho\Delta_j)[\tfrac{1}{2} - \rho]^2\Delta_j^2/2$$

$$\leq \pi^2 q(x_j-\rho\Delta_j)q^{-1}(x_j)[\tfrac{1}{2} - \rho]^2/2$$

and this last upper bound is less than ε if ρ is close to 1/2.

Remarks:

By the Borel-Nevanlinna-Biernacki result, the Wiman regularity assumption is true off a set of indices n_i for which $\sum_i (x_{n_i} - x_{n_i - 1}) < \infty$. This tempts one to hope that the assumption of Wiman regularity is not needed. It would be interesting to either prove that the Sansone and M^c^Shane conditions are equivalent, or produce an example of a function regular in the one sense but not the other.

Comparisons between some results on zero-tending solutions have been obtained by R. Wong [20]. A. Meir, D.W. Willett and J.S.W. Wong ([10],[11]) have obtained results which do not use regular growth - it would be interesting to see if their hypotheses imply S-regularity or M-regularity. Willett [18] has obtained a nice result which shows that a rate-of-growth condition on $q(x)$ will not suffice. For $q(x)$ not necessarily increasing, there are interesting results by D. Hinton [7] and Macki and Muldowney [8]. Finally, we mention a remarkable recent result of F.V. Atkinson [2] which should be compared with Theorem 2 above. His theorem states that the existence of a solution not tending to zero is equivalent to the existence of a second solution which goes to zero very rapidly.

Theorem 9 (Atkinson)

Let $q(x)$ *satisfy* (A). *Then there exists a solution of* (1) *with* $\limsup_{x\to\infty} |y(x)| > 0$ *if and only if there exists a solution* $y(x) = O(q^{-1/2}(x))$.

APPENDIX

Proof of Theorem 5

Let $y(x)$ be any solution of (1), with zeros $\{x_n\}$, $x_n \to +\infty$. We shall prove below that properties 4(i)(ii) hold always.

The function $V(x) = y^2(x) + [y'(x)]^2/q(x)$ is positive and decreasing, hence $\lim_{x\to\infty} V(x) = \alpha^2 \geq 0$ exists. Notice that at points z_n where $y'(z_n) \geq 0$, $|y(x)|$ is a maximum and $y_n \equiv |y(z_n)| = V^{1/2}(z_n)$. Thus $y(x) \to 0$ if and only if $\alpha = 0$.

We shall assume $\alpha > 0$ and conclude that the series in 4(iii) converges for any $0 < \rho < 1/2$. The convergence of this series is equivalent to the convergence of the infinite product

$\prod_1^\infty q(x_n+\rho\Delta_n)q^{-1}(x_n-\rho\Delta_n)$ which in turn is equivalent to

$$\sum_{n=1}^{\infty} [\log q(x_n+\rho\Delta_n) - \log q(x_n-\rho\Delta_n)] < \infty \quad . \tag{7}$$

We shall prove that 4(i)(ii) and (7) hold.

The inequalities 4(i) follow immediately from the Sturm Comparison Theorems. For 4(ii), assume that $\limsup_{n\to\infty} q(x_n)q^{-1}(x_{n+1}) = \ell < 1$. Then if we set $m^2 = \ell + \varepsilon < 1$ ($\varepsilon > 0$ sufficiently small) we have $q(x_n) < mq(x_{n+1})$ for n large, thus $q^{-1/2}(x_{n+1}) < mq^{-1/2}(x_n)$ for $n > N$. Then $\sum_N^\infty q^{-1/2}(x_n)$ will converge and by 4(i) so will $\sum_N^\infty (x_{n+1}-x_n)$. This is a contradiction, thus 4(ii) holds.

Turning to 4(iii), we first show that for any given $0 < \rho < 1/2$ we can choose $N(\rho)$ so large that

$$(8) \qquad z_{n-1} < x_n - \rho\Delta_n < x_n + \rho\Delta_n < z_n \ , \quad n > N,$$

where z_k is the first critical point to the right of x_k. Suppose not,then for some $0 < \rho < 1/2$ and some subsequence $\{x_{n_i}\}$ we would have one of

$$(9) \quad (a) \qquad x_{n_i} - z_{n_i-1} < \rho\pi q^{-1/2}(x_{n_i}) \qquad \text{or}$$

$$(b) \qquad z_{n_i} - x_{n_i} < \rho\pi q^{-1/2}(x_{n_i})$$

holding for $i = 1,2,\ldots$. Assume, to be specific, that 9(b) holds for all i (the proof when 9(a) holds is virtually identical). Writing (1) as

$$y'' + q(x_{n_i})y = [q(x_{n_i}) - q(x)]y, \ y(x_{n_i}) = 0,$$
$$y'(x_{n_i}) = q^{1/2}(x_{n_i})V^{1/2}(x_{n_i})$$

and using the appropriate Green's function, we see that

$$(10) \qquad y(x) = V^{1/2}(x_{n_i}) \sin q^{1/2}(x_{n_i})(x-x_{n_i}) -$$

$$- q^{-1/2}(x_{n_i}) \int_{x_{n_i}}^{x} y(x)[q(t)-q(x_{n_i})]\sin[q^{1/2}(x_{n_i})(x-t)]dt$$

for x $[x_{n_i}, z_{n_i}]$. Note that the integrand is nonnegative on this interval. Suppose, to be definite, that $y(x) > 0$ on $[x_{n_i}, z_{n_i}]$. Then (10) implies that

$$0 \leq y(x) \leq V^{1/2}(x_{n_i})\sin \rho\pi, \ x \ [x_{n_i}, z_{n_i}],$$
$$i = 1,2,\ldots,$$

and evaluating this at z_{n_i} we have (recall that V is decreasing)

$$V^{1/2}(x_{n_{i+1}}) \leq V^{1/2}(x_{n_i+1})$$

$$\leq V^{1/2}(z_{n_i}) \equiv |y(z_{n_i})| \leq V^{1/2}(x_{n_i})\sin \rho\pi.$$

This immediately implies that $\alpha = 0$, a contradiction. Thus (8) holds for all large n.

Given that (8) holds, we can again use a Green's function argument to show that

$$y(x) = V^{1/2}(x_n)\sin q^{1/2}(x_n)(x-x_n) -$$

$$- q^{-1/2}(x_n) \int_{x_n}^{x} y(t)[q(t) - q(x_n)] \sin[q^{1/2}(x_n)(x-t)]dt$$

for $n > N_\rho$, $x \quad [x_n - \rho\Delta_n, x_n + \rho\Delta_n]$. Because (8) holds, we can again conclude that

$$|y(x)| \leq V^{1/2}(x_n)\sin \rho\pi, \quad x \quad [x_n-\rho\Delta_n, x_n+\rho\Delta_n] \tag{11}$$

for n large. Since $V(x)$ decreases to α^2 as $x \to \infty$, for any $\varepsilon > 0$ we have $0 \leq V(x) - \alpha^2 < \varepsilon^2$ for x large. We choose $\varepsilon > 0$ so small that $\alpha^2\cos^2\rho\pi - \varepsilon^2\sin^2\rho\pi > 0$. Now $dV(x) = [V(x)-y^2(x)]d(\log q(x))$, so

$$V(x) - \alpha^2 = \int_x^\infty [V(t)-y^2(t)]d(\log q(t)) \quad \text{thus}$$

$$\varepsilon^2 > \int_x^\infty [\alpha^2-y^2(t)]d(\log q(t)) \quad \text{for } x \text{ large .} \tag{12}$$

Then, by (11)

$$\varepsilon^2 > \sum_{n>N_\rho} \int_{x_n-\rho\Delta_n}^{x_n+\rho\Delta_n} [\alpha^2-(\alpha^2+\varepsilon^2)\sin^2\rho\pi]\, d(\log q(t)) \; .$$

Now $\alpha^2 - (\alpha^2+\varepsilon^2)\sin^2\rho\pi = \alpha^2\cos^2\rho\pi - \varepsilon^2\sin^2\rho\pi$, so

$$\varepsilon^2[\alpha^2\cos^2\rho\pi - \varepsilon^2\sin^2\rho\pi] > \sum_{n>N_\rho} [\log q(x_n+\rho\Delta_n) - \log q(x_n-\rho\Delta_n)] \; ,$$

hence (7) holds.

Bibliography

1. G. Armellini, Sopra un'equazione differenziale della Dinamica, Rendiconti R. Accad. Naz. Lincei (6), 21 (1935), 111-116.

2. F.V. Atkinson, A stability problem with algebraic aspects, Proc. Royal Soc. Edinburgh, 78A (1978), 299-314.

3. M. Biernacki, Sur l'equation differentielle x" + A(t)x = 0, Prace Mat.-Fiz, 40 (1933), 163-171.

4. P. Hartman, Ordinary Differential Equations, Wiley, New York, 1964.

5. P. Hartman, On a theorem of Milloux, Amer. J. Math., 70 (1948), 395-399.

6. P. Hartman, The existence of large or small solutions of linear differential equations, Duke Math. J., 28 (1961), 421-430.

7. D. Hinton, Some stability conditions for y" + qy = 0, J. Math. Anal. Appl., 21 (1968), 126-132.

8. J. Macki and J. Muldowney, The asymptotic behaviour of solutions to linear systems of ordinary differential equations, Pac. J. Math., 33 (1970), 693-706.

9. E.J. McShane, On the solutions of the differential equation $y'' + p^2y = 0$, Proc. Amer. Math. Soc., 17 (1966), 55-61.

10. A. Meir, D.W. Willett and J.S.W. Wong, A stability condition for y" + p(x)y = 0, Mich. Math. J., 13 (1966), 169-170.

11. A. Meir, D.W. Willett and J.S.W. Wong, On the asymptotic behaviour of the solutions of $y'' + p(x)y = 0$, Mich. Math. J., 14 (1967), 47-52.

12. H. Milloux, Sur l'equation differentielle $x'' + A(t)x = 0$, Prace Mat.-Fiz., 41 (1934), 39-54.

13. R. Nevanlinna, Remarques sur les fonctions monotones, Bull. Sci. Math., 55 (1931), 140-144.

14. G. Prodi, Un'osservazione sugl'integrali dell'equazione $y'' + A(x)y = 0$ nel caso $A(x) \to \infty$, Atti Accad. Naz. Lincei Rend. Cl. Sci. Fis. Mat. Nat. (8), 8 (1950), 462-464.

15. G. Sansone, Sopra il comportamento asintotico delle soluzioni di un'equazione differenziale della dinamica, in Scritti Matematici offerti a Luigi Berzolari, Pavia, 1936, 385-403.

16. G. Sansone, Equazioni Differenziali nel Campo Reale, Vol. II, Zanichelli, vologna, 1948.

17. L. Tonelli, Estratto di lettera al prof. Giovanni Sansone, Scritti Matematici offerti a Luigi Berzolari, Pavia, 1936, 404-405.

18. D. Willett, On an example in second order linear ordinary differential equations, Proc. Amer. Math. Soc., 17 (1966), 1263-1266.

19. A. Wiman, Uber die reellen Losungen der linearen Differentialgleichungen zweiter Ordnung, Arkiv. for Mat. Astr. och Fys., 12 (1917), No. 14.

20. R. Wong, On a stability theorem of E.J. M[c]Shane, J. Math. Anal. Appl., 44 (1973), 215-217.

Asymptotic Distribution of the Eigenvalues of Non-Definite Sturm-Liouville Problems[1,2,3]

Angelo B. Mingarelli

1. The problem under consideration here is the derivation of asymptotic formulae for the *real* eigenvalues of the boundary problem

$$-y'' + q(t)y = \lambda r(t)y \tag{1.1}$$

where $\lambda \in C$ is a parameter and a non-trivial solution is required to satisfy

$$y(0)\cos\alpha - y'(0)\sin\alpha = 0 \tag{1.2}$$

$$y(b)\cos\beta + y'(b)\sin\beta = 0 \tag{1.3}$$

where $0 \le \alpha, \beta < \pi$. It is tacitly assumed that (1.1-2-3) gives rise to a *non-definite* Sturm-Liouville problem, that is, one which is neither left-definite nor right-definite in current terminology (cf., [3]).

---

1. This research is supported by the Natural Sciences and Engineering Research Council of Canada under grant UO 167.

2. This paper was presented at the Symposium on Differential Equations, Dundee, April 12, 1982.

3. Dedicated to the memory of a former teacher, Dr. R. Clive Moore.

Non-definite boundary problems of the above type may admit non-real eigenvalues. To see this consider the problem

$$-y'' + \exp(it)y = \lambda y \tag{1.3}$$

$$y(0) = y(\pi) = 0 \tag{1.4}$$

A glance at (1.3-4) shows that all of its eigenvalues must be strictly complex. Let $\lambda = \sigma + i\tau$ be one such non-real eigenvalue. Then a corresponding eigenfunction y will satisfy

$$-y'' + (\cos t - \alpha)y = i\,(\beta - \sin t)y. \tag{1.5}$$

Now (1.5) is of the form (1.1) with $q(t) = \cos t - \alpha$, $r(t) = \beta - \sin t$. Thus $\lambda = i$ is a non-real eigenvalue of (1.5-4). Since q,r will be assumed real, non-real eigenvalues will occur in complex-conjugate pairs.

Apart from an at most finite (though possibly empty) set of non-real eigenvalues the non-definite problem (1.1-2-3) has an infinity of real eigenvalues $\lambda_n^{\pm}$ with $\lambda_n^{\pm} \to \pm\infty$ as $n \to \infty$, (cf., [9] and [7]).

In the left-definite case, $q(t) \equiv 0$, of (1.1) it was essentially shown by Pleijel [8] that the positive eigenvalues λ_n^+ of (1.1) admit the asymptotic representation

$$\lambda_n^+ \sim n^2\pi^2 / \left(\int_0^b \sqrt{r_+(s)}\,ds\right)^2 \tag{1.6}$$

where $r_+(t) = \max\{r(t), 0\}$. Though Pleijel's form is not exactly (1.6), as it was derived for two-dimensional problems, its analog is (1.6) and we will show that (1.6) is satisfied under somewhat natural assumptions on q and r. The form (1.6) appears to have been conjectured by Jörgens [5, 5.16]. Since (1.1-2-3) also has negative eigenvalues $\lambda_n^- \to -\infty$ these will satisfy

$$\lambda_n^- \sim -n^2\pi^2/\left(\int_0^b \sqrt{r_-(s)}\, ds\right)^2 \tag{1.7}$$

where $r_-(t) = \max\{-r(t),0\}$, This can be easily seen upon replacing λ by $-\lambda$ and $r(t)$ by $-r(t)$ in (1.1) and in (1.6).

Let S^+ be a subinterval of $[0,b]$ on which $r(t) > 0$ a.e. (i.e. except for a set of zero Lebesgue measure). S^+ is said to be maximal if $r(t) > 0$ a.e. on an interval $S \supset S^+$ implies $S = S^+$. Similar definitions apply to those intervals S^- (S^0) of points in $[0,b]$ in which $r(t) < 0$ a.e. ($r(t) = 0$ a.e.).

Now let $[0,b] = S^+ \cup S^- \cup S^0$ where $S^\pm = \bigcup S_i^\pm$, $i = 1,2,\ldots,n^\pm$ is maximal, ($n^\pm \le \infty$). S^0 may or may not be empty. If $S^0 \ne \phi$ we suppose that $S^0 = \bigcup S_i^0$, $i = 1,2,\ldots,n^0$ ($\le \infty$) where each S_i^0 is maximal. If $S^0 = \phi$ we take it that $n^0 = 0$.

Since each non-trivial solution of the problem

$$-y'' + q(t)y = 0 \tag{1.8}$$

can admit an at most finite number of zeros in $[0,b]$ we will let n_1 (≥ 0) denote the maximum number of zeros that a given non-trivial solution may have in (0,b).

In the non-definite case it is necessary that r(t) changes its sign (cf., [7]) on some subset of positive Lebesgue measure and so we will assume that $n^+ \geq 1$, $n^- \geq 1$.

The following assumptions will be made at various parts in the sequel.

(H1) $q,r\colon [0,b] \to \mathbb{R}$; $q, r \in L(0,b)$, $\int_0^b |r| > 0$.

(H2) $0 \leq n^o < \infty$; $n^+ < \infty$, $n^- < \infty$.

(H3) i) The problem (1.1-2-3) admits a countable sequence or real eigenvalues $\lambda_n^+ > 0$, $\lambda_n^- < 0$ having no finite point of accumulation and such that $\lambda_n^{\pm} \to \pm\infty$ as $n \to \infty$.

ii) By relabeling the eigenvalues, if necessary, we assume that there exists $N_2 \geq 0$ such that for each $n \geq N_2$ the eigenfunctions $y(t, \lambda_n^{\pm})$ vanish precisely n times in (0,b).

In the non-definite case the eigenfunction corresponding to the smallest positive eigenvalue need not, necessarily, be of one sign in (0,b) as is the case where $r(t) > 0$ a.e. in (0,b). This is easily verified by setting $r(t) = 0$ a.e. on (0,c) where $0 < c < b$ and choosing q appropriately so that solutions of (1.8) actually vanish in (0,c) and a fortiori in (0,b). In any case the eigenvalues will be labeled according to the number of zeros of their corresponding eigenfunctions in (0,b), keeping in mind that the first positive eigenvalue may be denoted by λ_{N_2} where $N_2 > 0$ may be made as large as one wishes.

(H4) i) The boundary problem

$$-z'' + q(t)z = \mu\, r_+(t)z \tag{1.9}$$

where z satisfies (1.2-3) admits an infinite sequence of positive eigenvalues μ_n^+ for which $\mu_n^+ \to +\infty$ as $n \to \infty$ and having no finite point of accumulation.

ii) By relabeling these eigenvalues if necessary we will assume that corresponding eigenfunctions $z(t,\mu_n^+)$ vanish precisely n times in (0,b) for $n \geq N_3$.

Remarks. The first two assumptions in (H1) are natural, the third therein having been introduced so as to avoid cases where each $\lambda \in C$ may be an eigenvalue of (1.1-2-3).

The crucial assumption is (H2). Thus we will be essentially deriving (1.6) under the assumption that r(t) changes its sign finitely many times in the above-mentioned sense. Assumptions (H3) (i) and (ii) are verified in the event that $q,\ r \in C[0,b]$, (cf. [9]). It is most probable that (H3) is verified under the more general condition (H1). Recent work [2] and [4] has shown that (H4) is indeed verified as long as (H1) is satisfied, (cf., also [1, Chapter 8]).

For a given non-negative function $r_\varepsilon \in L(0,b)$, $\sqrt{r_\varepsilon} \in L(0,b)$, and $\int_0^b \sqrt{r_\varepsilon} \equiv K_\varepsilon$. Moreover if $r \in L(0,b)$, $r_\pm \in L(0,b)$ and $\int_0^b \sqrt{r_\pm} \equiv K_\pm$.

2. Lemma 2.1

Let (H1) and H4(i) be satisfied. Then the eigenvalues $\mu_n^+ \equiv \mu_n$ of (1.9) admit the asymptotic representation

$$\mu_n \sim n^2 \pi^2 / K_+^2 . \tag{2.1}$$

Proof. For $\varepsilon > 0$ define $r_\varepsilon(t) \equiv r_+(t) + \varepsilon$. Let $\mu_n(\varepsilon), z_n(t,\varepsilon)$ denote the eigenvalues and corresponding eigenfunctions of the problem

$$-z'' + q(t)z = \mu r_\varepsilon(t) z \tag{2.2}$$

where z satisfies (1.2-3). Then $q \in L(0,b)$, $r_\varepsilon \in L(0,b)$ and $r_\varepsilon > 0$ on $[0,b]$. Hence by Sturm-Liouville theory (cf., e.g. [1]) $\mu_n(\varepsilon)$ is real and for sufficiently large n, $\mu_n(\varepsilon) > 0$. Moreover if $\varepsilon_m \downarrow 0$ as $m \to \infty$, $\mu_n(\varepsilon_m) \le \mu_n(\varepsilon_{m+1})$ by variational principles [2, Chapter 6.2] since $r_{\varepsilon_m}(t) \ge r_{\varepsilon_{m+1}}(t)$. Thus since $r_\varepsilon(t) > r_+(t)$ we have $\mu_n(\varepsilon) \le \mu_n$. It now follows that if for $n \ge N_1$, $\mu_n(\varepsilon_1) > 0$ then $\mu_n(\varepsilon) > 0$ for each $0<\varepsilon<\varepsilon_1$ and $n \ge N_1$. Now as $\varepsilon_m \downarrow 0$ $\mu_n(\varepsilon_m) \to \mu_n$ by use of max-min characterization of $\mu_n(\varepsilon)$, [2, Chapter 6.1.4]. Use of the Lebesgue dominated convergence theorem first shows that $K_\varepsilon \to K_+$ as $\varepsilon \to 0^+$. Moreover $\mu_n(\varepsilon) \to \mu_n$ as $\varepsilon \to 0^+$. From Sturmian theory we also know that $\mu_n(\varepsilon)/n^2 \sim \pi^2/K_\varepsilon^2$ for each $\varepsilon > 0$ (cf., [6]) as $n \to \infty$. From this there follows (2.1).

Lemma 2.2

Let (H1-H4) be satisfied. Then for each n, $n \ge \max\{\gamma+N_2, 2+N_3\} \equiv N_4$ where $\gamma = n^- n_1 + n^0 n_1 + n^+ + 1$, we have

$$\mu^+_{n-2} \le \lambda^+_n \le \mu^+_{n+\gamma} \tag{2.3}$$

Proof. Let $t \in S^+_i$, some i, $1 \le i \le n^+$. Assume, on the contrary, that $\lambda^+_n > \mu^+_{n+\gamma}$ or that $\mu^+_n < \lambda^+_{n-\gamma}$ for some $n > N_4$. We will write $\mu^+_n \equiv \mu_n$, $\lambda^+_n \equiv \lambda_n$ for simplicity. On S^+_i, $r_+(t) = r(t) > 0$ a.e. and so $\mu_n r_+(t) - q(t) < \lambda_{n-\gamma} r(t) - q(t)$ a.e. We now apply Sturm's comparison theorem to the equations

$$y'' + (\lambda_{n-\gamma} r-q)y = 0 \tag{2.4}$$

$$z'' + (\mu_n r_+-q)z = 0 \tag{2.5}$$

over the interval S^+_i. It is then the case that between any two zeros of an eigenfunction $z_n(t,\mu_n)$ of (2.5) there is at least one zero of every non-trivial solution, i.e., an eigenfunction $y_{n-\gamma}(t,\lambda_{n-\gamma})$, of (2.4).

In the event that $n^0 > 0$ there follows from the above considerations that z_n has at most $n^0 n_1$ zeros in S^0(as z_n cannot vanish identically on subintervals). Similarly z_n can vanish at most $n^- n_1$ times in S^-. Since z_n has n zeros in $(0,b)$ (as $n > N_4$), z_n must vanish at least $n-n^0 n_1-n^- n_1$ times in S^+. Since S^+ is the union of n^+ distinct intervals, $y_{n-\gamma}$ must vanish at least $n-n^0 n_1-n^- n_1-n^+$ times in S^+ by a combinatorial argument and Sturm's comparison theorem. Thus $n-\gamma > n-n^0 n_1-n^- n_1-n^+$ because of H3(ii), i.e., $\gamma < n^0 n_1+n^- n_1+n^+$ which is impossible. This contradiction shows that for each $n > N_4$ we must have $\mu_n \ge \lambda_{n-\gamma}$ or $\mu_{n+\gamma} \ge \lambda_n$ for $n \ge N_4 - \gamma$.

In order to prove the left-side of (2.3) we assume on the contrary that there exists some $n > N_4$ such that $\lambda_n < \mu_{n-2}$. Since $r(t) \le r_+(t)$ a.e. on $[0,b]$ and $\lambda_n > 0$, $\lambda_n r(t) \le \lambda_n r_+(t)$ $< \mu_{n-2} r_+(t)$, i.e., $\lambda_n r(t) - q(t) < \mu_{n-2} r_+(t) - q(t)$ a.e. on $[0,b]$. An application of the Sturm comparison theorem to the equations

$$y'' + (\lambda_n r - q)y = 0$$

$$z'' + (\mu_{n-2} r_+ - q)z = 0$$

now shows that between any two zeros of $y_n(t,\lambda_n)$ there is at least one zero of $z_{n-2}(t,\mu_{n-2})$. This now forces Z_{n-2} to have at least n-1 zeros in $(0,b)$ and this is impossible as z_{n-2} vanishes precisely n-2 times in $(0,b)$ by H4(ii). This contradiction shows that for each $n > N_4$, $\lambda_n \ge \mu_{n-2}$ and this completes the proof of the lemma.

Combining these lemmata we obtain

Theorem 1

Let the assumptions (H1-H4) be satisfied for the boundary problem (1.1-2-3). Then the positive eigenvalues λ_n^+ of the problem admit the asymptotic representation (1.6) with a similar formula, (1.7), holding for the negative eigenvalues.

Proof. This follows from the above discussion upon dividing (2.3) by n^2 passing to the limit as $n \to \infty$ and applying lemma 2.1.

The author is grateful to Dr. C. Bennewitz (Uppsala) for having supplied the reference [8].

REFERENCES

[1] F.V. Atkinson, *Discrete and Continuous Boundary Problems*, Academic Press, New York, 1964.

[2] R. Courant and D. Hilbert, *Methods of Mathematical Physics*, Vol 1. Interscience, New York, 1953.

[3] W. N. Everitt, *On certain regular ordinary differential expressions and related differential operators*, in Spectral Theory of Differential Operators, I.W. Knowles and R.T. Lewis (eds) North-Holland, New York, 1981,115-167.

[4] W.N. Everitt, M.K. Kwong, A. Zettl, *Oscillation of eigenfunctions of weighted regular Sturm-Liouville problems*, To appear.

[5] K. Jörgens, *Spectral Theory of Second-Order Ordinary Differential Equations*, Matematisk Institut, Aarhus University, 1964.

[6] M.G. Kreĭn, *On the "indeterminate" case of the Sturm-Liouville boundary problem in the interval* $(0,\infty)$ Izvestiya Akad. Nauk. SSSR, Ser. Mat 16, (1954), 293-324, (Russian), (MR 14,558)

[7] A.B. Mingarelli, *Indefinite Sturm-Liouville problems* submitted.

[8] Å. Pleijel, *Sur la distribution des valeurs propres de problèmes régis par l'équation* $\Delta u + \lambda k(x,y) = 0$. Arkiv for Mat. Ast. o Fysik, 29B, (1942), 1-8.

[9] R.G.D. Richardson, *Contributions to the study of oscillation properties of the solutions of linear differential equations of the second order*, Amer. J. Math. 40, (1918), 283-316.

---

585 King-Edward Av, Ottawa, Ontario, Canada,
K1N 9B4.

SOME REMARKS ON THE ORDER OF AN ENTIRE FUNCTION ASSOCIATED WITH A SECOND ORDER DIFFERENTIAL EQUATION[1,2]

Angelo B. Mingarelli

1. This paper is motivated by a recent article of W.N. Everitt [2] wherein an in-depth study of the differential equation

$$-(p^{-1}(y'-ry))' - \overline{r}p^{-1}(y'-ry) + qy = \lambda\{i(\rho y)'+i\rho y'+wy\} \qquad (1.1)$$

was initiated. The following assumptions on the coefficients of (1.1) were made and will be assumed hereafter: $p : I \to \mathbb{R}$ and $p \in L_{loc}(I)$; $q : I \to \mathbb{R}$ and $q \in L_{loc}(I)$; $\rho : I \to \mathbb{R}$ and $\rho \in AC_{loc}(I)$; $w : I \to \mathbb{R}$ and $w \in L_{loc}(I)$; $r : I \to C$ and $r \in L_{loc}(I)$; $I = [a,b]$ is finite.

By a *solution* of (1.1) is meant a function $y \equiv y_\lambda^{[0]}(t,\lambda) \in AC_{loc}(I)$ with the property that $p^{-1}(y'-ry) + \lambda i\rho y \equiv y_\lambda^{[1]}(t,\lambda) \in AC_{loc}(I)$ and y satisfies (1.1) a.e. on I, [2, §2c]. It is known [2, p.120] that $y_\lambda^{[r]}(t,\cdot)$, $r = 0,1$, and the generalized Wronskian of two solutions y,z of (1.1) defined by

$$W(y,z)(t,\lambda) = (y_\lambda^{[0]}z_\lambda^{[1]} - y_\lambda^{[1]}z_\lambda^{[0]})(t,\lambda)$$

are all holomorphic functions on C for each t.

A boundary problem associated with (1.1) is defined once the boundary conditions [2, §3b, p.125]

---

1. This research is supported by the Natural Sciences and Engineering Research Council of Canada under Grant U0167.

2. Read before the Symposium on Differential Equations, Dundee, April 9, 1982.

$$y_\lambda^{[0]}(a)\cos\gamma - y_\lambda^{[1]}(a)\sin\alpha = 0$$

$$y_\lambda^{[0]}(b)\cos\delta + y_\lambda^{[1]}(b)\sin\delta = 0$$

are introduced $(\gamma,\delta\in[-\pi/2,\pi/2])$. Let ϕ,ψ be solutions of (1.1) satisfying

$$\phi_\lambda^{[0]}(a,\lambda) = \sin\gamma \quad \phi_\lambda^{[1]}(a,\lambda) = \cos\gamma \tag{1.2}$$

$$\psi_\lambda^{[0]}(b,\lambda) = -\sin\delta \quad \psi_\lambda^{[1]}(b,\lambda) = \cos\delta \tag{1.3}$$

and consider $W(\phi,\psi)(t,\lambda)$.

One of the questions which arises regards the order of W as an entire function [2, p.165 open problem (i)]. The aim of this paper will be to formulate some results in this direction. It should be noted that the order of $W(\phi,\psi)$ is the same as the order of $\phi_\lambda^{[0]}$ (or $\psi_\lambda^{[0]}$) which will in turn be shown to be the same as the order of $\phi_\lambda^{[1]}$ (or $\psi_\lambda^{[1]}$).

2. In the sequel will assume $r \equiv 0$ and rewrite (1.1) in the form

$$u' = pv \tag{2.1}$$

$$v' = -\lambda i(\rho u)' - \lambda i\rho u' + (q-\lambda w)u \tag{2.2}$$

The advantage of this approach lies in that we may allow p to vanish on sets of positive Lebesgue measure. When $p > 0$ a.e. we easily recover (1.1).

Theorem 2.1 Let u,v satisfy (2.1-2), $\lambda \in C$.

a) If for fixed $t \in I$

$$\int_0^t |\rho p| > 0 \tag{2.3}$$

then $u(t,\lambda)$ (and so $v(t,\lambda)$) is an entire function of λ of order not exceeding 1.

b) If for fixed $t \in I$,

$$\int_0^t |\rho p| = 0 \text{ and } \int_0^t (|p|+|w|) > 0 \tag{2.4}$$

then $u(t,\lambda)$ (and so $v(t,\lambda)$) is an entire function of λ of order not exceeding 1/2.

REMARK There is no sign restriction on any of p, q, ρ, w. Indeed these quantities need not even be real, (I am grateful to Professor F.V. Atkinson for pointing this out), as may be seen from the proofs which follow. The proof of part (a) and (b) may be obtained by adapting an argument of Atkinson [1, p.206] to our case. It should be noted that all that is required is that $u, v \in AC_{loc}(I)$ and that $u(t,\lambda)$, $v(t,\lambda)$ for fixed t, be entire functions of λ.

Proof of theorem 2.1. Let $\lambda \in C$, u, v as above. Then use of the relations (2.1-2) show that

$$\frac{d}{dt}\{|\lambda|u\bar{u}+v\bar{v}\} = 2|\lambda|p\ \mathrm{Re}(\bar{u}v) + 4\ \rho p|v|^2 \mathrm{Re}(\bar{\lambda}i) + 2\ \mathrm{Re}[(\bar{\lambda}i\rho'+q-\lambda w)\bar{u}v] \tag{2.5}$$

We now make use of the relation

$$2|uv| \le \{|\lambda||u|^2+|v|^2\}/\sqrt{|\lambda|}$$

in order to estimate the right-side of (2.5). Thus

$$\left|\frac{d}{dt}\{|\lambda|u\bar{u}+v\bar{v}\}\right| \le \sqrt{|\lambda|}\,|p|\{|\lambda||u|^2+|v|^2\} + 4|\rho p||\lambda|\{|\lambda||u|^2+|v|^2\}$$

$$+ |\bar{\lambda}i\rho'+q-\lambda w|\{|\lambda||u|^2+|v|^2\}/\sqrt{|\lambda|},$$

i.e.,

$$\le \{|\lambda||u|^2+|v|^2\}[|p|\sqrt{|\lambda|} + 4|\rho p||\lambda|+\sqrt{|\lambda|}\,|\rho'|$$

$$+ |q|/\sqrt{|\lambda|}+\sqrt{|\lambda|}\,|w|].$$

There now follows

$$\left|\frac{d}{dt}\log\{|\lambda||u|^2+|v|^2\}\right| \le \sqrt{|\lambda|}[|p|+|w|+|\rho'|] + 4|\lambda||\rho p| + |q|/\sqrt{|\lambda|} \qquad (2.6)$$

A simple observation now shows that, whenever (2.3) is satisfied, integration of (2.6) and exponentiation of the result will imply

$$|\lambda||u|^2 + |v|^2 = 0\,\{\exp(c|\lambda|)\} \qquad (2.7)$$

where $c > 0$ is a constant. Similarly whenever (2.4) is satisfied we must have $|\rho p| = 0$ a.e. on I and so the above observation applied to (2.6) will now yield

$$|\lambda||u|^2 + |v|^2 = 0\,\{\exp(c\sqrt{|\lambda|})\} \qquad (2.8)$$

where $c > 0$ is a constant and $|\lambda|$ sufficiently large.

Thus whenever (2.3) is satisfied (2.7) shows that each of u and v are of order not exceeding 1. Similarly whenever (2.4) is satisfied (2.8) implies that each of u and v are of order not exceeding 1/2.

REMARKS There is no loss of generality in assuming that $r \equiv 0$ in the above as if $r \not\equiv 0$ (2.1-2) may be modified so as to read

$$u' = ru + pv$$
$$v' = -\lambda i(\rho u)' - \lambda i\rho u' + (q-\lambda w)u - \bar{r}v.$$

The proof of theorem 2.1 then proceeds with straightforward changes to yield the same conclusions.

It may happen that p, q and w vanish on sets of positive Lebesgue measure in such a way that the problem (2.1-2) admits only finitely many eigenvalues, (cf. [1, §8.1]). In this case W is of order 0. Apart from these exceptional cases we will show below that under some natural assumptions the order of W is precisely 1, (1/2), if (2.3),((2.4)), is satisfied. This is supported by the examples described in [2, §4.11].

Theorem 2.2

Assume that the coefficients p, q, w, ρ are chosen so that (1.1) admits an infinity of eigenvalues λ_n.

a) If (2.3) is satisfied and $|\lambda_n| = 0(n)$ for all n sufficiently large, then $W(\phi,\psi)$ is precisely of order 1.

b) If (2.4) is satisfied and $|\lambda_n| = 0(n^2)$ for all n sufficiently large, then $W(\phi,\psi)$ is precisely of order 1/2.

Proof. We will only prove (a) as the proof of (b) is similar and so will be omitted.

Assume, if possible, that the order of $W(\phi,\psi)$ as a function of λ is of order $1-\delta$, $0 < \delta < 1$. The theory of functions of finite order then implies that

$$\sum_{|\lambda_r|\neq 0} |\lambda_r|^{-(1-\delta)-\varepsilon} < \infty$$

for each $\varepsilon > 0$, as the λ_r are the zeros of $W(\phi,\psi)$. Thus let $\varepsilon = \delta/2$. Then

$$\sum_{|\lambda_r|\neq 0} |\lambda_r|^{-1+\frac{\delta}{2}} < \infty$$

Now for r large, $|\lambda_r| < Ar$ and so $|\lambda_r|^{1-\frac{\delta}{2}} < (Ar)^{1-\frac{\delta}{2}}$

i.e., $$\sum |\lambda_r|^{-1+\frac{\delta}{2}} > A^{-1+\frac{\delta}{2}} \sum r^{-1+\frac{\delta}{2}}.$$

However the series on the right diverges by comparison with the harmonic series and so the series on the left must also diverge. This is a contradiction and so $W(\phi,\psi)$ must be of order 1.

The author is grateful to Professor F.V. Atkinson, Professor W.N. Everitt and Professor C. Bennewitz for some helpful discussions.

References

[1] F.V. Atkinson, *Discrete and Continuous Boundary Problems* Academic Press, New York, 1964.

[2] W.N. Everitt, *On certain regular ordinary differential expressions and related differential operators*, in Spectral Theory of Differential Operators, I.W. Knowles and R.T. Lewis (eds), North-Holland, New York, 1981, 115-165.

---

585 King-Edward Av. Ottawa, Ontario, Canada, K1N 9B4.

Steepest Descent for General Systems of Linear Differential Equations in Hilbert Space

J.W. Neuberger

1. Introduction

Many boundary value problems for linear differential equations may be posed as the problem of finding x so that

(*) $Tx = g$

where T is a continuous linear transformation from a real Hilbert space H to a real Hilbert space K and $g \in K$. A strategy for finding $x \in H$ satisfying (*) is the following: Find $x \in H$ so that $\phi(x)$ is minimum where

$$\phi(x) = \|Tx - g\|^2/2 \quad \text{for all } x \in H.$$

For $x \in H$ the Fréchet derivative $\phi'(x)$ satisfies

$$\phi'(x)h = \langle Th, Tx - g\rangle = \langle h, T^*(Tx - g)\rangle.$$

The element $T^*(Tx - g)$ is the gradient of ϕ at x and is denoted by $(\nabla\phi)(x)$.

Two related steepest descent schemes are

(I) Pick $x_0 \in H$ and define inductively

$x_{n+1} = x_n - \delta_n(\nabla\phi)(x_n)$, $n=0,1,2,...$

where δ_n is the number δ so that $\phi(x_n - \delta(\nabla\phi)(x_n))$ is minimum, i.e., $\delta = \|T^*(Tx - g)\|^2/\|TT^*(Tx - g)\|^2$. (The process of constructing $x_0, x_1, \ldots$ is understood to terminate at x_n if $(\nabla\phi)(x_n) = 0$).

(II) Pick $w \in H$ and define $z: [0,\infty) \to H$ such that $z(0) = w$ and $z'(t) = -(\nabla\phi)(x(t))$, $t \geq 0$.

We have in (I) and (II) discrete and continuous steepest descent processes.

It is mentioned in [4] that the idea of steepest descent goes back to Cauchy, [3]. Over the years many authors have made contributions to the subject. In the course of this paper we mention only a few authors - those whose work seems particularly relevant to our present development. References found in the listed papers authored or co-authored by Nashed [11], [12], [13], [14], [15], [16] are particularly extensive.

It is our purpose here to give a discussion of the use of steepest descent in connection with boundary value problems for linear systems of differential equations. Of particular interest is the choice of norms which lead to well-posed problems for numerical analysis. Some connections with Dirichlet spaces and finite dimensional potential theory [2] are indicated (see also [21]). We consider mainly the continuous steepest descent process II.

To begin we indicate a theorem which asserts that if there is a solution to (*), then a solution may be calculated by a steepest descent process. Such a theorem may be regarded as converting abstract existence results into algorithms for finding a solution.

Theorem 1. If there is a solution to (*) and z is defined as in (II), then

$$x \equiv \lim_{t\to\infty} z(t) \text{ exists}$$

and is a solution to (*).

This theorem may be derived from ([12], p. 386). For completeness we give an argument. In [5] there is an analogous result for a modification of process (I).

Proof of Theorem 1. From a standard variation of parameters formula

$$z(t) = e^{-tT^*T} z(0) + \int_0^t e^{-(t-j)T^*T} T^*g, \ t \geq 0$$

where $j(t) = t$, $t \in R$.

Suppose that for some $x \in H$, $Tx = g$. Then

$$z(t) = e^{-tT^*T} z(0) + \int_0^t e^{-(t-j)T^*T} T^*Tx$$

$$= e^{-tT^*T} z(0) + x - e^{-tT^*T} x, \quad t \geq 0 .$$

Now for any $v \in H$, $\lim_{t\to\infty} e^{-T^*T} v$ exists and equals the orthogonal projection of v onto $N(T^*T) = N(T)$ (cf, [25]). Hence $u \equiv \lim_{t\to\infty} z(t)$ exists and so

$$Tu = T(\lim_{t\to\infty} e^{-tT^*T} (z(0) - x)) + Tx = g$$

since the above limit is in $N(T)$.

In the application of the above theorem to differential equations certain orthogonal projections have a central role. To introduce these projections we examine a simple problem from perhaps a somewhat new point of view: We compute D^* where D: $H^1(R) \to L_2(R)$ is defined by $Dy = y'$, $y \in H^1(R)$ (Our general reference for Sobolev spaces is [1]). To this end denote $L_2(R) \times L_2(R)$ by H and denote by H' the subspace of H consisting of all $(y,y') \in H$. For $s \in H^1(R)$, $r \in K \equiv L_2(R)$,

$$\langle Ds, r\rangle_K = \langle s', r\rangle_K = \langle \binom{s}{s'}, \binom{0}{r} \rangle_H = \langle P\binom{s}{s'}, \binom{0}{r} \rangle_H$$

$$= \langle \binom{s}{s'}, P\binom{0}{r} \rangle_H = \langle s, u\rangle_{H^1}$$

where $u \equiv \pi P\binom{0}{r}$, $(\pi\binom{f}{g} \equiv f$, $\binom{f}{g} \in H)$ and P denotes the orthogonal projection of H onto H'. A formula for P is found in [18].

The use of such projections P are central to this author's study of differential equations as found in [18], [19], [20], [21], [22] for example. Roots of the theory are indicated in [17].

Next is presented an abstract theorem which follows from Theorem 1 and introduces some additional structure which will then be used in some concrete problems.

For this theorem denote by H and K two real Hilbert spaces, by H_0', H' closed subspaces of H so that $H_0' \subset H'$. Denote by g an element of K and by w a member of H'. Finally denote by A a member of $L(H,K)$ and consider the following problem:

Find $u \in H'$ such that $Au = g$ and $u - w \in R(P)$.

Of these last two expressions the first is an abstract expression of a linear differential equation and the second is a representation of boundary conditions. Examples are given in [18], [19], [21] on how to choose H, K, H', H_0', A, g, and w in concrete problems.

<u>Theorem 2</u>. Suppose that $w \in H'$ and there is $x \in H'$ such that $Ax = g$ and $x - w \in R(P)$. Suppose also that $z\colon\ [0,\infty) \to H$ so that $z(0) = w$, $z'(t) = -PA^*(Az(t) - g)$, $t \geq 0$. Then $u \equiv \lim_{t\to\infty} z(t)$ exists and satisfies

$$u \in H',\ Au = g,\ u - w \in R(P).$$

The following connects Theorem 2 with a variational problem:

Define ϕ: $H_0' \to R$ by

$$\phi(x) = \|A(x + w) - g\|^2/2,\ x \in H_0'.$$

<u>Theorem 3</u>. $(\nabla\phi)(x) = PA^*(A(x + w) - g)$, $x \in H_0'$.

<u>Proof of Theorem 3</u>. For x, $h \in H_0'$,

$$\phi(x + h) - \phi(x) = [\|A(x + w + h) - g\|^2 - \|A(x + w) - g\|^2]/2$$

$$= \langle Ah, A(x + w) - g\rangle + \langle Ah, Ah\rangle /2$$

so that

$$\phi'(x)h = \langle h, A^*(A(x + w) - g)\rangle$$

$$= \langle h, PA^*(A(x + w) - g)\rangle .$$

Since $PA^*(A(x + w) - g) \in H_0'$ it must be that

$$(\nabla\phi)(x) = PA^*(A(x + w) - g).$$

Proof of Theorem 2. Suppose that z is as in statement of Theorem 2. Denote AP by T. Then since $((I - P)z)'(t) = 0$, $t \geq 0$ and $z(0) = w$, it follows that $(I - P) z(t) = (I - P)w$, $t \geq 0$. Hence

$$\begin{aligned} z'(t) &= -PA^*(Az(t) - g) \\ &= -PA^*[A(Pz(t)) + (I - P)w - g] \\ &= -T^*(Tx(t) - h), \quad t \geq 0 \end{aligned}$$

where $h \equiv g - A(I - P)w$.

By hypothesis, there is $x \in H'$ such that $Ax = g$, $x - w \in R(P)$. Hence $Tx = APx + A(I - P)x - A(I - P)x = Ax - A(I - P)w = g - A(I - P)w = h$. Therefore by Theorem 1,

$$u \equiv \lim_{t\to\infty} z(t) \text{ exists and satisfies } Tu = h.$$

Now $Au = A(pu + (I - P)u) = Tu + A(I - P)u$. Since $u - w = \lim_{t\to\infty} (z(t) - w)$ and $z(t) - w \in R(p)$, $t \geq 0$, it follows that $u - w \in R(P)$. Hence $Au = Tu + A(I - P)w = h + A(I - P)w = g$. This completes a proof of Theorem 2.

The preceding structure gives a way to place steepest descent in Sobolev spaces. For example, the space H might be $L_2[0,1] \times L_2[0,1]$ and H' the subspace consisting of all elements (y,y') of H (derivatives in the L_2 sense, cf [1]). H' is clearly equivalent to the Sobolev space H^1. For problems based on a region Ω of R^2, H may be $L_2(\Omega)^3$ and H' the subspace of all elements (u, u_1, u_2) of H (u_1 partial derivative with respect to the first argument of u, etc.). Here H' is equivalent to the Sobolev space H^1 of Ω. H_0 might be all members of H' which are zero in $\delta\Omega$ in the trace sense. Concrete examples are found in [18], [19], [20], [21]. Various finite dimensional interpretations are indicated in the following section.

Minimization of functionals ϕ as above over all functions in H_0' is essentially minimization relative to a Sobolev norm. An example given later indicates important advantages of using a Sobolev norm rather than an L_2 norm. In [10] minimizations related to norms similar to Sobolev norms are considered. The effect of change of underlying norms is discussed in ([23], p. 245). The choice of proper norm for a minimization may be considered part of the topic of regularization (cf [13]).

Continuous steepest descent is studied in [24] and [12] as well as other authors mentioned in this last reference and [16]. Discrete steepest descent has been more widely considered. It is widely used in numerical analysis and it is the basis for many generalizations such as conjugate gradient methods (cf [7], [4], [23], [5], [11]) as well as many works in the field of operations research.

2. Finite Dimensional Problems

Here we consider some finite dimensional settings to which some of the preceding developments apply. The main object is to provide a setting for numerical analysis of a wide variety of linear systems. Such systems may be of interest for themselves or they may be considered as difference approximations to differential equations.

Suppose that each of m and n is a positive integer, Ω is a bounded region in R^m and $\delta > 0$. Denote by G'' a rectangular grid with even spacing δ such that the convex hull of G'' contains $\overline{\Omega}$. Denote $G'' \cap \overline{\Omega}$ by G and denote by K the vector space of all R^n-valued on G. Call an m-cube C minimal if (1) each edge of C has length δ and (2) each vertex of C is in G. Assume that every point of G is a vertex of at least one minimal m-cube. Denote by $\hat{G}$ the set of all centers of minimal m-cubes and denote by $\hat{K}$ the vector space of all R^n-valued functions on $\hat{G}$. If $u \in K$ define D_0: $K \to \hat{K}$ by

$$(D_0u)(p) = 2^{-m} \sum_{\substack{\varepsilon_j = \pm\delta/2 \\ j=1,\dots,m}} u(p+\varepsilon_1 e_1 + \dots + \varepsilon_m e_m)$$

and for i=1,...,m define $D_i: K \to \hat{K}$ by

$$(D_i u)(p) = (2^{-m+1}\delta^{-1}) \sum_{\substack{\varepsilon_j = \pm\delta/2 \\ j=1,\dots,m}} u(p+\varepsilon_1 e_1 + \dots + \varepsilon_m e_m)\,\mathrm{sgn}(\varepsilon_i) \quad .$$

Define D: $K \to \hat{K}^{m+1}$ so that if $u \in K$,

$$Du = \begin{pmatrix} D_0 u \\ D_1 u \\ \cdot \\ \cdot \\ D_m u \end{pmatrix}$$

for $u \in K$, $\|u\| \equiv (\sum_{p\in G} \|u(p)\|)^{\frac{1}{2}}$ and for $v \in \hat{K}$, $\|v\| \equiv (\sum_{q\in \hat{G}} \|v(q)\|^2)^{\frac{1}{2}}$

where $\|u(p)\|$, $\|v(q)\|$ denote Euclidean norms of u(p) and v(q) respectively, $p \in G$, $q \in \hat{G}$. Denote K^{m+1} by H and denote the range of D by H′. Denote by $B_1,\dots,B_n$ subsets of G and denote by K_0 the subspace of K consisting of all $u = (u^1,\dots,u^n) \in K$ such that if $i \in \{1,\dots,n\}$ then

$$u^i(p) = 0 \quad \text{if } p \in B_i.$$

Denote by π_0 the orthogonal projection of K onto K_0. Denote the image of K_0 under D by $H_0{}'$ and denote by P the orthogonal projection of H onto $H_0{}'$.

Theorem 4. If $E \equiv \pi_0 D^* D\big|_{K_0}$, then

$$P = DE^{-1}\pi_0 D^* \quad .$$

Proof. P is idempotent, symmetric, $R(P) \subset H_0{}'$ and $Pu = u$ if $u \in H_0{}'$. This characterizes the orthogonal of H onto $H_0{}'$.

We use this structure in the next section to make a comparison of gradients obtained using this structure and a more conventional structure.

3. Two Gradients

We use the notation of the previous section. Suppose $A \in L(H,\hat{K})$, $g \in \hat{K}$, $r \in K$. We compare two steepest descent schemes for solving for $u \in K$ such that

$$ADu = g, \quad Du - Dr \in R(P) .$$

For the first scheme, define ϕ: $H_0{}' \to R$ so that

$$\phi(x) = \|A(x + Dr) - g\|^2/2, \quad v \in H_0{}'$$

and for the second scheme define α: $K_0 \to R$ so that

$$\alpha(y) = \|AD(y + r) - g\|^2/2, \quad g \in K_0.$$

Denote by π the projection of H onto the first component of H. Hence if $u = (u^0, u^1, \ldots, u^m) \in H$, $\pi u = u^0$. Denote Dr by w.

Theorem 5. If $y \in K_0$.

$$\pi(\nabla\phi)(Dy) = E^{-1}(\nabla\alpha)(y) .$$

Proof. From Theorem 3,

$$(\nabla\phi)(x) = PA^*(A(x + w) - g) .$$

Using Theorem 4,

$$(\nabla\phi)(x) = DE^{-1}\pi_0 D^*A^*(A(x + w) - g)$$

and so

$$\pi(\nabla\phi)(Dy) = E^{-1}\pi_0 D^*A^*(AD(y + r) - g) .$$

As in the proof of Theorem 3,

$$\begin{aligned}\alpha'(y)k &= \langle ADk, AD(y + k) - g\rangle \\ &= \langle k, (AD)^*(AD(y + k) - g)\rangle \\ &= \langle k, \pi_0 D^*A^*(AD(y + k) - g)\rangle, \quad y,k \in K_0.\end{aligned}$$

Hence $(\nabla\alpha)(y) = \pi_0 D^*A^*(AD(y + k) - g)$ and the theorem follows.

Looking at the above in a slightly different way, if we regard α as being a function from K_0 to R with a new norm, $\|\ \|_D$ on K_0 define as $\|x\|_D = \|Dx\|_H$, then the gradient $\nabla_D\alpha$ calculated according to this new norm satisfies

$$(\nabla_D\alpha)(y) = \pi(\nabla\phi)(Dy), \quad y \in K_0.$$

To see this we start as above,

$$\begin{aligned}\alpha'(y)k &= \langle ADk, AD(y + k) - g\rangle \\ &= \langle Dk, A^*AD(y + k) - g\rangle \\ &= \langle PDk, A^*AD(y + k) - g\rangle \\ &= \langle Dk, PA^*(AD(y + k) - g)\rangle \\ &= \langle Dk, DE^{-1}\pi_0 D^*A^*(AD(y + k) - g)\rangle \\ &= \langle k, E^{-1}\pi_0(AD)^*(AD(y + k) - g)\rangle_D\end{aligned}$$

$$\begin{aligned}\text{so } (\nabla_D\alpha)(y) &= E^{-1}\pi_0(AD)^*(AD(y + k) - g) \\ &= \pi(\nabla\phi)(Dy)\ .\end{aligned}$$

4. Influence of Norm on Rate of Descent

In this section we are concerned with discrete steepest descent, i.e., that of type I in our introductory remarks. We choose discrete steepest descent to illustrate the effect of choice of norm since discrete steepest descent is more closely linked to computational practice than is continuous steepest descent.

Return briefly to a general setting in which each of H and K is a real Hilbert space, $T \in L(H,K)$, $T \neq 0$, $g \in K$ and define ϕ: $H \to R$ so that

$$\phi(x) = \|Tx - g\|^2/2 .$$

Then $(\nabla\phi)(x) = T^*(Tx - g)$, $x \in H$.

A way to measure the efficiency of a discrete steepest descent process is to consider rations

$$\phi(x - \delta(\nabla\phi)(x))/\phi(x)$$

defined for all $x \in H$ so that $\phi(x) \neq 0$ where δ is chosen optimally, i.e., $\delta = \|(\nabla\phi)(x)\|^2/\|T\nabla\phi(x)\|^2$.

Lemma 1. If $x \in H$ and $\phi(x) \neq 0$, then

$$\phi(x - \delta(\nabla\phi)(x))/\phi(x) = 1 - (\langle My,y\rangle/(\|y\|\|My\|))^2$$

where $y = Tx - g$ and $M \equiv TT^*$ and δ is chosen optimally.

This is an easy calculation based upon the fact that for δ optimally chosen,

$$\begin{aligned}\phi(x - \delta(\nabla\phi)(x)) &= \|Tx - g\|^2 - 2\delta\langle Tx - g, T(\nabla\phi)(x)\rangle \\ &\qquad + \delta^2\|T(\nabla\phi)(x)\|^2 \\ &= \|Tx - g\|^2 - \|T^*(Tx - g)\|^4/\|TT^*(Tx - g)\|^2 \\ &= \|y\|^2 - \|T^*y\|^4/\|TT^*y\|^2 \\ &= \|y\|^2 - (\langle My,y\rangle/\|My\|)^2, \quad y \equiv Tx - g .\end{aligned}$$

Lemma 2. Suppose $M \in L(K,K)$, $0 < \lambda < 1$ and $Mv = \lambda v$, $v \neq 0$. Suppose furthermore that $Mx = x$ if $\langle x,v\rangle = 0$. Then

$$1 - (\langle My,y\rangle/(\|y\|\|My\|))^2 \leq (1 - \lambda)^2/(1 + \lambda)^2 \quad \text{for all } y \in K.$$

Proof. Pick $y \in K$, $\|y\| = 1$ and write $y = x + r$ where r is a multiple of v and $\langle x,r\rangle = 0$. Note that $\|x\|^2 + \|r\|^2 = 1$. Then

$$(\langle My,y\rangle/\|My\|)^2 = (\langle x+\lambda v, x+v\rangle/\|x+\lambda v\|)^2 = (\|x\|^2+\lambda\|v^2\|)^2/(\|x\|^2+\lambda^2\|v\|^2).$$

Define $f(\theta) = (\sin^2\theta + \lambda\cos^2\theta)^2/(\sin^2\theta + \lambda^2\cos^2\theta)$, $0 \leq \theta \leq 2\pi$. Then f has its minimum at θ so that $\cos^2\theta = 1/(1+\lambda)$. Hence $(\langle My,y\rangle/\|My\|)^2 \geq 4\lambda/(1+\lambda)^2$ and the lemma follows.

We use the above to study convergence of discrete steepest descent for $\pi(\nabla\phi)$ and $\nabla\alpha$ indicated above. We consider a concrete example arising from numerical approximations to the initial value problem $y(0) = 1$, $y' = y$ on $[0,1]$. We believe the results given here to be typical of a wide class of linear problems even though arguments would be more complicated for more substantial differential equations.

Suppose n is a positive integer. Denote by G the grid $\{i/n\}_{i=0}^{n}$. Pick $m = 1$ and choose for this setting $\hat{G}$, H, H', K, $\hat{K}$ and D as in Section 2. Define $K_0 = \{y \in K \mid y(0) = 0\}$ and $A: H = \hat{K} \times \hat{K} \to \hat{K}$ by $A\binom{h}{g} = g - h$, $h, g \in \hat{K}$. Denote by P the orthogonal projection of H onto $H_0' \equiv DK_0$.

The problem of finding $u \in K$ such that $u(0) = 1$ and $ADu = 0$ is a finite difference approximation (using central differences) to the problem of finding y on $[0,1]$ such that $y(0) = 1$, $y' = y$. We compare the efficiency of the two approaches indicated above.

Choose $w \in K$ so that $w(p) = 1$, $p \in G$.

Theorem 6. Define $\phi: K_0 \to R$ by $\phi(x) = \|AD(x+w)\|^2/2$, $x \in K_0$. If $x \in K_0$ and $\phi(x) \neq 0$,

$$\phi(x-\delta(\nabla\phi)(x))/\phi(x) \leq (1-\lambda_n)^2/(1+\lambda_n)^2$$

where δ is chosen so that $\phi(x-\delta(\nabla\phi)(x))$ is minimum and

$$\lambda_n \equiv 1 - \frac{(p_n^{\,n} - q_n^{\,n})/2}{(p^n + q_n^{\,n})/2}, \quad p_n \equiv \frac{1+1/2n}{1-1/2n}, q_n \equiv 1/p_n .$$

(Note that $\lim_{n\to\infty} \lambda_n = 1 - \dfrac{\sinh 1}{\cosh 1}$ and $\lim_{n\to\infty} (1-\lambda_n)^2/(1+\lambda_n)^2 = .378199\ldots$.)

Sketch of Proof of Theorem 6. Define $M = APA^*$ where P is the orthogonal projection of H onto H_0'. Using Theorem 4, $M = ADE^{-1}\pi_0(AD)^*$ where π_0 is the orthogonal projection of K onto K_0 and $E \equiv \pi_0 D^*D\big|_{R(\pi_0)}$. Pick $g \in \hat{K}$. We wish to calculate an expression for Mg. To this end, define $f = E^{-1}\pi_0(AD)^*g$ and note that $Mg = ADf$.

Now $\pi_0 D^*Df = \pi_0(AD)^*g$. This expression is equivalent to the second order difference equation

$$c_n d_n f_{i+1} + (c_n^2 + d_n^2) f_i + c_n d_n f_{i-1} = d_n g_i + c_n g_{i+1},$$

$$i = 1, 2, \ldots, n-1$$

(where $c_n \equiv -(1/2n + \tfrac{1}{2})$, $d_n \equiv (1/2n - \tfrac{1}{2})$) together with the extra condition

$$2c_n d_n f_{n-1} + (c_n^2 + d_n^2) f_n = 2d_n g_n .$$

By ordinary means we may find the general solution to the second order difference equation and then use the extra condition together with the requirement $f_0 = 0$ to completely determine an expression for f. This done, we may calculate $Mg = ADf$ and find that

$$Mg = g - \frac{(p_n{}^n - q_n{}^n)/2}{(p_n{}^n + q_n{}^n)/2} \langle g, s \rangle s$$

where

$$s = (1/\|\gamma\|)\gamma , \quad \gamma = \begin{pmatrix} 1 \\ q \\ q^2 \\ \cdot \\ \cdot \\ q^{n-1} \end{pmatrix} \in \hat{K} .$$

Applying Lemmas 1 and 2 we have the conclusion to the theorem.

We now consider steepest descent using $\nabla\alpha$ instead of $\nabla\phi$. We retain the notation of the previous calculation. For n a positive integer define

$$\alpha_n(x) = \|AD(x+r)\|^2/2,\ x \in K_0,\ r(p) = 1,\quad p \in G.$$

Define grids $G_2, G_3, \ldots$, where for each positive integer n, $G_n = \{^i/n\}_{i=0}^n$. Denote by $\hat{G}_2, \hat{G}_3, \ldots$ the grids corresponding to $G_2, G_3, \ldots$ as in Section 2. Pick $y_2, y_3, \ldots$ in the following way: Fix $z \in C^2([0,1])$ so that $z(1) \neq 0$. Take y_n to be the function on $\hat{G}_n$ so that

$$y_n(p) = z(p)\ ,\quad p \in \hat{G}_n\ ,\quad n = 2,3,\ldots\ .$$

Pick $x_2, x_3, \ldots$ so that x_n is a function on G_n so that

$$y_n = ADx_n\ ,\quad n = 2,3,\ldots$$

where A and D are, for a particular n, defined as earlier in this section.

Theorem 7. $\lim_{n\to\infty} \alpha_n(x_n - \delta_n(\nabla\alpha_n)(x_n))/\alpha(x_n) = 1$

where for each n, δ_n is chosen so that $\alpha_n(x_n - \delta_n(\nabla\alpha_n)(x_n))$ is minimum.

Indication of Proof. For n a positive integer and $x \in \hat{K}_0$ (defined on G_n), $\alpha_n(x) = 0$ and δ chosen optimally, one has, using Lemma 1,

$$\alpha_n(x - \delta(\nabla\alpha_n)(x))/\alpha(x) = 1 - (\langle Ny,y\rangle/(\|y\|\,\|Ny\|))^2$$

where $N \equiv AD\pi_0(AD)^*$, $y \equiv AD(x+r)$, r is constant at 1.

Writing a matrix for N according to the natural basis for K_0, one arrives at the fact that

$$(\langle Ny,y\rangle/(\|y\|\,\|Ny\|))^2 = (f_n+h_n)^2/(i_n(a_n+b_n+e_n))$$

where

$$c_n = 1/(2n) + 1/2 \ , \ d_n = -1/(2n) - 1/2$$

$$a_n = (d_n{}^2 z(1/(2n)) + c_n d_n z(3/(2n)))^2$$

$$b_n = \Sigma_{i=1}^{n-2} (c_n d_n z((2i-1)/(2n)) + (c_n{}^2+d_n{}^2) z((2i+1)/(2n)) + c_n d_n z((2i+3)/(2n))^2$$

$$e_n = (c_n d_n z((2n-3)/(2n)) + (c_n{}^2+d_n{}^2) z((2n-1)/(2n))^2$$

$$f_n = \Sigma_{i=1}^{n-1} (d_n z((2i-1)/(2n)) + c_n z((2i+1)/(2n))^2$$

$$h_n = (d_n z((2n-1)/(2n)))^2$$

$$i_n = \Sigma_{i=1}^{n} \ z((2i-1)/(2n))^2$$

Careful examination yields the fact that $\lim_{n\to\infty} (f_n+h_n)^2/(i_n(a_n+b_n+e_n)) = 0$ and hence that the conclusion to the theorem holds.

Theorem 7 implies that as the mesh of G_n goes to 0, the proportional decrement of $\alpha(x)$ for the next discrete steepest descent step may be expected to be small. This is in sharp contrast to the use of the smoothed gradient of Theorem 6 in which there is always a large proportional improvement at each step - independent of n. Such evidence seems to indicate that the use of appropriate Sobolev norms is advantageous in steepest descent problems in differential equations.

References

[1] R.A. Adams, Sobolev Spaces, Academic Press (1975).

[2] A. Beurling et J. Deny, Espaces de Dirichlet I, le cas Elémentaire, Acta Math. 99 (1958), 203-224.

[3] A.L. Cauchy, Méthode générale pour la Resolution des Systemes d'Equations Simultanées, Comptes rendus, Ac. Sci. Paris 25 (1847), 536-538.

[4] Haskell B. Curry, The Method of Steepest Descent for Non-linear Minimization Problems, Quarterly App. Math 2 (1944), 258-261.

[5] V.M. Friedman, The Convergence of Steepest Descent Type Methods, Uspekhi Matematicheskikh Nauk XVII (1962), 201-204. (Russian)

[6] C.W. Groetsch, Steepest Descent and Least Squares Solvability, Canad. Math. Bull. 17 (1974), 275-276.

[7] M. Hestenes, Conjugate Direction Methods in Optimization, Springer-Verlag (1980).

[8] L.V. Kantorovitch, On an Effective Method of Solving Extremal Problems for Quadratic Functionals, C.R. (Doklady) Head. Sci. USSR 48 (1945), 455-460.

[9] ______, On the Method of Steepest Descent, C.R. (Doklady) Acad. Sci. USSR 56 (1947), 233-236.

[10] John Locker, Weak Steepest Descent for Linear Boundary Value Problems, Indiana J. Math 25 (1976), 525-530.

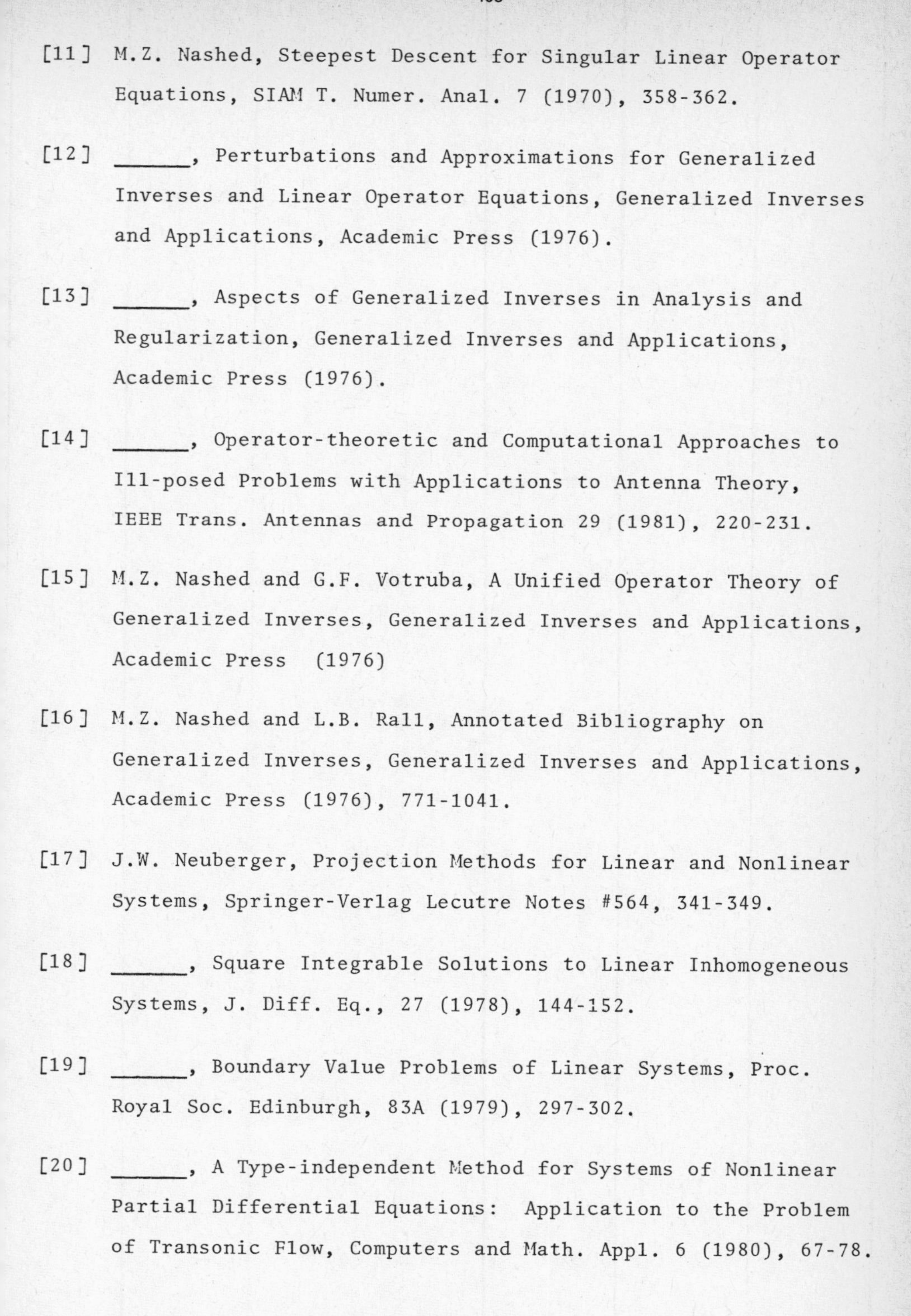

[11] M.Z. Nashed, Steepest Descent for Singular Linear Operator Equations, SIAM T. Numer. Anal. 7 (1970), 358-362.

[12] ______, Perturbations and Approximations for Generalized Inverses and Linear Operator Equations, Generalized Inverses and Applications, Academic Press (1976).

[13] ______, Aspects of Generalized Inverses in Analysis and Regularization, Generalized Inverses and Applications, Academic Press (1976).

[14] ______, Operator-theoretic and Computational Approaches to Ill-posed Problems with Applications to Antenna Theory, IEEE Trans. Antennas and Propagation 29 (1981), 220-231.

[15] M.Z. Nashed and G.F. Votruba, A Unified Operator Theory of Generalized Inverses, Generalized Inverses and Applications, Academic Press (1976)

[16] M.Z. Nashed and L.B. Rall, Annotated Bibliography on Generalized Inverses, Generalized Inverses and Applications, Academic Press (1976), 771-1041.

[17] J.W. Neuberger, Projection Methods for Linear and Nonlinear Systems, Springer-Verlag Lecutre Notes #564, 341-349.

[18] ______, Square Integrable Solutions to Linear Inhomogeneous Systems, J. Diff. Eq., 27 (1978), 144-152.

[19] ______, Boundary Value Problems of Linear Systems, Proc. Royal Soc. Edinburgh, 83A (1979), 297-302.

[20] ______, A Type-independent Method for Systems of Nonlinear Partial Differential Equations: Application to the Problem of Transonic Flow, Computers and Math. Appl. 6 (1980), 67-78.

[21] ______, Finite Dimensional Potential Theory Applied to Numerical Analysis of Linear Systems, Lin. Alg. Appl. 35 (1981), 193-202.

[22] ______, Steepest Descent for Systems of Nonlinear Partial Differential Equations, Oak Ridge National Laboratory, CSD/TM-161 (1981).

[23] J.M. Ortega and W.C. Rheinboldt, Iterative Solution of Nonlinear Equations in Several Variables, Academic Press (1970).

[24] P.C. Rosenblum, The Method of Steepest Descent, Proc. Symp. Appl. Math. Vol. VI, McGraw-Hill (1956).

[25] F. Riesz and B. Sz.-Nagy, Functional Analysis, Ungar (1955).

J.W. Neuberger
North Texas State University
Denton, Texas 76203
U.S.A.

On the Spectral Theory of Some Nonsymmetric Second Order Differential Operators

Thomas T. Read
Department of Mathematics
Western Washington University
Bellingham, Washington 98225

1. Introduction

In this paper we will develop some limit-point criteria for the nonsymmetric second order differential expression

$$L[y] = k^{-1}[-(py'+sy)' + ry' + qy] \tag{1}$$

on $[0,\infty)$, and will give some results on the location of the numerical range and the essential spectrum of m-sectorial extensions of the minimal operator under hypotheses which ensure that the minimal operator is sectorial.

The coefficients p, q, r, and s in (1) may in general all be complex-valued, although we will restrict some of them to be real-valued at times. For simplicity we assume that Re $p > 0$, that p and s are piecewise C^1, and that r and q are piecewise continuous. We assume also that the weight function k is positive and locally integrable. Associated with k is the weighted Hilbert space $\mathcal{L}_k^2(0,\infty) = \mathcal{L}_k^2$ of functions f such that $\int_0^\infty |f|^2 k < \infty$ with the inner product $(f,g) = \int_0^\infty f\bar{g}k$. The minimal operator T_0 defined by (1) is then the closure in $\mathcal{L}_k^2$ of the restriction of L to $C_0^\infty(0,\infty)$.

The representation of L in the form (1) is, of course, not unique. The choice of a suitable representation for L will be central to the discussion in Section 3 of the numerical range.

The expression (1) is formally symmetric if and only if p and q are both real valued and $s = \bar{r}$. In this case the following simple necessary and sufficient condition for the minimal operator T_0 to be positive definite was given in [13] for the case $k = 1$. The proof given there also holds in the general case without alteration.

THEOREM 1. *If L is formally symmetric, then these two properties are equivalent.*

(i) $(T_0 f,f) > 0$ *for all* f *in* C_0^∞.

(ii) L *has a representation* $L[y] = k^{-1}[-(py'+\bar{u}y)' + uy' + qy]$ *on* $(0,\infty)$ *where*

$$q \geq |u|^2/p \tag{2}$$

In Section 2 we give two limit-point criteria for nonsymmetric expressions L. Both are of the "Levinson type" of which many versions have appeared for formally symmetric expressions with real coefficients (for instance [1], [7, Theorem III 2.5], and [10]). The first together with Theorem 1 gives an immediate extension to this context of Hartman's result [5] that a nonoscillatory expression with $p^{-1/2}$ not absolutely integrable is limit-point. Corollaries 2 and 3 spell out some consequences of Theorem 2 for nonsymmetric expressions with real coefficients. We show that Theorem 3 contains a slight generalization of a theorem of Knowles and Race [9] on J-selfadjointness.

In Section 3 we investigate the numerical range of extensions of T_0. Here the analysis depends very heavily on Theorem 1 and on a slight strengthening of property (2) of that theorem.

2. Limit-point Criteria

We begin by reviewing briefly the notion of the mean deficiency index of a nonsymmetric expression introduced in [8]. See [7], Sections II4 and II5 for a more complete exposition.

DEFINITION. Let $L[y] = k^{-1}[-(py' + sy)' + ry' + qy]$ with associated minimal operator T_0 and maximal operator T_1. Then the *mean deficiency index*, d(L), of L is one-half the dimension of the quotient space domain T_1/domain T_0.

The mean deficiency index satisfies the inequality $1 \leq d(L) \leq 2$. We say L is *limit-point* if $d(L) = 1$.

If $T_0 - \lambda I$ has closed range for some complex number λ, then $2d(L) = \text{nullity}\ (T_1 - \lambda I) + \text{nullity}\ (T_1^+ - \bar{\lambda}I)$, where T_1^+ is the maximal operator associated with the formal adjoint L^+ of L. Thus d(L) coincides with the usual deficiency index for symmetric expressions with real coefficients. In any case, $2d(L) \geq$ nullity $(T_1 - \bar{\lambda}I)$ + nullity $(T_1^+ - \bar{\lambda}I)$ so that L cannot be limit-point if the equations $L[y] = 0$ and $L^+[y] = 0$ have at least 3 $\mathcal{L}^2$ solutions between them.

The expression L is limit-point if and only if $[f,g](x) \to 0$ as $x \to \infty$ for each f in domain T_1 and each g in domain T_1^+. Here $[f,g]$ is the Lagrange bilinear form. If $T_0 - \lambda I$ has closed range for some λ, then it is not difficult to see that $d(L) = 1$ if and only if $[f,g] = 0$ whenever $(T_1 - \lambda)f = (T_1^+ - \bar{\lambda})g = 0$.

We are now ready to state our first result.

THEOREM 2. *Let* $L[y] = k^{-1}[-py'+sy)' + ry' + qy]$ *with* $p > 0$ *Suppose that there is a uniformly bounded sequence* $\{w_n\}$ *of non-negative compactly supported absolutely continuous functions such that*

(i) $\int_0^\infty w_n(k/p)^{1/2} \to \infty$ as $n \to \infty$.

Suppose moreover that $\operatorname{Re} q = q_1 + q_2$ *and that for some* Q *with* $Q' = q_2$ *and some constant* K,

(ii) $p[w_n' + w_n \operatorname{Re}(r-s)/2p]^2 \leq [q_1 - |r + \bar{s} - 2Q|^2/4p]w_n^2 + Kk$.

Then L *is limit-point*.

Remarks. 1. For $k = 1$, q real-valued, and $r = s = 0$, this is a slight improvement on Theorem II 2.5 of [7]. (See also Remarks 2 and 3.)

2. If $T_0 - \lambda I$ has closed range for some λ, then the condition that the w_n be uniformly bounded may be omitted. This is achieved by modifying the proof very slightly to use the second characterization of L being limit-point given in the paragraph preceding Theorem 2.

3. If $\operatorname{Re} r = \operatorname{Re} s$ (in particular if L is formally symmetric), then the sequence $\{w_n\}$ may be replaced by a single bounded non-negative locally absolutely continuous function w such that

(i') $\int_0^\infty w(k/p)^{1/2}$

For, given such a w, let a(n) be defined by the equation $w(n) = \int_n^{a(n)} (k/p)^{1/2}$. Then the sequence $\{W_n\}$ defined by

$$W_n(x) = \begin{cases} w(x), & x \le n, \\ w(n) - \int_n^x (k/p)^{1/2}, & n \le x \le a(n), \\ 0, & a(n) \le x \end{cases}$$

satisfies the hypotheses of Theorem 2.

4. If q_1 is replaced by $q_1/(1 + \varepsilon)$ for some $\varepsilon > 0$ in (ii) then (i) may be replaced by

(i") $\int_0^\infty w_n^2 (q_1/p)^{1/2} \to \infty$ as $n \to \infty$.

(See Theorem 3 for a result of this sort.) This has the advantage that p may grow arbitrarily fast provided q_1 grows also.

Proof. The Lagrange bilinear form, for f in domain T_1 and g in domain T_1^+ is

$$[f,g] = p(\bar{g}'f - \bar{g}f') + (r - s)f\bar{g}$$

so that, abbreviating w_n to w,

$$\begin{aligned} w\sqrt{k/p}[f,g] = {} & \left(\sqrt{p}\ \bar{g}'w + \bar{g}w(r+\bar{s}-2Q)/2\sqrt{p} + \sqrt{p}\ \bar{g}w'\right)f\sqrt{k} \\ & - \left(\sqrt{p}\ f'w + fw(\bar{r}+s-2Q)/2\sqrt{p} + \sqrt{p}\ fw'\right)\bar{g}\sqrt{k} \\ & + (\bar{g}\sqrt{k})(fw/\sqrt{p})\operatorname{Re}(r-s). \end{aligned} \tag{3}$$

Here we have used the identity

$$((r+\bar{s})/2) - (\bar{r}+s)/2)f\bar{g} = f\bar{g}\ \operatorname{Im}(r+\bar{s}) = f\bar{g}\ \operatorname{Im}(r-s).$$

Now L is not limit-point if and only if for some f in domain T_1 and g in domain T_1^+, $[f,g] \to 1$ as $x \to \infty$. If so, then $\int_0^\infty w_n\sqrt{k/p}\ [f,g] \to \infty$ as $n \to \infty$. Thus in order to prove the theorem it suffices to show that each term on the right side of (3) is the product of two square-integrable functions whose $\mathcal{L}^2$ norm is bounded independently of n. To do this we note

$$\mathrm{Re}(L[f],fw^2) = \mathrm{Re}\int_0^\infty (pf'+sf)\bar{f}'w^2 + 2\mathrm{Re}\int_0^\infty (pf'+sf)\bar{f}ww' + \mathrm{Re}\int_0^\infty rf'\bar{f}w^2$$

$$+ \int_0^\infty q_1|f|^2w^2 - 2\mathrm{Re}\int_0^\infty Qf'\bar{f}w^2 - 2\int_0^\infty Q|f|^2ww'$$

$$= \int_0^\infty |\sqrt{p}f'w + fw(\bar{r}+s-2Q)/2\sqrt{p} + \sqrt{p}fw'|^2$$

$$+ \mathrm{Re}\int_0^\infty (s-\bar{r})|f|^2ww' + \int_0^\infty (q_1-|r+\bar{s}-2Q|^2 4p)|f|^2w^2 - \int_0^\infty p|f|^2w'^2$$

using $2\,\mathrm{Re}\int_0^\infty s|f|^2ww' = \mathrm{Re}\int_0^\infty (s+\bar{r})|f|^2ww' + \mathrm{Re}\int_0^\infty (s-\bar{r})|f|^2ww'$.
Now $\mathrm{Re}(s-\bar{r}) = \mathrm{Re}(s-r)$, so

$$\mathrm{Re}(L[f],fw^2) = \int_0^\infty |\sqrt{p}f'w+fw(r+\bar{s}-2Q)/2\sqrt{p}+\sqrt{p}fw'|^2 + \int_0^\infty (\mathrm{Re}(s-r))^2|f|^2w^2/4p$$

$$- \int_0^\infty p[w'+w\mathrm{Re}(r-s)/2p]^2|f|^2 + \int_0^\infty [q_1-|r+\bar{s}-2Q|^2/4p]|f|^2w^2.$$

By (ii), the sum of the last two terms on the right is bounded below by $-K\int_0^\infty |f|^2k$ for each $w = w_n$. It follows that each of the first two terms on the right is bounded above uniformly in n. Thus the second and third terms in (3) are each the products of two square-integrable functions whose $\mathcal{L}^2$ norm is bounded independently of n. The same conclusion can be shown to hold for the first term in (3) by repeating the argument with $\mathrm{Re}(L^+[g],gw^2)$. This completes the proof.

As an immediate corollary of Theorems 1 and 2, we can extend to complex coefficient symmetric expressions in a weighted Hilbert space, Hartman's theorem [5] that nonoscillatory expressions are limit-point. The proof is new even in the real coefficient case.

COROLLARY 1. _Let_ $L[y] = k^{-1}[-(py'+\bar{u}y)' + uy' + qy]$ _be formally symmetric_. _If_ $(T_0f,f) > 0$ _for_ f _in_ C_0^∞, _and if_ $\int_0^\infty \sqrt{k/p} = \infty$ _then_ L _is limit-point_.

Proof. By Theorem 1 we may assume that $q \geq |u|^2/p$. Hypothesis (ii) has the form

$$pw_n'^2 \leq [q - |u|^2/p]w_n^2 + Kk. \tag{4}$$

Moreover, by Remark 3 following the statement of Theorem 2, we may replace the sequence $\{w_n\}$ by the single function $w \equiv 1$ for which (4) is clearly true.

Remark. In general, L may fail to be limit-point if $\sqrt{k/p}$ is absolutely integrable. If, however, $q \geq (1+\varepsilon)|u|^2/p$, then it follows from Remark 4 above and a repetition of the proof of Corollary 1 that L is limit-point if p satisfies the weaker condition $\int_0^\infty \sqrt{kq/p} = \infty$. This condition is not sufficient if only $q \geq |u|^2/p$. For instance, $L[y] = -(x^4y')' - 2x^2y$ is limit-circle on $[1,\infty)$ with $k = 1$ although it can be rewritten in the form $L[y] = -(x^4y' + x^3y)' + x^3y' + x^2y$ where $q = x^2 = |u|^2/p$.

COROLLARY 2. *Let* $k = 1$ *and* $L[y] = a_2y'' + a_1y' + a_0y$ *where each* a_i *is real-valued* $(a_2 > 0)$ *and* a_2', a_1, *and* a_0 *are piecewise continuous. If there is a uniformly bounded sequence* $\{w_n\}$ *of non-negative compactly supported absolutely continuous functions such that*

(i) $\int_0^\infty w_n a_2^{-1/2} \to \infty$ as $n \to \infty$, *and*

(ii) $a_2[w_n' + w_n(a_2'-a_1)/2a_2]^2 \leq -a_0w_n^2 + K$,

then L *is limit-point*.

Proof. In Theorem 2 set $p_2 = -a_2$, $q_1 = -a_0$, and $r = -s = (a_2' - a_1)/2$.

When $a_2 = 1$ we can specialize this to

COROLLARY 3. $L[y] = y'' + a_1y' + a_0y$ *is limit-point if*

$$4a_1^2 - a_1/x \leq -4a_0 + Kx^2$$

on $[1,\infty)$ *and so, in particular, if* $|a_1| \leq C_1x$ *and* $a_0 \leq C_0x^2$.

Proof. Define the sequence $\{w_n\}$ by

$$w_n(x) = \begin{cases} 1/x, & 1 \leq x \leq n, \\ 1/n - x + n, & n \leq x \leq n + 1/n, \\ 0 & n + 1/n \leq x. \end{cases}$$

Then the result follows immediately from Corollary 2.

Remark. The expression $L[y] = y'' + a_1y' + a_0y$ may fail to be limit-point if $a_1(x) = x^{1+\varepsilon}$ for any $\varepsilon > 0$ even if a_0 is as in Corollary 3. If $L = (D + c/x)(D + x^{1+\varepsilon})$ where $D = d/dx$, then $L^+ = (-D+x^{1+\varepsilon})(-D+c/x)$. It is easy to verify that $L[y] = 0$ has 2 solutions in $\mathcal{L}^2$ and $L^+[y] = 0$ has 1 provided $c > 1/2$ and $c - \varepsilon < 1/2$. Clearly such a c can be chosen for any $\varepsilon > 0$, and it follows from the remarks preceding

Theorem 2 that such an L cannot be limit-point.

For our second limit-point criterion, we allow p to take values in a sector of the complex plane. We are still able to avoid any hypotheses on the imaginary part of q.

THEOREM 3. _Let_ $L]y] = k^{-1}[-(py')' + qy]$, _where_ $p \neq 0$ _and_

(i) $|p| \leq K_1 \operatorname{Re} p$.

Suppose that $\operatorname{Re} q = q_1 + q_2 + q_3$ _with_ $q_1 \geq 0$ _and that there is a uniformly bounded sequence_ $\{w_n\}$ _of compactly supported absolutely continuous functions such that_

(ii) $\int_0^\infty w_n^2 (q_1/\operatorname{Re} p)^{1/2} \to \infty$ as $n \to \infty$,

(iii) $-q_2 w_n^2 \leq K_2 k$,

and

(iv) $(1+\varepsilon)K_1\left(|p|w'^2 + Q^2w^2/|p| + 2Q(K_3 - \operatorname{Re}\sqrt{p/\bar{p}}\right)ww' \leq q_1 w^2 + K_4 k$.

Here $Q' = q_3$ _and_ $K_3 = (K_1\sqrt{1+\varepsilon})^{-1}$. _Then L is limit-point._

Remarks. 1. We have omitted the first order terms from (1) for simplicity. Such terms could be incorporated into the theorem without any major alteration in the proof either by making (iii) yet more complicated, or by bounding each of the terms in (iv) separately. It is desirable to avoid bounding the terms separately if possible since the "lumped" hypothesis can be genuinely useful, as in the proof of Corollary 1 or of Theorem 2 of [12].

2. As with Theorem 2, if $T_0 - \lambda I$ has closed range for some λ, then the requirement that the sequence $\{w_n\}$ be uniformly bounded can be dropped.

3. If p is real, then (iv) simplifies to

$$(1+\varepsilon)p\left(w_n' - Qw_n/p\right)^2 \leq q_1 w_n^2 + K_4 k.$$

Proof. We show again that $[f,g] \to 0$ whenever f is in domain T_1 and g is in domain T_1^+ by assuming that $[f,g] \to 1$ and reaching a contradiction. We have

$$w_n^2\sqrt{q_1/\operatorname{Re} p}[f,g] = (p\bar{g}'w_n/\sqrt{\operatorname{Re} p})(fw_n\sqrt{q_1}) - (pf'w_n/\sqrt{\operatorname{Re} p})(\bar{g}w_n\sqrt{q_1})$$

so that it suffices to show that for f in domain T_1,

$$\int_0^\infty \left((\mathrm{Re}\ p)|f'|^2 + q_1|f|^2\right)w_n^2 \le K \tag{5}$$

independently of n. Writing w in place of w_n, we have

$$\mathrm{Re}(L[f],fw^2) = \int_0^\infty (\mathrm{Re}\ p)|f'|^2 w^2 + \int_0^\infty (q_1 + q_2)|f|^2 w^2$$

$$+ 2\mathrm{Re}\int_0^\infty pf'\bar{f}ww' = 2\mathrm{Re}\int_0^\infty Qf'\bar{f}w^2 - 2\int_0^\infty Q|f|^2 ww'.$$

Choose $\delta > 0$ so that $\delta^2 K_1 = (1+\varepsilon)^{-1/2}$. We denote by $\sqrt{p}$ the branch of the square root such that $\sqrt{1} = 1$. Then

$$2\mathrm{Re}\int (pw'-Qw)f'\bar{f}w = \int_0^\infty |\delta\sqrt{p}f'w + \delta^{-1}\sqrt{\bar{p}}(fw' - Qfw/\bar{p})|^2$$

$$-\delta^2\int^\infty |p||f'|^2 w^2 - \delta^{-2}\int_0^\infty \left(|p|w'^2 + Q^2w^2/|p| - 2Q\mathrm{Re}\sqrt{p/\bar{p}}\ ww'\right)|f|^2.$$

Thus for each $w = w_n$,

$$\mathrm{Re}(L[f],fw^2) = \int_0^\infty (\mathrm{Re}\ p - \delta^2|p|)|f'|^2 w^2 +$$

$$+ \int_0^\infty |\delta\sqrt{p}f'w + \delta^{-1}\sqrt{\bar{p}}(fw'-Qfw/\bar{p})|^2$$

$$-\delta^{-2}\int_0^\infty \left(|p|w'^2 + Qw^2/|p| + 2Q(\delta^2 - \mathrm{Re}\sqrt{p/\bar{p}})ww'\right)|f|^2 + \int_0^\infty (q_1+q_2)|f|^2 w^2$$

$$\ge \lambda\int_0^\infty \left((\mathrm{Re}\ p)|f'|^2 + q_1|f|^2\right)w^2 - (K_2 + K_3(1+\varepsilon)^{-1/2})\int_0^\infty |f|^2 k$$

for some $\lambda > 0$. Thus (5) holds and the proof is complete.

As one application of Theorem 4 we easily obtain an extension of a result of Knowles and Race [9, Theorem 3.4].

THEOREM 5. (Knowles, Race) _Let_ $L[y] = k^{-1}[-(py')' + qy]$ _where_ p _takes values in a sector_ $\theta_1 \le \arg z \le \theta_2$, $\theta_2 - \theta_1 < \pi$, _which does not intersect the negative real axis. Suppose there is a positive locally absolutely continuous function_ w _on_ $[0,\infty)$ _such that_

(i) $\int_0^\infty w\sqrt{k/|p|} = \infty$,

and

(ii) $|p|(w')^2 \le K_0 k$,

and decompositions $\mathrm{Re}\ q = r_1 + r_2$, $\mathrm{Im}\ q = r_3 + r_4$ _such that_

(iii) $-r_i w^2 \le K_i k$, $i = 1, 3$,

(iv) $w(x)|\int_0^x r_i| \le K_i k(x)\sqrt{|p(x)|}$, $i = 2, 4$.

Suppose further that $T_0 - \lambda I$ _has closed range for some_ λ. _Then the extension_ T_γ _of_ T_0 _defined by_ L _on_

domain $T_\gamma = \{f \in \text{domain } T_1 \colon \gamma_1 f(0) + \gamma_2 p(0) f'(0) = 0\}$

for $\gamma = (\gamma_1, \gamma_2)$ *any fixed element of* $\mathcal{L}^2$ *is J-selfadjoint*.

Remark. The result of Knowles and Race [9] is stated for a unit weight function and requires the values of p to lie in the sector $0 \le \arg z \le \pi/2$. On the other hand, condition (iv) is stated in a slightly more general form.

Proof. We may assume that the function w is bounded, for if not then it follows from (i), (ii) and an argument in the proof of [11, Theorem 4] that

$$W(x) = w(x)/\left(1 + \int_0^x w\sqrt{k/|p|}\right)$$

is bounded and still satisfies (i) - (iv).

By assumption there is a constant θ, $|\theta| < \pi/2$, such that if $M[y] = \exp(-i\theta)L[y] = k^{-1}[-(p_0y')' + q_0y]$, then the values of p_0 lie in a sector $|\arg z| \le \phi$ for some $\phi < \pi/2$. Thus (i) of Theorem 4 holds with $K = (\cos\phi)^{-1}$. Also we have

$$\text{Re } q_0 = (r_1 + r_2)\cos\theta + (r_3 + r_4)\sin\theta.$$

Define a decomposition of Re q_0 by

$$q_1 = w^{-2}k,$$

$$q_2 = r_1 \cos\theta + r_3 \sin\theta - w^{-2}k,$$

$$q_3 = r_2 \cos\theta + r_4 \sin\theta.$$

Set $U(x) = w(x)\sqrt{k(x)/|p(x)|}/\left(1 + \int_0^x w\sqrt{k/|p|}\right)$.

Then $\int_0^x U = \log\left(1 + \int_0^x w\sqrt{k/|p|}\right) \to \infty$ as $x \to \infty$. Thus for each integer n there is $a(n) > n$ such that $w(n) = \int_n^{a(n)} U$.

Define a sequence $\{w_n\}$ by

$$w_n(x) = \begin{cases} w(x), & x \le n, \\ w(n) - \int_n^x U, & n \le x \le a(n), \\ 0, & a(n) \le x. \end{cases}$$

Then on $[0,n]$, $w_n^2\sqrt{q_1/\text{Re } p_0} = w\sqrt{k/\text{Re } p_0} \ge w\sqrt{k/|p|}$ so that (ii) of Theorem 4 holds. It is clear that (iii) holds and it follows from (iii) and (iv) of Theorem 5 that each term in (iv) of Theorem 4 is bounded above by a constant. Thus M, and hence L, is limit-point.

Finally, if $T_0 - \lambda I$ has closed range for some λ, then it follows from the remarks preceding Theorem 2 that

$$2 = 2d(T_0) = \text{nullity } (T_1 - \lambda I) + \text{nullity } (T_1^+ - \bar{\lambda} I)$$
$$= 2 \text{ nullity } (T_1 - \lambda I)$$

since $T_1 f = \lambda f$ if and only if $T_1^+ \bar{f} = \bar{\lambda}\bar{f}$. Thus each T_γ is J-selfadjoint by [9, Theorem 2.4].

3. Numerical Range

We turn now to an investigation of the numerical range and the essential spectrum of m-sectorial operators defined by (1). For this we require that the symmetric expression L_R, the <u>real part</u> of L, defined by

$$L_R[y] = k^{-1}[-(p_R y' + \bar{u} y)' + uy' + q_R y] \tag{5}$$

where $p_R = \text{Re } p$, and $u = (r+\bar{s})/2$ define a semibounded form on the C_0^∞ functions. If the minimal operator T_0 defined by L is quasi-sectorial, then by an adaptation of the construction of the Friedrichs extension of a symmetric operator, then T_0 and more generally the restriction of T_1 to the space $\mathcal{D}_\gamma$ of compactly supported functions in domain T_1 satisfying the boundary condition $\gamma_1 f(0)+\gamma_2 p(0)f'(0)=0$ $(\gamma = (\gamma_1,\gamma_2)$ has an m-sectorial extension T_γ. See Kato [6], p. 322 for the details of this construction. The domain of T_γ is the intersection of domain T_1 with the domain of the closure of the form h_γ defined on $\mathcal{D}_\gamma$ by

$$h_\gamma[f,g] = \int_0^\infty (p_R f'+uf)(p_R \bar{g}'+\bar{u}\bar{g}) + \int_0^\infty (q_R - |u|^2/p_R) f\bar{g}. \tag{6}$$

If condition (2) of Theorem 1 is strengthened slightly, then the domain of T_γ may be characterized more explicitly. We denote by $\mathcal{D}$ the class of functions

$$\mathcal{D} = \{f \in AC^{loc} : \int_0^\infty (p_R|f'|^2 + q_R|f|^2) < \infty\}. \tag{7}$$

THEOREM 6. <u>Suppose that</u> L_R <u>is as in</u> (5) <u>with</u>

(i) $q_R \geq k + (1+\varepsilon)|u|^2/p_R$ <u>for some</u> $\varepsilon > 0$,

(ii) $|(L[f],f)| \le K(L_R[f],f) = K \operatorname{Re}(L[f],f)$ _for all_ f _in_ C_0^∞.

Then for each ordered pair $\gamma = (\gamma_1,\gamma_1)$ _of complex numbers_ $(\gamma \neq (0,0))$, _the domain of the m-sectorial extension_ T_γ _of the restriction of_ T_1 _to_ $\mathcal{D}_\gamma$ _is_

$$\text{domain } T_\gamma = \{f \in \mathcal{D} \cap \text{domain } T_1 : \gamma_1 f(0) + \gamma_2 p(0) f'(0) = 0\}.$$

Remarks. 1. The k on the right side of (i) is there only for convenience since T_γ and $T_\gamma + Kk$ have the same domain for any constant K.

2. Each T_γ as in Theorem 6 has a square root with domain (or, if $\gamma_2 = 0$, $\mathcal{D}_0 = \{f \in \mathcal{D} : f(0) = 0\}$), but we shall not need this fact.

Proof. We suppose first that $\gamma_2 = 0$, that is, we consider the extension of the minimal operator T_0. It is shown in Theorem 4 of [13] that (i) implies that the set $\mathcal{D}$ defined by (7) also satisfies

$$\mathcal{D} = \{f \in AC^{loc} : ||f||^2 = \int_0^\infty |pf'+uf|^2 + \int_0^\infty (q_R - |u|^2/p)|f|^2 < \infty\} \quad (8)$$

and also that the equation $L_R[y] = 0$ has exactly one solution in $\mathcal{L}^2$. It then follows from an argument similar to that of Lemma 2.5 of [2] that the domain of the closure of the form h_0 defined by (6) on C_0^∞ is the set $\mathcal{D}_0 = \{f \in \mathcal{D} : f(0) = 0\}$. An argument as in Kato [6], p. 328 then shows that the domains of the m-sectorial extension of T_0 is as asserted in this case.

For the general case we note that the domain of the closure of the form h_γ defined by (6) is $\mathcal{D}$ as defined by (8). Also it is not difficult to see that there is $K > 0$ such that $|f(0)| \le K||f||_D$. Then the argument of [6], p. 328 may be used to give the result in this case also.

We shall give two results on the numerical range of L in which we make assumptions only about the average behavior of q. In the first of these we specialize to the case $p = k = 1$ in order to be in a position to make more accurate estimates. Corollaries 4 and 6 sharpen some results of Eastham [3] and Evans [4].

<u>Definition</u>. We shall say that a region R containing the origin in the complex plane is <u>appropriate</u> if for any $A \geq 0$ the set $R_A = \{z \in R : \text{Re } z \leq A\}$ is bounded and if z in R implies that $z + x$ is in R for each $x > 0$.

Given an appropriate region R and real constants a and A with $a > 0$, we shall define M_A, N_A, and I_A by

$$\arctan M_A = \sup\{\arg z : \text{Re } z = A,\ z \in R\},$$

$$-N = \inf\{a \text{ Re } z : z \in R\},$$

$$I_A = \sup\{|\text{Im}(z_1 - z_2)| : z_1, z_2 \in R_A,\ z_1 - z_2 \in R\}.$$

THEOREM 7. <u>Let</u> $L[y] = -y'' + qy$. <u>Suppose that for some appropriate region</u> R, <u>some constants</u> $a > 0$ <u>and</u> $A > 0$, <u>and each</u> $x \geq 0$,

(i) <u>there is</u> t, $x < t \leq x + a$ <u>such that</u>

$$(1/a) \text{ Re } \int_x^t q \geq A,$$

(ii) $$(1/a) \int_x^w q \in R$$

<u>whenever</u> $x \leq w \leq x + a$.

<u>Then the minimal operator</u> T_0 <u>is sectorial and the m-sectorial extension</u> T <u>of</u> T_0 <u>has the properties</u>

(i) domain $T = \{f \in \text{domain } T_1 : f(0) = 0$ and $\int_0^\infty (|f'|^2 + (q_1+1)|f|^2) < \infty\}$.

(ii) $\text{Re } T \geq A - (Aa+N)^2/4 = x_0$.

(iii) <u>The numerical range of</u> T <u>is contained in the set</u>

$\{x+iy : x \geq x_0,$

$$|y| \leq \begin{cases} 2I_A(x-x_0)^{1/2} + y_1, & x - x_0 \leq (I_A/M_A)^2, \\ M_A(x-x_0) + y_2, & x - x_0 > (I_A/M_A)^2, \end{cases}$$

where $y_1 = I_A(Aa+N) + AM_A$ and $y_2 = y_1 + I_A^2/M_A$.

<u>Remarks</u>. 1. In general it is not possible to restrict the numerical range of T to a sector with semiangle smaller than $\theta = \arctan M_A$, for the hypotheses allow $q(x) = e^{i\theta} f(x)$ where f is any real valued function such that $f \geq A/\cos\theta$.

2. Hypothesis (ii) may be regarded as a Brinck condition for a complex valued function. We shall use a more general version in

Theorem 8.

Proof. We define a new representation of L of the form (1) so that (2) holds for the real part of $L - x_0$ as follows. Suppose first that $A = 0$. Fix $C > 0$. Set $x_1 = 0$ and if x_n has already been defined set

$$x_{n+1} = \inf\{x \le x_n + a : \mathrm{Re} \int_{x_n}^{x} q \ge Ca\}$$

if such an x_{n+1} exists and

$$x_{n+1} = \sup\{x \le x_n + a : \mathrm{Re} \int_{x_n}^{x} q \ge 0\}$$

otherwise.

Define q_1 to be the real valued step function whose value on $[x_n, x_{n+1})$ is

$$q_1(x) = (x_{n+1} - x_n)^{-1} \mathrm{Re} \int_{x_n}^{x_{n+1}} q .$$

Then $0 \le \int_{x_n}^{x} q_1 \le Ca$ for $x \le x \le x_{n+1}$.

Define v_1 to be the function given on $[x_n, x_{n+1})$ by

$$v_1(x) = (1/2)N + \mathrm{Re} \int_{x_n}^{x} (q - q_1).$$

Note that $v_1(x_n) = (1/2)N$ for each n, so that v_1 is continuous. Moreover, since

$$-N - Ca \le \mathrm{Re} \int_{x_n}^{x} (q - q_1) \le \mathrm{Re} \int_{x_n}^{x} q \le Ca,$$

we have $|v_1(x)| \le N/2 + Ca$ for all x.

For f in C_0^∞,

$$\begin{aligned}
(\mathrm{Re}\, T_0 f, f) &= \int_0^\infty (|f'|^2 + (\mathrm{Re}\, q)|f|^2) = \int_0^\infty (|f'|^2 + q_1|f|^2 + v_1'|f|^2) \\
&= \int_0^\infty (|f'|^2 + q_1|f|^2) - \int_0^\infty v_1(f'\bar{f} + f\bar{f}') \\
&\ge \int_0^\infty (|f'|^2 + q_1|f|^2) - 2\left(\int_0^\infty |f'|^2 \int_0^\infty v_1^2|f|^2\right)^{1/2} \\
&\ge \int_0^\infty (q_1 - v_1^2)|f|^2 \\
&\ge -(Ca + N/2)^2 \int_0^\infty |f|^2.
\end{aligned}$$

Thus $\mathrm{Re}\, T_0 \ge -(Ca+N/2)^2$. Since $C > 0$ is arbitrary, $\mathrm{Re}\, T_0 \ge -N^2/4$. The same inequality holds for the m-sectorial extension T of T_0 so that (ii) is proved when $A = 0$.

If $A > 0$, then $L_A = L - A$ satisfies the hypotheses of Theorem 7 with N replaced by $Aa + N$. Thus it follows from the case just proved that $\mathrm{Re}\, T - A \geq -(Aa+N)^2/4$ and (ii) is established.

To establish (i) and (iii), let $C > 0$ and let $\{x_n\}$, q_1 and v_1 be chosen as in the first part of the proof for the expression $L_A + L - A$. Thus $\mathrm{Re}\, q = A + q_1 + v_1'$ where $q_1 \geq 0$, $|v_1| \leq Ca + (Aa+N)/2$, and

$$\mathrm{Re}\int_{x_n}^{x_{n+1}} q = A(x_{n+1}-x_n) + \int_{x_n}^{x_{n+1}} q_1 .$$

Define a step function q_2 and a continuous function v_2 by

$$q_2(x) = (x_{n+1}-x_n)^{-1}\ \mathrm{Im}\int_{x_n}^{x_{n+1}} q$$

and

$$v_2(x) = \mathrm{Im}\int_{x_n}^{x} (q-q_2)$$

on $[x_n, x_{n+1})$. Then

$$(x_{n+1}-x_n)|q_2(x)| \leq M_{A+C}\ \mathrm{Re}\int_{x_{n+1}}^{x_{n+1}} q = M_{A+C}(A+q_1)(x_{n+1}-x_n)$$

so that $|q_2| \leq M_{A+C}(q_1+A)$.

On the interval $[x_n, x_{n+1})$, the numbers $\mathrm{Im}\int_{x_n}^{x} q$ lie in an interval of length I_{A+C} containing 0 since for any x and x',

$$|\mathrm{Im}\int_{x_n}^{x'} q - \mathrm{Im}\int_{x_n}^{x} q| = |\mathrm{Im}\int_{x}^{x'} q| \leq I_{A+C}.$$

Since for any x in $[x_n, x_{n+1})$,

$$v_2(x) = \mathrm{Im}\int_{x_n}^{x} q - \left[(x-x_n)/(x_{n+1}-x_n)\right]\mathrm{Im}\int_{x_n}^{x_{n+1}} q = t_1 - \alpha t_2$$

where t_1 and t_2 are in I_{A+C} and $0 \leq \alpha \leq 1$, we have $|v_2| \leq I_{A+C}$.

If f is in C_0^∞ with $||f|| = 1$, then for $i = 1, 2$ and any $\varepsilon_i > 0$,

$$|\int^{\infty} v_i'|f|^2| = |\int_0^{\infty} v_i(f'\bar{f}+f\bar{f}')| \leq \varepsilon_i\int_0^{\infty}|f'|^2 + \varepsilon_i^{-1}\int_0^{\infty}|v_i|^2|f|^2.$$

Thus for any $\delta > 0$,

$$\operatorname{Re}(T_0f,f) - \delta|\operatorname{Im}(T_0f,f)| = \int_0^\infty |f'|^2 + \int_0^\infty (q_1+A+v_1')|f|^2$$
$$-\delta\left|\int_0^\infty (q_2+v_2')|f|^2\right|$$
$$\geq (1-\varepsilon_1-\delta\varepsilon_2)\int_0^\infty |f'|^2 + \int_0^\infty (q_1+A)(1-\delta M_{A+C})|f|^2$$
$$- \int_0^\infty (v_1^2/\varepsilon_1 + \delta v_2^2/\varepsilon_2)|f|^2.$$

By setting $\varepsilon_2 = (1-\varepsilon_1)/\delta$ and using the estimates for v_1 and v_2 with the abbreviation $J = Ca + (Aa+N)/2$,

$$\operatorname{Re}(T_0f,f) - \delta|\operatorname{Im}(T_0f,f)| \geq \int_0^\infty (q_1+A)(1-\delta M_{A+C})|f|^2$$
$$- (J^2/\varepsilon_1 + \delta I_{A+C}^2/(1-\varepsilon_1)).$$

For fixed δ, the minimum value of the last term is $(J+\delta I_{A+C})^2$. If $\delta \leq M_{A+C}^{-1}$, then for $x = \operatorname{Re}(T_0f,f)$ and $y = \operatorname{Im}(T_0f,f)$ we have

$$|y| \leq (x+J^2)/\delta + 2JI_{A+C} + \delta I_{A+C}^2 + \int_0^\infty (q_1+A)(M_{A+C}-\delta^{-1})|f|^2$$
$$\leq (x-A+J^2)/\delta + \delta I_{A+C}^2 + 2JI_{A+C} + AM_{A+C}.$$

Thus, minimizing over δ,

$$|y| \leq 2I_{A+C}\left[J+(x-A+J^2)^{1/2}\right] + AM_{A+C}$$

if $x - A + J^2 \leq (I_{A+C}/M_{A+C})^2$ and

$$|y| \leq M_{A+C}(x-A+J^2) + 2I_{A+C}J + AM_{A+C} + I_{A+C}^2/M_{A+C}$$

if $x - A + J^2 > (I_{A+C}/M_{A+C})^2$.

These inequalities hold for each $C > 0$. Thus, since $I_{A+C} \to I_A$ and $J \to (Aa+N)/2$ as $C \to 0$, (iii) is established.

Finally, for any $C > 0$ we have $q = q_1 + A + v_1' + i(q_2+v_2')$ which corresponds to the representation

$$L[y] = -\left[y'-(v_1+iv_2)y\right]' - (v_1+iv_2)y' + (q_1+A+iq_2)y$$

with

$$L_R[y] = -(y'-v_1y)' - v_1y' + (q_1+A)y$$

Since v_1 is bounded, it is clear that $q_1 + A + K \geq (1+\varepsilon)v_1^2$ for some positive constant K. Hypothesis (ii) of Theorem 6 is clear for L + K from the estimates used to establish (iii) of Theorem 7. Since T and

T + K have the same domain, (i) follows from Theorem 6 and the proof of Theorem 7 is complete.

COROLLARY 4. *Suppose that* q *is real valued and that for some* $a > 0$, $A \geq 0$, $C > 0$, *and each* $x \geq 0$,

(i) *there is* t, $x < t \leq x + a$ *such that* $(1/a) \int_x^t q \geq A$,

(ii) $\int_x^{x+a} |q| \leq C$.

Then $T \geq A - (C+Aa)^2/16 \geq A - C^2/4$.

Remark. With similar hypotheses, Eastham [3] obtained the estimate $T \geq A - C^2$.

Proof. It suffices to show that the hypotheses of Theorem 7 hold with $N \leq (C - Aa)/2$. If $\int_x^w q = -K$, then with t as in (i), $\int_x^w q \geq K + Aa$ so that $C \geq \int_x^w |q| + \int_w^t |q| \geq 2K + Aa$. Hence $K \leq (C-Aa)/2$ so that N is as asserted.

COROLLARY 5. *If for some real constants* c_1 and c_2, *and* all $x \geq 0$

$$\left|\int_0^x (\text{Re } q - c_1)\right| \leq K; \quad \left|\int_0^x (\text{Im } q - c_2)\right| \leq K,$$

for some constant $K > 0$, *then the numerical range of* T *is contained in*

$$\{x+iy: x \geq c_1 - s_1^2/4 = x_0, \ |y-c_2| \leq s_2(\sqrt{x-x_0} + s_1/2)\}$$

where $s_1 = \sup_{a,b} \left|\int_a^b (\text{Re } q - c_1)\right|$, $s_2 = \sup_{a,b} \left|\int_a^b (\text{Im } q - c_2)\right|$.

Remarks. 1. The hypotheses hold with $c_1 = c_2 = 0$ if q is in $\mathcal{L}^1$. Then $s_1 \leq \int_0^\infty |\text{Re } q| = K_1$, $s_2 \leq \int_0^\infty |\text{Im } q| = K_2$ and s_1 and s_2 may be much smaller than K_1 and K_2 if Re q and Im q are oscillatory. This result was proved by Evans [4] with $2K_1$ and $2K_2$ in place of s_1 and s_2.

2. The hypotheses also hold if q is periodic with period P if $c_1 = \int_0^P \text{Re } q/P$, $c_2 = \int_0^P \text{Im } q/P$. More generally, they hold if q is almost periodic and $q-c$ has an almost periodic primitive for some complex constant c.

Proof. We may assume $c_1 = c_2 = 0$. Set

$$M = \sup_x \int_0^x \text{Re } q, \quad m = \inf_x \int_0^x \text{Re } q, \quad v_1(x) = -(M+m)/2 + \int_0^x \text{Re } q.$$

Then $|v_1| \leq (M-m)/2 = s_1/2$. Similarly we may define $v_2(x)$ with $v_2' = \text{Im } q$ and $|v_2| \leq s_2/2$. Now a repetition of the arguments in the proof of Theorem 7 leads to

$x = \text{Re}(Tf,f) \geq -s_1^2/4$ and $x - \delta|y| \geq -(s_1^2/4\varepsilon_1 + \delta^2 s_2^2/4(1-\varepsilon_1))$ for any $\varepsilon_1 > 0$ and $\delta > 0$. The desired inequality for y is then obtained as in the proof of Theorem 7.

COROLLARY 6. <u>Let</u> $B(X) = \sup\{|\text{Im} \int_x^y q| : x \geq X, x \leq y \leq x + a\}$. <u>If</u> (i) <u>of</u> Theorem 7 <u>holds and</u> $B(X) \to 0$ as $X \to \infty$, <u>then the essential spectrum of</u> T <u>is real</u>.

<u>Remark</u>. The condition holds if $\text{Im } \int_0^x q$ approaches a finite limit as $x \to \infty$. This is true, for instance if $\text{Im}q(x) = x^\alpha \sin x^\beta$ with $\alpha < \beta - 1$. In particular it holds if Im q is in $\mathcal{L}^1$. If also $\lim_{x\to\infty} \int_0^x \text{Re } q$ exists (or Re q is in $\mathcal{L}^1$) then by applying Corollary 5 to L[X] we obtain the result that the essential spectrum of T is $[0,\infty)$.

<u>Proof</u>. Replacing q by q + a if necessary, we may assume $A \geq 1$. Then the constants $M_A(X)$, $I_A(X)$ defined as in Theorem 7 for the restriction L(X) of L to $[X,\infty)$ approach 0 as $X \to \infty$. It then follows from Theorem 7 that any point not on the real axis will fail to be in the numerical range of T(X) for X sufficiently large. Thus it is not in the essential spectrum of T = T(0).

We now extend Theorem 7 to an operator in a weighted Hilbert space whose potential q may oscillate in an arbitrarily wild fashion provided the average behavior of Re q is sufficiently positive. We do this by using the device of weighted averages introduced in [10]. However the function h is not the same as in [10].

THEOREM 8. <u>Let</u> $L[y] = k^{-1}[-(py')' + qy]$. <u>Let</u> g <u>and</u> h <u>be positive left continuous functions such that</u> $h(x) \geq c_1 k(w)$ <u>and</u> $(g(x))^2 h(x) \leq p(w)$ <u>whenever</u> $x \leq w \leq x + g(x)$. <u>Suppose that for some appropriate</u>

region R, *some constants* $A > 0$ *and* $\theta \le 1$, *and each* $x \ge 0$,

(i) *there is* t, $x < t \le x + \theta g(x)$ *such that* $(1/gh(x))\mathrm{Re}\int_x^t q \ge A$,

(ii) $A/\theta - (A+N)^2 > 0$,

(iii) $(1/gh(x))\int_x^w q \in R$ *whenever* $x \le w \le x + g(x)$.

Then $\mathrm{Re}\, T \ge c_1[(A/\theta) - (A+N)^2] = x_0$ *and the numerical range of* T *is contained in*

$$\{x+iy : x \ge x_0,\quad |y| \le \begin{cases} 2I_A\sqrt{c_1(x-x_0)} + y_1, & x-x_0 \le c_1(\delta_0 I_A)^2, \\ \delta_0^{-1}x, & x-x_0 > c_1(\delta_0 I_A)^2. \end{cases}\}$$

Here $y_1 = c_1[2I_A(A+N) + AM_A/\theta]$, N is now defined by $-N = \inf\{\mathrm{Re}\, z : z \in R\}$, and δ_0 is the positive root of $(A+N+\delta I_A)^2 + (\delta M_A - 1)(A/\theta) = 0$.

Proof. Set $x_1 = 0$ and if x_n has already been defined, set

$$x_{n+1} = \inf\{x > x_n : \mathrm{Re}\int_{x_n}^x q \ge Agh(x_n)\}.$$

Define q_1, q_2, v_1, and v_2 as in the proof of Theorem 7. Then on $[x_n, x_{n+1})$, $q_1(x) \ge (A/\theta)h(x_n)$, $|q_2(x)| \le M_A\, q_1(x)$, $|v_1(x)| \le (A+N)gh(x_n)$, and $|v_2(x)| \le I_A\, gh(x_n)$. Proceeding as in the proof of Theorem 7 with f in C_0^∞, $\int_0^\infty |f|^2 k = 1$,

$$\mathrm{Re}(T_0 f, f) \ge \int_0^\infty (q_1 - v_1^2/p)|f|^2 = \int_0^\infty k^{-1}(q_1 - v_1^2/p)|f|^2 k.$$

Now $k^{-1}(q_1 - v_1^2/p) \ge k^{-1}h\,[(A/\theta) - (A+N)^2] \ge x_0$ so the first assertion is established. For the second, we have as in the proof of Theorem 7 that if $x = \mathrm{Re}\,(T_0 f, f)$, $y = \mathrm{Im}(T_0 f, f)$, then

$$\begin{aligned} x - \delta|y| &\ge \int_0^\infty [(A/\theta)(1-\delta M_A) - (A+N+\delta I_A)^2](h/k)|f|^2 k \\ &\ge c_1[(A/\theta)(1-\delta M_A) - (A+N+\delta I_A)^2] \end{aligned}$$

as long as the right side is nonnegative, that is, as long as $\delta \le \delta_0$. The inequalities for $|y|$ now follow as before by choosing the optimal δ.

If we restrict the behavior of the imaginary part of q somewhat more, we can contain the numerical range of T within a parabolic region.

THEOREM 9. *Let* L, g, *and* h *be as in Theorem* 8 *with also* $h(x) \le c_2 k(w)$ *when* $x \le w \le x + g(x)$. *Suppose that for some* $A > 0$, $B > 0$, and $\theta \le 1$, (i) *of Theorem* 8 *holds and also*

$$\text{(ii')} \quad \left[1/gh(x)\right] \operatorname{Re}\int_x^w q \ge -N, \quad \left[1/gh(x)\right] \left|\operatorname{Im}\int_x^w q\right| \le B$$

whenever $x \le w \le x + g(x)$.

Then $\operatorname{Re} T \ge x_0' = c[(A/\theta) - (A+N)^2]$ *where* $c = c_1$ *if this expression is nonnegative and* $c = c_2$ *otherwise. The numerical range of* T *is contained in*

$$\{x+iy : x \ge x_0',\ |y| \le 2B\sqrt{2c(x-x_0^c)} + Bc[4(A+N)+1]\}.$$

Here $c = c_1$ if $A/\theta - (A+N)^2 > 0$ and $x - x_0' \le 4c_1 B^2 \delta_0$ where δ_0 is the positive root of $(A+N + 2\delta B)^2 + \delta B - A/\theta = 0$, $c = c_2$ otherwise, and x_0^c is the value of x_0' associated with that value of c.

Proof. Choose $\{x_n\}$, q_1, and v_1 as in the proof of Theorem 8. The inequality $\operatorname{Re} T \ge x_0'$ follows just as in Theorem 8.

Set $x_1' = 0$, and $x_{n+1}' = x_n' + g(x_n')$ and on $[x_n', x_{n+1}')$ define

$$q_2(x) = (x_{n+1}' - x_n')^{-1} \int_{x_n}^{x_{n+1}} \operatorname{Im} q, \text{ and } v_2(x) = \operatorname{Im}\int_{x_n}^{x} (q-q_2). \quad \text{Then}$$

$|q_2(x)| \le Bh(x_n)$ and $|v_2(x)| \le 2g(x_n)h(x_n)$. We obtain

$$x - \delta|y| \ge \int_0^\infty [(A/\theta) - \delta B - (A+N + 2\delta B)^2](h/k)|f|^2 k$$

$$\ge c[(A/\theta) - \delta B - (A+N+2\delta B)^2]$$

where $c = c_1$ or c_2 depending on whether the expression is positive or negative. The inequality for $|y|$ then follows upon choosing the optimal value of δ as before.

References

1. F. V. Atkinson and W. D. Evans, On solutions of a differential equation which are not of integrable square, Math. Z. 127 (1972), 323-332.
2. J. S. Bradley, D. B. Hinton and R. M. Kauffman, On the minimization of singular quadratic functionals, Proc. Roy. Soc. Edinburgh 87A(1981) 193-208.
3. M.S.P. Eastham, Semi-bounded second order differential operators, Proc. Roy. Soc. Edinburgh 72A (1978), 9-16.
4. W. D. Evans, On the spectra of Schrödinger operators with a complex potential, Math. Ann. 255 (1981), 57-76.
5. P. Hartman, Differential equations with non-oscillatory eigenfunctions, Duke Math. J. 15 (1948), 697-709.
6. T. Kato, Perturbation Theory for Linear Operators, second edition (Berlin, Springer-Verlag, 1980).
7. R. M. Kauffman, T. T. Read and A. Zettl, The Deficiency Index Problem for Powers of Ordinary Differential Expressions (Lecture Notes in Mathematics No. 621, Springer-Verlag, 1977).
8. R. M. Kauffman, Polynomials and the limit-point condition, Trans. Amer. Math. Soc. 201 (1975), 347-366.
9. I. Knowles and D. Race, On the point spectra of complex Sturm-Liouville operators, Proc. Roy. Soc. Edinburgh 85A (1980), 263-290.
10. T. T. Read, Factorization and discrete spectra for second-order differential expressions, J. Differential Equations 35 (1980) 388-406.
11. T. T. Read, A limit-point criterion for expressions with intermittently positive coefficients, J. London Math. Soc. (2) 15 (1977), 271-276.
12. T. T. Read, A limit-point criterion for expressions with oscillatory coefficients, Pacific J. Math. 66 (1976) 243-255.
13. T. T. Read, Sectorial second order differential operators, to appear in Proceedings of the 1982 Dundee Conference on Ordinary and Partial Differential Equations (Lecture Notes in Mathematics, Springer-Verlag).

The Cauchy and Backward-Cauchy Problem for a
Nonlinearly Hyperelastic/Viscoelastic Infinite Rod

R. Saxton

In this paper we consider both simpler and more general forms of a nonlinear partial differential equation arising from a particular model of a rod vibrating longitudinally (cf. Jaunzemis [4], Love [6]). This is the Euler equation derived from a Lagrangian.

$$L \equiv \int_{t_1}^{t_2} \int_{\Omega} [\tfrac{1}{2}u_t^2 + \tfrac{1}{2}f(u_x)u_{xt}^2 - \hat{W}(u_{x,} f)] \, dx \; dt, \quad \Omega \subseteq \mathbb{R} . \qquad 1.$$

Setting $\delta L = 0$ provides the equation of motion. Here $f, \hat{W}$ are generally nonlinear functions ($\hat{W}$ is the strain-energy density of the material and f is a function corresponding to the product of a non-linear Poisson's ratio with a variable radius of gyration about the central axis of the rod). Elsewhere ([9]) we consider $f(u_x)$ to become singular as $u_x \to -1$ for physical reasons of material inversion, but now we generalize in another direction and let $f(\phi) = \phi$.
If we define

$$\hat{W}(u_{x,} u_x) = W(u_x), \; \frac{dW(u_x)}{du_x} = \sigma(u_x), \qquad 2.$$

then the equation of motion obtained is

$$u_{tt} - u_{xxtt} - \sigma_x(u_x) = 0. \qquad 3.$$

We may consider more generally the equation

$$u_{tt} - u_{xxtt} - T_x(x,t,u_x,u_{xt},u,u_t) = 0 \qquad 4.$$

(see [8]) where the nonlinear function T depends on the two dependent and independent variables as well as on the first spatial derivatives of u and u_t. However in the case of elasticity it is appropriate for T to have dependence only on x, u_x and u_{xt}, and we consider primarily homogeneous materials for which T has the form

$$T = \sigma(u_x) + \tau(u_{xt}) . \qquad 5.$$

In places, we indicate also how one treats the more general problem 4. The paper proves local and global existence results for various spaces with $x \in \mathbb{R}$, $t \geq 0$. Some of these spaces are chosen to demonstrate possible behaviour of solutions as $x \to \pm \infty$, and are interesting in view of solitary wave solutions (see [10]) which are known to exist.

First several restrictions are imposed on σ, τ and W to permit sufficient tractability in investigating local and global existence.

Presence or absence of viscous terms does not affect the question of global existence since the equation will be shown to possess globally unique solutions even in the hyperelastic case, when $\tau \equiv 0$. Any assumptions of viscous damping such as $\phi\tau(\phi) > 0 \ \forall \ \phi \in \mathbb{R}$ only help to improve matters by letting solutions decay to equilibrium as $t \to + \infty$. On the other hand when $\phi\tau(\phi) < 0$ is permitted, this may produce solutions growing as $t \to + \infty$ and is equivalent to making the change of independent variable $t \to -t$ and looking at the case of $\phi\tau(\phi) > 0$, as $t \to - \infty$. In certain cases this may lead to blow-up of solutions in finite time, as in the undamped nonlinear string (Lax [5]). Here we consider three cases, generally grouped together for ease of presentation, in which blow-up does not occur i.e. for which solutions exist for all time $t > 0$.

Hypothesis H1) Let $\tau(.)$ be locally Lipschitz - continuous, i.e. $\forall \ \phi,\psi \in \mathbb{R}$ such that for some $R>0$ $|\phi|<R, |\psi|<R$, there exists a constant $\Gamma(R) > 0$, $\Gamma(R) \to +\infty$ as $R \to \infty$ with $|\tau(\phi) - \tau(\psi)| \leq \Gamma(R)|\phi - \psi|$.

H1)i) $\phi\tau(\phi) > 0, \ \forall \phi \in \mathbb{R}$,

H1)ii) $\tau(\phi) \equiv 0$,

H1)iii) $\phi\tau(\phi) = -\frac{\alpha\phi}{2}$, some $\alpha > 0$.

It will be seen later that the appropriate 'energy' estimate becomes less useful as we proceed from Hi) through Hii) to Hiii). Further hypotheses required for $\sigma(.)$, $W(.)$ are

H2) Let $\sigma(.)$ be locally Lipschitz -continuous (without loss of generality we may take the same Lipschitz constant $\Gamma(R)$ as in H1), and set $\sigma(o) \equiv 0$.

H3) Assume $W(\phi) \geq 0$, $\forall\, \phi \in \mathbb{R}$.

We provide some notation.

$L^p(\mathbb{R})$ denotes the space of measurable real-valued functions on $\mathbb{R}$ for which

$$\| f \|_p \equiv \left(\int_{-\infty}^{\infty} |f(x)|^p \, dx \right)^{\frac{1}{p}} < \infty , \quad 1 \leq p < \infty ,$$

or

$$\| f \|_\infty \equiv \operatorname{ess}_{x \in \mathbb{R}} \sup |f(x)| < \infty, \quad \text{when } p = \infty .$$

The Sobolev spaces $W_o^{m,p}(\mathbb{R})$, $1 \leq p < \infty$, $W^{m,\infty}(\mathbb{R})$, $m \in \mathbb{N}$, consist of those functions in $L^p(\mathbb{R})$ all of whose generalised derivatives up to and including order m belong to $L^p(\mathbb{R})$.

We define norms on $W_o^{m,p}(\mathbb{R})$, $W^{m,\infty}(\mathbb{R})$ by

$$\| f \|_{m,p} = \left(\sum_{j=0}^{m} \| \frac{d^j f}{dx^j} \|_p^p \right)^{\frac{1}{p}} , \quad 1 \leq p < \infty,$$

or

$$\| f \|_{m,\infty} = \sum_{j=0}^{m} \| \frac{d^j f}{dx^j} \|_\infty ,$$

respectively.

Each of the spaces $W_o^{m,p}(\mathbb{R})$, $W^{m,\infty}(\mathbb{R})$ is a Banach space under the given norm.

let Y be a Banach space and let A be a bounded or semibounded set in $\mathbb{R}$. Then the Banach space $C^k(A;Y)$, $k \in \mathbb{N}\cup\{0\}$ is the class of k-times continuously differentiable mappings $\frac{d^j u}{dt^j}(t) : A \to Y$, $0 \leq j \leq k$.

$C^k(A ; Y)$ has the norm

$$|u|_{k,Y} \equiv \sum_{j=0}^{k} \sup_{t \in A} \left\| \frac{d^j u}{dt^j}(t) \right\|_Y .$$

Similarly, we define the Banach space $L^1(A ; Y)$ to be the class of measurable mappings $u(t) : A \to Y$ such that for $u \in L^1(A ; Y)$

$$\|u\|_{1,Y} \equiv \int_A \|u(t)\|_Y \, dt < \infty .$$

Now we want to consider local existence of solutions to equation 4. with τ given by 5. The domain of x will be taken to be the real line over which the following Cauchy data are given -

$$u_o(x) \equiv u(x,0) , \quad u_1(x) \equiv u_t(x,0) , \quad x \in \mathbb{R} . \qquad 6.$$

We present the proof in two stages. The Theorem applies to the space $x \in W^{1,\infty}(\mathbb{R}) \cap W_o^{1,2}(\mathbb{R})$ which is the broadest to which our method applies. The possibility of working in a larger space $W_o^{1,p}(\mathbb{R})$, $p<\infty$, using other techniques such as a monotonicity approach is not considered here, due partly to the restrictions thereby imposed on σ but also because there appear to be difficulties in the conservative case ($\tau = 0$). The Corollary that follows the Theorem is concerned with restricting the initial data to special classes of functions which decay at various rates as $|x| \to \infty$ and showing that the solution to 4., 5., 6. behaves likewise (this is independent of the choice of subsidiary H1i), H1ii) or H1iii)).

<u>Theorem 1</u> Let hypotheses H1) and H2) be satisfied, and suppose that $u_o(x), u_1(x) \in W^{1,\infty}(\mathbb{R}) \cap W_o^{1,2}(\mathbb{R})$. Then there exists a solution $u(x,t)$ to 4., 5. and 6. belonging to $C^1([o,\tau[; W^{1,\infty}(\mathbb{R}) \cap W_o^{1,2}(\mathbb{R}))$

defined on a maximal interval $[o,\tau[$, $\tau \leq \infty$. If $\tau<\infty$, then

$$\|u(.,t)\|_{1,\infty} + \|u_t(.,t)\|_{1,\infty} \to \infty \text{ as } t \to \tau-,$$

Remark 1 The proof of the Theorem is divided into two steps, one of which we postpone until later. Here we show that equation 4. is formally equivalent to an integro-differential equation which we solve by contraction mapping. The Theorem will be proved when it is shown that the solution found in this way solves 4. in a weak sense. Henceforth we consider this to be true.

Proof of Theorem 1

We write 4., 5., as

$$(1 - \partial_x^2)u_{tt} = \partial_x (\sigma + \tau). \qquad 7.$$

and we assume $u(\pm\infty,t) = 0$, $\forall\, t > 0$.

Formally operating on 7. with $(1 - \partial_x^2)^{-1}$ we obtain

$$u_{tt}(x,t) = \int_{-\infty}^{\infty} G(x,\xi)\partial_\xi(\sigma + \tau)\, d\xi$$

$$= - \int_{-\infty}^{\infty} G_\xi(x,\xi)\, (\sigma + \tau)\, d\xi \qquad 8.$$

where

$$G(x,\xi) = \tfrac{1}{2}e^{-|x-\xi|} , \qquad 9.$$

and, on integrating twice with respect to time, 8. becomes

$$u(x,t) = u_o(x) + tu_1(x) - \int_o^t \int_{-\infty}^{\infty} (t - \eta)\, G_\xi(x,\xi)\, (\sigma + \tau)\, d\xi\, d\eta. \qquad 10.$$

The derivative of 10. with respect to x is given by

$$u_x(x,t) = u_o'(x) + tu'_1(x) - \int_o^t\int_{-\infty}^{\infty} (t-\eta)\, G_{\xi x}(x,\xi)(\sigma+\tau)\, d\xi d\eta$$

$$- \int_o^t (t - \eta)\, (\sigma + \tau)\, d\eta \qquad 11.$$

We try to find a fixed point in $W^{1,\infty}(\mathbb{R}) \cap W_o^{1,2}(\mathbb{R})$ for the operator equation

$$(Au)(x,t) = u_o(x) + tu_1(x) - \int_o^t \int_{-\infty}^{\infty} (t-\eta)\, G_\xi(x,\xi)(\sigma+\tau)\, d\xi\, d\eta \qquad 12.$$

using a standard application of the contraction mapping principle (see e.g. [1]), which then implies the existence of a unique solution for 10. To show that the fixed point exists we need to demonstrate that A maps the ball

$$B(R) \equiv \{u(x,t) \in C^1([o,T]\ ;\ W^{1,\infty}(\mathbb{R}) \cap W_o^{1,2}\ (\mathbb{R}) :$$
$$:\ |u|_{1,\ W^{1,\infty}} + |u|_{1,\ W^{1,2}} \leq R\ \} \qquad 13.$$

into itself for T sufficiently small, and that A is a contraction, i.e. that for some $R, T > 0$, $0 < \theta < 1$ and all $u,\ v \in B(R)$,

$$|Au|_{1,x} \leq |u|_{1,x} \text{ and } |Au - Av|_{1,x} \leq \theta\ |u-v|_{1,x}\ , \qquad 14.$$

where $|\ |_{1,x}$ denotes the norm inside 13.

To satisfy 14. it may in fact be shown ([8]) that because of the similarity in the steps of the calculations it is sufficient to verify that

$$|(Au)_x|_D \leq |u_x|_D \text{ and } |(Au)_x - (Av)_x|_D \leq \theta |u_x - v_x|_D \qquad 15.$$

where

$$|\phi(.,.)|_D \equiv \sup_{t \in [o,T]} (\|\phi(.,t)\|_\infty + \|\phi(.,t)\|_2 + \|\phi_t(.,t)\|_\infty + \|\phi_t(.,t)\|_2) \qquad 16.$$

is the norm associated with $C^1([0,T];\ L^\infty(\mathbb{R}) \cap L^2(\mathbb{R}))$.

Writing

$$|\psi(.)|_E \equiv \|\psi(.)\|_\infty + \|\psi(.)\|_2 \qquad 17.$$

and noticing that $(Au)_x$ is given by the right side of 11., we find that

$$|(Au)_x|_D \leq |u_o'|_E + (1+T)|u_1'|_E +$$

$$+ \left| \int_{o}^{t} \int_{-\infty}^{\infty} (t - \eta)\, G_{\xi x}(\sigma + \tau)\, d\xi d\eta \right|_{D}$$

$$+ \left| \int_{o}^{t} (t - \eta)(\sigma + \tau)\, d\eta \right|_{D}$$

$$\leq |u_o'|_E + (1 + T)|u_1'|_E + C\left| \int_{o}^{t} (t - \eta)(\sigma + \tau) d\eta \right|_D$$

where $C > 0$ depends only on the Greens function. Thus by 13., H1. and H2.,

$$|(Au)_x|_D \leq |u_o'|_E + (1 + T)|u_1'|_E + C\Gamma(R)(1 + T)\int_o^t |u_x|_D\, d\eta \quad ,$$

that is

$$|(Au)_x|_D \leq |u_o'|_E + (1 + T)|u_1'|_E + CR\Gamma(R)T(1 + T) \quad , \qquad 18.$$

where the inequalities follow from the definition of the norms and properties of the Bochner integral (Yosida [ ]). Hence 18. shows that provided there exists some $R > 0$, $T > 0$ such that there holds

$$|u_o'|_E + (1 + T)|u_1'|_E + CR\Gamma(R)T(1 + T) \leq R \qquad 19.$$

then $AB(R) \subset B(R)$

In a similar way, A can be shown to be a contraction provided $R, T > 0$ exist such that for some finite constant $C' > 0$

$$C'\,\Gamma(R)\, T\, (1 + T) \leq \theta < 1 \quad . \qquad 20.$$

Conditions 19. and 20. may be simultaneously satisfied by choosing large enough R and small enough $T > 0$, which thereby proves the existence of a unique fixed point for 12. and hence the existence of a unique solution $u(x,t)$ to the integral equation 10., with $u \in C^1([0,T]\, ;\, W^{1,\infty}(\mathbb{R}) \cap W_o^{1,2}(\mathbb{R}))$.

The last part of the Theorem is proved by a routine continuation argument as found in Reed ([7]), replacing the interval of existence $[0,T]$ by $[0,\tau[$, where τ is the supremum of the T over which existence holds.

For the Corollary to this Theorem we are interested in finding whether if the initial data approach zero at a certain rate as $x \to \pm\infty$ then the solution to the problem does so also. More clearly, by way of example, we might ask that given $e^{\alpha|x|}(|u_o(x)| + |u_1(x)|) \to 0$ as $|x| \to \infty$, α some positive number, does $e^{\alpha|x|}(|u(x,t)| + |u_t(x,t)|) \to 0$, $t>0$ as $|x| \to \infty$? To treat this question more fully we define a general class of function, together with associated Banach spaces.

Let $\psi = \psi(x)$ satisfy the conditions

$$1 \leqq \psi(x) < e^{\alpha|x|}, \quad 0 \leqq \alpha < \tfrac{1}{2} \quad , \tag{21.}$$

$$\psi(x + y) \leqq \psi(x)\psi(y) \quad \forall\, x, y \in \mathbb{R} \quad , \tag{22.}$$

$$\psi(x) \text{ continuous on } \mathbb{R} \tag{23.}$$

and

$$-K < \psi'(0-) \leqq \psi'(0+) < K \quad \text{for some } K > 0 \, . \tag{24.}$$

Further let $J = J(x) > 0$, let $Z = C^1([0,T];\ W^{1,\infty}(\mathbb{R}) \cap W_o^{1,2}(\mathbb{R}))$ and

$$Z(J) = \{u \in Z : Ju \text{ and } Ju_x \in C^1([0,T] ; L^\infty(\mathbb{R}) \cap L^2(\mathbb{R}))\} \tag{25.}$$

with corresponding norm (see 14.)

$$\|u\|_{Z(J)} \equiv |u|_{1,x} + |Ju|_D + |Ju_x|_D \quad . \tag{26.}$$

Note that when $J(x) = 1\ \forall x \in \mathbb{R}$, $Z(J)$ reduces to Z since the norms are then equivalent.

We now prove

<u>Corollary 1</u>. Let $\psi(x)$ satisfy 21. - 24. Provided ψu_o, $\psi u_o'$, ψu_1 and $\psi u_1'$ belong to $L^2(\mathbb{R}) \cap L^\infty(\mathbb{R})$, there exists a unique solution $u(x,t)$ to

4. under the conditions of Theorem 1, with $u(x,t) \in Z(\psi)$ defined over a maximal interval of existence $[0,\tau[$. If $\tau < \infty$, then $\|u\|_{Z(\psi)} \to \infty$ as $t \to \tau-$.

<u>Proof</u> The procedure of Theorem 1 can be repeated with the ball $B(R)$ now replaced by

$$B_Z(R) \equiv \{u \in Z(\psi) \; : \; \|u\|_{Z(\psi)} \leq R\} \quad . \tag{27.}$$

It is sufficient to show that if $u \in Z(\psi)$ then $Au \in Z(\psi)$ since the fixed point argument then completes the proof as before. We therefore establish estimates corresponding to those of Theorem 1 by considering the representative term

$$\psi(x)(Au)_x(x,t) = \psi(x)u_o'(x) + t\psi(x)u_1'(x)$$
$$- \int_0^t (t - \eta) \; \psi(x) \int_{-\infty}^{\infty} G_{\xi x}(x,\xi)(\sigma + \tau)d\xi d\eta$$
$$- \int_0^t (t - \eta) \; \psi(x)(\sigma + \tau) \; d\eta \; . \tag{28.}$$

We note that since $\psi u_x \in B_Z\,(R)$ then $\psi(\sigma + \tau) \in L^1(0, T \; ; L^2\,(\mathbb{R}) \cap L^\infty(\mathbb{R}))$ for appropriate $T>0$. This can be seen using hypotheses H1., H2. :-

$$\int_0^t (t - \eta)\,\|\psi\,(.)(\sigma(.,\eta)+\tau(.,\eta))\|_2\,d\eta$$

$$\leq \int_0^t (t - \eta)\,\|\psi\,(.)\Gamma(R)(|u_x(.,\eta)|+|u_{x\eta}(.,\eta)|)\|_2\,d\eta$$

for $\sup_{t\in[0,T]} \{\|u_x\|_2 + \|u_{xt}\|_2\} \leq R.$

Thus

$$\int_0^t (t - \eta) \| \psi(.)(\sigma(.,\eta) + \tau(.,\eta)) \|_2 d\eta \le 2T^2\Gamma(R)\ R. \qquad 29.$$

A similar result may likewise be established for

$$\int_0^t (t - \eta) \| \psi(.)(\sigma(.,\eta) + \tau(.,\eta)) \|_\infty d\eta ,$$

and so the last term of 28. is bounded in $C^1([0,T];L^2(\mathbb{R})\cap L^\infty(\mathbb{R}))$ for $0\le t\le T$. We need to obtain the same type of estimates as 29. for the second last term also. These are obtained as follows. Since $\psi(\sigma + \tau) \in L^2(\mathbb{R}) \cap L^\infty(\mathbb{R})$ for almost all t, it is only necessary to show there exists a finite constant $C > 0$ such that

$$\| \psi(.) \int_{\infty}^{\infty} G_{\xi x}(.,\xi) f(\xi) d\xi \|_\infty < C \| \psi(.) f(.) \|_\infty , \qquad 30.$$

and

$$\| \psi(.) \int_{\infty}^{\infty} G_{\xi x}(.,\xi) f(\xi) d\xi \|_2 < C \| \psi(.) f(.) \|_2 , \qquad 31.$$

for all $f(x)$ with $\psi f \in L^2(\mathbb{R}) \cap L^\infty(\mathbb{R})$.

We denote by K and M, $\|\psi(.)f(.)\|_\infty$ and $\|\psi(.)f(.)\|_2$ respectively. Hence, for almost all $\xi, |f(\xi)| \le \frac{K}{\psi(\xi)}$.

Thus

$$\tfrac{1}{2}\psi(x) \int_{-\infty}^{\infty} e^{-|x-\xi|} |f(\xi)| d\xi \le \frac{K}{2}\int_{-\infty}^{\infty} e^{-|x-\xi|} \frac{\psi(x)}{\psi(\xi)} d\xi$$

$$\le \frac{K}{2}\int_{\infty}^{\infty} e^{-|x-\xi|} \psi(x-\xi) d\xi \quad , \text{ by 22.} ,$$

$$\le \frac{K}{2}\int e^{(\alpha-1)|x-\xi|} d\xi \quad , \text{ by 21.} ,$$

$$\le C \| \psi(.)f(.) \|_\infty$$

where $C = C(\alpha) < \infty$ for $\alpha < 1$, and so certainly for $0\le\alpha<\frac{1}{2}$. Hence 30. follows at once on taking the essential supremum of the first and last parts of the inequality.

Next we have,

$$\| \psi(.)\int_{-\infty}^{\infty} |G_{\xi x}(.,\xi)f(\xi)|d\xi \|_2^2$$

$$\leq \int_{-\infty}^{\infty} \psi^2(x)\left[\tfrac{1}{2}\int_{-\infty}^{\infty} e^{-|x-\xi|}|f(\xi)|d\xi\right]^2 dx$$

$$\leq \tfrac{1}{4}\int_{-\infty}^{\infty} \psi^2(x)\int_{-\infty}^{\infty} e^{-|x-\xi|}d\xi \int_{-\infty}^{\infty} e^{-|x-\xi|}f^2(\xi)d\xi dx$$

$$\leq \tfrac{1}{2}\int_{-\infty}^{\infty}\int_{-\infty}^{\infty}\psi^2(x)e^{-|x-\xi|}f^2(\xi)dxd\xi \quad \equiv \quad I,$$

where we used the Cauchy Schwartz inequality for the second last line, and in the last line interchanged the order of integration by applying Fubini's Theorem. On noting that by 22. ,

$$I \leq \tfrac{1}{2}\int_{-\infty}^{\infty} \psi^2(\xi)f^2(\xi)\int_{-\infty}^{\infty} \psi^2(\xi-x)e^{-|x-\xi|}dx\, d\xi$$

$$\leq \tfrac{1}{2}\int_{-\infty}^{\infty} \psi^2(\xi)f^2(\xi)\int_{-\infty}^{\infty} e^{(2\alpha-1)|x-\xi|}dxd\xi$$

$$\leq C^2\| \psi(.)f(.)\|_2^2 \quad ,$$

we obtain 31., where $C = C(\alpha) < \infty$ for $0 \leq \alpha < \frac{1}{2}$.

It is now straightforward to complete the proof of the Corollary using again the first part and then continuing as outlined in the proof of the Theorem. Fuller details are in [8], Yosida [11], Elcrat and MacLean [3].

We make the further remark that the above procedure may also be applied in some other equations - here for example, when $T = T(x,t,u_x,u_{xt},u,u_t)$ - but in this case the proof becomes slightly more elaborate involving an intermediate space $Y(\psi)$ of function pairs $[u,v]$,

$$Y(\psi) \equiv \{[u,v] \in \{C([0,T]; W^{1,\infty}(\mathbb{R}) \cap W_o^{1,2}(\mathbb{R})\}^2 \quad :$$

$$: [\psi u, \psi v], [\psi u_x, \psi u_x] \in \{C([0,T]; L^\infty(\mathbb{R}) \cap L^2(\mathbb{R})\}^2 \}$$

which reduces to $Z(\psi)$ once it is shown that $v = u_t$.

To finally complete the proof of Theorem 1 it is only necessary to verify that the solution to the integral equation 10. solves the differential equation 4. (5., 6.). As is evident, we have so far considered solutions which can generally only be interpreted in a weak sense for 4. Thus we have the following definitions and a Lemma to Theorem 1.

Definition Let $\phi = \phi(x,t)$, $\psi = \psi(x,t) \in L^2(]0,T[\times \mathbb{R})$, $0 \le t_1 < t_2 \le T$

and $$\langle \phi(.,.), \psi(.,.) \rangle \equiv \int_{t_1}^{t_2} \int_{-\infty}^{\infty} \phi(x,t)\psi(x,t)\, dxdt \qquad 32.$$

denote the inner product on $L^2(]0,T[\times \mathbb{R})$.

Let $f =(x)$, $g = g(x) \in L^2(\mathbb{R})$

and $$(f(.), g(.) \equiv \int_{-\infty}^{\infty} f(x)\, g(x)\, dx \qquad 33.$$

denote the inner product on $L^2(\mathbb{R})$

Lemma 1 The unique solution $u(x,t)$ to equation 10. satisfies

$$u(x,t) \in C^2([0,\tau[; W^{1,\infty}(\mathbb{R}) \cap W_o^{1,2}(\mathbb{R}))$$

and for every $\phi(x,t) \in C^1([0,T] ; W_o^{1,1}(\mathbb{R}))$ there holds

$$\langle u_t, \phi_t \rangle + \langle u_{xt}, \phi_{xt} \rangle - \langle \sigma + \tau, \phi_x \rangle$$

$$= (u_t, \phi)\Big|_{t_1}^{t_2} + (u_{xt}, \phi_x)\Big|_{t_1}^{t_2} \quad , \forall T < \tau \quad . \qquad 34.$$

Proof The first part may be seen by differentiating the right side of 10. twice with respect to t and noticing that all the terms are continuous in t by H1., H2. and Theorem 1.

The second part follows on substituting u_t and u_{xt} from 10. into the first two terms of 34. and integrating by parts (see [8]) for $\phi(x,t) \in C_o^\infty(]0,T[\times\mathbb{R})$, then using density to include all $\phi(x,t)$.

Remark 2 Results on regularity may now be found when hypotheses H1. and H2. are strengthened and initial data are made smoother but since these are straightforward reapplications of the contraction mapping principle or simple 'bootstrap' arguments for time dependence, we avoid stating them and again refer to [8] for details.

Remark 3 It is possible to prove by a continuation argument that the interval of existence $[0,\tau[$ of the regular solutions in Remark 2 is the same as that in Theorem 1 under the same hypotheses on the initial data and $\sigma(.)$ and $\tau(.)$, and the same may hold for solutions corresponding to data as given in Corollary 1. Therefore it is only necessary to obtain global existence (i.e. $\tau = \infty$) of solutions in $W^{1,\infty}(\mathbb{R}) \cap W_o^{1,2}(\mathbb{R})$ to infer global existence in any of the other spaces alluded to above under the appropriate conditions.

Lemma 1 leads immediately to a result concerning continuous dependence of solutions on their initial data :-

Lemma 2 Suppose $u(x,t)$ and $u_{mn}(x,t)$ are solutions of 5.. 6. corresponding to initial data 7., and $u_{om}(x)$, $u_{1n}(x)$ respectively, where $\{u_{om}\}$, $\{u_{1n}\}$ are bounded sequences in $W^{1,\infty}(\mathbb{R})$ such that

$$u_{om}(.) \to u_o(.) \quad \text{in} \quad W_o^{1,2}(\mathbb{R}) \ , \qquad 35.$$

$$u_{1n}(.) \to u_1(.) \quad \text{in} \quad W_o^{1,2}(\mathbb{R}) \ , \qquad 36.$$

as $m,n \to \infty$.

Then for some $\tau > 0$,

$$u(x,t),\ u_{mn}(x,t) \in C^2([0,\tau[;\ W_o^{1,2}(\mathbb{R}) \cap W^{1,\infty}(\mathbb{R})).$$

Further, for all $t \in [0,T]$, $T < \tau$, as $m,n \to \infty$

$$u_{mn}(.,t) \to u(.,t) \text{ in } W_o^{1,2}(\mathbb{R}) \ , \qquad 37.$$

$$u_{mnt}(.,t) \to u_t(.,t) \text{ in } W_o^{1,2}(\mathbb{R}) \ , \qquad 38.$$

$$u_{mn}(.,t) \xrightarrow{*} u(.,t) \text{ weak-star in } W^{1,\infty}(\mathbb{R}) \quad , \tag{39.}$$

$$u_{mnt}(.,t) \xrightarrow{*} u_t(.,t) \text{ weak-star in } W^{1,\infty}(\mathbb{R}) \quad , \tag{40.}$$

<u>Proof</u> Every bounded sequence in $W^{1,\infty}(\mathbb{R})$ contains a subsequence which converges weak-* to a member of $W^{1,\infty}(\mathbb{R})$ (e.g. [11]). If a sequence converges to a member in the norm topology of $W_o^{1,2}(\mathbb{R})$ and is bounded in $W^{1,\infty}(\mathbb{R})$ then we may show that the entire sequence converges weak-* to that element in $W^{1,\infty}(\mathbb{R})$ ([2], P.76).

Thus 35. and 36. imply

$$u_{om}(.) \xrightarrow{*} u_o(.) \quad \text{weak-* in } W^{1,\infty}(\mathbb{R})$$

and

$$u_{1n}(.) \xrightarrow{*} u_1(.) \quad \text{weak-* in } W^{1,\infty}(\mathbb{R}) .$$

Similarly 39. and 40. follow when 37. and 38. are proved. These latter are obtained by setting

$$W_{mn}(x,t) = u_{mn}(x,t) - u(x,t). \tag{41.}$$

Then by Lemma 1 with $<,>$ defined in $L^2(]0,T[\times \mathbb{R})$,

$$<W_{mn_t}, \phi_t> + <W_{mn_{xt}}, \phi_{xt}> - <\sigma_{mn} + \tau_{mn} - \sigma - \tau, \phi_x>$$

$$= (W_{mn_t}, \phi)\Big|_o^T + (W_{mn_{xt}}, \phi_x)\Big|_o^T \tag{42}$$

where the meaning of σ_{mn}, τ_{mn} is evident. We may substitute $\phi = W_{mn_t}$ in 42. to obtain

$$\| W_{mn_t}(.,t)\|^2_{1,2} \Big|_o^T + 2 <\sigma_{mn} + \tau_{mn} - \sigma - \tau, W_{mn_{xt}}> = 0 . \tag{43.}$$

Letting

$$\| u_{om} - u_o\|^2_{1,2} = \delta_{om} , \tag{44.}$$

$$\| u_{1n} - u_1\|^2_{1,2} = \delta_{1n} \tag{45.}$$

and using H1., H2., 43. shows

$$\| W_{mn_t}(.,T)\|^2_{1,2} \le \delta_{1n} + C\int_0^T [\|W_{mn}(.,\eta)\|^2_{1,2} + \|W_{mn\eta}(.,\eta)\|^2_{1,2}]\, d\eta \qquad 47.$$

where $C = C(\Gamma(R))$ (see the proof of Theorem 1).

But $W^2_{mn}(x,t) = W^2_{mn}(x,0) + 2\int_0^t W_{mn}(x,\eta)\, W_{mn\eta}(x,\eta)\, d\eta$

and so

$$\| W_{mn}(.,T)\|^2_{1,2} \le \delta_{om} + \int_0^T [\| W_{mn}(.,\eta)\|^2_{1,2} + \| W_{mn\eta}(.,\eta)\|^2_{1,2}]d\eta \qquad 48.$$

Adding 47. and 48. finally gives an inequality to which Grönwall's Lemma may be applied, and so

$$\| W_{mn}(.,T)\|^2_{1,2} + \| W_{mn_t}(.,T)\|^2_{1,2} \le (\delta_{om} + \delta_{1n}) \exp((C+1)T) \quad , \qquad 49.$$

giving the desired results, since $\delta_{om}, \delta_{1n} \to 0$ as $m,n \to \infty$.

Before turning to the final question of global existence, we investigate a simple property concerning the propagation of initial discontinuities in the first derivatives of the given data.

We define the 'jump' $[\phi(.,.)](x)$ in $\phi(x.,)$ at a point x by the relation

$$[\phi(.,.)](x) = \phi(x+,.) - \phi(x-,.) \;. \qquad 50.$$

Lemma 3 Let $u(x,t)$ be the solution of 4. - 6. and suppose $u_o(x) \in C^1(\mathbb{R} \setminus \chi_{om})$, $u_1(x) \in C^1(\mathbb{R} \setminus \chi_{1n})$ where $\chi_{om}, \chi_{1n} \subset \mathbb{R}$ are arbitrary sets of $m,n \in N \cup \{o\}$ points at which $u_o'(x)$, respectively $u_1'(x)$ has a discontinuity. Then the only points at which discontinuities in $u_x(x,t)$, $u_{xt}(x,t)$ may occur for $t>0$ are contained in $\chi_{om} \cup \chi_{1n}$.

Proof By Theorem 1, the solution $u(x,t)$ for 4. - 6. satisfies

$$u_x(x,t) = u_o'(x) + tu_1'(x) - \int_0^t\int_{-\infty}^{\infty}(t-\eta)G_{\xi x}(x,\xi)(\sigma(u_\xi) + \tau(u_{\xi\eta}))d\xi d\eta$$

$$- \int_0^t (t-\eta)(\sigma(u_x) + \tau(u_{x\eta}))\, d\eta \qquad 51.$$

for almost every x. It is easy to show that the map

$\phi(x) \to \int_{-\infty}^{\infty} G_{\xi x}(x,\xi)\phi(\xi)d\xi$ takes $L^p(\mathbb{R})$ into $C_B(\mathbb{R})$ $\quad 1 \le p \le \infty$,

and so the third term in 51. is continuous. Thus

$$[u_x(.,t)](x) = [u_o'(.)](x) + t[u_1'(.)](x) - \int_o^t (t-\eta)([\sigma(u_x(.,\eta))](x) + [\tau(u_{x\eta}(.,\eta))](x))d\eta. \quad 52.$$

Similarly,

$$[u_{xt}(.,t)](x) = [u_1'(.)](x) - \int_o^t ([\sigma(u_x(.,\eta))](x) + [\tau(u_{x\eta}(.,\eta))](x))d\eta. \quad 53.$$

By hypotheses H1., H2., adding 52., 53. implies that for $t \in [0,T] \subset [0,\tau[$,

$$\begin{aligned} &|[u_x(.,t)](x)| + |[u_{xt}(.,t)](x)| \\ &\le |[u_o'(.)](x)| + (1+T)|[u\ '(.)](x)| \\ &\quad + (1+T)\Gamma(R)\int_o^t (|[u_x(.,\eta)](x)| + |[u_{x\eta}(.,\eta)](x)|) \\ &\le \Big(|[u_o'(.)](x)| + (1+T)|[u_1'(.)](x)|\Big) \exp(T(1+T)\Gamma(R)) \end{aligned} \quad 54.$$

where the last inequality is a consequence of Grönwall's lemma.

Hence for $x \in \chi_{om} \cup \chi_{1n}$ the right side is zero, and the result follows immediately.

Finally we show that under mild additional hypotheses there appear solutions to 4. - 6. which exist globally in time.

Theorem 2 In addition to the conditions of Theorem 1 let hypotheses H3. and H1 i), ii) or iii) hold. Then the solution $u(x,t)$ of 4. - 6. belongs to the class $C^2([0,\tau[;W^{1,\infty}(\mathbb{R}) \cap W_o^{1,2}(\mathbb{R}))$ for every finite $\tau > 0$.

Proof We may use Lemma 1 to derive an 'energy' estimate for $u(x,t)$ on replacing $\phi(x,t)$ by $u_t(x,t)$ in the expression 34. This delivers

$$\frac{1}{2}\int_{-\infty}^{\infty} (u_t^2 + u_{xt}^2)dx + \int_{-\infty}^{\infty} W(u_x)dx + \int_0^t \int_{-\infty}^{\infty} u_{xt}\tau(u_{xt})dxdt$$

$$= \frac{1}{2}\int_{-\infty}^{\infty} (u_1^2 + u_1'^2)dx + \int_{-\infty}^{\infty} W(u_o')dx \quad \equiv \quad E_o \ , \qquad 55.$$

where we have let $t_1 \to 0$ and $t_2 = t < \tau$ in 34.

In cases H1.i), ii) we therefore have the a priori bounds

$$\frac{1}{2}\| u_t \|_{1,2}^2 (t) + \int_{\infty}^{\infty} W(u_x)dx(t) < E_o \qquad 55\ i)$$

and

$$\frac{1}{2}\| u_t \|_{1,2}^2 (t) + \int_{-\infty}^{\infty} W(u_x)\ dx(t) \quad = E_o\ . \qquad 55\ ii)$$

In cases H1.iii) we rewrite 55. and use Grönwall's lemma.

$$\frac{1}{2}\| u_t \|_{1\,2}^2 (t) + \int_{-\infty}^{\infty} W(u_x)dx(t) = E_o + \frac{\alpha}{2}\int_0^t \int_{-\infty}^{\infty} u_{xt}^2\ dx\ dt.$$

implies

$$\| u_{xt} \|_2^2 (t) \quad \le E_o e^{\alpha t}$$

and, in turn, therefore

$$\frac{1}{2}\| u_t \|_{1,2}^2 (t) + \int_{\infty}^{\infty} W(u_x)dx(t) \le E_o(1 + \tfrac{1}{2}\ e^{\alpha t}) \quad . \qquad 55\ iii)$$

The bounds 55. are in themselves insufficient except in a special case when σ is uniformly Lipschitz continuous. Here we must supplement them by pointwise bounds on u_x, u_{xt} valid almost everywhere. To do this most conveniently, we add two further hypotheses which may however be relaxed, or disposed of entirely when the rod is finite.

H4. Let $0 \le K < \infty$ be such that σ, W are assumed to satisfy, for $|\ \phi\ | \ge K$,

$$|\ \phi\ | \le |\ \sigma\ (\phi)| \le W(\phi)\ .$$

H5. (case H1.i)) Let $p \in \mathbb{N}$, $0 \le K < \infty$ be such that for $|\phi| \ge K$,

$$\tau(\phi) = \alpha\phi^{2p-1}, \quad \text{some } \alpha > 0 .$$

Now we differentiate 10. twice with respect to t and once with respect to x, giving

$$u_{ttx} + \sigma(u_x) + \tau(u_{xt}) = -\int_{-\infty}^{\infty} G_{\xi x}(\sigma(u_\xi) + \tau(u_{\xi t}))\, d\xi \ . \qquad 56.$$

Thus in case H1. i), using 55.,

$$\begin{aligned} |u_{ttx} + \sigma + \tau| &\le \left\{ \int_{-\infty}^{\infty} G_{\xi x}(|\sigma|+|\tau|)d\xi \right\}_{|u_\xi|,|u_{\xi t}|<K} \\ &\quad + \left\{ \int_{-\infty}^{\infty} G_{\xi x}(|\sigma|+|\tau|)d\xi \right\}_{|u_\xi|,|u_{\xi t}|\ge K} \\ &\le 2\Gamma(K)\int_{-\infty}^{\infty} G_{\xi x}(x,\xi)d\xi \\ &\quad + \int_{-\infty}^{\infty} W(u_\xi)d\xi + \alpha\int_{-\infty}^{\infty} |u_{\xi t}|^{2p-1}\, d\xi \\ &\le 2\Gamma(K) + E_o + \alpha\int_{-\infty}^{\infty} |u_{\xi t}|^{2p-1}\, d\xi \end{aligned} \qquad 57.$$

Next we multiply both sides of 57. by $|u_{xt}|$ and integrate with respect to time to obtain

$$\begin{aligned} &\tfrac{1}{2}\, u_{xt}^2(x,t) + W(u_x(x,t)) + \int_o^t u_{x\eta}\, \tau\, d\eta(x) \\ &\le \tfrac{1}{2}\, u_1'^2\,(x) + W(u_o'(x)) + \int_o^t |u_{x\eta}|(2\Gamma(K) + E_o + \alpha\int_{-\infty}^{\infty} |u_{\xi\eta}|^{2p-1}d\xi)d\eta(x). \end{aligned} \qquad 58.$$

In particular therefore, for $t \in [0,T]$,

$$\int_0^t u_{x\eta}^{2p}\, d\eta(x) \le E_1(x) + (2\Gamma(K) + E_0)\int_0^t |u_{x\eta}|\, d\eta(x)$$

$$+ \alpha\int_0^t |u_{x\eta}| \int_{-\infty}^{\infty} |u_{\xi\eta}|^{2p-1} d\xi\, d\eta(x)$$

where $E_1(x) \equiv \frac{1}{2}u_1'^2(x) + W(u_0'(x))$.

So by Hölder's inequality and 55.,

$$\int_0^t u_{x\eta}^{2p}\, d\eta(x) \le E_1(x) + (2\Gamma(K) + E_0)T\left\{\int_0^t |u_{x\eta}|^{2p} d\eta(x)\right\}^{\frac{1}{2p}}$$

$$+ \alpha\left\{\int_0^t |u_{x\eta}|^{2p}\, d\eta(x)\right\}^{\frac{1}{2p}} \left\{\int_0^t \int_{-\infty}^{\infty} |u_{\xi\eta}|^{2p}\, d\xi\, d\eta\right\}^{1-\frac{1}{2p}}$$

$$\le E_1(x) + (2\Gamma(K) + E_0)T\left\{\int_0^t |u_{x\eta}|^{2p}\, d\eta(x)\right\}^{\frac{1}{2p}}$$

$$+ \alpha\left\{\int_0^t |u_{x\eta}|^{2p}\, d\eta(x)\right\}^{\frac{1}{2p}} E_0^{1-\frac{1}{2p}} . \qquad 59.$$

It follows that the term on the left side of 59. is uniformly bounded for almost every x and each $t \in [0,T]$. By 58. and H4., it is immediate that also $u_x(x,t)$ and $u_{xt}(x,t)$ remain uniformly bounded, i.e. we have that for some J, $0<J<\infty$,

$$\| u_x(.,t)\|_\infty ,\ \| u_{xt}(.,t)\|_\infty ,\ \left\| \int_0^t u_{x\eta}\ \tau\ d\eta\right\|_\infty \le J ,\ \forall t\in[0,T]. \qquad 60.$$

Using these estimates in 10. shows

$$\| u(.,t)\|_{1,\infty} ,\ \| u_t(.,t)\|_{1,\infty} \le f(t) < \infty \qquad 61.$$

where f(t) is uniformly bounded on every finite interval [0,T]. Similarly, in case H1. ii) when $\tau = 0$, 56. leads to

$$|u_{ttx} + \sigma| \le \Gamma(K) + \int_{-\infty}^{\infty} W(u_\xi)d\xi$$

$$\le \Gamma(K) + E_0 \qquad 62.$$

and 58. becomes

$$\tfrac{1}{2}u_{xt}^2(x,t) + W(u_x(x,t)) \le E_1(x) + (\Gamma(K)+E_o)(\tfrac{1}{2}\int_o^t (1 + u_{x\eta}^2)\, d\eta(x)) \quad . \qquad 63.$$

Therefore, by Grönwall's inequality, for $t \in [0,T]$,

$$u_{xt}^2(x,t) \le \left\{2E_1(x) + [\Gamma(K) + E_o]T\right\} \exp[(\Gamma(K) + E_o)t] \qquad 64.$$

and we obtain, as before,

$$\| u(.,t)\|_{1,\infty},\ \|u_t(.,t)\|_{1,\infty} \le g(t) < \infty \qquad 65.$$

where $g(t)$ grows at most exponentially in time.

In case H1. iii) when $\tau = -\frac{\alpha\phi}{2}$, $\alpha > 0$ we have

$$|u_{ttx} + \sigma| \le \left| \int_{-\infty}^{\infty} G_{\xi x}(\sigma - \frac{\alpha u_{\xi t}}{2})d\xi \right| + \frac{\alpha}{2}|u_{xt}|$$

$$\le \Gamma(K) + \int_{-\infty}^{\infty} W(u_\xi)d\xi + \frac{\alpha}{2}\| G_{\xi x}(x.,) \|_2 \| u_{\xi t}(.,t)\|_2 + \frac{\alpha}{2}|u_{xt}|$$

$$\le \Gamma(K) + E_o + \frac{\alpha}{2} E_o^{\frac{1}{2}} e^{\frac{\alpha}{2}t} + \frac{\alpha}{2}|u_{xt}|. \qquad 66$$

where we used the inequality preceding 55.iii). Multiplying by $|u_{xt}|$ and integrating,

$$\tfrac{1}{2}u_{xt}^2(x,t) + W(u_x) \le E_1(x) + \int_o^t \left(\Gamma(K) + E_o + \frac{\alpha}{2} E_o^{\frac{1}{2}} e^{\frac{\alpha}{2}t}\right)|u_{x\eta}(x,\eta)|d\eta + \frac{\alpha}{2}\int_o^t u_{x\eta}^2(x,\eta)\, d\eta$$

$$\le E_1(x) + \tfrac{1}{2}\left(\Gamma(K)T + E_oT + E_o^{\frac{1}{2}} e^{\frac{\alpha}{2}T}\right) + \tfrac{1}{2}\left(\Gamma(K) + E_o + \alpha + \frac{\alpha}{2} E_o^{\frac{1}{2}} e^{\frac{\alpha}{2}T}\right) \int_o^t u_{x\eta}^2(x,\eta)d\eta \qquad 67.$$

from which, again by Grönwall's lemma

$$u_{xt}^2(x,t) \leq \left(2E_1(x)+\Gamma(K)T+E_oT+E_o^{\frac{1}{2}}e^{\frac{\alpha}{2}T}\right) \exp\left\{\left(\Gamma(K)+E_o+\alpha+\frac{\alpha}{2}E_o^{\frac{1}{2}}\ e^{\frac{\alpha}{2}T}\right)t\right\} \qquad 68.$$

and we have

$$\| u(.,t)\|_{1,\infty}\ ,\quad \| u_t(.,t)\|_{1,\infty} \leq h(t)\ < \infty \qquad 69.$$

where $h(t)$ is exponentially bounded on every finite interval $[0,T]$.

We have therefore found in each case that an a priori bound for $|u|_{1},_{W^{1,\infty}} + |u|_{1},_{W^{1,2}}$ (see Theorem 1) exists on every finite interval of the form $[0,T]$. A standard continuation argument therefore proves that these solutions exist for all time $t \in [0,\infty[$ (see, for example, Reed [7], or [8]).

REFERENCES

1. T. B. Benjamin, J.L. Bona and J.J. Mahony, Model Equations for long waves in nonlinear dispersive systems, Phil. Trans. Roy. Soc. Lond. *272*, 47-78, 1220.

2. R.W. Carroll, "Abstract Methods in Partial Differential Equations", Harper and Row, New York (1969).

3. A.R. Elcrat and H.A. Maclean, Weighted Wirtinger and Poincaré inequalities on unbounded domains, Indiana University Math. J., *29*, 3 (1980).

4. W. Jaunzemis, "Continuum Mechanics", New York Macmillan (1967).

5. P.D. Lax, Development of singularities of solutions of nonlinear hyperbolic partial differential equations, J. Math. Phys., *5* (1964).

6. A.E.H. Love, "Theory of Elasticity" (4th Ed.) §278, Cambridge University Press (1927).

7. M. Reed, "Abstract Nonlinear Wave Equations", Springer-Verlag, Berlin (1976).

8. R. Saxton, Ph.D. Thesis, Heriot-Watt University (1981).

9. R. Saxton, in preparation

10. R. Saxton, to appear in the Proceedings of the Conference on Ordinary and Partial Differential Equations, Dundee 1982, Springer-Verlag, Berlin.

11. K. Yosida, "Functional Analysis", 3rd ed., Springer-Verlag, Berlin.

ACKNOWLEDGEMENTS

This research was carried out with the assistance of a British SERC research studentship and SERC grant GR/B/78083.

Floquet Theory for Doubly-Periodic Differential Equations and a Number Theory Conjecture

by

B D Sleeman

and

P D Smith

§0 Introduction

The classic Floquet theory is fundamental to the study of ordinary differential equations with periodic coefficients. Roughly speaking the theory states that if the coefficients of an ordinary differential equation are periodic with period π, then under rather mild conditions there always exists at least one multiplicative solution $u(x)$; that is, $u(x)$ has the property:-

$$u(x+\pi) = su(x) , \tag{0.1}$$

for some constant $s \in \mathbb{C}$. From this it follows that if we write $s = \exp(\mu\pi)$ then

$$u(x) = \exp \mu x . P(x) , \tag{0.2}$$

where $P(x)$ is periodic of period π.

The theory we develop here is in answer to the natural question: how far does Floquet theory extend to ordinary differential equations whose coefficients are doubly periodic with periods w, w' say?

More than a hundred years ago, Hermite [1], established that if the general solution of such an ordinary differential equation is regular in the complex plane, then there exists at least one doubly multiplicative solution $u(z)$ such that

$$u(z+w) = su(z) \quad , \quad u(z+w') = s'u(z)$$

for some constants $s,s' \in \mathbb{C}$. Furthermore such a solution can be expressed in the form

$$u(z) = \exp \mu z \frac{\Theta\,(z-a)}{\Theta\ (z)} P(z)$$

where $\Theta(z)$ is the Jacobian theta function, μ and a are constants and $P(z)$ is doubly periodic with periods w, w'.

From 1877 until about the mid 1960's the problem remained dormant. In 1968 Arscott and Sleeman [2] obtained some results, in an algebraic setting, which have some bearing on the subject (see also Sleeman [3]). The first major advance was achieved by Arscott and Wright [4] and Wright [5]. In these works Arscott and Wright laid the foundations for a very general extension of the Floquet theory, and illustrated their ideas with a number of significant examples which in turn pointed towards a quite extensive theory. This paper, which is expository in part, is intended to re-evaluate the theory proposed by Arscott and Wright, to bring their advances to some degree of completion and for one of us (B.D.S.) is offered as a tribute to inspiring teaching and guidance of Felix Arscott.

To appreciate the essential difficulty in extending the Floquet theory recall that an ordinary differential equation with singly-periodic coefficients can normally be cast into a form in which the equation has no singularities in a strip of the complex plane which includes all the real axis. Consequently there is no difficulty in continuing any solution analytically throughout this strip and the complete analytic function so obtained is single valued there. As an example we note that Mathieu's equation has no finite singularities at all.

In the case of equations with doubly periodic coefficients the situation is entirely different. First of all, a doubly periodic function with no singularities is just a constant and so any doubly-periodic equation which is not trivial must have an infinite number of singularities in the finite part of the complex plane. Secondly these singularities cannot be by-passed because double periodicity is essentially a property which involves the whole complex plane. Thus in order to make progress we restrict the discussion to equations having only one singularity in each fundamental period parallelogram. The theory which we develop depends in a fundamental way on the "exponent" ν of the equation, defined in the following section. The Hermite theory mentioned above applies essentially to the case when ν is an integer. Here in this paper we develop the Arscott-Wright theory when ν is rational. A fairly complete theory rests on the resolution of a conjecture raised by Arscott and Wright [4] and tested by them in a number of non-trivial cases. The conjecture is too involved to state here, without prior motivation, but will emerge naturally in our development. It rests essentially on whether a certain trigometrical equation has integer solutions and is proved in the later sections of the paper.

§1 The characteristic exponent ν

Here and throughout the paper we find it convenient to adapt the notation of Jacobian elliptic functions. To begin then, consider the ordinary differential equation

$$\frac{d^2 w}{dz^2} + \Phi(z)w = 0 , \tag{1.1}$$

where $\Phi(z) = \Phi(-z)$ and is periodic with periods $2K$, $2iK'$ and analytic except at points congruent to iK' mod $(2K, 2iK')$. It follows that $\Phi(z)$ is integrable and expressible as an absolutely and uniformly convergent series of the form

$$\Phi(z) = \sum_{m=0}^{\infty} A_m \, \mathrm{sn}^{2m} z \quad , \qquad |z| < \infty . \tag{1.2}$$

Typical examples of (1.1) are Lamé's equation and the ellipsoidal wave equation.

Following a standard argument, [6, p.162], it is easy to show that there exists at least one solution $w(z)$ of (1.1) valid in a neighbourhood N of iK' $\left(N = \{z : |z-iK'| < \min(2K,2K')\}\right)$ which on being analytically continued along a negative half circuit about iK' is multiplied by an appropriate constant σ say. We often write this property as

$$w\left(iK' + (z-iK')e^{-i\pi}\right) = \sigma w(z) . \tag{1.3}$$

Clearly $w(z)$ is only determined up to a multiplicative constant. The constant σ, appearing in 1.3, is determined as a root of a quadratic equation of the form

$$\sigma^2 - 2A\sigma - 1 = 0 , \tag{1.4}$$

so that if σ is a root then so is $-\sigma^{-1}$. The coefficient A depends on the normalisation or initial values of $W(z)$. From (1.4) it follows that there is also a solution $w(z)$ of (1.1) for which

$$\hat{w}\left(iK' + (z-iK')e^{-i\pi}\right) = -\sigma^{-1}\hat{w}(z) . \tag{1.5}$$

We now define the exponent ν by the relations

$$\begin{aligned} \sigma &= \exp i\nu\pi \\ -\sigma^{-1} &= \exp -i(\nu+1)\pi . \end{aligned} \tag{1.6}$$

It is important to note that ν may or may not appear explicitly in the ordinary differential equation (Lamé's equation is an exception). However ν and σ are determined by A and thus in turn is determined by the differential equation. For $z \in N$ the solutions $w(z)$, $\hat{w}(z)$ are expressible as Laurent series, viz:-

$$\begin{aligned} w(z) &= (z-iK')^{-\nu} \sum_{-\infty}^{\infty} C_n (z-iK')^{2n} , \\ \hat{w}(z) &= (z-iK')^{\nu+1} \sum_{-\infty}^{\infty} \hat{C}_n (z-iK')^{2n} . \end{aligned} \tag{1.7}$$

These solutions are linearly independent if 2ν is not an odd integer and in order to be precise we restrict ν to the semi-open interval

$$-\tfrac{1}{2} < \nu \leq \tfrac{1}{2} . \tag{1.8}$$

The case $\nu = \frac{1}{2}$ is exceptional, since then $w(z)$, $\hat{w}(z)$ may not be independent and the general solution valid near iK' may involve a logarithmic term. For a detailed study of this case see Wright [5].

§2 Hermite Theory (ν=0)

If we regard (1.1) as an ordinary differential equation with a singly periodic coefficient of period $2K$, then the classic Floquet theory shows that there always exists a multiplicative solution $u(z)$ analytic throughout the strip $|\text{Im } z| < K'$ such that

$$u(z + 2K) = s\, u(z) , \tag{2.1}$$

for some constant $s \in C$. Now define

$$\hat{u}(z) = u(z + 2iK'). \tag{2.2}$$

If $\hat{u}(z)$ is a constant multiple of $u(z)$, i.e.

$$\hat{u}(z) = s'\, u(z),$$

for some $s' \in \mathbb{C}$ then $u(z)$ is doubly multiplicative and there is nothing more to prove. If, however, $\hat{u}(z)$ is not a multiple of $u(z)$ then $\hat{u}(z)$ and $u(z)$ are linearly independent and so the general solution of (1.1) is expressible in the form

$$v(z) = c\, u(z) + \hat{c}\, \hat{u}(z) . \tag{2.3}$$

Now

$$\begin{aligned} v(z+2K) &= c\, u(z+2K) + \hat{c}\, \hat{u}(z+2K) \\ &= c\, s\, u(z) + \hat{c}\, u(z+2K+2iK') \\ &= c\, s\, u(z) + \hat{c}\, s\, u(z+2iK') \\ &= c\, s\, u(z) + \hat{c}\, s\, \hat{u}(z) \\ &= s\, v(z). \end{aligned}$$

Next, since $\Phi(z+2iK') = \Phi(z)$, $\hat{u}(z+2iK')$ is a solution of the differential equation (1.1) and so

$$\hat{u}(z+2iK') = d\, u(z) + \hat{d}\, \hat{u}(z).$$

Consequently

$$\begin{aligned} v(z+2iK') &= c\, u(z+2iK') + \hat{c}\, \hat{u}(z+2iK') \\ &= c\, \hat{u}(z) + \hat{c}\, d\, u(z) + \hat{c}\, \hat{d}\, \hat{u}(z) \\ &= \hat{c}\, d\, u(z) + (c+\hat{c}\hat{d})\hat{u}(z) . \end{aligned}$$

Thus

$$v(z+2iK') = s'v(z) \quad \text{if and only if}$$

$$\hat{c}d = s'c , \quad (c+\hat{c}\hat{d}) = s'\hat{c} ,$$

and so $c, \hat{c}$ are determined non-trivially if and only if

$$s'^2 - \hat{d}\, s' - d = 0 . \tag{2.4}$$

With s' satisfying (2.4) we see that $v(z)$ is doubly multiplicative and Hermite's theory is essentially established.

However we can say more than this; we have

$$u(z+4iK') = d\, u(z) + \hat{d}\, u(z+2iK')$$

and so

$$u'(z+4iK') = d\, u'(z) + \hat{d}\, u'(z+2iK').$$

On multiplying these by $u'(z+2iK')$ and $u(z+2iK')$ respectively and subtracting we obtain

$$\begin{aligned} &u(z+4iK')u'(z+2ik') - u'(z+4iK')u(z+2iK') \\ &= d\{u(z)u'(z+2iK') - u'(z)u(z+2iK')\} \end{aligned}$$

or, in the notation of the Wronskian,

$$W\{u(z+4iK'), u(z+2iK')\} = - dW\{u(z+2iK'), u(z)\} .$$

Now by the Abel identity both Wronskians in this equality must be a constant k say, and so we conclude

$$k = -dk , \quad \text{or} \quad d = -1.$$

Hence from (2.4) we find

$$s'^2 - \hat{d}s' + 1 = 0 ,$$

and this means that if s' is a root then so is s'^{-1}.

§3 Notation

Throughout the rest of this paper we adopt the following notation

$$\nu = \ell/m \quad (-\tfrac{1}{2} < \nu < \tfrac{1}{2}), \quad \sigma = \exp i\nu\pi .$$

From the solutions $w(z)$, $\hat{w}(z)$ defined in Section 1 we further write

$$w(z-2miK') = w_m(z) ,$$

$$\hat{w}(z-2miK') = \hat{w}_m(z) ,$$

where $w_m(z)$, $\hat{w}_m(z)$ are defined in

$$0 < |z - (2m+1)iK'| < R = \min(2K, 2iK').$$

Let $W_m(z)$ denote the solution column vector

$$W_m(z) = \{w_m(z), \hat{w}_m(z)\} \quad \text{for} \quad |m| \geq 1 ,$$

when $m = 0$, $W_0(z) = \{w(z), \hat{w}(z)\}$.

Note that $w_m(z)$, $\hat{w}_m(z)$ are determined up to an arbitrary multiplicative constant. Finally we use M to denote the diagonal matrix

$$M = \begin{pmatrix} \sigma & 0 \\ 0 & -\sigma^{-1} \end{pmatrix} \equiv \begin{pmatrix} \exp i\nu\pi & 0 \\ 0 & \exp -i(\nu+1)\pi \end{pmatrix} . \tag{3.1}$$

§4 The Shift Matrix T

Here we introduce a further matrix T which is fundamental to the development of the theory. That T plays such a central role comes about from the following observations. The solution vector $W_m(z)$ which is defined for $0 < |z - (2m+1)iK'| < R$ is not periodic in general and is not the same as $W_{m+1}(z)$, valid near $(2m+3)iK'$. Nevertheless these solution vectors do have a common region of validity; so there is a constant <u>shift matrix</u> T such that

$$W_{m+1}(z) = TW_m(z) . \tag{4.1}$$

Because of the periodic nature of the coefficient $\Phi(z)$ in (1.1) it is clear that T is independent of m. Furthermore, because of the symmetry of $\pm iK'$ with respect to the origin in the z-plane and the fact that $\Phi(z)$ is even, $W(-z) = \{w(-z), \hat{w}(-z)\}$ is also a solution column vector valid for $0 < |z + iK'| < R$. From the series expansions (1.7) we observe that

$$w(-z) = (-z-iK')^{-\nu} \sum_{-\infty}^{\infty} c_n (-z-iK')^{2n}$$

$$= \exp -i\nu\pi \, (z+iK')^{-\nu} \sum_{-\infty}^{\infty} c_n (z+iK')^{2n}$$

$$= \exp -i\nu\pi \;\; w_{-1}(z).$$

Similarly

$$\hat{w}(-z) = \exp i(\nu+1)\pi \, \hat{w}_{-1}(z) \, .$$

But

$$W(z) = T\, W_{-1}(z) = T\, M\, W(-z) \, ,$$

by definition of T and M above. In addition we have

$$W(-z) = T\, M\, W(z) = (TM)^2 W(-z)$$

and so we arrive at the first fundamental property of T, namely

$$(TM)^2 = I \, , \tag{4.2}$$

where I is the 2×2 matrix identity.

Now suppose we set

$$T = \begin{pmatrix} \alpha & \beta \\ \gamma & \delta \end{pmatrix} , \tag{4.3}$$

then from (4.2) the following identities must hold

$$\begin{gathered} \sigma^2\alpha^2 - \beta\,\gamma = \sigma^{-2}\delta^2 - \beta\,\gamma = 1 \, , \\ \beta(\sigma^2\alpha - \delta) = \gamma(\sigma^2\alpha - \delta) = 0 \, . \end{gathered} \tag{4.4}$$

There are two cases to be considered.

Case I $\sigma^2\alpha - \delta \neq 0$

In this case we must have $\beta = \gamma = 0$ and so T takes the diagonal form

$$T = \pm\begin{pmatrix}\sigma^{-1} & 0\\ 0 & -\sigma\end{pmatrix}. \tag{4.5}$$

Case II $\sigma^2\alpha - \delta = 0$

This case leads to the following possibilities for T.

i) $$T = \pm\begin{pmatrix}\sigma^{-1} & \beta\\ 0 & \sigma\end{pmatrix}, \quad \gamma = 0$$

ii) $$T = \pm\begin{pmatrix}\sigma^{-1} & 0\\ \gamma & \sigma\end{pmatrix}, \quad \beta = 0$$

iii) $$T = \begin{pmatrix}\alpha & (\sigma^2\alpha^2-1)^{\frac{1}{2}}\\ (\sigma^2\alpha^2-1)^{\frac{1}{2}} & \sigma^2\alpha\end{pmatrix}, \quad \alpha \neq \pm\sigma^{-1}.$$

The possibility iii) comes about from a simple normalisation bearing in mind the indeterminancy of $W(z)$.

Case I can be disposed of quickly, the analysis of case II forms a substantial part of the theory.

If T is of the form (4.5) then in terms of the matrix M we have

$$T = M^{-1} \quad \text{or} \quad T = -M^{-1}.$$

The former case implies, $W(z) = W(-z)$, or $w(z) = w(-z)$ and $\hat{w}(z) = \hat{w}(-z)$ In other words (1.1) has a pair of linearly independent even solutions. But since $\Phi(z)$ is even in z and $z = 0$ is an ordinary point of (1.1) the usual theory of ordinary differential equations shows that there cannot be two even non-trivial solutions. Similarly if $T = -M^{-1}$ then $w(z)$ and $\hat{w}(z)$ must be odd and again this is impossible unless both solutions are trivial.

§5 The Uniformity condition and an extension of the Hermite theory

We have seen that the Hermite theory asserting the existence of doubly multiplicative solutions to (1.1) applies only when the general solution of (1.1) is uniform and this is the case if and only if $\nu = 0$.

Suppose $\nu = \ell/m$ is rational, then in order to make progress we cut the z-plane in such a manner that a closed circuit, if it surrounds any singularity at all, can surround only a multiple of m singularities.

In this cut plane, described below, we find that the general solution is uniform if a certain uniformity condition is satisfied.

When this condition is satisfied the Hermite theory is easily extended to establish the existence of doubly multiplicative solutions in the cut plane.

Let a cut be made in the z-plane from iK' to $(2m-1)iK'$ thus joining a chain of m singularities and let congruent cuts (modulo $2K$, $2miK'$) be made also as shown in Figure 1. We consider the cut joining iK' to $(2m-1)iK'$ and a path which makes a positive circuit about it. To assist the analysis we deform both the cut and the path slightly, so that the latter consists only of finite shifts of magnitude $2iK'$ parallel to the imaginary axis and circuits about individual singularities as shown in Figure 2. Such deformations do not affect the results.

First, we observe that the solution vector $W_m(z)$ on making a positive circuit about $(2m+1)iK'$ becomes multiplied by

$$\begin{pmatrix} \exp -2\nu\pi i & 0 \\ 0 & \exp 2\pi i\nu \end{pmatrix} = \begin{pmatrix} \sigma^{-2} & 0 \\ 0 & \sigma^{2} \end{pmatrix} = M^{-2} .$$

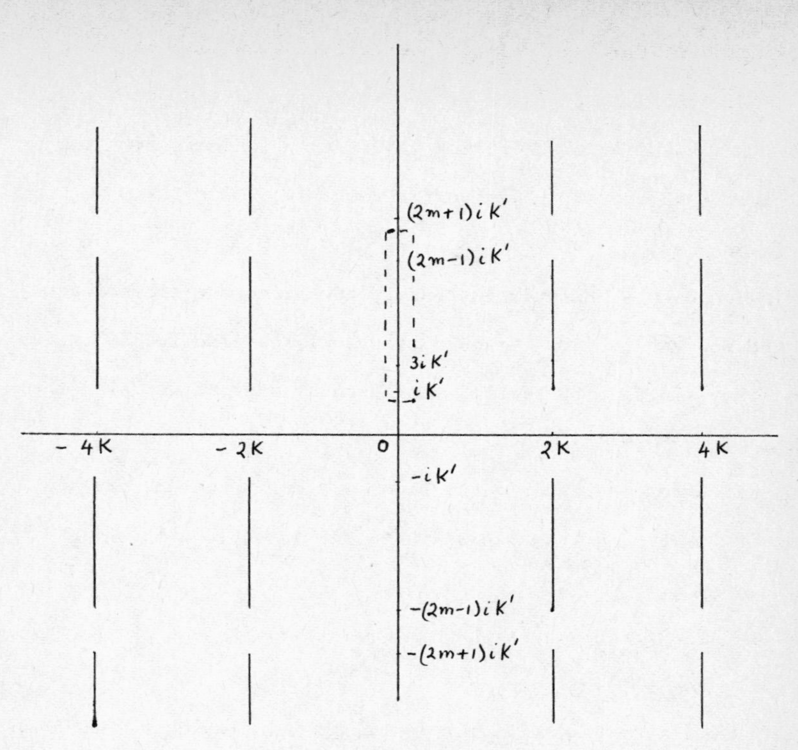

Figure 1

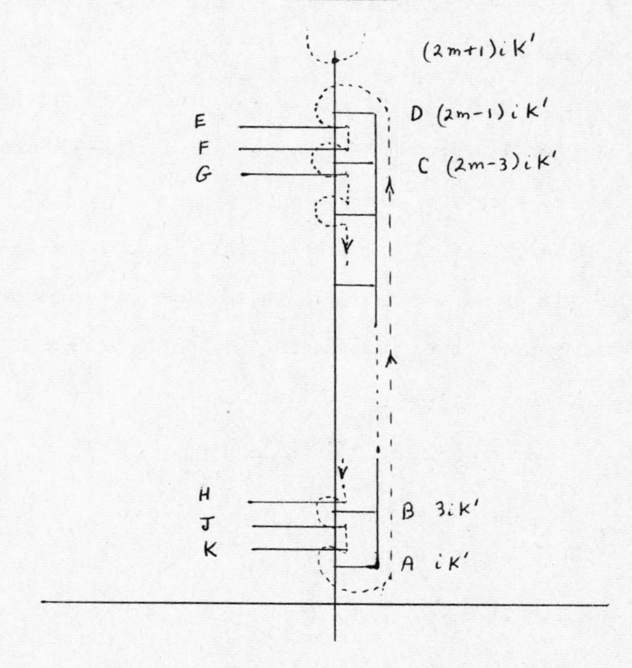

Figure 2

Now with $W(z) = W_0(z)$ at A, its analytic continuation at B is $T^{-1}W_1$ (c.f. 4.1) and at C is $T^{-(m-2)}W_{m-2}$ and at D is $T^{-(m-1)}W_{m-1}$. From the above remark we see that if we continue this solution vector about $(2m-1)iK'$ it becomes $T^{-(m-1)}M^{-2}W_{m-1}$ at E. At F it becomes $T^{-(m-1)}M^{-2}TW_{m-2}$, and $T^{-(m-1)}M^{-2}TM^{-2}W_{m-2}$ at G. Proceeding with this analytic continuation we see that the solution vector becomes $T^{-(m-1)}(M^{-2}T)^{m-2}W_1$ at H, $T^{-(m-1)}(M^{-2}T)^{m-2}M^{-2}W_1$ at J, $T^{-(m-1)}(M^{-2}T)^{m-1}W_0$ at K and finally returns to A with the value $T^{-(m-1)}(M^{-2}T)^{m-1}M^{-2}W_0$.

Thus for the general solution $W \equiv W_0$ to be uniform in the cut plane we must have the necessary and sufficient condition

$$T^{-(m-1)}(M^{-2}T)^{m-1}M^{-2} = I$$

$$\text{or} \quad \underline{T^{(m-1)}M^2 = (M^{-2}T)^{m-1}} \,. \tag{5.1}$$

This is the <u>uniformity condition</u> referred to above which together with case II defining T in section 4 is crucial to a satisfactory extension of the Hermite theory.

The cases $m = 3,4,5,6,7,8$ were discussed in some detail in the pioneering work of Arscott and Wright [4] and which led them to the conjecture stated and resolved later in this paper.

The analysis of (5.1) falls essentially into two parts depending on the form of T (case II). Thus we need to consider the case when T is upper or lower triangular and the case when T is symmetric.

<u>Part I T - triangular</u>

Suppose $S = \begin{pmatrix} a & 0 \\ c & d \end{pmatrix}$ then, provided $a \neq d$,

$$S^n = \begin{pmatrix} a^n & 0 \\ c\,\dfrac{(a^n-d^n)}{a-d} & d^n \end{pmatrix} .$$

Thus if $T = \begin{pmatrix} \sigma^{-1} & 0 \\ \gamma & \sigma \end{pmatrix}$ then

$$T^{m-1} = \begin{pmatrix} \sigma^{-(m-1)} & 0 \\ \dfrac{\gamma\left(\sigma^{-(m-1)} - \sigma^{(m-1)}\right)}{\sigma^{-1} - \sigma} & \sigma^{m-1} \end{pmatrix} \tag{5.2}$$

provided $\sigma^2 \neq 1$. However since $\nu = \ell/m \neq 0$ this condition holds. Also

$$(M^{-2}T)^{m-1} = \begin{pmatrix} \sigma^{-3(m-1)} & 0 \\ \dfrac{\sigma^2\gamma\left(\sigma^{-3(m-1)} - \sigma^{3(m-1)}\right)}{\sigma^{-3} - \sigma^3} & \sigma^{3(m-1)} \end{pmatrix} \tag{5.3}$$

provided $\sigma^6 \neq 1$, that is provided $\exp 6i\nu\pi \equiv \exp 6i\,\frac{\ell}{m}\,\pi \neq 1$. Thus (5.3) holds provided $m \neq 3$. The case $m = 3$ calls for separate treatement and will be considered later.

With the condition $m > 3$ the uniformity condition (5.1) reads

$$\begin{pmatrix} \sigma^{-(m-3)} & 0 \\ \dfrac{\sigma^2\gamma\left(\sigma^{-(m-1)} - \sigma^{(m-1)}\right)}{\sigma^{-1} - \sigma} & \sigma^{m-3} \end{pmatrix} = \begin{pmatrix} \sigma^{-3(m-1)} & 0 \\ \dfrac{\sigma^2\gamma\left(\sigma^{-3(m-1)} - \sigma^{3(m-1)}\right)}{\sigma^{-3} - \sigma^3} & \sigma^{3(m-1)} \end{pmatrix}$$

which results in the equations

$$\sigma^{-(m-3)} = \sigma^{-3(m-1)},$$

$$\frac{\gamma\left(\sigma^{-(m-1)} - \sigma^{(m-1)}\right)}{\sigma^{+1} - \sigma} = \frac{\gamma\left(\sigma^{-3(m-1)} - \sigma^{3(m-1)}\right)}{\sigma^{-3} - \sigma^3}, \tag{5.4}$$

$$\sigma^{m-3} = \sigma^{3m-3}.$$

Since $\sigma \equiv \exp i\nu\pi$ and $\nu = \ell/m$ we see that $\sigma^{2m} = 1$ and so the first and third equation (5.4) are satisfied identically. To satisfy the second equation we require

$$\gamma\left(\sigma^{m-2} - \sigma^{m-4} - \sigma^{m+4} + \sigma^{m+2}\right) = \gamma\left(\sigma^{m+2} - \sigma^{m-4} - \sigma^{m+4} + \sigma^{m-2}\right)$$

and this holds identically for any γ provided $m > 3$. Now consider the special case when $m = 3$. Here we require

$$T^2M^2 = (M^{-2}T)^2$$

$$\text{or} \qquad \gamma(\sigma^{-1} + \sigma^5) = \gamma(\sigma + \sigma^3)$$

$$\text{i.e.} \qquad \gamma\sigma(1 - \sigma^2)(1 + 2\sigma^2) = 0 \ ,$$

and this is only satisfied if $\gamma = 0$.

To conclude then we see that for $T = \begin{pmatrix} \sigma^{-1} & 0 \\ \gamma & \sigma \end{pmatrix}$, the uniformity condition is satisfied for any γ provided $m > 3$ and is satisfied for $m = 3$ if and only if $\gamma = 0$.

In a similar manner, it is easily checked that if T has the form

$$T = \begin{pmatrix} -\sigma^{-1} & 0 \\ \gamma & -\sigma \end{pmatrix}$$

then the uniformity condition is satisfied for any γ provided $m > 3$ and is only satisfied for $m = 3$ when $\gamma = 0$. Next, if

$$T = \pm \begin{pmatrix} \sigma^{-1} & \beta \\ 0 & \sigma \end{pmatrix}$$

the same results prevail; that is the uniformity condition holds identically for any β provided $m > 3$ and is only satisfied for $m = 3$ when $\beta = 0$.

To summarise these calculations we deduce the following: If T is lower triangular then from (4.1) we see that the solution $w(z)$ is multiplicative for an increase of $2iK'$ with periodicity factor $\pm\sigma^{-1}$. Furthermore since

$$TM = \begin{pmatrix} \pm 1 & 0 \\ \sigma\gamma & \mp 1 \end{pmatrix}$$

and recalling that $W(-z) = TMW(z)$, $w(z)$ will be either even or odd about $z = 0$.

Similarly if T is upper triangular $\hat{w}(z)$ will be multiplicative for an increase of $2iK'$ with periodicity factor $\pm\sigma$ and is either odd or even respectively about $z = 0$.

For an increase of amount $2miK'$, then depending whether T is lower or upper triangular $w(z)$ or $\hat{w}(z)$ respectively will be multiplicative and in each case the periodicity factor is ∓ 1. When $m = 3$, T is diagonal and so $w(z)$ and $\hat{w}(z)$ are both multiplicative solutions for the period $2iK'$ with periodicity factors $\pm\sigma^{-1}$, $\pm\sigma$ respectively. Since in this case $TM = \begin{pmatrix} \pm 1 & 0 \\ 0 & \mp 1 \end{pmatrix}$, $w(z)$ is even (odd) and $\hat{w}(z)$ is odd (even). For the period $6iK'$ they are of course multiplicative with the same periodicity factor (∓ 1) and consequently the general solution is multiplicative for the period $6iK'$.

Part 2 T - symmetric

Here we consider the problem of satisfying the uniformity condition (5.1) when T is symmetric and of the form

$$T = \begin{pmatrix} \alpha & (\sigma^2\alpha^2-1)^{\frac{1}{2}} \\ (\sigma^2\alpha^2-1)^{\frac{1}{2}} & \sigma^2\alpha \end{pmatrix} \tag{5.5}$$

where $\alpha^2 \neq \sigma^{-2}$. In order to analyse this case we first reduce T and $M^{-2}T$ to Jordan canonical form and this of course depends on whether the eigenvalues of T and/or $M^{-2}T$ are distinct or not.

To begin with let $\alpha = \sigma^{-1}\cos\theta$ where $\theta \in \mathbb{C}$ and $\mathrm{Re}\,\theta \in [0,\pi]$. Then in terms of θ, T is given by

$$T = \begin{pmatrix} \sigma^{-1}\cos\theta & \pm i \sin\theta \\ \pm i \sin\theta & \sigma\cos\theta \end{pmatrix} . \tag{5.6}$$

The eigenvalues of T are given by the roots of the equation

$$\lambda^2 - (\sigma+\sigma^{-1})\lambda\cos\theta + 1 = 0$$

$$\text{or} \quad \lambda^2 - 2\lambda\cos\nu\pi\cos\theta + 1 = 0 . \tag{5.7}$$

If we further write

$$\cos\phi = \cos\nu\pi\cos\theta , \quad \mathrm{Re}\,\phi \in [0,\pi] ,$$

then the roots λ_1, λ_2 are given by

$$\lambda_1 = \exp i\phi , \qquad \lambda_2 = \exp(-i\phi) \quad \text{respectively}$$

and are distinct provided $\phi \neq 0,\pi$.

Consider the choice

$$T_{+} = \begin{pmatrix} \sigma^{-1}\cos\theta & i \sin\theta \\ i \sin\theta & \sigma\cos\theta \end{pmatrix}$$

then

$$M^{-2}T_{+} = \begin{pmatrix} \sigma^{-3}\cos\theta & i\,\sigma^{-2}\sin\theta \\ i\,\sigma^{2}\sin\theta & \sigma^{3}\cos\theta \end{pmatrix} . \tag{5.8}$$

Now if we write

$$\cos\psi = \cos 3\nu\pi\cos\theta , \ \mathrm{Re}\,\psi \in [0,\pi]$$

then the eigenvalues of $M^{-2}T_{+}$ are given by

$$\tilde{\lambda}_1 = \exp i\psi , \qquad \tilde{\lambda}_2 = \exp(-i\psi)$$

and these are distinct provided $\psi \neq 0,\pi$. We note that if $\alpha = 0$ so that T is diagonal then by definition of θ, $\cos\theta = 0$, $\lambda_1 = -\lambda_2 = i$ and $\tilde{\lambda}_1 = -\tilde{\lambda}_2 = i$ and so in this case the eigenvalues of T_{+} and $M^{-2}T_{+}$ are distinct.

We now prove that the eigenvalues of T_+ and $M^{-2}T_+$ are distinct in general. Suppose this is not the case: then by definition of ψ and ϕ we must have

$$\cos 3\nu\pi \cos\theta = \pm 1,$$

$$\text{and} \qquad \cos \nu\pi \cos\theta = \pm 1, \tag{5.9}$$

where the $\pm$ signs are independent. From (5.9) we conclude that either

$$\cos\nu\pi = \cos 3\nu\pi \tag{5.10a}$$

$$\text{or} \qquad \cos\nu\pi = -\cos 3\nu\pi\ . \tag{5.10b}$$

Now (5.10a) implies

$$\cos\nu\pi = 4\cos^3\nu\pi - 3\cos\nu\pi$$

$$\text{i.e.} \qquad \cos\nu\pi \sin^2\nu\pi = 0\ .$$

But if $\nu = \ell/m$ then this equation implies either $\ell\pi/m = (2n+1)\ \pi/2$ or $\ell\pi/m = p\pi$ for some integers n and p. However both these possibilities are impossible since $\ell/m \in (-\frac{1}{2},\frac{1}{2}) \setminus \{0\}$. The same conclusion holds for the possible choice (5.10b).

Now suppose the eigenvalues of T_+ are equal and those of $M^{-2}T_+$ are distinct. In particular suppose $\lambda_1 = \lambda_2 = 1$. Then

$$\cos\theta = \sec\nu\pi = 2\sigma(1+\sigma^2)^{-1}. \tag{5.11}$$

If we introduce the notation

$$t = \frac{1-\sigma^2}{1+\sigma^2} \tag{5.12}$$

then on taking that branch of $(\sigma^2\alpha^2-1)^{\frac{1}{2}}$ which takes the value $i\sin\theta$ when $\alpha = \sigma^{-1}\cos\theta$ we can write T_+ in the form

$$T_+ = \begin{pmatrix} \dfrac{2}{1+\sigma^2} & it \\ it & \dfrac{2\sigma^2}{1+\sigma^2} \end{pmatrix} \tag{5.13}$$

and $M^{-2}T_+$ as

$$M^{-2}T_{+} = \begin{pmatrix} \frac{2}{\sigma^2(1+\sigma^2)} & \frac{it}{\sigma^2} \\ i\sigma^2 t & \frac{2\sigma^4}{1+\sigma^4} \end{pmatrix} . \qquad (5.14)$$

Since the eigenvalues of (5.13) are 1,1 there is a non-singular matrix U such that

$$U^{-1}T_{+}U = \begin{pmatrix} 1 & 1 \\ 0 & 1 \end{pmatrix} .$$

In fact $U = \begin{pmatrix} -i & 0 \\ 1 & -t^{-1} \end{pmatrix}$ and $U^{-1} = \begin{pmatrix} i & 0 \\ it & -t \end{pmatrix}$.

From these results we find that

$$T_{+}^{m-1} = U\begin{pmatrix} 1 & m-1 \\ 0 & 1 \end{pmatrix} U^{-1} = \begin{pmatrix} 1 + (m-1)t & i(m-1)t \\ i(m-1)t & 1 - (m-1)t \end{pmatrix} . \qquad (5.15)$$

We now wish to express $(M^{-2}T_{+})^{m-1}$ in a suitable form bearing in mind that the eigenvalues of $M^{-2}T_{+}$ are distinct. The Jordan matrix for $M^{-2}T_{+}$ is

$$L = \begin{pmatrix} \exp i\psi & 0 \\ 0 & \exp(-i\psi) \end{pmatrix}$$

and the eigenvectors of $M^{-2}T_{+}$ are

$$\begin{pmatrix} -\frac{2\sigma^4}{1+\sigma^2} + \exp i\psi \\ i\sigma^2\tau \end{pmatrix} \quad \text{and} \quad \begin{pmatrix} -\frac{2\sigma^4}{1+\sigma^2} + \exp(-i\psi) \\ i\sigma^2 t \end{pmatrix} .$$

Fruthermore the matrix U^* such that

$$U^{*-1}(M^{-2}T_{+})U^* = L$$

is given by

$$U^* = \begin{pmatrix} -\dfrac{2\sigma^4}{1+\sigma^2} + \exp i\psi & -\dfrac{2\sigma^4}{1+\sigma^2} + \exp(-i\psi) \\ i\sigma^2 t & i\sigma^2 t \end{pmatrix}$$

and so

$$U^{*-1} = \frac{1}{2\sigma^2 t \sin\psi} \begin{pmatrix} i\sigma^2 t & \dfrac{2\sigma^4}{1+\sigma^2} - \exp(i\psi) \\ -i\sigma^2 t & -\dfrac{2\sigma^4}{1+\sigma^2} + \exp i\psi \end{pmatrix} .$$

These equations now give

$$(M^{-2}T_{+})^{m-1} = -\frac{1}{2\sigma^2 t \sin\psi} \begin{pmatrix} -\dfrac{2\sigma^4}{1+\sigma^2} + \exp i\psi & -\dfrac{2\sigma^4}{1+\sigma^2} + \exp(-i\psi) \\ i\sigma^2 t & i\sigma^2 t \end{pmatrix}$$

$$\times \begin{pmatrix} \exp i(m-1)\psi & 0 \\ 0 & \exp -i(m-1)\psi \end{pmatrix} \begin{pmatrix} i\sigma^2 t & \dfrac{2\sigma^4}{1+\sigma^2} - \exp(-i\psi) \\ -i\sigma^2 t & -\dfrac{2\sigma^4}{1+\sigma^2} + \exp i\psi \end{pmatrix}$$

$$= -\frac{1}{2\sigma^2 t \sin\psi} \begin{pmatrix} \dfrac{4\sigma^6 t}{1+\sigma^2} \sin(m-1)\psi - 2\sigma^2 t \sin m\psi & R \\ -2i\sigma^4 t^2 \sin(m-1)\psi & -\dfrac{4\sigma^6 t}{1+\sigma^2} \sin(m-1)\psi + 2\sigma^2 t \sin(m-2)\psi \end{pmatrix} \tag{5.16}$$

where

$$R = 2i\left[-\frac{4\sigma^8}{(1+\sigma^2)^2} \sin(m-1)\psi - \sin(m-1)\psi + \frac{2\sigma^4}{1+\sigma^2}\left(\sin m\psi + \sin(m-2)\psi\right)\right] .$$

Now in order to simplify matters a little we make use of the Tchebycheff polynomial $U_n(z)$ defined by

$$U_n(z) = (1-z^2)^{\frac{1}{2}}\sin[(n+1)\cos^{-1}z] ,$$

$$\text{or} \qquad U_n(\cos\psi) = \frac{\sin(n+1)\psi}{\sin\psi} . \tag{5.17}$$

In this notation

$$(M^{-2}T_+)^{m-1} = -\frac{1}{2\sigma^2 t}\begin{pmatrix} \frac{4\sigma^2 t}{1+\sigma^2}U_{m-2} - 2\sigma^2 t\, U_{m-1} & R \\ -2i\sigma^4 t^2 U_{m-2} & -\frac{4\sigma^6 t}{1+\sigma^2}U_{m-2} + 2\sigma^2 t\, U_{m-3} \end{pmatrix}$$

$$\text{where} \quad R = 2i[-\frac{4\sigma^8}{(1+\sigma^2)^2}U_{m-2} - U_{m-2} + \frac{2\sigma^4}{1+\sigma^2}(U_{m-1}+U_{m-3})] . \tag{5.18}$$

If the uniformity condition is to be satisfied then we deduce from (5.15) and (5.18)

$$\sigma^2\left(1+(m-1)t\right) = U_{m-1} - \frac{2\sigma^4}{1+\sigma^2}U_{m-2} ,$$

$$(m-1)t = \frac{4\sigma^8}{1-\sigma^4}U_{m-2} + t^{-1}U_{m-2} - \frac{2\sigma^4}{1-\sigma^2}(U_{m-1} + U_{m-3})$$

$$(m-1)t = tU_{m-2} , \tag{5.19}$$

$$1 - (m-1)t = \frac{2\sigma^6}{1+\sigma^2}U_{m-2} - \sigma^2 U_{m-3} .$$

Note - If we had taken the other possible branch of $(\sigma^2\alpha^2-1)^{\frac{1}{2}}$ in T the same set of equations would result.

Since $\sigma^2 \neq 1$, $t \neq 0$ and immediately we must have

$$U_{m-2} = m-1 . \tag{5.20}$$

Furthermore from the recurrence relation

$$U_{m-1} - 2\cos\psi\, U_{m-2} + U_{m-3} = 0$$

for Tchebycheff polynomials we also require

$$1 - (m-1)t = \sigma^2 U_{m-1} - \frac{2}{1+\sigma^2}U_{m-2} . \tag{5.21}$$

Next we deduce

$$U_{m-1} = m\sigma^2 , \tag{5.22}$$

and using this in (5.21) implies

$$1 - (m-1)\left(\frac{1-\sigma^2}{1+\sigma^2}\right) = m\sigma^4 - \frac{2(m-1)}{1+\sigma^2}$$

$$\text{or} \quad m = \sigma^4 m \ .$$

Since $m \neq 0$ then $\sigma^4 = 1$ which is impossible since $\nu \in (-\frac{1}{2},\frac{1}{2})$. We conclude therefore that if the eigenvalues of T are both +1 and those of $M^{-2}T$ distinct then the uniformity condition cannot be satisfied for any non-trivial symmetric matrix T. The same conclusion is reached if the eigenvalues of T are -1, -1. In the same way it can be shown, see Wright [5], that if the eigenvalues of $M^{-2}T$ are equal and those of T are distinct then the uniformity condition cannot be satisfied by a non trivial symmetric matrix T.

It remains then to consider the case when T and $M^{-2}T$ each have <u>distinct</u> eigenvalues.

Recall

$$T = T_{+} = \begin{pmatrix} \sigma^{-1}\cos\theta & i\sin\theta \\ i\sin\theta & \sigma\cos\theta \end{pmatrix}, \tag{5.23}$$

and

$$M^{-2}T = \begin{pmatrix} \sigma^{-3}\cos\theta & i\,\sigma^{-2}\sin\theta \\ i\,\sigma^{2}\sin\theta & \sigma^{3}\cos\theta \end{pmatrix} \tag{5.24}$$

where $\sin\theta \neq 0$ and so $\operatorname{Re}\theta \in (0,\pi)$.

Note:- our analysis is the same if we consider $T = T_{-}$. [c.f.(5.6)]

Again setting

$$\begin{aligned} \cos\phi &= \cos\nu\pi\cos\theta \\ \cos\psi &= \cos 3\nu\pi\cos\theta \end{aligned} \tag{5.25}$$

we know that the eigenvalues of T are $\exp \pm i\phi$ and those of $M^{-2}T$ are $\exp \pm i\psi$ and since they are assumed distinct

$$\operatorname{Re}\phi \in (0,\pi)\ , \quad \operatorname{Re}\psi \in (0,\pi)\ .$$

Now reduce T and $M^{-2}T$ to Jordan canonical form, that is write

$$T = ULU^{-1}\ , \quad M^{-2}T = U^{*}L^{*}U^{*-1}$$

where

$$U = \begin{pmatrix} i\sin\theta & i\sin\theta \\ \exp i\phi - \sigma^{-1}\cos\theta & \exp(-i\phi) - \sigma^{-1}\cos\theta \end{pmatrix},$$

$$U^{*} = \begin{pmatrix} -\sigma^{3}\cos\theta + \exp i\psi & -\sigma^{3}\cos\theta + \exp(-i\psi) \\ i\sigma^{2}\sin\theta & i\sigma^{2}\sin\theta \end{pmatrix}$$

$$L = \begin{pmatrix} \exp i\phi & 0 \\ 0 & \exp(-i\phi) \end{pmatrix} \qquad L^{*} = \begin{pmatrix} \exp i\psi & 0 \\ 0 & \exp -i\psi \end{pmatrix}.$$

From this we easily deduce that

$$T^{m-1} = U \begin{pmatrix} \exp(m-1)i\phi & 0 \\ 0 & \exp -(m-1)i\phi \end{pmatrix} U^{-1}$$

and

$$(M^{-2}T)^{m-1} = U^{*} \begin{pmatrix} \exp i(m-1)\psi & 0 \\ 0 & \exp -(m-1)i\psi \end{pmatrix} U^{*-1}.$$

If we again introduce the Tchebycheff polynomials $U_m(\cos\theta) \equiv U_m$, $U_m(\cos\psi) \equiv \tilde{U}_m$ and form the uniformity condition then the following equations require to be satisfied:

$$\sigma^{2}U_{m-1} - \sigma^{3}\cos\theta\ U_{m-2} = \tilde{U}_{m-1} - \sigma^{3}\cos\theta\ \tilde{U}_{m-2}\ , \tag{5.26a}$$

$$\sin^{2}\theta\ U_{m-2} = \sin^{2}\theta\ \tilde{U}_{m-2}\ , \tag{5.26b}$$

$$\sin^{2}\theta\ U_{m-2} = \sin^{2}\theta\ \tilde{U}_{m-2}\ , \tag{5.26c}$$

$$\sigma\ U_{m-1} - \cos\theta\ U_{m-2} = \sigma^{3}\tilde{U}_{m-1} - \cos\ \tilde{U}_{m-2}\ . \tag{5.26d}$$

From the fact that $\sin\theta \neq 0$ these reduce to the set

$$U_{m-2} = \tilde{U}_{m-2} \tag{5.27}$$

$$\sigma^2 U_{m-1} = \tilde{U}_{m-1}$$

$$\sigma U_{m-1} = \sigma^3 \tilde{U}_{m-1} .$$

Eliminating $\tilde{U}_{m-1}$ and recalling that $m > 2$, and so $\sigma^4 \neq 1$, we require

$$U_{m-1} = 0 \quad \text{and} \quad \tilde{U}_{m-1} = 0 .$$

Now since $\text{Re}\ \phi$, $\text{Re}\ \psi \in (0,\pi)$ we deduce from these equations that

$$\sin m\phi = \sin m\psi = 0$$

and from (5.27) we deduce that

$$\cos m\phi = \cos m\psi .$$

Thus if the uniformity condition is to be satisfied for suitable θ, ϕ and ψ then the following equations must hold

$$\cos\phi = \cos \ell/m\ \pi \cos\theta , \tag{α}$$

$$\cos\psi = \cos 3\ell/m\ \pi \cos\theta, \tag{β}$$

$$\sin m\phi = \sin m\psi = 0 , \tag{γ}$$

$$\cos m\phi = \cos m\psi . \tag{δ}$$

First of all we see that if $\theta = \phi = \psi = \pi/2$ then equations (α), (β) and (δ) are satisfied, but (γ) is only satisfied in this case if m is <u>even</u>. This yields

$$T = \pm \begin{pmatrix} 0 & i \\ i & 0 \end{pmatrix} \overset{\text{def}}{\equiv} \pm\Omega . \tag{5.28}$$

More generally we can reduce equations (α) - (δ) still further. From (γ) we have

$$\phi = \frac{p\pi}{m} , \quad \psi = \frac{q\pi}{m} \tag{5.29}$$

for integers $p,q = 1,2,\ldots,m-1$ and so from (δ) we must have

$$p \equiv q (\text{mod } 2).$$

With the choice (5.29) equations (α), (β) take the form

$$\cos p\pi/m = \cos \ell\pi/m \cos\theta \ ,$$

$$\cos q\pi/m = \cos 3\ell\pi/m \cos\theta,$$

respectively and which are solvable for $\cos\theta \neq 0$ if and only if

$$\cos p\pi/m \cos 3\ell\pi/m = \cos \ell\pi/m \cos q\pi/m \ . \tag{5.30}$$

Now ℓ/m is rational and $\ell/m \in (-\frac{1}{2},0) \cup (0,\frac{1}{2})$ and since (5.30) is unaffected if ℓ/m is replaced by $\ell/m + 1$ we may consider $\ell/m \in (0,\frac{1}{2}) \cup (\frac{1}{2},1)$. Thus ℓ can take the values $1,2,\ldots,m-1$ with ℓ,m co-prime. Equation (5.30) can be solved for certain trivial values of p and q but a brief calculation shows that they must in fact be excluded and we must attach the further conditions

$$p \not\equiv \pm\ell \ , \quad q \not\equiv \pm 3\ell \pmod m, \quad p/m \neq \tfrac{1}{2} \quad , \quad q/m \neq \tfrac{1}{2} \ .$$

These calculations lead us to propose the following theorem.

<u>Theorem 1</u> (Arscott-Wright Conjecture)

Let Q_m be the set of positive integers r such that

$$1 \leq r \leq m-1 \quad \text{and} \quad r/m \neq \tfrac{1}{2} \ .$$

Choose $\ell \in Q_m$ so that ℓ and m are co-prime. Then there are no integers $p, q \in Q_m$ with

$$p \equiv q \pmod 2), \ p \equiv \pm \ \ell \pmod m), \ q \not\equiv \pm \ 3\ell(m),$$

satisfying the equation

$$\cos p\pi/m \cos 3\ell\pi/m - \cos \ell\pi/m \cos q\pi/m = 0 \ .$$

This result was proposed, as we have indicated, as a conjecture by Arscott and Wright more than 10 years ago who also verified its truth for values of $m = 3$ to 7.

In order not to interrupt the flow of our development we defer the proof of Theorem 1 until section 7.

On the basis of Theorem 1 we conclude

Theorem 2

Let m be an integer greater than 2. If m is odd then there is no symmetric non-diagonal matrix T which satisfies the uniformity condition. If m is even then $T = \pm \Omega$ are the only matrices satisfying the uniformity condition.

Finally the analysis of this section can be summarized in the following theorem.

Theorem 3

The uniformity condition

$$T^{m-1}M^2 = (M^{-2}T)^{m-1}$$

where M is defined by (3.1) is satisfied by matrices T of the following form

i) $T = \pm M^{-1}J$ where $J = \begin{pmatrix} 1 & 0 \\ 0 & -1 \end{pmatrix}$

ii) $T = \begin{pmatrix} \pm \sigma^{-1} & 0 \\ \gamma & \pm \sigma \end{pmatrix}$ and $T = \begin{pmatrix} \pm \sigma^{-1} & \beta \\ 0 & \pm \sigma \end{pmatrix}$

for arbitrary β and γ but where $\beta = \gamma = 0$ if $m = 3$.

iii) $T = \pm\begin{pmatrix} 0 & i \\ i & 0 \end{pmatrix} = \pm \Omega$ provided m is even.

§6 Doubly-multiplicative solutions in the cut plane

If $\nu = \ell/m$ is rational, the z-plane is cut as described in section 5 and T is such that the uniformity condition then it is natural to investigate the extension of the Hermite theory outlined in section 2 to the cut plane.

To begin with we define what is meant by addition of a period. If $u(z)$ is a solution of (1.1) valid in a region including the point z, then $u(z+2K)$, say, is to be interpreted as the analytic continuation of $u(z)$ to a region including the point $z + 2K$ by a path which avoids the cuts. Because of the uniformity of the general solution the analytic continuation is unique and does not depend on the particular path followed so long as the path does not cross a cut.

The analysis of section 2 applies virtually unchanged to this situation and we establish the existence of at least one doubly-multiplicative solution with pseudo-periods $2K$, $2iK'$. That is, a solution $u(z)$ with constants s, s' such that

$$u(z + 2K) = s\,u(z), \quad u(z + 2iK') = s'u(z). \tag{6.1}$$

We now turn to the question of analysing those doubly-multiplicative solutions for pseudo-periods $2K$, $2iK'$ and for $2K$, $2miK'$ when the matrix T takes one of the possible forms listed in Theorem 3. Throughout we shall retain the notation of section 5.

Suppose $T = M^{-1}J$, then $w(z)$ and $\hat{w}(z)$ are both multiplicative for the period $2iK'$ but with the periodicity factors σ^{-1} and σ respectively. For $T = -M^{-1}J$, $w(z)$ and $\hat{w}(z)$ are again multiplicative for the period $2iK'$ but this time the periodicity factors are $-\sigma^{-1}$ and $-\sigma$ respectively. Thus if $T = \pm M^{-1}J$ no linear combination of $w(z)$, $\hat{w}(z)$ can be multiplicative for the period $2iK'$ and the only possible doubly-multiplicative solutions are therefore w and $\hat{w}$ themselves. Now if $T = M^{-1}J$ then from (4.1) we have $W(z) = JW(-z)$ so that $w(z)$ is even and $\hat{w}(z)$ is odd. Similarly if $T = -M^{-1}J$ then $w(z)$ is odd and $\hat{w}(z)$ is even. We now show that w and $\hat{w}$ cannot both be multiplicative for the period $2K$ without one of them being identically zero. Thus suppose $T = M^{-1}J$ and

$$w(z + 2K) = s\,w(z)\ , \quad \hat{w}(z + 2K) = \hat{s}\,\hat{w}(z)\ . \tag{6.2}$$

With $z = -2K$ we see that $w(0) = s\,w(-2K) = s\,w(2K) = s^2w(0)$. Similarly on differentiating $\hat{w}(z)$ we find that $\hat{w}'(0) = \hat{s}^2\hat{w}'(0)$. Now $z = 0$ is an ordinary point of the differential equation (1.1), so no non-trivial solution can have a double zero. Thus $w(0) \neq 0$ and so $s^2 = 1$, $\hat{s}^2 = 1$. The solution $w(z)$ therefore has the properties

$$w(z) = w(-z), \quad w(z + 2K) = s\,w(z) \quad (s = \pm 1)$$

$$w(z + 2iK') = s'w(z) \quad (s' = \sigma^{-1}) .$$

Then

$$w(K + iK') = s\,w(-K + iK') = ss'w(-K-iK') = ss'w(K + iK')$$

and so $w(K + iK') = 0$, since $ss' \neq 1$.

Similar calculation shows that $\hat{w}(K + iK') = 0$, but this is impossible since $K + iK'$ is an ordinary point of the doubly-periodic equation (1.1) and the general solution cannot be zero there. The same conclusion holds if $T = -M^{-1}J$.

In summary then, if $T = \pm M^{-1}J$ only one of $w, \hat{w}$ is doubly multiplicative for pseudo periods $2K$, $2iK'$, and we have either

i) $w(z)$ is even and doubly-multiplicative $(2K, 2iK')$ with periodicity factors $s = \pm 1$, $s' = \sigma^{-1}$ or is odd and doubly-multiplicative $(2K, 2iK')$ with factors $s = \pm 1$, $s' = -\sigma^{-1}$,

or

ii) $\hat{w}(z)$ is odd and doubly-multiplicative $(2K, 2iK')$ with periodicity factors $\hat{s} = \pm 1$, $s' = \sigma$ or even and doubly-multiplicative $(2K, 2iK')$ with factors $\hat{s} = \pm 1$, $s' = -\sigma$.

In the case $T = \begin{pmatrix} \pm\sigma^{-1} & 0 \\ \gamma & \pm\sigma \end{pmatrix}$, $w(z)$ is even or odd and doubly-multiplicative with pseudo periods $2K$, $2iK'$ and periodicity factors ± 1, $\pm\sigma^{-1}$, while $\hat{w}(z)$ has no particular properties.

Similarly if $T = \begin{pmatrix} \pm\sigma^{-1} & \beta \\ 0 & \pm\sigma \end{pmatrix}$ then $\hat{w}(z)$ is odd or even and doubly multiplicative $(2K, 2iK')$ with periodicity factors ± 1, $\pm\sigma$, while this time $w(z)$ has no particular properties.

Next if m is even $T = \pm \Omega$ will satisfy the uniformity condition. For $T = \Omega$ we find that $w(z) \pm \hat{w}(z)$ are both multiplicative solutions for the period $2iK'$ with periodicity factors $\pm i$ respectively, while for $T = -\Omega$, $w(z) \pm \hat{w}(z)$ are both multiplicative for the period $2iK'$ with periodicity factors $\mp i$ respectively.

If we consider the possibility of doubly-multiplicative solutions with pseudo-periods $2K$, $2miK'$ then the reasoning of section 2 shows the existence of at least one doubly-multiplicative solution.

If $T = \pm M^{-1}J$ the general solution is multiplicative for $2miK'$ with periodicity factor ∓ 1 so that a doubly-multiplicative solution is not necessarily w or $\hat{w}$ but may be a linear combination of these and the restriction $s = \pm 1$ no longer holds. If

$$T = \begin{pmatrix} \pm \sigma^{-1} & 0 \\ 0 & \pm \sigma \end{pmatrix} \quad \text{or} \quad T = \begin{pmatrix} \pm \sigma^{-1} & \beta \\ 0 & \pm \sigma \end{pmatrix}$$

then respectively $w(z)$ is doubly multiplicative $(2K, 2miK')$ and $\hat{w}(z)$ is not, or, $\hat{w}(z)$ is doubly multiplicative $(2K, 2miK')$ while $w(z)$ is not.

The restriction $s = \pm 1$ here does apply.

Finally if $T = \pm \Omega$ the solutions $w(z) \pm \hat{w}(z)$ are both multiplicative for $2miK'$ but in all cases, since m must be even, the periodicity factor is the same, namely $(-1)^{m/2}$.

§7 Proof of the Arscott-Wright Conjecture (Theorem 1)

In this and subsequent sections we shall prove theorem 1; in fact, it is convenient to prove the following slightly stronger result, which does not consider the cases $m = 4$ or 6.

Theorem 4

Let m be a positive integer, not equal to 1,2,4 or 6. The only solutions in integers p, q, ℓ of the equation

$$\cos p\pi/m \cos 3\ell\pi/m = \cos \ell\pi/m \cos q\pi/m \tag{7.1}$$

subject to the constraints (i) p and q are both even or odd, (ii) $\cos p\pi/m \cos q\pi/m \neq 0$, and (iii) ℓ is coprime to m, are $p \equiv \pm \ell(m)$, $q \equiv \pm 3\ell(m)$.

When $m = 4$, the solutions of (7.1), subject to the same constraints, include also the solutions $p \equiv 4(8)$, $q \equiv 0(8)$ and $p \equiv 0(8)$, $q \equiv 4(8)$; these additional solutions are excluded by the slightly stronger hypotheses of theorem 1. When $m = 6$, the solutions of (7.1) subject to constraint (iii) only, are given by $\cos q\pi/6 = 0$ i.e. $q \equiv 3(6)$; but again all such solutions are excluded under the stronger hypotheses of theorem 1. The next section has a brief discussion on arithmetic in fields of roots of unity; this will provide the tools necessary to establish the solutions of (7.1).

§8 Arithmetic in fields of roots of unity

For a positive integer m, let μ_m denote the group of the mth roots of unity. Let Q denote the field of rational numbers, and $F_m = Q(\mu_m)$ denote the smallest field containing Q and μ_m. When m is odd, F_m contains the $2m^{th}$ roots of unity; when m is even, F_m contains no more roots of unity than those in μ_m. Henceforth we shall suppose that either m is odd or divisible by 4 (equation (7.1) will be suitably modified, as explained later).

The field F_m is a vector space over Q, of dimension $\phi(m)$, the number of coprime residue classes modulo m. Setting $\zeta_m = e^{2\pi i/m}$, a basis for this vector space is $1, \zeta_m, \zeta_m^2, \ldots, \zeta_m^{\phi(m)-1}$. Let $G(F_m/Q)$ denote the set of automorphisms of the field F_m which leave the rational field Q fixed. This set is a group, known as the Galois group of F_m over Q ; it has order $\phi(m)$ and is isomorphic to the group of coprime residue classes modulo m i.e. the invertible elements of the ring $\mathbb{Z}/m\mathbb{Z}$, which we denote by $(\mathbb{Z}/m\mathbb{Z})^\times$. The isomorphism is obtained as follows - an automorphism $\sigma \in G(F_m/Q)$ is determined completely by its action on ζ_m, and since $\zeta_m^{\ \sigma}$ is obviously another primitive m^{th} root of unity, there is an integer a such that $\zeta_m^{\ \sigma} = \zeta_m^{\ a}$; a is well defined modulo m and is invertible in $\mathbb{Z}/m\mathbb{Z}$, and so we obtain the map which is an isomorphism : $G(F_m/Q) \xrightarrow{\sim} (\mathbb{Z}/m\mathbb{Z})^\times$. Denote the element corresponding to a by σ_a.

An important subfield of F_m is its maximal real subfield denoted by $F_m^{\ +} = Q(\mu_m)^+$. This is generated by the values $\zeta_m + \zeta_m^{\ -1}$, $\zeta_m^{\ 2} + \zeta_m^{\ -2}$,... over Q, or equivalently by the values $\cos 2\pi/m$, $\cos 4\pi/m$,... over Q. Its Galois group $G(F_m^{\ +}/Q)$ is isomorphic to $(\mathbb{Z}/m\mathbb{Z})^\times/(\pm 1)$, the original group with ± 1 factored out. The action of an element $\sigma_a \in G(F_m/Q)$ on a value $\cos 2\pi b/m$ is given by $(\cos 2\pi b/m)^{\sigma_a} = \cos 2\pi ab/m$ (whether b is coprime to m or not). If b is coprime to m, then σ_a fixes $\cos 2\pi b/m$ only when $a \equiv \pm 1\ (m)$.

F_m contains other subfields, particularly the fields $F_{m_1} = Q(\mu_{m_1})$, where m_1 is a divisor of m (also odd or divisible by 4). Such a field F_m is characterized as the fixed field of the subgroup $\{\sigma_a \in G(F_m/Q) \mid a \equiv 1(m_1)\}$; i.e. $F_{m_1} = \{x \in F_m \mid x^{\sigma_a} = x$ for all $\sigma_a \in G(F_m/Q)$ with $a \equiv 1(m_1)\}$. We define two maps from F_m into such a subfield F_{m_1}, namely the trace and the norm. The trace, denoted by $\mathrm{Tr}_{F_m/F_{m_1}}$, maps $\alpha \to \sum_{a\equiv 1(m_1)} \alpha^{\sigma_a}$; the norm, denoted by $N_{F_m/F_{m_1}}$, maps $\alpha \to \prod_{a\equiv 1(m_1)} \alpha^{\sigma_a}$; here $\alpha \in F_m$, and the sum or product is taken over all $\sigma_a \in G(F_m/Q)$ with $a \equiv 1(m_1)$. It is easily checked that $\alpha^{\sigma_a} = \alpha$ whenever $a \equiv 1(m_1)$, so the values $\mathrm{Tr}_{F_m/F_{m_1}}(\alpha)$ and

$N_{F_m/F_{m_1}}(\alpha)$ do indeed belong to F_{m_1}.

Notice that if $\alpha \in F_{m_1}$, then $Tr_{F_m/F_{m_1}}(\alpha) = \frac{\phi(m)}{\phi(m_1)}\alpha$, since $\phi(m)/\phi(m_1)$ is the order of the subgroup which fixes F_{m_1}; similarly $N_{F_m/F_m}(\alpha) = \alpha^{\phi(m)/\phi(m_1)}$.

We need the following result to compute norms and traces later (see Lang [7]).

Theorem 5

Suppose m_1, m_2 are coprime integers, odd or divisible by 4. The field of $m_1m_2^{th}$ roots of unity, $F_{m_1m_2}$, is the compositum the fields F_{m_1} and F_{m_2}, and $F_{m_1} \cap F_{m_2} = Q$. Furthermore

$G(F_{m_1m_2}/Q) \overset{\sim}{\rightarrow} G(F_{m_1}/Q) \times G(F_{m_2}/Q)$.

N.B. In these circumstances, we also have $G(F_{m_1}/Q) \overset{\sim}{\rightarrow} G(F_{m_1m_2}/F_{m_2})$; this enables us to "trace out" the m_1-part of $\zeta_{m_1m_2}$ and obtain an m_2^{th} root of unity.

We apply this theorem to obtain the following results about tracs and norms.

Lemma 8.1 Let m be an integer, odd or divisible by 4, and let p be an odd prime

a) If p does not divide m, $Tr_{F_{mp}/F_m}(\zeta_{mp}) = -\zeta_m{}^y$, where y

is uniquely determined (mod m) by $yp \equiv 1(m)$.

b) If p divides m, then $Tr_{F_{mp}/F_m}(\zeta_{mp}) = 0$.

Proof Suppose first that $p \nmid m$. Then there exist integers x,y such that $mx + py = 1$; thus $\zeta_{mp} = \zeta_p{}^x.\zeta_m{}^y$. Note that y is determined uniquely mod m, and that x is not divisible by p, so that $\zeta_p{}^x$ is a primitive p^{th} root of unity. Applying the theorem above,

$$\mathrm{Tr}_{F_{mp}/F_m}(\zeta_{mp}) = \mathrm{Tr}_{F_{mp}/F_m}(\zeta_p^{\,x}.\zeta_m^{\,y}) = \zeta_m^{\,y}.\mathrm{Tr}_{F_{mp}/F_m}(\zeta_p^{\,x}) = \zeta_m^{\,y}.\mathrm{Tr}_{F_p/Q}(\zeta_p^{\,x}).$$

Finally $\mathrm{Tr}_{F_p/Q}(\zeta_p^{\,x}) = \sum_{Q=1}^{p-1} \zeta_p^{\,a} = -1$, which yields result a).

Turning to the case in which $p|m$, let $m = p^e m_1$ where $e \geq 1$ and m_1 is not divisible by p. Again there exist integers x, y such that $m_1 x + p^{e+1}y = 1$, thus $\zeta_{mp} = \zeta_{p^{e+1}}^{\,x}\,\zeta_{m_1}^{\,y}$. Then applying the theorem above,

$$\mathrm{Tr}_{F_{mp}/F_m}(\zeta_{mp}) = \mathrm{Tr}_{F_{mp}/F_m}(\zeta_{p^{e+1}}^{\,x}\cdot\zeta_{m_1}^{\,y}) = \zeta_{m_1}^{\,y}.\mathrm{Tr}_{F_{mp}/F_m}(\zeta_{p^{e+1}}^{\,x})$$

$$= \zeta_{m_1}^{\,y}.\mathrm{Tr}_{F_{p^{e+1}}/F_{p^e}}(\zeta_{p^{e+1}}^{\,x}).$$

Finally

$$\mathrm{Tr}_{F_{p^{e+1}}/F_{p^e}}(\zeta_{p^{e+1}}^{\,x}) = \sum_{b=0}^{p-1} \zeta_{p^{e+1}}^{\,x(1+bp^e)} = \zeta_{p^{e+1}}^{\,x}.\sum_{b=0}^{p-1} \zeta_p^{\,xb} = 0 .$$

<u>Corollary 8.2</u> Let m be a square free odd integer, and let m_1 divide m, and $m_2 = m/m_1$. Let r be the number of prime factors of m_2 and let a be an integer.

a) If a is co prime to m_2, $\mathrm{Tr}_{F_m/F_{m_1}}(\cos 2\pi a/m) = (-1)^r \cos 2\pi ay/m_1$ where $ym_2 \equiv 1(m_1)$.

b) If a is not co prime to m_2, $\mathrm{Tr}_{F_m/F_{m_1}}(\cos 2\pi a/m) = A \cos 2\pi ay/m_1$ where $ym_2 \equiv 1(m_1)$, and A is an integer, determined as follows Suppose a and m_2 have s prime factors in common, say $q_1,\ldots,q_s$. Then $A = (-1)^{r-s}\phi(q_1 \ldots q_s)$.

<u>Proof</u> The proof of lemma 8.1 is valid if ζ_{mp} is replaced by $\zeta_{mp}^{\,a}$ (where a is co prime to mp). The trace has the property $\mathrm{Tr}_{F_n/F_{n_2}}(\alpha) = \mathrm{Tr}_{F_{n_1}/F_{n_2}}\left(\mathrm{Tr}_{F_n/F_{n_1}}(\alpha)\right)$ where $n_1|n$ and $n_2|n$; from

this part a) follows by induction. As for part b), let $a_1 = a/q_1\ldots q_s$, and let $m_3 = m_2/q_1\ldots q_s$. Then $Tr_{F_m/F_{m_1}}(\cos 2\pi a/m) = Tr_{F_{m_1m_3}/F_{m_1}}(Tr_{F_m/F_{m_1m_3}} \cos 2\pi a/m_1m_3) = \phi(q_1\ldots q_s)Tr_{F_{m_1m_3}/F_{m_1}}(\cos 2\pi a/m_1m_3)$

$= \phi(q_1\ldots q_s)(-1)^{r-s} \cos 2\pi a_1z/m_1$, (by part a)), where z is determined by $zm_3 \equiv 1\ (m_1)$. It is easily checked that $a_1z \equiv ay(m_1)$ with $ym_2 \equiv 1(m_1)$. This proves the corollary.

Lemma 8.3 Let m be an integer, odd or divisible by 4, and let p be an odd prime dividing m. If p does not divide a,

$$N_{F_{mp}/F_m}(2 \cos 2\pi a/mp) = 2 \cos 2\pi a/m \quad .$$

Proof Write $m = p^e m_1$ where p does not divide m_1, and $e \geq 1$, choose integers x, y such that $m_1x + p^{e+1}y = 1$, and so $\zeta_{mp} = \zeta_{p^{e+1}}{}^x.\zeta_{m_1}{}^y$. We see that $2 \cos 2\pi a/mp = \zeta_{mp}{}^a + \zeta_{mp}{}^{-a} = \zeta_{mp}{}^{-a}(\zeta_{mp}{}^{4a}-1)/(\zeta_{mp}{}^{2a}-1)$; we shall determine the norm of each factor and multiply the results. Let $\alpha = \zeta_{mp}{}^a$ and $n = 2$ or 4. Now $N_{F_{mp}/F_m}(\zeta_{mp}{}^{an}-1)$

$$= \prod_{b=0}^{p-1} (\zeta_{p^{e+1}}{}^{nx(1+bp^e)a}\zeta_{m_1}{}^{nya} - 1) = \prod_{b=0}^{p-1} (\alpha^n.\zeta_p{}^{bxna} - 1) = \alpha^{np} - 1 \ ;$$

also $N_{F_{mp}/F_m}\zeta_{mp}{}^{-a} = \prod_{b=0}^{p-1} \zeta_{p^{e+1}}{}^{-ax(1+bp^e)}\zeta_{m_1}{}^{-ay} = \prod_{b=0}^{p-1} \alpha^{-1}.\zeta_p{}^{-bxa} = \alpha^{-p}$.

Thus $N_{F_{mp}/F_m}(2 \cos 2\pi a/mp) = \alpha^{-p}\left(\frac{\alpha^{4p}-1}{\alpha^{2p}-1}\right) = \zeta_m{}^{-a}\left(\frac{\zeta_m{}^{4a}-1}{\alpha_m{}^{2a}-1}\right) = 2 \cos 2\pi a/m$

as required.

We need one more result on traces when m is divisible by 4.

Lemma 8.4 Let m be divisible by 4, and a be co prime to m. Then if $m' = \frac{1}{2}m$,

$$Tr_{F_m/F_{m'}}(\cos 2\pi a/m) = 0 \ .$$

[If m is an odd multiple of 4, $F_{m'} = F_{\frac{1}{2}m'}$.]

<u>Proof</u> Let 2^{n+2} be the exact power of 2 dividing m, and $m_1 = m/2^{n+2}$. Choose integers x, y such that $xm_1 + 2^{n+2}y = 1$. Then $\zeta_m = \zeta_{2^{n+2}}{}^x \zeta_{m_1}{}^y$. Thus $Tr_{F_m/F_{m'}}(\zeta_m{}^a) = \zeta_{m_1}{}^{ay}.Tr_{F_{2^{n+2}}/F_{2^{n+1}}}(\zeta_{2^{n+2}}{}^a)$ $= \zeta_{m_1}{}^{ay}(\zeta_{2^{n+2}}{}^a + \zeta_{2^{n+2}}{}^{a(1+2^{n+1})}) = 0$. The result follows.

Finally, we need the following basis result when m is a square free odd integer. See Long [8], p. 167.

<u>Theorem 6</u>

Let m be a square free odd integer. Let A be a complete set of inequivalent representatives for $(\mathbb{Z}/m\mathbb{Z})^\times$. The set $A_m = \{\zeta_m{}^a | a \in A\}$ is a basis for F_m/Q.

Let us return to the basic equation

$$\cos p\pi/m \cos 3\pi/m = \cos \pi/m \cos q\pi/m \ . \tag{8.1}$$

When m is odd, constraint (ii) holds automatically, furthermore we can suppose that p, q, ℓ are even - for there exist integers ℓ', p', q' such that $\ell \equiv 2\ell'$, $p \equiv 2p'$, $q \equiv 2q'$ modulo m, and using the condition $p \equiv q(2)$ we obtain

$$\cos 2\pi p'/m \cos 6\pi\ell'/m = \cos 2\pi\ell'/m \cos 2\pi q'/m \ . \tag{8.2}$$

Thus when m is odd, we aim to prove that the only solutions of (8.2) - subject to the one constraint that ℓ' is comprime to m - are $p' \equiv \pm\ell'(m)$, $q' \equiv \pm 3\ell'(m)$. When m is even, we can again obtain equation (8.2) upon replacing m by 2m; henceforth we will assume that if m is even, it is divisible by 4, and we aim to prove that the only solutions of (8.2) - subject to the three constraints (i) $p' \equiv q'(2)$, (ii) $\cos 2p'\pi/m \cos 2q'\pi/m \neq 0$, (iii) $(\ell',m) = 1$ - are $p' \equiv \pm\ell'(\frac{1}{2}m)$, $q' \equiv \pm 3\ell'(\frac{1}{2}m)$.

Applying the automorphism $\sigma_{\ell'}{}^{-1} \in G(F_m/Q)$ to equation (8.2) yields

$$\cos 2p''\pi/m \cos 6\pi/m = \cos 2\pi/m \cos 2q''\pi/m \tag{8.3}$$

where p'', q'' are determined by $p' \equiv p''\ell'$, $q' \equiv q''\ell'$ (mod m); the solution $p' \equiv \pm\ell'$, $q' \equiv \pm 3\ell'$ (m) of equation (8.2) corresponds to the solution $p'' \equiv \pm 1$, $q'' \equiv \pm 3$ (m) of equation (8.3).

It is clearly sufficient to prove the following.

Let m be a positive integer, either odd or divisible by 4, $m \neq 1,4,8,12$. Consider the equation in integers p,q :

$$\cos 2\pi p/m \cos 6\pi/m = \cos 2\pi/m \cos 2\pi q/m \quad . \tag{8.4}$$

If m is odd, (8.4) has only the solutions $p \equiv \pm 1$ (m), $q \equiv \pm 3$ (m). If m is even, and provided (i) $p \equiv q(2)$ and (ii) $\cos 2\pi p/m \cos 2\pi q/m \neq 0$, equation (8.4) has only the solutions $p \equiv \pm 1$ (m/2), $q \equiv \pm 3$ (m/2).

We will establish this result in the following four sections by considering four separate cases

a) m a square free odd integer (section 9)

b) m four times a square free odd integer (section 10)

c) m an odd integer or four times an odd integer (section 11)

d) m divisible by 8. (section 12)

Trigonometric manipulation shows that (8.4) yields the equivalent identities

$$\cos \frac{2\pi(p-3)}{m} + \cos \frac{2\pi(p+3)}{m} = \cos \frac{2\pi(q-1)}{m} + \cos \frac{2\pi(q+1)}{m} \tag{8.5}$$

$$\cos \frac{2\pi(p-2)}{m} + \cos \frac{2\pi(p+2)}{m} = \cos \frac{2\pi p}{m} + \cos \frac{2\pi q}{m} \tag{8.6}$$

$$2 \cos \frac{4\pi}{m} = 1 + \frac{\cos \frac{2\pi q}{m}}{\cos \frac{2\pi p}{m}} \tag{8.7}$$;

we recall in (8.7) that $\cos 2\pi p/m \neq 0$ by condition (ii).

We make two observations from (8.7).

Suppose $m \neq 3$ or 12.

Lemma 8.5

a) Let p_1 be an odd prime divisor of m. Then p_1 does not divide both p and q.

b) Write $p/m = p'/m_1$, $q/m = q'/m_2$ in lowest terms.

If m is odd, then m_1 and m_2 are not co prime.

If m is four times an odd integer, and p,q are even, then m_1 and m_2 are not coprime, nor is $(m_1,m_2) = 2$.

Proof a) If p_1 divides both p and q, then (8.7) shows that $\cos 4\pi/m \in Q(\mu_{m/p})^+$, and hence $Q(\mu_m)^+ \subseteq Q(\mu_{m/p})^+$; this is impossible unless $m = 3$ or 12.

b) Suppose m is odd, and suppose, to the contrary, $(m_1,m_2) = 1$. Equation (8.7) shows $Q(\mu_m)^+ = Q(\mu_{m_1})^+ Q(\mu_{m_1})^+$. Since the degree of $F_{m_1}^+ . F_{m_2}^+$ is $\phi(m_1)/2.\phi(m_2)/2 = \phi(m_1m_2)/4$, whereas that of $F_{m_1m_2}^+$ is $\phi(m_1m_2)/2$, we obtain a contradiction.

Suppose m is four times an odd integer m_0. Equation (8.7) then shows $\cos \pi/m_0 \in Q(\cos \frac{\pi\, q/2}{m_0}, \cos \frac{\pi\, p/2}{m_0})$ and hence $Q(\mu_{m_0})^+ \subseteq Q(\mu_{m_1})^+ . Q(\mu_{m_2})^+$; this is again impossible by considering the degrees.

§9 The odd square free case

Throughout this section we shall suppose that m is a square-free odd integer, not equal to 1. Using equations (8.5) and (8.6), together with a suitable basis for $Q(\mu_m)^+$, we shall establish the main result that the equation $\cos 2\pi p/m \cos 6\pi/m = \cos 2\pi/m \cos 2\pi q/m$ in integers p and q has only the solutions $p \equiv \pm 1\ (m)$, $q \equiv \pm 3\ (m)$.

Let B be the set $\{a \in \mathbb{Z} \mid 1 \le a \le \frac{m-1}{2}$ and $(a,m) = 1\}$. Then the set $A = \{\pm a \mid a \in B\}$ is a complete set of inequivalent residue classes (mod m) which are coprime to m. The set $\mathcal{A}_m = \{\zeta_m^a \mid a \in A\}$ is a basis for $Q(\mu_m)/Q$; the set $\mathcal{B}_m = \{\cos 2\pi a/m\ ;\ a \in B\}$ is a basis for $Q(\mu_m)^+/Q$. Examining equation (8.5) we see that if m is coprime to the four numerators $p \pm 3$, $q \pm 1$, then each term in (8.5) belongs to basis $\mathcal{B}_m$, so that either $p + 3 \equiv \pm(q-1)$ and $p - 3 \equiv \pm(q+1)$, or $p + 3 \equiv \pm(q+1)$ and $p - 3 \equiv \pm(q-1)$ (mod m); hence

$p \equiv \pm 1\ (m)$ and $q \equiv \pm 3\ (m)$. Thus if we can show that m is coprime to the four numerators $p \pm 3$, $q \pm 1$, the result is established; the remainder of this section is devoted to proving this.

Before proving the next two lemmas, note that in basis B_m, 1 is uniquely expressible as

$$1 = \sum_{a \in B} -2 \cos \frac{2\pi a}{m} .$$

<u>Lemma 9.1</u> Every prime divisor of m, not equal to 3 or 5, divides at most one of the four numbers $p \pm 3$, $q \pm 1$.

<u>Proof</u> Suppose the lemma is false, and that some prime divisor p_1 ($\neq 3,5$) of m divides at least two of $p \pm 3$, $q \pm 1$. Hence p_1 divides exactly one of $p \pm 3$, and one of $q \pm 1$. Applying the trace map $Tr_{F_m/F_{p_1}}$ to equation (8.6), we obtain an equation of the form

$$A + B \cos \frac{2\pi . 6y}{p_1} = C + D \cos \frac{2\pi . 2y}{p_1}$$

where A, B, C, D are non-zero integers, and y is determined by $y(\frac{m}{p_1}) \equiv 1 \pmod{p_1}$; applying the automorphism $\sigma_{2y}^{-1} \in G(F_{p_1}/Q)$ to this produces

$$A + B \cos \frac{6\pi}{p_1} = C + D \cos \frac{2\pi}{p_1}$$

i.e.
$$D \cos \frac{2\pi}{p_1} - B \cos \frac{6\pi}{p_1} = (A - C) . \qquad \text{(i)}$$

If $A \neq C$, then by the remark above on the unique expression of 1 in the basis B_{p_1}, equation (i) is impossible (since $p_1 \geq 7$); if $A = C$, equation (i) is equally impossible. This proves the lemma.

<u>Lemma 9.2</u> Every prime divisor of m, not equal to 3 or 5, is coprime to all of the four numbers $p \pm 3$, $q \pm 1$.

<u>Proof</u> Suppose the lemma is false, and that some prime divisor p_1 ($\neq 3,5$) of m divides at least one of these four numbers; by lemma 9.1 it divides exactly one of them.

First suppose that $p_1 | p + 3$ or $p_1 | p - 3$. Applying the trace $Tr_{F_m/F_{p_1}}$ to equation (8.5) yields an equation of the form

$$A = B\cos\frac{2\pi(q+1)y}{p_1} + C\cos\frac{2\pi(q-1)y}{p_1} - D\cos\frac{2\pi.6y}{p_1}$$ with non-zero

integers A, B, C, D and y determined by $y(m/p_1) \equiv 1(p_1)$; applying the automorphism $\sigma_y^{-1} \in G(F_{p_1}/Q)$ gives

$$A = B\cos\frac{2\pi(q+1)}{p_1} + C\cos\frac{2\pi(q-1)}{p_1} - D\cos\frac{12\pi}{p_1}. \qquad \text{(ii)}$$

If $p_1 \geq 11$, our remark on the unique expression of 1 in basis B_{p_1} shows that (ii) is impossible. When $p_1 = 7$, this remark forces either $(q-1) \equiv \pm 2$ and $(q+1) \equiv \pm 3$, or $(q+1) \equiv \pm 2$ and $(q-1) \equiv \pm 3$ (mod 7), so that $q \equiv \pm 3$ (7); applying the trace map Tr_{F_m/F_7} and a suitable automorphism of $G(F_7/Q)$ to equation (8.6) yields an equation of the form

$$A'\cos 2\pi/7 + B'\cos 4\pi/7 = C'\cos 6\pi/7$$

(where A', B', C' are integers, A', B' nonzero); and this is clearly impossible in the basis B_7.

Turning to the case in which $p_1 | q+1$ or $p_1 | q-1$, we obtain (in a similar fashion) from equation (8.5) the following :

$$A = B\cos\frac{2\pi(p+3)}{p_1} + C\cos\frac{2\pi(p-3)}{p_1} - D\cos\frac{4\pi}{p_1},$$ with A, B, C, D non-zero

integers. If $p_1 \geq 11$, our remark above shows this is impossible; when $p_1 = 7$, this remark forces either $p + 3 \equiv \pm 1$ and $p - 3 \equiv \pm 3$, or $p + 3 \equiv \pm 3$ and $p - 3 \equiv \pm 1$; but it is easily checked that no such p exists. This proves the lemma.

<u>Lemma 9.3</u> Suppose m = 3 or 5 or 15. Then equation (8.4) has only the solution $p \equiv \pm 1$ (m), $q \equiv \pm 3$ (m).

<u>Proof</u> With m = 3, equation (8.4) becomes $\cos 2\pi p/3 = -\frac{1}{2}\cos 2\pi q/3$ which has only the solution $p \equiv \pm 1$ (3), $q \equiv \pm 3$ (3) as stated.
With m = 5, if $p \equiv \pm 1$ (5) then equation (8.4) immediately implies

$q \equiv \pm 3\ (5)$; so suppose $p \not\equiv \pm 1\ (5)$. Equation (8.6) yields

$$2 \cos 4\pi/5 - \cos 2\pi q/5 = 1 \qquad \big(\text{if } p \equiv 0\ (5)\big) \quad \text{or}$$

$$\cos 4\pi/5 - \cos 2\pi/5 + \cos 2\pi q/5 = 1 \quad \big(\text{if } p \equiv \pm 2\ (5)\big)\ ;$$

clearly $q \not\equiv 0\ (5)$, and recalling the basis B_5 of $Q(\mu_5)$ both these equations are seen to be impossible.

It remains to consider the case $m = 15$: we consider in turn the possibilities of 5 dividing none, one or two of the numerators $p \pm 3$, $q \pm 1$. If 5 divides none, applying Tr_{F_{15}/F_3} to equation (8.5) yields $2 \cos \frac{2\pi p}{3} = \cos \frac{2\pi(q-1)}{3} + \cos \frac{2\pi(q+1)}{3}$ from which we conclude $p \not\equiv 0\ (3)$, $q \equiv 0\ (3)$; thus 3, 5 and 15 are coprime to the four numerators, and the result immediately follows. If 5 divides exactly one numerator, applying the trace Tr_{F_{15}/F_3} gives

$$\pm 3 \cos 2\pi p/3 = -\left(\cos \frac{2\pi(q-1)}{3} + \cos \frac{2\pi(q+1)}{3}\right) \quad \big(\text{if } p \equiv \pm 3\ (5)\big)$$

or $$-2 \cos 2\pi p/3 = 4 \cos \frac{2\pi(q\mp 1)}{3} - \cos \frac{2\pi(q\pm 1)}{3} \qquad \big(\text{if } q \equiv \pm 1\ (5)\big)$$

each of which are easily checked to have no solutions. If 5 divides exactly two numerators, applying trace Tr_{F_{15}/F_3} yields

(i) $$3 \cos 2\pi p/3 = 4 \cos \frac{2\pi(q-1)}{3} - \cos \frac{2\pi(q+1)}{3} \qquad (p \equiv 3,\ q \equiv 1)$$

(ii) $$-3 \cos 2\pi p/3 = 4 \cos \frac{2\pi(q-1)}{3} - \cos \frac{2\pi(q+1)}{3} \qquad (p \equiv -3,\ q \equiv 1)$$

(iii) $$3 \cos 2\pi p/3 = -\cos \frac{2\pi(q-1)}{3} + 4 \cos \frac{2\pi(q+1)}{3} \qquad (p \equiv 3,\ q \equiv -1)$$

(iv) $$-3 \cos 2\pi p/3 = -\cos \frac{2\pi(q-1)}{3} + 4 \cos \frac{2\pi(q+1)}{3} \qquad (p \equiv -3,\ q \equiv -1),$$

(i) and (iii) possess solutions $p \not\equiv 0\ (3)$, $q \equiv 0\ (3)$, so that $p \equiv -2, 8\ (15)$, $q \equiv 6\ (15)$, whilst (ii) has solutions $p \equiv 0\ (3)$,

$q \equiv -1$ (3) i.e. $p \equiv 3$ (15), $q \equiv -4$ (15), and (iv) has solution $p \equiv 0$ (3), $q \equiv 1$ (3) i.e. $p \equiv 3$ (15), $q \equiv 4$ (15). It is easily checked that on substituting these congruence solutions in equation (8.5), we obtain equations about elements of B_{15} or B_5 which are impossible. This concludes the proof of the lemma.

We are now able to prove the main result of this section.

Theorem 7

Let m be a square free odd integer, not equal to 1. The only solutions in integers p, q to the equation $\cos 2\pi p/m \cos 6\pi/m = \cos 2\pi/m \cos 2\pi q/m$ are $p \equiv \pm 1$ (m), $q \equiv \pm 1$ (m).

Proof Let $m = m_1 . m_2$ where $m_1 | 15$, and $(m_2, 15) = 1$. By lemma 9.2, m_2 is prime to each of $p \pm 3$, $q \pm 1$. We show that m is prime to each of these numerators also. If $m_1 = 1$, there is nothing to prove; if $m_1 \neq 1$, consider the trace $Tr_{F_m/F_{m_1}}$ of equation (8.5) :

$$\cos \frac{2\pi(p-3)y}{m_1} + \cos \frac{2\pi(p+3)y}{m_1} = \cos \frac{2\pi(q+1)y}{m_1} + \cos \frac{2\pi(q-1)y}{m_1}$$

where y is determined by $ym_2 \equiv 1$ (m_1). Applying the automorphism $\sigma_y^{-1} \in G(F_{m_1}/Q)$ to this equation, and using lemma 9.3, we conclude that $p \equiv \pm 1$ (m_2), $q \equiv \pm 3$ (m_2), thus m_2 and m are coprime to $p \pm 3$, $q \pm 1$. The result now follows as explained at the beginning of this section.

§10 The case with m equal to four times an odd square free integer

Throughout this section, we shall suppose that m is of the form $4m_1$, where m_1 is an odd square free integer. We shall establish the result that, provided $m_1 \neq 1,3$, the equation $\cos 2\pi p/m \cos 6\pi/m = \cos 2\pi/m \cos 2\pi q/m$ (in integers p,q) subject to the restrictions that (i) $p \equiv q$ (2) and (ii) $p,q \not\equiv m_1$ $(2m_1)$, has only the solutions

$p \equiv \pm 1\ (2m_1)$, $q \equiv \pm 3\ (2m_1)$. When $m_1 = 3$ the equation (subject to no restrictions on p,q) has the solutions $q \equiv \pm 3\ (2m_1)$, with p unrestricted; henceforth we shall suppose that $m_1 \neq 1,3$.

The first part of this section considers the case when p,q are both even, and a contradiction is obtained from equation (8.6), so that no solutions (with p,q even) exist; the second part considers the case when p,q both odd, and the result is obtained from equation (8.5).

a) The case with p,q even. Throughout this subsection, we will suppose that p,q are both even : we rewrite the terms of equation (8.6), in order to use the properties of the basis B_{m_1} of $Q(\mu_{m_1})^+$. Note that restriction (ii) holds automatically. Since m_1 is odd, there exist integers x,y such that $4x + m_1 y = 1$ and thus $\zeta_{4m_1} = \zeta_{m_1}^{\,x}.i^y$. Thus if a is even, $\cos 2\pi a/4m_1 = (-1)^{a/2} \cos 2\pi ax/m_1$; equation (8.6) becomes, after applying the automorphism $\sigma_a^{-1} \in G(F_{m_1}/Q)$,

$$\cos \frac{2\pi(p+2)}{m_1} + \cos \frac{2\pi(p-2)}{m_2} + \cos \frac{2\pi p}{m_1} = \pm \cos \frac{2\pi q}{m_1} \qquad (10.1)$$

with the plus (resp. minus) sign if p and q are incongruent (resp. congruent) modulo 4.

Lemma 10.1 Every prime divisor of m_1, exceeding 5, divides at most one of the four numerators p, $p \pm 2$, q.

Proof Suppose the lemma is false : so there exists a prime $p_1 \geq 7$ which divides at least two of the four numerators. By lemma 8.5, p_1 does not divide both p and q, so $p \equiv \pm 2\ (p_1)$, $q \equiv 0\ (p_1)$. Applying the trace $Tr_{F_{m_1}/F_{p_1}}$ to equation (10.1) and a suitable automorphism of $G(F_{p_1}/Q)$ yields an equation of form

$$A + B \cos 8\pi/p_1 + C \cos 4\pi/p_1 = D ,$$

with nonzero integers A, B, C, D. Recalling that $\mathcal{B}_{p_1}$ is a basis for $Q(\mu_{p_1})^+$, this last identity is impossible, and the lemma is proved.

Lemma 10.2 Every prime divisor of m_1, exceeding 5, divides none of the four numerators p, $p \pm 2$, q.

Proof Suppose the result is false: then there is a prime divisor p_1 of m_1 dividing exactly one of the four numerators. Applying trace $Tr_{F_{m_1}/F_{p_1}}$, and a suitable automorphism, as in lemma 10.1, we obtain equations of form

$$\text{(i)} \qquad A \cos \frac{2\pi(p-2)}{p_1} + B \cos \frac{2\pi(p+2)}{p_1} + C \cos \frac{2\pi p}{p_1} = \pm D \quad \left(\text{if } q \equiv 0 \ (p_1)\right)$$

$$\text{(ii)} \qquad A \cos 4\pi/p_1 + B \cos 2\pi q/p_1 = \pm D \quad \left(\text{if } p \equiv 0 \ (p_1)\right)$$

$$\text{(iii)} \qquad A \cos 4\pi/p_1 + B \cos 8\pi/p_1 + C \cos 2\pi q/p_1 = \pm D \quad \left(\text{if } p \equiv \pm 2 \ (p_1)\right)$$

(A, B, C, D nonzero integers).

If $p_1 > 7$, these identities are untenable, upon recalling the basis $\mathcal{B}_{p_1}$. If $p_1 = 7$, identity (ii) is equally untenable; whilst identity (i) requires $p \equiv \pm 3 \ (7)$, $q \equiv 0 \ (7)$ and identity (iii) requires $p \equiv \pm 2 \ (7)$, $q \equiv \pm 1 \ (7)$.

We now show that when $p_1 = 7$, these last congruences produce contradictions, by rearranging equation (10.1). Using addition formulae and the identity $\sin 3\theta = \sin\theta\,(1 + 2\cos 2\theta)$, we obtain

$$i \sin \frac{2\pi(p+3)}{m_1} - i \sin \frac{2\pi(p-3)}{m_1} = \pm \left(i \sin \frac{2\pi(q+1)}{m_1} - i \sin \frac{2\pi(q-1)}{m_1}\right);$$

applying the trace $Tr_{F_{m_1}/F_7}$ and a suitable automorphism of $G(F_7/Q)$ yields an equation of form

$$i A \sin \frac{2\pi(p+3)}{7} + i B \sin \frac{2\pi(p-3)}{7} = \pm\left(i C \sin \frac{2\pi(q+1)}{7} - i D \sin \frac{2\pi(q-1)}{7}\right).$$

Substituting the congruences $p \equiv \pm 3 \ (7)$, $q \equiv 0 \ (7)$ and $p \equiv \pm 2 \ (7)$, $q \equiv \pm 1 \ (7)$ gives identities of form $A' \sin 2\pi/7 = B'$, and

$A'\sin 4\pi/7 + B'\sin 2\pi/7 = C'$ (A',B',C' integers) which are impossible with the basis A_7. This establishes the lemma.

<u>Lemma 10.3</u> Suppose m_1 is coprime to 15. Then equation (8.4) has no solutions in even integers p,q.

<u>Proof</u> Since m_1 is coprime to the numerators p, $p \pm 2$, q, equation (10.1) is clearly impossible in the basis B_{m_1}.

<u>Lemma 10.4</u> Suppose $m_1 = 3m_2$ where m_2 is coprime to 15. Then equation (10.1) has no solutions in even integers p,q.

<u>Proof</u> Since m_2 is coprime to the numerators p, $p \pm 2$, q, applying the trace $Tr_{F_{m_1}/F_3}$ to equation (10.1) yields

$$\cos\frac{2\pi(p-2)}{3} + \cos\frac{2\pi(p+2)}{3} + \cos\frac{2\pi p}{3} = \pm\cos\frac{2\pi q}{3},$$

which is impossible, since the left hand side is zero. Thus the equation has no solutions.

<u>Lemma 10.5</u> Suppose $m = 5m_2$ where m_2 is coprime to 15. Then equation (8.4) has no solutions in even integers p,q.

<u>Proof</u> Applying trace $Tr_{F_{m_1}/F_5}$ and a suitable automorphism to equation (10.1) yields $\cos\frac{2\pi(p-2)}{5} + \cos\frac{2\pi(p+2)}{5} + \cos\frac{2\pi p}{5} = \pm\cos\frac{2\pi q}{5}$.

According as $p \equiv 0, \pm 1, \pm 2 \pmod 5$ we obtain the equations

$$2\cos 4\pi/5 \pm \cos 2\pi q/5 = -1,$$

$$2\cos 2\pi/5 + \cos 4\pi/5 = \pm\cos 2\pi q/5,$$

$$\cos 2\pi/5 + \cos 8\pi/5 = -1 \pm \cos 2\pi q/5,$$

each of which are seen to be impossible, given the basis B_5 of $Q(\mu_5)^+$.

<u>Lemma 10.6</u> Suppose $m_1 = 15m_2$, where m_2 is coprime to 15. Then equation (8.4) has no solutions in even integers p,q.

<u>Proof</u> Since m_2 is coprime to each of p, $p \pm 2$, q, applying the trace $Tr_{F_{m_1}/F_{15}}$ and a suitable automorphism of $G(F_{15}/Q)$ to equation (10.1), yields

$$\cos\frac{2\pi(p-2)}{15} + \cos\frac{2\pi(p+2)}{15} + \cos\frac{2\pi p}{15} = \pm\cos\frac{2\pi q}{15}. \qquad \text{(i)}$$

We claim that this last equation has no solutions, and this in turn proves the lemma.

Clearly, 5 divides exactly one of p, $p \pm 2$; for otherwise applying the trace Tr_{F_{15}/F_3} yields $\cos 2\pi q/3 = 0$ which is absurd. Also obviously, 3 divides exactly one of p, $p \pm 2$.

Furthermore $(q,15) \neq 1$, for otherwise the application of the trace $Tr_{F_{15}/Q}$ to (i) yields one of the two impossible identities $1 - \phi(3) - \phi(5) = \pm 1$ or $1 + 1 + \phi(15) = \pm 1$; Lemma 8.5 then shows that $(p,15) = 1$. We claim that $3 \mid q$, but $5 \nmid q$. For, if $q \equiv 0\ (5)$, then $p \equiv \pm 2\ (15)$ $\left(\text{else equation (10.1) would imply } \cos 4\pi/15 \in Q\right)$; but then equation (10.1) yields the impossible identity $\cos 4\pi/15 + \cos 8\pi/15 = -1 \pm \frac{1}{2}$; thus $5 \nmid q$ and hence $3 \mid q$. Considering equation (10.1) again,

$$\cos\frac{2\pi(p-2)}{15} + \cos\frac{2\pi(p+2)}{15} + \cos\frac{2\pi p}{15} = \cos\frac{2\pi\, q/3}{5},$$

we now see that $p \equiv \pm 2\ (15)$ [for otherwise $\cos 4\pi/15 \in Q(\mu_5)^+$] and taking trace $Tr_{F_{15}/Q}$ yields the impossible identity $\phi(15) + 1 + 1 = \pm\phi(3)$. This proves the lemma.

We gather the results of lemmas 10.3-10.6 in the following.

<u>Theorem 8</u>

Suppose m_1 is a square free odd integer, $\neq 1$ or 3. The equation $\cos 2\pi p/4m_1 \cos 6\pi/4m_1 = \cos 2\pi/4m_1 \cos 2\pi q/4m_1$ has no solutions in even integers p,q.

We can now turn to:

b) The case with p,q odd

Suppose henceforth in this subsection that p and q are odd. We rewrite the terms of equation (8.5) in order to utilize the properties of the basis B_{m_1}. We obtain in a similar fashion to the above,

$$\cos\frac{2\pi(p-3)}{m_1} - \cos\frac{2\pi(p+3)}{m_1} = \pm\left(\cos\frac{2\pi(q-1)}{m_1} - \cos\frac{2\pi(q+1)}{m_1}\right) \qquad (10.2)$$

with the plus (resp. minus) sign if p and q are incongruent (resp. congruent) modulo 4.

Lemma 10.7 Every prime divisor of m_1 exceeding 5 divides at most one of the four numerators $p \pm 3$, $q \pm 1$.

Proof Suppose the result is false : so there exists a prime $p_1 \geq 7$ which divides exactly one of $p \pm 3$ and exactly one of $q \pm 1$. Applying the trace $Tr_{F_{m_1}/F_{p_1}}$ to equation (10.2) and a suitable automorphism of $G(F_{p_1}/Q)$ yields an equation of form $A + B\cos 2\pi/p_1 = C + D\cos 4\pi/p_1$ with nonzero integers A, B, C, D. But this is clearly untenable, recalling that B_{p_1} is a basis for $Q(\mu_{p_1})^+$; the lemma is thus proved.

Lemma 10.8 Every prime divisor of m_1, exceeding 5, divides none of the four numerators $p \pm 3$, $q \pm 1$.

Proof Suppose the result is false then there is a prime divisor p_1 of m_1 dividing exactly one of the four numerators. Applying $Tr_{F_{m_1}/F_{p_1}}$, and a suitable automorphism, as in lemma 10.7, we obtain equations of the form

$$\text{(i)} \quad A\cos\frac{2\pi(p-3)}{p_1} + B\cos\frac{2\pi(p+3)}{p_1} + C\cos\frac{4\pi}{p_1} = D \quad \left(\text{if } q \equiv \pm 1 \ (p_1)\right),$$

$$\text{or} \quad \text{(ii)} \quad A\cos\frac{2\pi(q-1)}{p_1} + B\cos\frac{2\pi(q+1)}{p_1} + C\cos\frac{12\pi}{p_1} = D \quad \left(\text{if } p \equiv \pm 3 \ (p_1)\right),$$

with nonzero integers A, B, C, D. If $p_1 > 7$, these identities are untenable, upon recalling that B_{p_1} is a basis. If $p_1 = 7$, the first equation requires $p + 3 \equiv \pm 1$, $p - 3 \equiv \pm 3$ (or vice-versa) and this is impossible; if $p_1 = 7$, the second equation is tenable only with $p \equiv \pm 3$ (7), $q \equiv \pm 3$ (7). By rearranging equation (10.2) we now show this last congruence produces a contradiction. Using addition formula and the identity $\sin 3\theta = \sin\theta\,(1 + 2\cos 2\theta)$, we obtain

$$i \sin \frac{2\pi(p-2)}{m_1} + i \sin \frac{2\pi(p+2)}{m_1} + i \sin \frac{2\pi p}{m_1} = \pm\, i \sin \frac{2\pi q}{m_1} ;$$

applying the trace $Tr_{F_{m_1}/F_7}$ and a suitable automorphism of $G(F_7/Q)$ yields an equation of the form

$$i\,A \sin \frac{2\pi(p-2)}{7} + i\,B \sin \frac{2\pi(p+2)}{7} + i\,C \sin \frac{2\pi p}{7} = \pm i\,D \sin \frac{2\pi q}{7} ,$$

with nonzero A, B, C, D. Substituting the congruences $p \equiv \pm 3$ (7), $q \equiv \pm 3$ (7) gives the identity $i\,A \sin 2\pi/7 + i\,B \sin 4\pi/7 + i(C \pm D) \sin 6\pi/7 = ($ which is clearly impossible, upon recalling the basis A_7 of $Q(\mu_7)$. This proves the lemma.

<u>Lemma 10.9</u> Suppose p, q are odd integers, and $m_1 = 5$ or 15 (so $m = 20$ or 60). If $p,q \not\equiv m_1$ $(2m_1)$, then the equation $\cos 2\pi p/4m_1 \cos 6\pi/4m_1 = \cos 2\pi/4m_1 \cos 2\pi q/4m_1$ has only the solutions $p \equiv \pm 1$ $(2m_1)$, $q \equiv \pm 3$ $(2m_1)$.

<u>Proof</u> Since p, q are odd, it suffices to suppose $p,q \not\equiv 0$ (m_1), and to seek solutions $p \equiv \pm 1$ (m_1), $q \equiv \pm 3$ (m_1).

If $m_1 = 5$, we have $p,q \not\equiv 0$ (5), and if $p \equiv \pm 1$ (5) or $q \equiv \pm 3$ (5) we immediately obtain the stated solution; it remains to check the infeasibility of the case $p \equiv \pm 2$ (5) and $q \equiv \pm 1$ (5), which is evident from equation (10.2).

Turning to the case $m_1 = 15$, we consider in turn the cases of 5 dividing none, one or two of the numerators $p \pm 3$, $q \pm 1$. If 5 divides none, the application of Tr_{F_{15}/F_3} to equation (10.2) yields

$0 = \cos\frac{2\pi(q-1)}{3} - \cos\frac{2\pi(q+1)}{3}$; thus $q \equiv 0\ (3)$, and (from equation (8.7)) we have $p \not\equiv 0\ (3)$, so that 3, 5 and 15 are coprime to the four numerators $p \pm 3$, $q \pm 1$: recalling that B_{15} is a basis of $Q(\mu_{15})^+$, we deduce $p \equiv \pm 1\ (15)$, $q \equiv \pm 3\ (15)$ as stated.

We now show that if 5 divides at least one of the numerators, contradictions ensue. Suppose 5 divides at least two, and hence exactly two, of these numerators. Applying Tr_{F_{15}/F_3} to equation (10.2) yields, in the case $p \equiv 3\ (5)$, $q \equiv 1\ (5)$

$$5\cos 2\pi p/3 = 4\cos\frac{2\pi(p-3)}{3} + \cos\frac{2\pi(p+3)}{3} = \pm\left(4\cos\frac{2\pi(q-1)}{3} + \cos\frac{2\pi(q+1)}{3}\right)$$

and this is possible only with $p \not\equiv 0\ (3)$, $q \equiv 0\ (3)$; thus $q \equiv 6\ (15)$, $p \equiv -2$ or $8\ (15)$, and each of these is seen to be impossible in equation (10.2). A similar argument in the remaining three cases ($p \equiv \pm 3$, $q \equiv \pm 1$) yields similar contradictions, hence 5 cannot divide at least two of the four numerators.

Finally suppose 5 divides exactly one numerator. First, suppose $p \equiv \pm 3\ (5)$. Applying trace Tr_{F_{15}/F_3} to equation (10.2) yields

$5\cos 2\pi p/3 = \pm\left(\cos\frac{2\pi(q-1)}{3} - \cos\frac{2\pi(q+1)}{3}\right)$ which is impossible, since the right hand side has value 0 or $\pm\frac{3}{2}$. Thus $p \not\equiv \pm 3\ (5)$ and suppose $q \equiv \pm 1\ (5)$. Applying Tr_{F_{15}/F_3} now yields

$0 = 4\cos\frac{2\pi(q\pm 1)}{3} + \cos\frac{2\pi(q\mp 1)}{3}$, which is impossible.

The lemma is now completely proved.

We can now establish the main result of this section.

Theorem 9

Suppose m_1 is a square free odd integer, not equal to 1 or 3. The equation $\cos 2\pi p/4m_1 \cos 6\pi/4m_1 = \cos 2\pi/4m_1 \cos 2\pi q/4m_1$ in integers p, q subject to the restrictions (i) $p \equiv q\ (2)$ and (ii) $p, q \not\equiv m_1\ (2m_1)$ has only the solutions $p \equiv \pm 1\ (2m_1)$, $q \equiv \pm 3\ (2m_1)$.

<u>Proof</u> When p, q are both even, this has already been established, so suppose p, q are odd. Write $m_1 = m_2m_3$, where $m_2 | 15$ and $(m_3,15) = 1$. When $m_3 = 1$, $m_1 = 5$ or 15 and the result has been proved. So suppose $m_3 \neq 1$. Lemma 10.8 shows that m_3 is coprime to $p \pm 3, q \pm 1$. If $m_2 = 1$, since $\mathcal{B}_{m_3}$ is a basis for $Q(\mu_{m_3})^+$, we immediately obtain $p \equiv \pm 1\ (m_1)$, $q \equiv \pm 3\ (m_1)$ and since p, q are odd, we conclude $p \equiv \pm 1\ (2m_1)$, $q \equiv \pm 3\ (2m_1)$. If $m_2 = 3$, applying the trace $Tr_{F_{m_1}/F_3}$ to equation (10.2) yields $0 = \cos \frac{2\pi(q-1)}{3} - \cos \frac{2\pi(q+1)}{3}$; thus $q \equiv 0\ (3)$ and (from lemma 8.5) $p \not\equiv 0\ (3)$, so that $3m_2$ is coprime to the four numerators $p \pm 3$, $q \pm 1$; by the same basis argument as above we obtain $p \equiv \pm 1\ (2m_1)$, $q \equiv \pm 3\ (2m_1)$. If $m_2 = 5$ or 15, applying the trace $Tr_{F_{m_1}/F_{m_2}}$ to equation (10.2), followed by a suitable automorphism of $G(F_{m_2}/Q)$, yields the equation

$$\cos \frac{2\pi(p-3)}{m_2} - \cos \frac{2\pi(p+3)}{m_2} = \pm\left(\cos \frac{2\pi(q-1)}{m_2} - \cos \frac{2\pi(q+1)}{m_2}\right) .$$

By lemma 10.9, this equation has either the solutions $p \equiv \pm 1\ (m_2)$, $q \equiv \pm 3\ (m_2)$ or else $p \equiv q \equiv 0\ (m_2)$. Since $p,q \not\equiv 0\ (m_1)$, equation (8.7) shows that it is not possible for $p \equiv q \equiv 0\ (m_2)$, hence m_2, and also m_1, are coprime to $p \pm 3$, $q \pm 1$. The result now follows, using the usual basis argument.

We have thus established the result when m is four times an odd square free integer.

§11 <u>The case of m odd or four times an odd integer</u>

Throughout this section we shall suppose that m is either an odd integer or four times an odd integer : denote this odd integer m_1, so that $m = m_1$ or $m = 4m_1$. We shall, in addition, suppose that $m \neq 1$, 4 or 12, and establish the following result.

Theorem 10

a) If m is odd, equation (8.4) has only the solutions $p \equiv \pm 1$ (m), $q \equiv \pm 3$ (m).

b) If m is even, equation (8.4), subject to the two restrictions (i) $p \equiv q$ (2) and (ii) $p,q \not\equiv m_1$ $(2m_1)$, has only the solutions $p \equiv \pm 1$ $(2m_1)$, $q \equiv \pm 3$ $(2m_1)$.

Lemma 11.1 Let p_1 be an (odd) prime divisor of m_1 such that $p_1^2 | m_1$. Then a) p is not divisible by p_1,
and b) if $p_1 \neq 3$, q is not divisible by p_1.

Proof First we note that p_1 does not divide both p and q - for otherwise equation (8.7) would show $\cos 4\pi/m \in Q(\mu_{m/p_1})^+$, which is impossible. Suppose, to the contrary p were divisible by p_1. Then $p_1 \nmid q$ and $p_1 \nmid p \pm 2$, and applying the trace $Tr_{F_m/F_{m/p_1}}$ to equation (8.6) yields the impossible equation $\cos 2\pi p/m = 0$; thus $p_1 \nmid p$. Now suppose $p_1 | q$. If p_1 divides neither of $p \pm 2$, the same trace argument shows $\cos 2\pi q/m = 0$, which is impossible; thus p_1 divides exactly one of $p \pm 2$. But under this trace map equation (8.6) then shows that $\cos \frac{2\pi(p\pm 2)}{m} = \cos \frac{2\pi q}{m}$ and hence $\cos \frac{2\pi p}{m} = \cos \frac{2\pi(p\mp 2)}{m}$; but this implies that $p \equiv \pm 1$ (m), $q \equiv \pm 3$ (m) which is impossible since $p_1 \neq 3$. The lemma is now proved.

Lemma 11.2 Suppose $9 | m$. Then theorem 10 is true.

Proof First suppose that m is odd. Lemma 11.1 shows that $3 \nmid p$. Suppose that $3 \nmid q$: then 3 divides exactly one of $q \pm 1$, say $q + 1$. Let $m' = \frac{1}{3}m$ and apply the trace $Tr_{F_m/F_{m'}}$ to equation (8.5) : we obtain $\cos \frac{2\pi(q+1)}{m'} = 0$, an obvious impossibility. Thus $3 | q$. Following the same argument as lemma 11.1 now shows that $p \equiv \pm 1$ (m), $q \equiv \pm 3$ (m). This establishes case a) of theorem 10.

We now turn to the case when m is even. Let $m_1' = \frac{1}{3}m_1$. We first show that both p and q cannot be even, by applying the trace $Tr_{F_{m_1}/F_{m_1'}}$ to equation (10.1). Clearly $3|q$, (for otherwise (10.1) yields $\cos\frac{2(p\pm 2)}{m_1'} = 0$) and so we obtain $\cos\frac{2\pi p}{m_1} = -\cos\frac{2\pi(p\pm 2)}{m_1}$. But then $2p \equiv m_1 \pm 2(p\pm 2)$ modulo $2m_1$, which is impossible. Thus p and q must be odd. Lemma 11.1 shows that $3 \nmid (p\pm 3)$; we apply the trace $Tr_{F_{m_1}/F_{m_1'}}$ to equation (10.2), and deduce that $q \equiv 0$ (3). Application of the trace $Tr_{F_{4m_1}/F_{4m_1'}}$ to equation (8.6) now yields $\cos\frac{2\pi p}{m} = \cos\frac{2\pi(p\pm 2)}{m}$, so that $p \equiv \pm 1$ $(2m_1)$ and $q \equiv \pm 3$ $(2m_1)$. This proves the lemma.

<u>Lemma 11.3</u> Let p_1 be a prime divisor of m, not equal to 3. Suppose that theorem 4 is true. Then theorem 10 is also true when m is replaced by mp_1.

<u>Proof</u> Consider the equation $\cos\frac{2\pi p}{p_1 m}\cos\frac{6\pi}{p_1 m} = \cos\frac{2\pi}{p_1 m}\cos\frac{2\pi q}{p_1 m}$. If m is odd, no restrictions are imposed on p and q; if m is even, we require $p \equiv q$ (2) and $p,q \not\equiv p_1 m_1$ $(2p_1 m_1)$ [recall $m = 4m_1$]. By lemma 11.1, p_1 does not divide p or q. Applying the norm map $N_{F_{p_1 m}/F_m}$ to equation (8.4) gives

$$\cos\frac{2\pi p}{m}\cos\frac{6\pi}{m} = \cos\frac{2\pi}{m}\cos\frac{2\pi q}{m}\ .$$

If m is odd, we deduce that $p \equiv \pm 1$ (m), $q \equiv \pm 3$ (m); if m is even, we deduce that either $p \equiv \pm 1$ $(2m_1)$, $q \equiv \pm 3$ $(2m_1)$ or $p \equiv 0$ (m_1) or $q \equiv 0$ (m_1). But lemma 11.1 shows that $p \not\equiv 0$ (m_1), $q \not\equiv 0$ (m_1), so that $p \equiv \pm 1$, $q \equiv \pm 3$ modulo m or modulo $2m_1$, accordingly as m is odd or even. For simplicity we shall suppose $p \equiv 1$, $q \equiv 3$; the argument below holds equally well with the other three possibilities. Thus for some integers a, b we have $p = 1 + am_1$, $q = 3 + bm_1$; when m is even, a and b must be even. We will show that $a \equiv b \equiv 0\ (p_1)$ and this proves the lemma.

From equation (8.5) we have

$$\cos\frac{2\pi(3+am)}{p_1m} + \cos\frac{2\pi(-1+am)}{p_1m} = \cos\frac{2\pi(1+am)}{p_1m} + \cos\frac{2\pi(3+bm)}{p_1m},$$

or in terms of ζ_{p_1m} and ζ_{p_1}:

$$\zeta_{p_1}{}^a\zeta_{mp_1}{}^3 + \zeta_{p_1}{}^{-a}\zeta_{mp_1}{}^{-3} + \zeta_{p_1}{}^a\zeta_{mp_1}{}^{-1} + \zeta_{p_1}{}^{-a}\zeta_{mp_1}$$

$$= \zeta_{p_1}{}^a\zeta_{mp_1} + \zeta_{p_1}{}^{-a}\zeta_{mp_1}{}^{-1} + \zeta_{p_1}{}^b\zeta_{mp_1}{}^3 + \zeta_{p_1}{}^{-b}\zeta_{mp_1}{}^{-3}.$$

Writing $\alpha = \zeta_{p_1}{}^a - \zeta_{p_1}{}^{-a}$ and $\beta = \zeta_{p_1}{}^b - \zeta_{p_1}{}^a$ and $\bar\beta$ for its complex conjugate, we obtain $\alpha\zeta_{mp_1} - \alpha\zeta_{mp_1}{}^{-1} + \beta\zeta_{mp_1}{}^3 + \bar\beta\zeta_{mp_1}{}^{-3} = 0$. Thus

$$\bar\beta - \alpha\zeta_{mp_1}{}^2 + \alpha\zeta_{mp_1}{}^4 + \beta\zeta_{mp_1}{}^6 = 0 . \qquad \text{(i)}$$

First consider the case with m odd. The set $S = \{1, \zeta_{mp_1}{}^2, \zeta_{mp_1}{}^4, \ldots, \zeta_{mp_1}{}^{2(p_1-1)}\}$ is a basis for $Q(\mu_{mp_1})/Q(\mu_m)$, and since $\alpha,\beta,\bar\beta \in Q(\mu_m)$, we conclude, as $p_1 \geq 5$, that $\alpha = \beta = 0$ and hence $a \equiv b \equiv 0\ (p_1)$. Turning to the case with m even $(m = 4m_1)$, S is a basis for $Q(\mu_{m_1p_1})/Q(\mu_{m_1})$, and again we conclude $a \equiv b \equiv 0\ (p_1)$. The lemma is now completely proven.

Theorem 10 now follows either from lemma 11.2 (if $9 \mid m$) or lemma 11.3 by induction, beginning with the cases established in section 10.

§12 The case $m \equiv 0\ (8)$

Throughout this section we shall suppose that m is divisible by 8, so that $m = 2^{n+2}m_1$ where $n \geq 1$ and m_1 is an odd integer; also $m \neq 8$. Let $m' = \frac{1}{2}m$, $m'' = \frac{1}{4}m$. We shall establish the result:

Theorem 11

The equation in integers p, q

$$\cos 2\pi p/m \cos 6\pi/m = \cos 2\pi/m \cos 2\pi q/m$$

subject to the restrictions (i) $p \equiv q\ (2)$, (ii) $p,q \not\equiv 2^n m_1\ (2^{n+1}m_1)$ has only the solutions $p \equiv \pm 1\ (2^{n+1}m_1)$, $q \equiv \pm 3\ (2^{n+1}m_1)$.

[If $m = 8$ the equation subject to these restrictions has solutions $p \equiv \pm 1\ (4)$, $q \equiv \pm 3\ (4)$ (if p,q odd) and $p \equiv 4\ (8)$, $q \equiv 0\ (8)$ or $p \equiv 0\ (8)$, $q \equiv 4\ (8)$ (if p,q even).]

Lemma 12.1 Suppose p, q are even integers. Equation (8.4) subject to restrictions (i) and (ii) above has no solutions.

Proof We consider equation (8.6). Suppose first that $p \equiv 0\ (4)$. Then $\mathrm{Tr}_{F_{m'}/F_{m''}} \cos \frac{2\pi(p\pm 2)}{m} = 0$. Clearly $q \not\equiv 2\ (4)$ for otherwise this trace map would imply $\cos 2\pi p/m = 0$, i.e. $p \equiv 2^n m_1\ (2^{n+1}m_1)$. Thus $q \equiv 0\ (4)$, and we deduce $\cos \frac{2\pi(p+2)}{m} + \cos \frac{2\pi(p-2)}{m} = 0$. Since $m \neq 8$, this implies that $\cos 2\pi p/m = 0$, contrary to the restriction $p \not\equiv m'\ (m'')$.

On the other hand, suppose $p \equiv 2\ (4)$. Again it is clear that $q \not\equiv 0\ (4)$ (for the trace $\mathrm{Tr}_{F_{m'}/F_{m''}}$ would show $\cos 2\pi p/m = 0$); applying the trace $\mathrm{Tr}_{F_{m'}/F_{m''}}$ again shows $\cos \frac{2\pi(p+2)}{m} + \cos \frac{2\pi(p-2)}{m} = 0$, which as above is impossible. Thus equation (8.4) has no solutions subject to restrictions (i) and (ii).

Lemma 12.2 Suppose p, q are odd integers. Equation (8.4) has only the solutions $p \equiv \pm 1\ (m')$, $q \equiv \pm 3\ (m')$.

Proof Since p, q are odd, $p - 3 \equiv q - 1$ or $p - 3 \equiv q + 1\ (4)$. Applying the trace $\mathrm{Tr}_{F_{m'}/F_{m''}}$ to equation (5) yields

$$\cos \frac{2\pi(p-3)}{m} = \cos \frac{2\pi(q-1)}{m} \quad \text{or} \quad \cos \frac{2\pi(q+1)}{m} .$$

Thus $p - 3 \equiv \pm(q-1)$ and $p + 3 \equiv \pm(q+1) \pmod m$

or $p + 3 \equiv \pm(q-1)$ and $p - 3 \equiv \pm(q+1) \pmod m$;

these congruences have only the solutions $p \equiv \pm 1$, $q \equiv \pm 3 \pmod{m'}$.

Theorem 11 follows immediately from Lemmas 12.1 and 12.2. Theorems 7, 9, 10, 11 establish the truth of theorem 4 and hence of the Arscott-Wright conjecture.

REFERENCES

1. Ch Hermite : Sur Quelques Applications des Functions Elliptiques, Oeuvres Vol III, 264-428.

2. F M Arscott and B D Sleeman : Multiplicative solutions of linear differential equations. J. London Math. Soc. 43 (1968), 263-270.

3. B D Sleeman : Multiparameter periodic differential equations. Proc. Conf. on the Theory of Ordinary and Partial Differential Equations (1978). Lecture Notes in Mathematics Vol 827, 229-250, (1980).

4. F M Arscott and G P Wright : Floquet Theory for Doubly-periodic Differential Equations. Spisy Přírodov fak Univ. J E Purkyně v Brně T5 (1969) 111-124.

5. G P Wright : Floquet Theory for Doubly-periodic Differential Equations. Ph.D. Thesis, University of Surrey, 1970.

6. F M Arscott : Periodic Differential Equations. Pergamon Press, Oxford, 1964.

7. S Lang : Algebraic Number Theory. Addison Wesley, 1970.

8. R L Long : Algebraic Number Theory. Monographs and textbooks in pure and applied maths, No 41. Marcel Dekker, 1977.

POINCARÉ'S TYPE OF EQUATIONS
in the study of limit cycles

Tung Chin-Chu
(Graduate School, Chinese Academy of Sciences)

1. We shall say that equationsof the form

$$\dot{x} = F_y G(x,y) + F_x H_1(x,y) , \quad \dot{y} = -F_x G(x,y) + F_y H_2(x,y) , \tag{1}$$

are of Poincaré type. A great number of the examples used by Poincaré in his memoir for studying the limit cylces are of this form [1]. If the functions on the right hand side satisfy:

(I) F = constant denotes a family of closed curves around $(0,0)$, and $(F_x, F_y) = (0,0)$ only at $(x,y) = (0,0)$;

(II) $H(x,y)$ are bounded on a bounded region of (x,y), and $H_i(x,y) = 0$ denote simple closed curves with $(H_{ix}, H_{iy}) \neq (0,0)$ on $H_i = 0$ for $i = 1, 2$;

(III) $G(x,y) \neq 0$, $H_1 H_2 + G^2 \neq 0$ on a region which contains the curves $H_i = 0$, $i - 1, 2$,

then we shall say that the equations are of the first class; otherwise they are of the second class.

This type of equation yields naturally a topographical structure F = constant and a set of contact curves determined by F and H_i, $i = 1, 2$. It seems that the global phase portrait is intimately connected with such a structure.

2. For the equations of the first class it is easy to establish the following

Theorem 1. If $H_1 = H_2 = H$, $H(0,0) \neq 0$; and $H(x,y) = 0$ is a simple closed curve enclosing $(0,0)$, then the system has a limit cycle which is stable for $H(0,0) < 0$, and is unstable for $H(0,0) > 0$.

Proof. That the singular point of the system is determined by $(F_x, F_y) = 0$ follows from the condition (III); that $(x,y) = (0,0)$ is the only singular point follows from the condition (I).

Since $H(0,0) \neq 0$, $H \neq 0$ in the neighbourhood of $(0,0)$, and therefore $(0,0)$ is either a node or a focus which is stable for $H(0,0) < 0$ and unstable for $H(0,0) > 0$.

From condition (II), the sign of $H(x,y)$ changes as (x,y) goes across $H(x,y) = 0$. Let

$k_1 = \min F(x,y)$, $k_2 = \max F(x,y)$, for (x,y) on $H(x,y) = 0$.

Then $F = c_1$ with $c_1 < k_1$ and $F = c_2$ with $c_2 > k_2$ together form the boundary of the ring-shaped region needed by application of the Poincare-Bendixson theorem, since they are cycles without contact. Therefore the existence of a limit cycle with a stability opposite to that of the singular point $(0,0)$ is insured.

Concrete examples are often constructed with $F = x^2 + y^2$ and a convex curve $H = 0$. We do not require $H = 0$ to be convex here. The following is an example:

Let $F = x^4 + 6x^2y^2 + y^4$, $H = (x^2 + y^2) + x^2 - y^2 - 2$

then the system has a limit cycle in the region $1 < F < 100$.

The uniqueness of the limit cycle can be shown by using the following theorem or Poincare (Theorem XIX on page 77 or [1]).

Theorem. Let the system of equations be written in polar coordinates as

$$\frac{d\rho}{d\omega} = \phi(\rho,\omega) \tag{2}$$

and let $\psi(\rho)$ be a single-valued function which takes finite values

<u>for finite ρ. Then there must be a point in the ring-shaped region bounded by two limit cycles of (2) on which one of the following is true:</u>

$$\frac{d\psi}{d\rho} = \infty \; ; \quad \phi(\rho,\omega) = \infty \; ; \quad \frac{\partial}{\partial\rho}(\phi \frac{\partial\psi}{\partial\rho}) = 0 \; ; \quad \frac{\partial\phi}{\partial\rho} = \infty \; ; \quad \frac{\partial^2\psi}{\partial\rho^2} = \infty \, . \tag{3}$$

In the original statement of Poincare the last two items are missing. They were supplied by Chin [2]. Another amendment in the proof of it can be remarked as follows.

In proving the theorem, $\psi(\rho(\omega))$ has been taken to be single-valued in ω; while, however, the function ρ expressing the limit cycle should not be necessarily single-valued of ω unless we restrict it to be convex. Now suppose that $\rho(\omega)$ is not single-valued. Since the limit cycle is a closed curve enclosing the origin, there must be a half ray $\omega = \omega_o$ which has contact with it, so that $\phi(\rho(\omega_o),\omega_o) = \infty$. And that means we have already a point <u>on</u> the limit cycle where one of conditions in (3) is true.

Application of the theorem to a region where none of the conditions holds shows that there is not more than one limit cycle, and when there is a limit cycle, it must have a single-valued expression.

A generalized version of this theorem is the following

<u>Theorem.</u> <u>There is a topological transformation</u> $\xi = \phi(x,y)$, $\eta = \psi(x,y)$ <u>which transforms</u> $\frac{dy}{dx} = G(x,y)$ <u>into</u> $\frac{d\xi}{d\eta} = \textcircled{H}$ <u>such that for the new equation there must be a point in the region bounded by two limit cycles for which either</u> $\textcircled{H} = \infty$ <u>or</u> $\frac{\partial \textcircled{H}}{\partial\xi} = 0$.

Now let us apply this new version of the theorem to show the uniqueness of the limit cycle for the system (1) as follows.

Theorem 2. In spite of the conditions in theorem 1, suppose $M(x,y) > \varepsilon > 0$ for a constant ε, where $M(x,y)$ is an integrating factor of the orthogonal family

$\Phi(x,y) = $ constant of the curves $F(x,y) = $ constant i.e.

$$d\Phi = M(x,y)[\frac{\partial F}{\partial y}\,dx - \frac{\partial F}{\partial x}\,dy] = 0\ ;$$

and further

$$\frac{\partial}{\partial x}\left(\frac{H}{MG}\right)\frac{\partial F}{\partial x} + \frac{\partial}{\partial y}\left(\frac{H}{MG}\right)\frac{\partial F}{\partial y} \neq 0\ . \tag{4}$$

Then (1) has a unique limit cycle.

Proof. The transformation $\xi = F(x,y)$, $\eta = \Phi(x,y)$ with inverse $x = f(\xi,\eta)$, $y = \phi(\xi,\eta)$ transforms the system into

$$\frac{d\xi}{d\eta} = \frac{H(f,\phi)}{MG(f,\phi)} = \textcircled{H}(\xi,\eta)$$

That $\textcircled{H} \neq \infty$ in the region $k_1 < F < k_2$ follows from the conditions (I), (II), (III) of section 1, and $\frac{\partial \textcircled{H}}{\partial \xi} \neq 0$ follows from (4) since

$$\frac{\partial \textcircled{H}}{\partial \xi} = \frac{\partial}{\partial x}\left(\frac{H}{MG}\right)\frac{\partial f}{\partial \xi} + \frac{\partial}{\partial y}\left(\frac{H}{MG}\right)\frac{\partial \phi}{\partial \xi} = \frac{M}{\Delta}\left[\frac{\partial}{\partial x}\left(\frac{H}{MG}\right)\frac{\partial F}{\partial x} + \frac{\partial}{\partial y}\left(\frac{H}{MG}\right)\frac{\partial F}{\partial y}\right] \neq 0$$

where $\Delta = \frac{\partial(F,\Phi)}{\partial(x,y)}$. Therefore the theorem follows from the above generalized theorem of Poincare.

We have made H_1 equal to H_2 in the above discussions. For the general case of $H_1 \neq H_2$, we have the following

Theorem 3. If

$$H(x,y) \equiv -\frac{F_x^2 H_1 + F_y^2 H_2}{G(F_x^2 + F_y^2) + F_x F_y (H_1 - H_2)}\,G$$

satisfies the conditions given in theorems 1 and 2, then (1) has one limit cycle which is stable for $H(0,0) < 0$ and unstable for $H(0,0) > 0$.

Proof. The equations (1) can be reduced to that of the case $H_1 = H_2$ by introducing a new H given by (5).

It is easy to verify

$$\frac{F_y G + F_x H}{-F_x G + F_y H} = \frac{F_y G + F_x H_1}{-F_x G + F_y H_2}$$

Hence the theorem is true.

3. Equations of the second class are too complicated to make a systematic study even for the simplest case where F, G, H are polynomials. The question of distribution of a family of algebraic curves which gives the topographical structure at the start is already difficult enough. Let us be content with a few examples.

Example 1. The system

$$\frac{dx}{dt} = -yG + xK_1K_2K_3 \qquad \frac{dy}{dt} = xG + yK_1K_2K_3$$

where

$G = x^2 + y^2 - 2x - 8$; $K_1 = x^2 + y^2 - 1$; $K_2 = x^2 + y^2 - 9$;

$K_3 = x^2 + y^2 - 25$,

has two limit cycles $K_1 = 0$ and $K_3 = 0$ on the finite plane. The first cycle encloses one singular point, while the second encloses three singular points.

Example 2. The system

$$\frac{dx}{dt} = \frac{\partial F}{\partial y} + \frac{\partial F}{\partial x} F \qquad \frac{dy}{dt} = -\frac{\partial F}{\partial x} + \frac{\partial F}{\partial y} F$$

where

$$2F \equiv (x^2 + y^2) - 4x^2y^2 + \varepsilon ,$$

has five singular points: $(0,0)$; $(\pm\frac{1}{\sqrt{3}}, \pm\frac{1}{\sqrt{3}})$; $(\pm\frac{1}{\sqrt{3}}, \mp\frac{1}{\sqrt{3}})$ in which $(0,0)$ is of higher order.

Let $\varepsilon = -0{\cdot}1$, then the system has a non-convex limit cycle L . Let the value of ε increase. Then L turns into four separatrix cycles with $(0,0)$ as their common point for $\varepsilon = 0$. They break into four limit cycles for $\varepsilon \in (0,4/27)$ and vanish at the four singular points $(\pm\frac{1}{\sqrt{3}}, \pm\frac{1}{\sqrt{3}})$; $(\pm\frac{1}{\sqrt{3}}, \mp\frac{1}{\sqrt{3}})$ respectively at $\varepsilon = 4/27$.

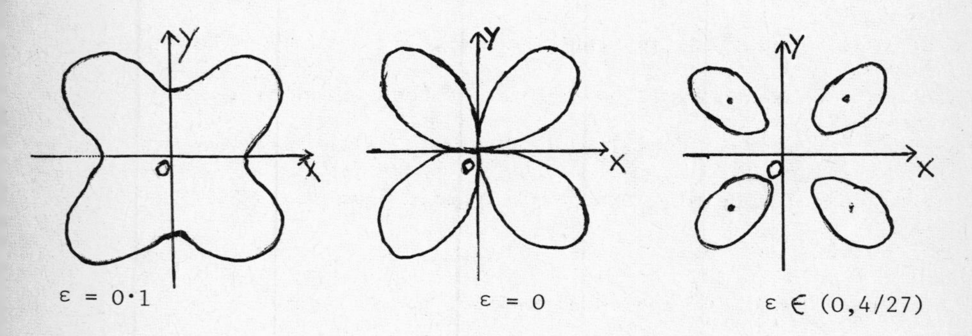

$\varepsilon = 0{\cdot}1$ $\varepsilon = 0$ $\varepsilon \in (0,4/27)$

Example 3. (T.M. Cherry) [3]. The system

$$\frac{dx}{dt} = \frac{\partial F}{\partial y} + \frac{\partial F}{\partial x} H \qquad \frac{dy}{dt} = -\frac{\partial F}{\partial x} + \frac{\partial F}{\partial y} H$$

where $F = (x^2 + y^2)^2 + 2(y^2 - x^2)$; $H = xF$, takes the lemniscate $F = 0$ as a separatrix-cycle. There are three singular points: a saddle $(0,0)$, a stable focus $(1,0)$ and an unstable focus $(-1,0)$. The curve $F = 0$ is an inner boundary of a family of closed trajectories; while within its two eyes the trajectories are spirals.

The last example shows that systems with a polynomial right-hand side may have phase-portraits of unexpected feature. If we rotate the vector field of this example slightly through a positive angle, the system is turned into a system with one limit cycle round $(-1,0)$.

In such a process of continuous variation of the vector field a limit cycle comes into existence from one eye of the lemniscate, while the other eye breaks and vanishes together with the family of closed curves outside of it.

In conclusion, let us remark that Poincaré's type of equation has received little attention though many questions may be raised from the consideration of it. For instance, how far does a topographical structure which appears in the system determine the phase-portrait? Or, more exactly, under what conditions may two topologically equivalent structures determine the same phase-portrait? And in what connection is it related to the theory of structural stability? Relatively easier problems at students' level can also be made from this type of equation. For instance, one may discuss the case where $H(x,y) = 0$ does not enclose the singular point $(0,0)$ aiming at a proof of non-existence of a limit cycle. One may discuss the phase-portrait of equations with a simpler F but more complicated disposition of a number of simple closed curves $H(x,y) = 0$.

References

1. H. Poincaré: Mémoire sur les courbes definies par une équation differéntielle, Oeuvres de Henri Poincaré, Vol. 1, Paris, 1892.
2. Chin Yuan-Shin: Curves defined by differential equations, Science Press (in Chinese).
3. T.M. Cherry: The pathology of differential equations, The Journal of Australian Mathematical Society, Vol. 1, pp. 1-16.

Das Gesetz von Maxwell-Clausius in der phaenomenologischen Thermodynamik.

Johann Walter

§ 1 Der kritische Zustand

Als Entdecker der kritischen Phaenomene bei realen Gasen muß Cagniard de la Tour [11] gelten. Er erwärmte 1822 verschiedene Substanzen (Äther etc.) in einem hermetisch abgeschlossenen Stück eines Gewehrlaufs, wobei die Flüssigkeit zu Beginn des Versuches etwa die Hälfte des Rohres einnahm, während die andere Hälfte nur Dämpfe derselben Substanz enthielt. Am Klang eines kleinen miteingeschlossenen Kieselsteines, den er während der Erwärmung des Rohres in diesem herumrollen ließ, glaubt er das Nichtvorhandensein von Flüssigkeit oberhalb einer gewissen Temperatur feststellen zu können. Weitere Versuche, nun mit aus Glasröhren gefertigten Behältern, zeigten bei geeigneter Wahl des Ausgangsverhältnisses von Flüssigkeit und Dampf auch optische Phaenomene: Ist ursprünglich etwa 2/5 des Volumens mit Flüssigkeit, der Rest mit Dampf gefüllt, so steigt bei vorsichtigem Erwärmen das von der Flüssigkeit eingenommene Volumen auf das Doppelte. Jedoch bei einer für die Substanz charakteristischen Temperatur (von Mendelejev als absoluter Siedepunkt bezeichnet; vergl. [4] p.341) verschwindet plötzlich der die Flüssigkeit und das Gas voneinander trennende Meniskus. Erniedrigt man hierauf die Temperatur wieder, so wird im unteren Teil des Rohres wieder eine Flüssigkeitssäule sichtbar, welche oben von einem deutlich wahrnehmbaren Meniskus begrenzt ist.

Im Jahr 1869 verallgemeinert Thomas Andrews [1] die Versuchsanordnung von Cagniard de la Tour, indem er die Fixierung des Volumens aufhebt, dieses also während des Versuches variieren kann. Er untersucht mit dieser Apparatur Kohlendioxyd und stellt seine Versuchsergebnisse als ein Diagramm von Isothermen in der v,p-Ebene dar.

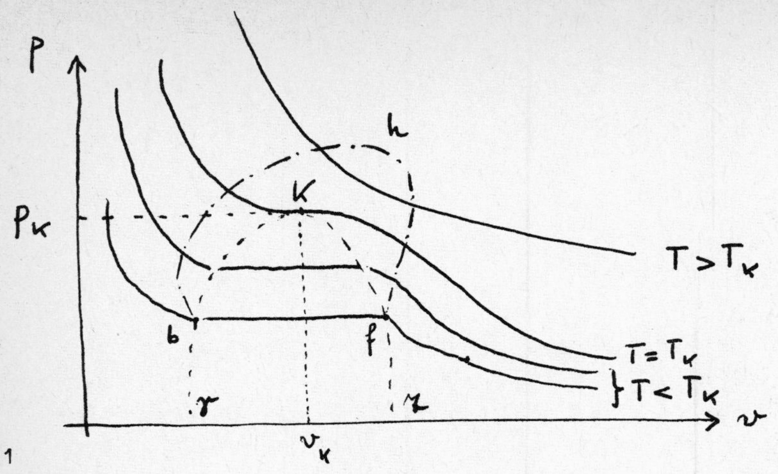

Fig. 1

Während für $T > T_K$ die Isothermen streng monoton fallende Funktionen von v sind und für $T >> T_K$ den Isothermen eines idealen Gases gleichen, enthalten sie für $T < T_K$ horizontale Strecken, die von der Siedelinie γ und der Taulinie τ begrenzt werden und ein Zweiphasensystem beschreiben. Den Übergang bildet die Isotherme $T = T_K$ mit in K horizontaler Tangente. K heißt kritischer Punkt, das zwischen γ und τ gelegene Gebiet Koexistenzgebiet.

Die Versuche von Cagniard de la Tour entsprechen in diesem Diagramm offensichtlich einer zum Teil im Koexistenzgebiet verlaufenden Zustandsänderung längs einer Vertikalen $v \approx v_K$ derart, daß beim Überqueren der kritischen Isotherme das oben beschriebene Verschwinden bzw. Wiederauftauchen des Meniskus erfolgt.

Andrews folgerte aus seinen Ergebnissen im wesentlichen zweierlei:

1.) Oberhalb der kritischen Temperatur T_K kann ein Gas auch bei noch so hohem Druck nicht verflüssigt werden.

2.) "Der gewöhnliche Gas- und gewöhnliche Flüssigkeitszustand sind nur weit voneinander getrennte Formen eines und desselben Aggregatzustandes" [1] , da eine im Zustand b (vergl. Fig. 1) befindliche eindeutig als Flüssigkeit anzusprechende Substanzmenge bei der den kritischen Punkt oben umrundenden Zustandsänderung b h f ohne Meniskusbildung - kontinuierlich - in den gasförmigen Zustand f derselben Temperatur und desselben Druckes überführt werden kann (Kontinuitätshypothese).

Während nun von den Andrewsschen Ergebnissen viele für die weitere Erforschung der realen Gase förderliche Impulse ausgingen, wurde seine Kontinuitätshypothese, die ja keinesfalls eine reine Tatsachenbehauptung darstellt, zum Ausgangspunkt einer bis heute umstrittenen Spekulation, mit der wir uns im folgenden kritisch auseinandersetzen wollen.

§ 2 Der Kontinuitätsbegriff von Thomson und van der Waals

Schon seit dem 18. Jahrhundert ist das Phaenomen des Siedevorzuges bekannt. Bei genügender Vorsicht kann man eine Flüssigkeit weit über ihren Siedepunkt erhitzen, ohne daß Sieden eintritt, und sie so in einen Zustand bringen, in dem ihr Druck geringer ist als der Druck des gesättigten Dampfes bei derselben Temperatur. Diese sogenannten überhitzten Zustände einer Flüssigkeit sind zwar mit einiger Sicherheit reproduzierbar aber nicht stabil; sie gehen bisweilen unter explosionsartiger Dampfbildung in andere Zustände über.

Thomson [10] trägt 1871 als erster die der Überhitzung entsprechenden homogenen Zustände in das Isothermendiagramm von Andrews ein (Fig. 2) und erhält auf diese Weise eine über γ hinweg ins Koexistenzgebiet hineinragende Fortsetzung b c des Flüssigkeitsastes a b einer unterkritischen Isotherme $T = T_1$.

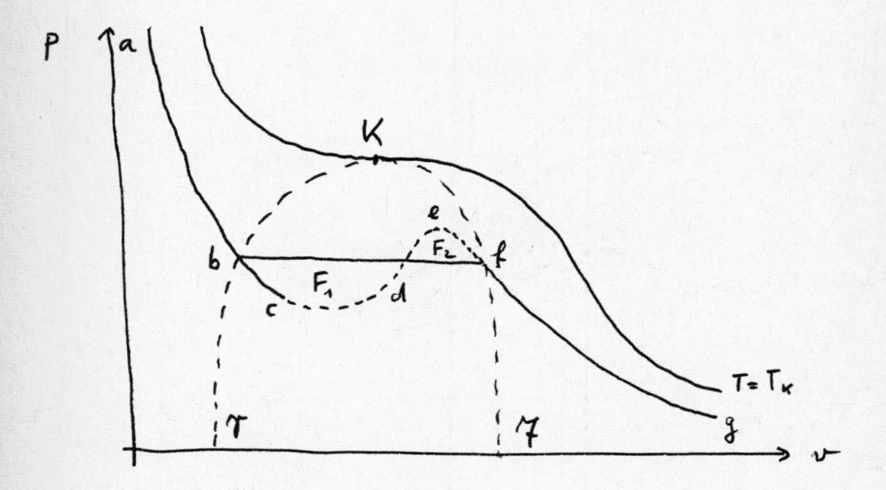

Fig. 2

Man könnte dies auch so ausdrücken, daß sich die Zustandsfläche $T = T(v,p)$ längs der Siedelinie γ verzweigt, wobei die obere Fläche homogene, die untere zweiphasige Zustände beschreibt[1].

Durch das derart ergänzte Diagramm und einige naheliegende, geometrisch-qualitative Betrachtungen wird nun Thomson dazu angeregt, - sozusagen spielerisch - den Flüssigkeitsast a b c der Isotherme $T = T_1$ mit ihrem Dampfast f g durch die (in Fig. 2 gepünktelte) Linie c d e f zu verbinden. Die so entstehende bruchlose Linie a b c d e f g wird von Thomson als Isotherme die Zustände auf dem Stück c d e f analog denen auf a b c und f g als homogen, der längs dieser hypothetischen Isotherme vollzogene Übergang vom flüssigen zum gasförmigen Zustand als neuartiger Ausdruck der Andrewsschen Kontinuitätshypothese aufgefaßt.

Freilich beeilt sich Thomson hinzuzufügen, daß insbesondere die dem Stück c d e zugeordneten Zustände labil und deshalb keineswegs realisierbar seien.

Nennt man Thomas Andrews den Tycho Brahe der realen Gase, so darf van der Waals als ihr Kepler gelten: Er veröffentlicht im Jahr 1873 in seiner Dissertation [12] die berühmte Zustandsgleichung

$$(1)\quad \left(P + \frac{a}{v^2}\right)(v - b) = R \cdot T .$$

Das zu (1) gehörende Isothermendiagramm zeigt Fig. 3:

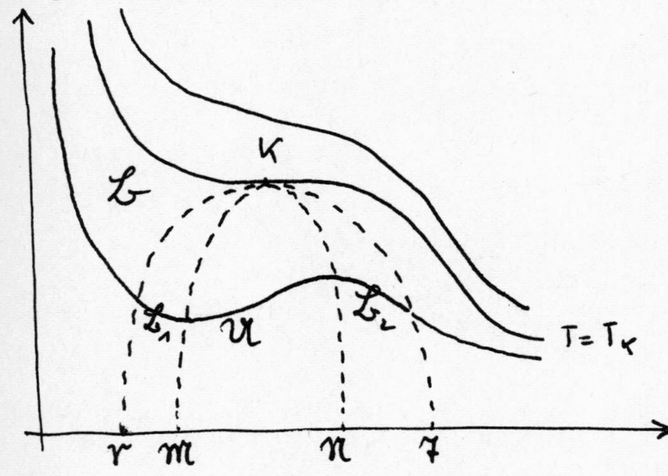

1) Eine solche Verzweigung der Zustandsfläche findet auch längs der Taulinie τ statt. Es handelt sich hierbei um das Phaenomen des übersättigten Dampfes, welches jedoch Thomson noch nicht bekannt war.

Es existiert eine kritische Isotherme $T = T_K$. Die überkritischen Isothermen sind streng monoton fallend. Die unterkritischen Isothermen sind, wie wir dies oben bei Thomson kennengelernt haben, nach Art eines Polynoms dritten Grades geschwungen.

Seien $\mathfrak{M}$ bzw. $\mathfrak{N}$ die Lagenorte der Minima bzw. Maxima der unterkritischen Isothermen. Dann erstreckt sich der experimentelle gesicherte Geltungsbereich $\mathfrak{L}$ der Gleichung (1) auf das Außere des von den Kurven $\mathfrak{M}$ und $\mathfrak{N}$ begrenzten Gebietes $\mathfrak{U}$. Zwischen $\mathfrak{r}$ und $\mathfrak{M}$ bzw. zwischen $\mathfrak{N}$ und $\mathfrak{f}$ beschreibt (1) die Zustände der überhitzten Flüssigkeit bzw. des übersättigten Dampfes. Homogene Zustände zwischen $\mathfrak{M}$ und $\mathfrak{N}$ sind nicht realisierbar. Über die Lage der Kurven $\mathfrak{r}$ und $\mathfrak{f}$, d. h. über die Lage der die zweiphasigen Zustände beschreibenden horizontalen Isothermenstücke, vermag die Gleichung (1) zunächst nichts auszusagen. Mit der Entdeckung von (1) kann vielmehr erst die (alsbald von van der Waals aufgeworfene) Frage entstehen, ob die Lage von $\mathfrak{r}$ und $\mathfrak{f}$ aus der Zustandsgleichung erschlossen werden könne oder ob sie eine zusätzliche empirische Gegebenheit sei.

Die eigentliche Leistung der Zustandsgleichung (1) sah van der Waals darin, daß sie sowohl den flüssigen wie auch den gasförmigen Zustand einer Substanz zu beschreiben gestattet. Er deutete diese Tatsache als eine Bestätigung der Andrewsschen Kontinuitätshypothese. Und es fällt selbst heute noch schwer, sich dem van der Waalsschen Gedanken zu entziehen, daß der zwar prinzipiell nicht verifizierbare Isothermenverlauf zwischen $\mathfrak{M}$ und $\mathfrak{N}$ dennoch einer - wie immer gearteten - Bedeutung nicht ganz und gar entbehrt.

§ 3 Das Gesetz von Maxwell-Clausius

Eine solche Bedeutung erwuchs dem fiktiven Isothermenverlauf wenige Jahre später dadurch, daß er von Maxwell [7] in einer suggestiven Weise zur Festlegung von $\mathfrak{r}$ und $\mathfrak{f}$ verwendet wurde.

Maxwell war 1875 unabhängig von Thomson und van der Waals, ohne sich wie jener auf geometrische Intuition oder wie dieser auf analytische Fortsetzung einer geeigneten Zustandsgleichung zu stützen, durch statistische Betrachtungen auf den merkwürdigen Verlauf aufmerksam geworden, den eine unterkritische Isotherme haben müßte, wenn sie homogene Zustände beschreiben sollte.

Er betrachtet dann einen fiktiven, isothermen Kreisprozess $\mathcal{L} = \mathcal{L}_1 + \mathcal{L}_2$ bei welchem die Substanz zunächst als Zweiphasensystem (Fig. 2) das horizontale Stück $\mathcal{L}_1 =$ b f, danach als homogenes System das teilweise fiktive Stück $\mathcal{L}_2 =$ f e d c b der Isotherme $T = T_1$ durchläuft. Seien F_1 und F_2 die Flächeninhalte, welche die Kurve $\mathcal{L}_2$ mit der Strecke $\mathcal{L}_1$ einschließt. Bei geeigneter Orientierung von $\mathcal{L}$ gilt dann

$$\int_{\mathcal{L}} p\,dv = |F_1 - F_2| . \tag{2}$$

Die linke Seite von (2) stellt die bei dem Prozess $\mathcal{L}$ gewonnene mechanische Energie dar. Diese Energie wird einem Wärmereservoir der Temperatur T_1 entnommen. Wäre nun $|F_1 - F_2| > 0$, so wäre $\mathcal{L}$ ein sogenanntes Perpetuum mobile zweiter Art (PM_{II}), d. h. eine periodisch wirkende Maschine, die bei jedem Arbeitsgang ein Quantum Wärme der Temperatur T_1 vollständig in mechanische Energie überführt. Der zweite Hauptsatz der Thermodynamik verbietet aber die Existenz solcher Maschinen. Hieraus folgert Maxwell die Gleichheit von F_1 und F_2 und gelangt so zu einer Festlegung von γ und τ unter alleiniger Verwendung des zweiten Hauptsatzes und der van der Waalsschen Zustandsgleichung.

Unabhängig von Maxwell erhält 1880 Clausius [5] dieselben Ergebnisse.

Wir zitieren nun der Vollständigkeit halber aus den "Vorlesungen über Thermodynamik" von Planck [8] einen von dem obigen etwas abweichenden, ursprünglich wohl auf Gibbs zurückgehenden Beweis des Gesetzes von Maxwell-Clausius. Durch allgemeine Überlegungen, auf deren Darlegung wir hier verzichten können, gelangt Planck zunächst zu der Gleichgewichtsbedingung

$$s_1 - s_2 = \frac{u_1 - u_2 + p(v_1 - v_2)}{T} \tag{3}$$

für die Koexistenz von Flüssigkeit und Gas, die sich unter Verwendung des Gibbsschen Potentials $g = u + pv - Ts$ auch in der Form $g_1 = g_2$ schreiben läßt. Der Index 1 bezieht sich hierbei auf die Flüssigkeit, der Index 2 auf den Dampf. s, u, v bedeuten spezifische Entropie, spezifische innere Energie und spezifisches Volumen der Substanz. Planck setzt nun voraus, daß s, p, u in der Umgebung des kritischen

Punktes K eindeutige Funktionen der Variablen T und v sind [2],
und integriert in der Darstellung

$$S_1 - S_2 = \int_2^1 \frac{du + p\,dv}{T} ,$$

deren rechte Seite bekanntlich von der Wahl des den Zustand 1 mit dem Zustand 2 verbindenden Integrationsweges unabhängig ist, längs der durch die Zustände 1 und 2 gehenden fiktiven Isotherme.
Es ergibt sich

$$S_1 - S_2 = \frac{u_1 - u_2}{T} + \frac{1}{T}\int_2^1 p\,dv ,$$

woraus mit (3) folgt

$$\int_2^1 p\,dv = p(v_1 - v_2). \qquad (4)$$

Dies bedeutet wiederum die Gleichheit von F_1 und F_2.

§ 4 Kritische Stimmen zu den Beweisen des Gesetzes von Maxwell - Clausius

Ein Bewußtsein des schwankenden Grundes, auf dem die Beweise des Maxwellschen Gesetztes fußen, findet sich bei Maxwell, Clausius und allen Autoren, die in der Folgezeit dies Gesetz in ihre Darstellung der Thermodynamik aufgenommen haben. Dies Bewußtsein erscheint jedoch konsequenzenlos abgedrängt in die stereotype Formel, mit der die - sogar prinzipielle - Nichtrealisierbarkeit des Isothermenstücks c d e (Fig. 2) eingeräumt wird. Dennoch fehlt es nicht an kritischen Stimmen.

In der 1881 erschienenen deutschen Übersetzung seiner Dissertation [13] sagt van der Waals zu den Beweisen von Maxwell und Clausius in einer Anmerkung "Beide Beweise benutzen einen Erfahrungssatz in einem Fall, der durch das Experiment nicht verwirklicht werden kann, und sind daher nicht über jeden Zweifel erhaben", läßt diesen Zweifel jedoch nicht in den eigentlichen Text seiner Arbeit übergehen.

---

2) $p = p(v, T)$ kann z. B. die nach p aufgelöste Form der van der Waalsschen Zustandsgleichung (1) sein.

Richards beruhigt sich in [9] mit der Erklärung: "The fact that the dotted loops in the figure (vergl. unsere Fig. 2) are not physically attainable is immaterial. If g (Gibbssches Potential, vgl. § 3) is related to p and v by an <u>analytic</u> relation like the van der Waals equation, $g_1 - g_2$ may then be <u>evaluated</u> by using such mathematical relations even though the "path" involved has no physical meaning".

Guggenheim [6] entzieht sich, trotz starken Mißtrauens gegen das Maxwellsche Gesetz, einer klaren Stellungnahme mit den Worten: "Since the portion c d e (unsere Fig. 2) of the curve cannot be realized experimentally, instead of saying that the two phase aquilibrium is determined by the condition of equality of the two areas F_1 and F_2 it is perhaps more correct to say that the connecting portion c d e of the curve must be sketched in such a manner as to make the two areas equal".

Am weitesten vorgedrungen in Richtung auf eine Verwerfung des Maxwellschen Gesetzes scheint Baehr in seiner Dissertation [2]. Er schreibt dort p. 271 in einem klein gedruckten Absatz: "Aus Stetigkeitsgründen rein mathematischer Natur müssen die Isothermen im Zweiphasengebiet die bekannte Schleifenform aufweisen, den dabei auftretenden instabilen Zuständen kann keine physikalische Bedeutung zukommen, denn sie können in der Natur nicht auftreten. Der Isothermenverlauf im heterogenen Gebiet hat nur formale Bedeutung: Soll die Zustandsgleichung $P=P(v,T)$ die Stabilitätsgrenzen (γ u. $\mathcal{F}$, vergl. Fig. 2) des homogenen Gebietes wiedergeben, so muß die Extrapolation der Isotherme ins Zweiphasengebiet so erfolgen, daß $\int_2^1 p\,dv = p(v_1 - v_2)$ ist".

In sein Lehrbuch [3] hat Baehr das Gesetz von Maxwell nicht mehr aufgenommen. Freilich hätte man sich in Anbetracht der Tatsache, daß alle anderen uns bekannten Lehrbücher an diesem Gesetz noch festhalten, gerade von diesem Autor eine gewisse Polemik gewünscht.

§ 5 Widerlegung der Beweise des Gesetzes von Maxwell-Clausius

Die Schwäche der beiden in § 3 vorgeführten Beweise des Maxwellschen Gesetzes besteht, wie wir gesehen haben, in der Verwendung des prinzipiell nicht realisierbaren Isothermenverlaufes im Gebiet (Fig. 3). Um eine präzise Vorstellung von den hiermit implizierten

Schlußfehlern zu gewinnen, ist es nützlich, den genauen Inhalt der durch die Zustandsgleichung $T=T(\vartheta, p)$ über das Verhalten eines realen Gases gemachten Aussagen zu untersuchen. Dabei denken wir uns die Gleichung $T=T(\vartheta,p)$ im Sinn von Thomson und van der Waals ins Gebiet $\mathfrak{V}$ irgendwie eindeutig forgesetzt.

Die wichtigste Aussage, die aus jeder in diesem Sinn fortgesetzten Zustandsgleichung folgt, besteht darin, daß die ϑ,p-Ebene in der Umgebung des kritischen Punktes K in zwei Gebiete $\mathfrak{L}$ und $\mathfrak{V}$ zerfällt, derart daß die Inklusionen $\{\vartheta,p\}\varepsilon\mathfrak{L}$ bzw. $\{\vartheta,p\}\varepsilon\mathfrak{V}$ über Existenz bzw. Nichtexistenz eines entsprechenden homogenen Zustandes der Substanz entscheiden. Allerdings vermag die Zustandsgleichung nichts über die genaue Form von $\mathfrak{L}$ bzw. $\mathfrak{V}$ auszusagen.

Wir bilden nun die in folgenden mit $A(\vartheta,p)$ bzw. $B(\vartheta,p)$ bezeichneten Aussagen "Es existiert ein homogener Zustand mit dem Volumen ϑ und dem Druck p" bzw. "Ein homogener Zustand mit dem Volumen ϑ und dem Druck p hat die Temperatur $T(\vartheta,p)$".
Die in der Zustandsgleichung $T=T(\vartheta,p)$ zusammengefaßten Informationen formulieren wir wie folgt: Die Aussage $A(\vartheta,p)$ ist wahr oder falsch jenachdem $\{\vartheta,p\}\varepsilon\mathfrak{L}$ oder $\{\vartheta,p\}\varepsilon\mathfrak{V}$ ist. Die Aussage $B(\vartheta,p)$ ist wahr, wenn $\{\vartheta,p\}\varepsilon\mathfrak{L}$ ist. Für $\{\vartheta,p\}\varepsilon\mathfrak{V}$ ist jedoch $B(\vartheta,p)$ weder wahr noch falsch; denn man kann nicht nachprüfen, ob ein nichtrealisierbarer Zustand $\{\vartheta,p\}$ die Temperatur $T(\vartheta,p)$ hat oder nicht. Aussagen, deren Verifikation prinzipiell unmöglich ist, pflegt man als sinnlos (im technischen, nicht im wertenden Sinn) zu bezeichnen.

1) Widerlegung des Maxwellschen Beweises.
Der erste Beweis von § 3 wird indirekt geführt. Er beginnt mit der Annahme $F_1 \neq F_2$ und schließt mit der Behauptung, daß der (geeignet orientierte) Kreisprozess $\mathcal{L}$ ein PM_{II} darstelle. Dies steht im Widerspruch zum 2. Hauptsatz. Also ist doch $F_1=F_2$. Es kommt bei der Beurteilung dieses Beweises alles auf die Bedeutung der (im folgenden mit C abgekürzten) Aussage " $\mathcal{L}$ ist ein PM_{II}" an. Sieht man in ihr lediglich Aussagen der Form $B(\vartheta,p)$ verwendet, verzichtet man also auf eine Aussage über die Realisierbarkeit von $\mathcal{L}$, so ist C sinnlos, da sie Aussagen der Form $B(\vartheta,p)$ für $\{\vartheta,p\}\varepsilon\ \mathcal{L}_2\cap\mathfrak{V}$, d. h. sinnlose Aussagen, voraussetzt. Die Sinnlosigkeit von C zeigt sich insbesondere darin, daß nicht entschieden werden kann, ob C zum 2. Hauptsatz im Widerspruch steht oder nicht. Der 2. Hauptsatz verbietet die Realisierbarkeit eines PM_{II}. Gerade über die Realisierbarkeit von $\mathcal{L}$ wird aber nichts gesagt.

Interpretiert man die Aussage C als " $\mathcal{L}$ ist ein realisierbarer Prozess und $\mathcal{L}$ ist ein PM_{II}", so steht sie zwar in klarem Widerspruch zum 2. Hauptsatz, ist aber selbst falsch, da sie Aussagen der Form $A(v,p)$ für $\{v,p\} \varepsilon\ \mathcal{L}_2 \cap \mathfrak{V}$, d. h. falsche Aussagen, zur Voraussetzung hat. Also kann $F_1 = F_2$ nicht erschlossen werden.

Sieht man schließlich in C die Aussage "Falls $\mathcal{L}$ realisierbar wäre, so wäre $\mathcal{L}$ ein PM_{II}", so ist diese Aussage zwar richtig, steht aber nicht mehr im Widerspruch zum 2. Hauptsatz. Also kann wiederum $F_1 = F_2$ nicht geschlossen werden.

Etwas drastisch könnte man den Beweis von Maxwell so entkräften: Der Prozess $\mathcal{L}$ ist ein fiktiver nur in Gedanken bestehender Prozess. Ein gedachter Prozess darf aber ein PM_{II} sein. Denn der 2. Hauptsatz verbietet nicht die Denkbarkeit, sondern bloß die Realisierbarkeit dieser so brauchbaren Maschine.

2) Widerlegung des Planckschen Beweises.

Der Plancksche Beweis des Gesetzes von Maxwell geht aus von der Gleichgewichtsbedingung (3). Hiernach hat man zunächst die Funktionen $s(v,T)$, $u(v,T)$, $p(v,T)$ im Bereich $\mathcal{L}$ empirisch zu bestimmen und danach durch Einsetzen in (3) die Gleichungen von γ und $\mathfrak{F}$ zu ermitteln. Von der Zustandsgleichung $p=p(v,T)$, die man sich als die van der Waalsche vorstellen mag, sei nun eine eindeutige Fortsetzung ins Gebiet $\mathfrak{V}$ hinein bekannt. Die Zurückführung von (3) auf das Maxwellsche Gesetz beruht dann auf der Annahme, daß auch die Funktionen $s(v,T)$ und $u(v,T)$, sogar unter Erhaltung der Beziehung $ds = \frac{du + p \cdot dv}{T}$ ins Gebiet $\mathfrak{V}$ hinein fortgesetzt werden können.

Der Beweis von Planck hat an dieser Stelle eine Lücke, da die Erlaubtheit dieser Annahme keinesfalls erwiesen ist. Wir wollen die Annahme im folgenden etwas umformen, um ihre Bedeutung besser beurteilen zu können.

Für die Vollständigkeit des Differentials $\frac{du + p\, dv}{T}$ ist notwendig und hinreichend das Verschwinden des Skalarproduktes $(\mathfrak{w}, \mathfrak{m})$, wobei $\mathfrak{w} = \{T_p, -T_v, T - pT_p\}$ und $\mathfrak{m} = \{u_v, u_p, -1\}$ gesetzt wurde (Integrabilitätsbedingung). $\mathfrak{w}$ ist aus der Zustandsgleichung bekannt, $\mathfrak{m}$ ist ein Normalenvektor zur Fläche $u=u(v,p)$. Die Gleichung $(\mathfrak{w}, \mathfrak{m}) = 0$ besagt dann geometrisch, daß die Fläche $u=u(v,p)$ sich aus den Lösungen (Charakteristiken) des Gleichungssystems

$$(5)\quad \dot{v} = T_p\,,\quad \dot{p} = -T_v\,,\quad \dot{u} = T - pT_p$$

zusammensetzen läßt. Von (5) spalten wir das u nicht enthaltende System

(6) $\dot{\mathfrak{v}} = T_p, \; \dot{p} = - T_{\mathfrak{v}}$

ab, das als Lösungen die Isothermen der Zustandsgleichung $T=T(\mathfrak{v},p)$ besitzt. Aus der letzten Gleichung von (5) ergibt sich, daß die innere Energie $u(\mathfrak{v},p)$ durch Quadraturen in der ganzen $\mathfrak{v}$,p-Ebene berechnet werden kann, wenn sie etwa längs einer Vertikalen $\mathfrak{v}$= const empirisch bekannt ist.

Eine Komplikation topologischer Natur tritt jedoch für $T < T_K$ ein. Der unterkritische Teil von $\mathcal{L}$ wird nämlich durch $\mathfrak{U}$ in zwei nicht zusammenhängende Komponenten $\mathcal{L}_1$ (Flüssigkeit) und $\mathcal{L}_2$ (Dampf) zerlegt (Fig. 3). Seien $\mathfrak{v}_1$ und $\mathfrak{v}_2$ Volumina mit $\mathfrak{v}_1 < \mathfrak{v}_K < \mathfrak{v}_2$ und sei $\mathfrak{u}(\mathfrak{v},p)$ auf den Vertikalen $\mathfrak{v} = \mathfrak{v}_1$ bzw. $\mathfrak{v} = \mathfrak{v}_2$, soweit sie in $\mathcal{L}_1$ bzw. $\mathcal{L}_2$ liegen, empirisch bestimmt. Durch Quadraturen erhält man dann Funktionen $\mathfrak{u}_1(\mathfrak{v},p)$ und $\mathfrak{u}_2(\mathfrak{v},p)$, die in $\mathcal{L}_1$ bzw. $\mathcal{L}_2$ die innere Energie des realen Gases beschreiben. Erstreckt man nun die von $\mathfrak{v} = \mathfrak{v}_1$ ausgehende Integration der letzten Gleichung von (5) durch $\mathfrak{U}$ hindurch ins Gebiet $\mathcal{L}_2$ hinein (dies ist mathematisch möglich, da die Zustandsgleichung auch in $\mathfrak{U}$ erklärt ist), so entsteht die Möglichkeit in $\mathcal{L}_2$ die Funktionen $\mathfrak{u}_1(\mathfrak{v},p)$ und $\mathfrak{u}_2(\mathfrak{v},p)$ zu vergleichen. Die Plancksche Annahme läuft darauf hinaus, daß in $\mathcal{L}_2$ $\mathfrak{u}_1(\mathfrak{v},p) = \mathfrak{u}_2(\mathfrak{v},p)$ gilt. Man kann aber zunächst nur sagen, daß $\mathfrak{u}_1(\mathfrak{v},p) - \mathfrak{u}_2(\mathfrak{v},p)$ eine Funktion von T allein ist, die für $T=T_K$ verschwindet. Würde nun für eine gewisse Zustandsgleichung $\mathfrak{u}_1(\mathfrak{v},p) = \mathfrak{u}_2(\mathfrak{v},p)$ tatsächlich gelten, so kann die Gleichheit aufgehoben werden durch eine Abänderung des Isothermenverlaufs in $\mathfrak{U}$. Diese Beobachtung leitet uns zu einem letzten Argument, das das Gesetz von Maxwell an der Wurzel trifft: Seien $T=T_1(\mathfrak{v},p)$ und $T=T_2(\mathfrak{v},p)$ zwei Zustandsgleichungen, die in $\mathcal{L}$ übereinstimmen und sich nur in $\mathfrak{U}$ unterscheiden. Da der Verlauf der Isotherme in $\mathfrak{U}$ für das meßbare Verhalten des Systems völlig irrelevant ist, können wir uns vorstellen, daß beide Zustandsgleichungen dasselbe System beschreiben. Insbesondere kann nach (3) die Lage von γ und $\mathcal{T}$ ohne Benutzung des Verhaltens der Isothermen in $\mathfrak{U}$ festgelegt werden. Die nach dem Maxwellschen Gesetz berechneten Kurven γ und $\mathcal{T}$ sind aber im allgemeinen bei verschiedenen Zustandsgleichungen verschieden. Ob also mit Hilfe des Maxwellschen Gesetzes die Lage der Kurven γ und $\mathcal{T}$ richtig vorhergesagt werden kann, hängt von der "richtigen" Wahl der Zustandsgleichung im Gebiet $\mathfrak{U}$ ab. Da man aber dort, wie wir bereits vielfach betont haben, die Isothermen nicht messen kann, bleibt als einziges Kriterium für die "Richtigkeit" dieser Wahl die Gültigkeit des Maxwellschen Gesetzes. Dieses aber sollte ursprünglich bewiesen werden.

§ 6 Bemerkungen

Wir hoffen, im § 5 die Unhaltbarkeit der Beweise des Gesetzes von Maxwell überzeugend nachgewiesen zu haben. Trotzdem bleibt natürlich unbestritten, daß das Maxwellsche Gesetz, angewandt auf geeignete Zustandsgleichungen, ggf. zu einer befriedigenden Festlegung von $\mathfrak{r}$ und $\mathfrak{z}$ führen könnte. In der Tat zeigt Baehr in seiner Dissertation [2], daß bei geeigneter Wahl einer bei K analytischen Zustandsgleichung in einer allerdings sehr kleinen Umgebung des kritischen Punktes K die gemessenen Werte von $\mathfrak{r}$ und $\mathfrak{z}$ mit den nach dem Maxwellschen Gesetz berechneten verträglich sind. Ob sich in dieser Tatsache ein physikalisches Phaenomen oder einfach die Anpassungsfähigkeit der analytischen Funktionen widerspiegelt, vermag ich nicht zu sagen. Jedenfalls wird man, falls das erstere der Fall ist, die Erklärung nicht in der phaenomenologischen Thermodynamik suchen dürfen.

Zusätzliche Bemerkung (April 1982):

Die vorliegende Arbeit ist in der voranstehenden Form bereits im Sommer 1966 zu Papier gebracht worden und dokumentiert einen stark von neopositivistischen Ansätzen *) beeinflußten Zugriff, wie ihn sich der Verfasser während seiner Assistentenzeit 1960 - 1966 an der Technischen Universität in Berlin-Charlottenburg in weitgehender fachlicher Isolation erarbeitet hat. Der Verfasser gedenkt an dieser Stelle gerne seines Lehrers in Thermodynamik, Professor Gerhard Schubert, in dessen Vorlesung an der Johannes-Gutenberg-Universität Mainz im Jahr 1955 ihm die der vorliegenden Arbeit zugrundeliegende Frage zuerst aufging. Während der "Fourth Conference on Ordinary and Partial Differential Equations held at Dundee, Scotland 1976" hat Professor J. Serrin von der University of Minnesota dem Verfasser freundlicherweise mitgeteilt, daß er dieselben Fragen bearbeitet hat und im wesentlichen zu denselben Ergebnissen gelangt ist, und ihm 1978 sein Manuskript "The Elementary van der Waals Fluid and Maxwell's Rule" zugänglich gemacht. Diese Informationen stellten sich dem Verfasser dar als Hinweis auf ein Anwachsen des Interesses an von Mathematikern mit mathematischen Methoden gewonnenen thermodynamischen Aussagen **) und hat ihm Mut gemacht, aus der oben erwähnten Isolation herauszustreben und diese Arbeit der Öffentlichkeit vorzulegen.

*) Vergl. z.B. unseren ca. 1965 entstandenen Dialog: "Gespräch zwischen einem Positivisten und einem Thermodynamiker über die Gültigkeit der sogenannten Maxwellschen Regel" (Manuskript).

**) Vergl. etwa unsere Arbeit in Proc. Royal Soc. Edinburth 82 A, 87 - 94, 1978 und die dort erwähnte Literatur.

Literatur

1. Andrews, Th.: On the continuity of the gaseous and liquid states of matter. Phil. Trans. Roy. Soc. London 159,575-589 (1869). (oder: Ostwalds Klassiker, Nr. 132 Leipzig 1902)

2. Baehr, H.D.: Der kritische Zustand und seine Darstellung durch eine Zustandsgleichung. Abhandl. Mainzer Akad., Math. Natw. Klasse Nr. 6, 233-333 (1953).

3. Baehr, H.D.: Thermodynamik, Berlin 1962

4. Chwolson, O.D.: Lehrbuch der Physik, III. 2, zweite deutsche Auflage, Braunschweig 1922

5. Clausius, R.: Über das Verhalten der Kohlensäure in Bezug auf Druck, Volumen und Temperatur. Ann. Phys. Chemie Neue Folge IX, 337-357 (1880).

6. Guggenheim, E.A.: Thermodynamics, classical and statistical. Handbuch der Physik III, 2, 1-118, Berlin 1959

7. Maxwell, J.C.: On the dynamical evidence of the molecular constitution of bodies. Nature 11, 357-360 (März 1875).

8. Planck, M.: Vorlesungen über Thermodynamik. 10. Auflage, Berlin 1954

9. Richards, P.J.: Manual of Mathematical Physics. New York 1959

10. Thomson, J.: Considerations on the abrupt change at boiling or condensing in reference to the continuity of the fluid state of matter. Proc. Roy. Soc. London 19, 1-10 (Nov. 1871).

11. Tour, C. de: Exposé de quelques résultats obtenue par l'action combinée de la chaleur et de la compression sur certains liquides, tel que l'eau, l'alcool, l'éther sulphurique et l'essence de pétrole rectifiée. Ann. Chim. Phys. 21, 127-132 (Paris 1822).

12. Waals, J.D. van: Over de continuiteit van den gas en vloeistoftoestand. Diss. Leiden 1873.

13. Waals, J.D. van: Die Kontinuität des gasförmigen und flüssigen Zustandes. (Übersetzt von F.Roth) Leipzig 1881.